De Gruyter Expositions in Mathematics 53

De Gruyter Expositions in Mathematics

20 Semigroups in Algebra, Geometry and Analysis, *K. H. Hofmann, J. D. Lawson, E. B. Vinberg* (Eds.)

21 Compact Projective Planes, *H. Salzmann, D. Betten, T. Grundhöfer, H. Hähl, R. Löwen, M. Stroppel*

22 An Introduction to Lorentz Surfaces, *T. Weinstein*

23 Lectures in Real Geometry, *F. Broglia* (Ed.)

24 Evolution Equations and Lagrangian Coordinates, *A. M. Meirmanov, V. V. Pukhnachov, S. I. Shmarev*

25 Character Theory of Finite Groups, *B. Huppert*

26 Positivity in Lie Theory: Open Problems, *J. Hilgert, J. D. Lawson, K.-H. Neeb, E. B. Vinberg* (Eds.)

27 Algebra in the Stone-Čech Compactification, *N. Hindman, D. Strauss*

28 Holomorphy and Convexity in Lie Theory, *K.-H. Neeb*

29 Monoids, Acts and Categories, *M. Kilp, U. Knauer, A. V. Mikhalev*

30 Relative Homological Algebra, *E. E. Enochs, O. M. G. Jenda*

31 Nonlinear Wave Equations Perturbed by Viscous Terms, *V. P. Maslov, P. P. Mosolov*

32 Conformal Geometry of Discrete Groups and Manifolds, *B. N. Apanasov*

33 Compositions of Quadratic Forms, *D. B. Shapiro*

34 Extension of Holomorphic Functions, *M. Jarnicki, P. Pflug*

35 Loops in Group Theory and Lie Theory, *P. T. Nagy, K. Strambach*

36 Automatic Sequences, *F. v. Haeseler*

37 Error Calculus for Finance and Physics, *N. Bouleau*

38 Simple Lie Algebras over Fields of Positive Characteristic. I. Structure Theory, *H. Strade*

39 Trigonometric Sums in Number Theory and Analysis, *G. I. Arkhipov, V. N. Chubarikov, A. A. Karatsuba*

40 Embedding Problems in Symplectic Geometry, *F. Schlenk*

41 Approximations and Endomorphism Algebras of Modules, *R. Göbel, J. Trlifaj*

42 Simple Lie Algebras over Fields of Positive Characteristic. II. Classifying the Absolute Toral Rank Two Case, *H. Strade*

43 Modules over Discrete Valuation Domains, *P. A. Krylov, A. A. Tuganbaev*

44 Quasi-Stationary Phenomena in Nonlinearly Perturbed Stochastic Systems, *M. Gyllenberg, D. S. Silvestrov*

45 Distribution Theory of Algebraic Numbers, *P.-C. Hu, C.-C. Yang*

46 Groups of Prime Power Order. Volume 1, *Y. Berkovich*

47 Groups of Prime Power Order. Volume 2, *Y. Berkovich, Z. Janko*

48 Stochastic Dynamics and Boltzmann Hierarchy, *D. Ya. Petrina*

49 Applied Algebraic Dynamics, *V. Anashin, A. Khrennikov*

50 Ultrafilters and Topologies on Groups, *Y. G. Zelenyuk*

51 State Observers for Linear Systems with Uncertainty, *S. K. Korovin, V. V. Fomichev*

52 Stability Analysis of Impulsive Functional Differential Equations, *I. Stamova*

Products of Finite Groups

by

Adolfo Ballester-Bolinches
Ramón Esteban-Romero
Mohamed Asaad

De Gruyter

Mathematics Subject Classification 2010
Primary 20D40, 20D10, 20D35;
Secondary 20D30, 20E15, 20F16, 20F17

ISBN 978-3-11-020417-9
e-ISBN 978-3-11-022061-2
ISSN 0938-6572

Library of Congress Cataloging-in-Publication Data

Asaad, Mohamed.
Products of finite groups / by Mohamed Asaad, Adolfo Ballester-Bolinches, Ramón Esteban-Romero.
p. cm. – (De Gruyter expositions in mathematics ; 53)
Includes bibliographical references and index.
ISBN 978-3-11-020417-9 (alk. paper)
1. Finite groups. I. Ballester-Bolinches, Adolfo. II. Esteban-Romero, Ramón. III. Title.
QA177.A78 2010
512'.23–dc22
2010031029

Bibliographic information published by the Deutsche Nationalbibliothek

The Deutsche Nationalbibliothek lists this publication in the Deutsche Nationalbibliografie; detailed bibliographic data are available in the Internet at http://dnb.d-nb.de.

Typesetting: Da-TeX Gerd Blumenstein, Leipzig, www.da-tex.de
Printing: Hubert & Co. GmbH & Co. KG, Göttingen
∞ Printed on acid-free paper

Printed in Germany

www.degruyter.com

For the ones we love:

Fran
Anita
Ramón
Ana
Juan Miguel
José Luis
Beatriz
Asaad (in memoriam)
Aisha (in memoriam)
Fatma (in memoriam)

Preface

Debajo de las multiplicaciones
hay una gota de sangre de pato.
Debajo de las divisiones
hay una gota de sangre de marinero.
Debajo de las sumas, un río de sangre tierna;
...

FEDERICO GARCÍA LORCA
New York (oficina y denuncia),
Poeta en Nueva York

The central theme of this book is the structural study of a finite group which can be factorised as a product of finitely many pairwise permutable subgroups. The origin of the theory may be traced back to the year 1903, when Burnside published his well-known p^a-lemma; he discovered that a finite group cannot be simple if it factorises as the product of a Sylow subgroup and the centraliser of a non-trivial element. One year later, he used this result to prove his celebrated $p^a q^q$-theorem about the solubility of finite groups whose order is divisible by at most two primes, which is also a theorem about factorised groups. Burnside's early work was followed by Hall in the decade 1928–1937 in a great sequence of papers which established the basic theory of finite soluble groups: this work determined the direction of research in finite soluble groups for many years, and may have well provided the motivation for a new field of research in the theory of factorised groups; he discovered that a finite group is soluble if and only if it is the product of pairwise permutable Sylow subgroups. But it was a short paper of Itô in 1955 which set the theory of products in motion. He shows that any (not necessarily finite) group is metabelian whenever it is the product of two abelian subgroups. After appearance of Itô's theorem, and motivated by the results of Burnside and Hall, attention is shifted to finite groups that are product of nilpotent subgroups, the conjecture being that such groups are soluble. In 1958, Wielandt confirms this conjecture in the coprime case and Kegel in the general one in 1962. The outcome of this research is known nowadays as the theorem of Kegel and Wielandt: every finite group which is factorised as a product of pairwise permutable nilpotent subgroups is soluble. However it is not known whether its derived length can be bounded in terms of the nilpotency classes of the factors. Nor it is known whether the Kegel and Wielandt theorem can be extended to infinite groups. In the much more special case when the factors are normal and nilpotent, then the product is nilpotent. This is a well-known result of Fitting. However, the product of pairwise normal supersoluble subgroups need not be supersoluble even in the finite case. A natural path of inquiry opens up when one asks how the structure of the factors affects the structure of the factorised group when they are connected by certain permutability conditions: totally and mutually permutable products naturally emerge. It was Shunkov in 1964 who first studied pairwise totally permutable subgroups of prime power order. Shunkov's early work was followed in 1989 by a seminal paper of Asaad and Shaalan which established the basic theory of totally and mutually permutable products. Since that time there was gradually emerging a body of techniques and a series of general methods for studying these kind of products. Although the entire field is presently in a state of ferment and fluidity, a degree of stability appears to be settling over certain aspects of the subject.

Apart from the relevant monographs of N. S. Chernikov [95] (in Russian) of 1987 and the one by Amberg, Franciosi, and de Giovanni [9] of 1992, which contain many beautiful results about products of groups, especially in the infinite case, none of the existing books on group theory describes more recent developments on products of finite groups. The present volume has therefore the modest object of presenting in a unified form a comprehensible account of more than twenty years of work on this subject, focusing the attention on products of finite nilpotent groups and products of mutually and totally permutable finite groups. There are some omissions that have been caused rather by the necessity of keeping the project within reasonable limits. Thus, for example, the systematic treatment of the finite products of connected subgroups with respect to a class of groups, apart from a brief outline without proofs in Chapter 5, has been omitted. This is an area that has been the subject recently of great deal of investigation. We have not dealt with factorisations of finite simple groups nor trifactorised groups either. Nevertheless, we hope that this book will serve as a basic reference in the subject area, as a text for postgraduate studies, and also as a source of research ideas.

All of our work is concerned with finite groups. Therefore our unspoken rule is that *all groups are finite*. Chapter 1 is foundational material, and it is written on the assumption that the reader has a basic knowledge in group theory, representation theory, and theory of classes of groups, such as which may be obtained from reading Chapters A, B, I, and II of the book of Doerk and Hawkes [119], and Chapters 2 and 4 of the book of Ballester-Bolinches and Ezquerro [47]. In selecting or rejecting material for this chapter, we have held steadfast to a single-minded purpose: to present only those results deemed essential for applications in the rest of the book. In Chapter 2 classes of groups defined by means of certain permutability conditions which have been proved useful in questions involving products are discussed. A major part of the chapter is devoted to characterisation theorems. Chapter 3 provides a comprehensive look at products of two nilpotent groups, focusing the attention on products of abelian groups, structure theorems, and Hall–Higman type results. The study of the subgroup structure of pairwise mutually and totally permutable products is the main aim of the Chapter 4. The last chapter of the book is dedicated to the study of the behaviour of such products with respect to classes of groups.

We would like to acknowledge our gratitude to James C. Beidleman, Jonas Bochtler, Ángel Carocca, Nikolay S. Chernikov, John Cossey, Luis M. Ezquerro, Andrew Fransman, Michiel Hazewinkel, Hermann Heineken, Lev S. Kazarin, Leonid A. Kurdachenko, Thomas Masiwa, Victor D. Mazurov, María Carmen Pedraza-Aguilera, and Alexander N. Skiba. We wish to express our thanks to Robert Plato and Simon Albroscheit for their continuous support from Walter de Gruyter during the production of this book. Finally, we thank the *Ministerio de Educación y Ciencia* (Spanish Government) and *FEDER* (European Union) for their financial support via the research grant MTM2007-68010-C03-02.

Adolfo Ballester-Bolinches, Ramón Esteban-Romero, and Mohamed Asaad
Torres-Torres, València, and Cairo, August 2010

Contents

Chapter 1

Prerequisites

In this chapter we consider several primarily unrelated basic topics of finite group theory that we shall need in varying degree throughout the book. These include essential preliminary properties of factorised groups in Section 1.1, and general facts about permutability, Sylow permutability, subnormality, pronormality, and related subgroup embedding properties in Sections 1.2 and 1.5. Working on factorised groups whose factors are connected by certain permutability conditions, and on classes of groups defined by certain permutability properties, it is important to have properties of power automorphisms and Dedekind groups at our disposal. This is the reason why we have included them in Sections 1.3 and 1.4. In Section 1.6 we deal with an interesting extension of the class of all supersoluble groups: the class of all SC-groups or groups whose chief factors are simple. We finish the chapter establishing results on some numerical invariants which will be applied to give information about products of nilpotent groups. Later chapters will treat various aspects of these subjects in great detail and depth. Here we shall be interested in deriving elementary properties, in introducing auxiliary concepts, and in developing some important notational conventions.

1.1 Factorised groups – basic concepts

If A and B are subgroups of a group G, the *Frobenius product* AB of A and B is defined to be the subset of all elements of G with the form ab, where $a \in A, b \in B$. It is well known that AB is a subgroup of G if and only if $AB = BA$, that is, if the subgroups A and B permute. Should it happen that AB coincides with the group G, with the result that $G = AB = BA$, then G is said to be *factorised* by its subgroups A and B. More generally, a group G is said to be the *product of its pairwise permutable subgroups* $G_1, G_2, \dots, G_r$ if $G = G_1G_2 \cdots G_r$ and $G_iG_j = G_jG_i$ for all integers i and j with $i, j \in \{1, 2, \dots, r\}$. This implies that for every choice of indices $1 \le i_1 \le i_2 \le \cdots \le i_k \le r$, the product $G_{i_1}G_{i_2} \cdots G_{i_k}$ is a subgroup of G. In this case we also say that G is *factorised* by the subgroups $G_1, G_2, \dots, G_r$. It is clear that every epimorphic image $G/N = (G_1N/N)(G_2N/N) \cdots (G_rN/N)$ of G is also a product of the pairwise permutable epimorphic images $G_1N/N, G_2N/N, \dots, G_rN/N$ of G_1, $G_2, \dots, G_r$ respectively.

The following lemma will prove to be useful.

Lemma 1.1.1. *Let G be a group. If A, B, and C are subsets of G such that $C^{-1} = \{c^{-1} : c \in C\} \subseteq C$, then:*

1. *if $AC \subseteq A$, then $(A \cap B)C = A \cap BC$, and*
2. *if $CA \subseteq A$, then $C(A \cap B) = A \cap CB$.*

Proof. We will prove only the part 1, because the proof of the part 2 is similar. Clearly, $(A \cap B)C \subseteq AC \cap BC \subseteq A \cap BC$. Now let $a \in A \cap BC$ and write $a = bc$ with $b \in B$ and $c \in C$. Then $ac^{-1} \in AC \cap B \subseteq A \cap B$ and therefore $a = (ac^{-1})c \in (A \cap B)C$. □

As a consequence of Lemma 1.1.1, we obtain the well-known *Dedekind identity* (the modular law).

Theorem 1.1.2. *Let G be a group. Let A, B, and C be subgroups of G. If $C \leq A$, then $(A \cap B)C = A \cap BC$ and $C(A \cap B) = A \cap CB$.*

Theorem 1.1.2 will be used in the sequel without further reference.

If A and B are subgroups of a group G, then the set AB is a union of right cosets of A; moreover, for $b_1, b_2 \in B$, Ab_1 and Ab_2 are equal if and only if $b_1 b_2^{-1} \in A \cap B$, that is, $(A \cap B)b_1 = (A \cap B)b_2$, and hence AB contains exactly $|B : (A \cap B)|$ right cosets of A. Therefore we obtain the following *product formula*.

Lemma 1.1.3. *Let A and B subgroups of a group G. Then $|AB|\,|A \cap B| = |A|\,|B|$.*

The following elementary result will be used in Section 5.1.

Theorem 1.1.4. *Let $G = AB$ be the product of the subgroups A and B. If W is a subgroup of G which permutes with both A and B and such that $W \cap A$ and $W \cap B$ permute, then $|G : (W \cap A)(W \cap B)|$ divides $|G : W|^2$.*

Proof. Applying Lemma 1.1.3

$$|G| = \frac{|A|\,|B|}{|A \cap B|} \quad \text{and} \quad |(W \cap A)(W \cap B)| = \frac{|W \cap A|\,|W \cap B|}{|W \cap A \cap B|}.$$

Hence

$$\begin{aligned} |G : (W \cap A)(W \cap B)| &= \frac{|A|}{|W \cap A|}\frac{|B|}{|W \cap B|}\frac{|W \cap A \cap B|}{|A \cap B|} \\ &= \frac{|AW : W|\,|BW : W|}{|A \cap B : W \cap A \cap B|}, \end{aligned}$$

which is a divisor of $|G : W|^2$. □

The next lemma, due to Ballester-Bolinches and Pérez-Ramos [55, Lemma 1], finds many applications in the study of totally permutable products.

Lemma 1.1.5. *Let the group $G = NB$ be the product of two subgroups N and B. Suppose that N is normal in G. Since B acts by conjugation on N, we can construct the semidirect product $X = [N]B$ with respect to this action. Then the natural map $\alpha\colon X \to G$ given by $(nb)\alpha = nb$ for every $n \in N$ and every $b \in B$ is an epimorphism,* $\operatorname{Ker}\alpha \cap N = 1$, *and* $\operatorname{Ker}\alpha \leq \mathrm{C}_X(N)$.

Proof. This is obtained by direct checking. □

In working with permutable subgroups and factorised groups it is important to have the concept of a subnormal subgroup and some of its properties at our disposal. We know that in general normality is not a transitive relation on the set of subgroups of a group. Following Wielandt, this lack of transitivity is repaired by defining a new subgroup embedding property, the subnormality, which is the transitive closure of the relation of normality.

Definition 1.1.6. Let G be a group. A subgroup K of G is said to be *subnormal* in G if either $K = G$ or there are a non-negative integer r and a series

$$K = K_0 \trianglelefteq K_1 \trianglelefteq K_2 \trianglelefteq \cdots \trianglelefteq K_r = G$$

of subgroups of G.

If H is subnormal in G, then $H \cap L$ is subnormal in L for every subgroup L of G and, if N is a normal subgroup of G, then HN/N is subnormal in G/N. Furthermore, the set $\mathrm{sn}(G)$ of all subnormal subgroups of a group G is a sublattice of the subgroup lattice of G (see [119, A, 14.1, 14.2, and 14.4]).

We now record without proof two further useful properties of subnormal subgroups: the first result is due to Maier and Wielandt and the second one was proved by Wielandt. For detailed proofs, see [189, 7.7.1 and 7.2.5].

Theorem 1.1.7. *Let the group $G = AB$ be the product of the subgroups A and B. If $X \leq A \cap B$ is subnormal in both A and B, then X is subnormal in G.*

The following consequence of Theorem 1.1.7 can be proved by using an inductive argument.

Corollary 1.1.8. *Let the group $G = G_1 G_2 \cdots G_r$ be the product of the pairwise permutable subgroups $G_1, G_2, \dots, G_r$. If X is a subgroup of $\bigcap_{i=1}^{r} G_i$ which is subnormal in G_i for all $i \in \{1, 2, \dots, r\}$, then X is subnormal in G.*

If X is a subgroup of a group G and Y is a subset of G, we denote by $\langle X^Y \rangle$ the subgroup $\langle X^y : y \in Y \rangle$. Hence $\langle X^G \rangle$ is the *normal closure* of X in G, that is, smallest normal subgroup of G containing X.

Theorem 1.1.9. *Let G be a group and let A and B be two subgroups of G such that $AB^g = B^g A$ for every $g \in G$. Then the following properties hold:*

1. *If $G = A\langle B^G \rangle = B\langle A^G \rangle$, then $G = AB$.*
2. *$\langle A^B \rangle \cap \langle B^A \rangle$ is a subnormal subgroup of G.*
3. *If $AB \leq H \leq G$, then $\langle A^H \rangle \cap \langle B^H \rangle$ is a subnormal subgroup of G.*
4. *If X and Y are two subsets of G, then the commutator $[\langle A^X \rangle, \langle B^Y \rangle]$ is a subnormal subgroup of G.*

Let π be a set of primes. Recall that $\mathrm{O}_\pi(G)$ is the unique largest normal π-subgroup of the group G, which may, of course, be the trivial subgroup. When $\pi = \{p\}$, p a prime, we write $\mathrm{O}_{\{p\}}(G) = \mathrm{O}_p(G)$. We define the *Fitting subgroup* of G, denoted by $\mathrm{F}(G)$, to be the product of the subgroups $\mathrm{O}_p(G)$ as p runs over the prime divisors of the order of G. We know, by [157, III, 4.3], that $\mathrm{F}(G)$ is the intersection of the centralisers of the chief factors of G. If G is soluble, it can be also proved that $\mathrm{F}(G)$ is the intersection of all H appearing in complemented self-centralising chief factors H/K.

For the maximal normal p-subgroup of a factorised group, p a prime, and its Fitting subgroup the following consequence of Theorem 1.1.7 was proved by Kazarin [173] and Maier [192] (see also Monakhov [199, 200]) in the case of two factors.

Corollary 1.1.10. *Let the group $G = G_1 G_2 \cdots G_r$ be the product of the pairwise permutable subgroups $G_1, G_2, \dots, G_r$. Then $\bigcap_{i=1}^r \mathrm{O}_p(G_i)$ is contained in $\mathrm{O}_p(G)$ and $\bigcap_{i=1}^r \mathrm{F}(G_i)$ is contained in $\mathrm{F}(G)$.*

Proof. Clearly $\bigcap_{i=1}^r \mathrm{O}_p(G_i)$ is a subnormal subgroup of G_j for all $j \in \{1, 2, \dots, r\}$ and thus $\bigcap_{i=1}^r \mathrm{O}_p(G_i)$ is a subnormal subgroup of G by Corollary 1.1.8. Therefore $\bigcap_{i=1}^r \mathrm{O}_p(G_i)$ is contained in $\mathrm{O}_p(G)$. The proof for the Fitting subgroup is analogous. □

Given a set of primes π and a group G, let $\mathrm{O}^\pi(G)$ denote the unique smallest normal subgroup of G for which the corresponding factor group is a π-group. The following simple observation is also important.

Lemma 1.1.11. *Let G a group and let π be a set of primes. If H is a subnormal subgroup of G and $|G : H|$ is a π-number, then $\mathrm{O}^\pi(G) = \mathrm{O}^\pi(H)$.*

Proof. Arguing by induction on $|G : H|$, it is enough to assume that H is a normal subgroup of G. Then $\mathrm{O}^\pi(H)$ is a characteristic subgroup of H. Hence $\mathrm{O}^\pi(H)$ is a normal subgroup of G and $|G : \mathrm{O}^\pi(H)|$ is a π-number. Therefore $\mathrm{O}^\pi(G) \leq \mathrm{O}^\pi(H)$. It follows that $\mathrm{O}^\pi(G)$ is a normal subgroup of H of index a π-number, which implies that $\mathrm{O}^\pi(H) \leq \mathrm{O}^\pi(G)$. We conclude that $\mathrm{O}^\pi(G) = \mathrm{O}^\pi(H)$. □

In the structural study of factorised groups it is interesting to know which subgroups inherit the factorisation. In the following, we shall be concerned with this property, which we now formalise with appropriated definitions.

Definition 1.1.12. A subgroup S of a group $G = G_1G_2\cdots G_r$ which is the product of the pairwise permutable subgroups $G_1, G_2, \dots, G_r$, is called *prefactorised* (with respect to this factorisation) if $S = (S \cap G_1)(S \cap G_2)\cdots(S \cap G_r)$, or, equivalently, if every $s \in S$ can be written as $s = s_1s_2\cdots s_r$, with $s_i \in S \cap G_i$.

Definition 1.1.13. Let the group $G = G_1G_2\cdots G_r$ be the product of the pairwise permutable subgroups $G_1, G_2, \dots, G_r$. A subgroup S of G is said to be *factorised* (with respect to the factorisation $G = G_1G_2\cdots G_r$) when it satisfies the following condition: if $a_{i_1}a_{i_2}\cdots a_{i_r} \in S$ with $a_{i_j} \in G_{i_j}$, where $\{i_1, i_2, \dots, i_r\} = \{1, 2, \dots, r\}$, then a_{i_j} belongs to S for all $j \in \{1, 2, \dots, r\}$.

The concept of factorised subgroup of a product of two subgroups was first introduced by Wielandt [256].

Lemma 1.1.14. *Let the group $G = G_1G_2\cdots G_r$ be product of the pairwise permutable subgroups $G_1, G_2, \dots, G_r$. If S is a factorised subgroup of G, then S is prefactorised and $G_i \cap \prod_{j\neq i} G_i \leq S$ for all $i \in \{1, 2, \dots, r\}$. The converse is true provided that $(S \cap G_i)$ permutes with $(S \cap G_j)$ for all i, j.*

Proof. If $x = a_1a_2\cdots a_r$ is an element of S with $a_i \in G_i$, then by the hypothesis a_i belongs to $S \cap G_i$ for all $i \in \{1, 2, \dots, r\}$. This shows that $S = (S \cap G_1)(S \cap G_2)\cdots(S \cap G_r)$. Moreover, if $x \in G_i \cap \prod_{j\neq i} G_j$ for some $i \in \{1, 2, \dots, r\}$, then $x^{-1} \in \prod_{j\neq i} G_j$ and $xx^{-1} = 1 \in S$. Thus x lies in S because S is factorised.

Conversely, assume that $S = (S \cap G_1)(S \cap G_2)\cdots(S \cap G_r)$ is the product of the pairwise permutable subgroups $S\cap G_1, S\cap G_2, \cdots, S\cap G_r$ and $G_i\cap\prod_{j\neq i} G_i \leq S$ for all i. Let $x = a_{i_1}a_{i_2}\cdots a_{i_r}$ be an element of S with $a_{i_j} \in G_{i_j}$ where $\{i_1, i_2, \dots, i_r\} = \{1, 2, \dots, r\}$. Since $S = (S \cap G_1)(S \cap G_2)\cdots(S \cap G_r)$ and the factors are pairwise permutable, we have also that $x = b_{i_1}b_{i_2}\cdots b_{i_r}$ with $b_{i_j} \in S \cap G_{i_j}$. Therefore $b_{i_1}^{-1}a_{i_1} = (b_{i_2}\cdots b_{i_r})(a_{i_2}\cdots a_{i_r})^{-1} \in G_{i_1} \cap G_{i_2}\cdots G_{i_r} \leq S$ and so $a_{i_1} \in S$. Now we deduce that $a_{i_2}\cdots a_{i_r}$ belongs to S. Arguing as before, $a_{i_2}\cdots a_{i_r} = c_{i_1}c_{i_2}\cdots c_{i_r} = d_{i_2}d_{i_1}c_{i_3}\cdots c_{i_r}$ with $c_{i_j} \in S \cap G_{i_j}$ and $d_{i_j} \in S \cap G_{i_j}$. Therefore $d_{i_2}^{-1}a_{i_2}$ belongs to $G_{i_2} \cap \prod_{j\neq i_2} G_{i_j}$, which is contained in S. Consequently we obtain that $a_{i_2} \in S$ and similarly $a_{i_3}, \dots, a_{i_r}$ belong to S and S is factorised. □

Proposition 1.1.15. *Let the group $G = G_1G_2\cdots G_r$ be the product of the pairwise permutable subgroups $G_1, G_2, \dots, G_r$. Every subgroup of G containing all factors but one is factorised.*

Proof. Assume that $G_2 \cdots G_r \leq S \leq G$. Then

$$\begin{aligned} S &= S \cap G_1 G_2 \cdots G_r \\ &= (S \cap G_1)(G_2 \cdots G_r) = (S \cap G_1)(S \cap G_2) \cdots (S \cap G_r) \end{aligned}$$

and so S is the product of the pairwise permutable subgroups $S \cap G_1, G_2, \ldots, G_r$. In particular, S is prefactorised. On the other hand, the product of $r - 1$ subgroups of the family $\{S \cap G_1, G_2, \ldots, G_r\}$ is contained in S. Moreover, they are pairwise permutable. By Lemma 1.1.14, S is factorised. □

Corollary 1.1.16. *Let the group $G = AB$ be the product of the subgroups A and B.*

1. *Every factorised subgroup of G is prefactorised.*
2. *A subgroup S is factorised if and only if it is prefactorised and $A \cap B$ is contained in S.*
3. *Every subgroup of G containing A or B is factorised.*

The next lemma studies the behaviour of prefactorised and factorised subgroups in epimorphic images. It also shows that the intersection of factorised subgroups of groups which are the products of pairwise permutable subgroups are factorised (cf. [9, Lemma 1.1.2]).

Lemma 1.1.17. *Let the group $G = G_1 G_2 \cdots G_r$ be the product of the pairwise permutable subgroups $G_1, G_2, \ldots, G_r$. Then:*

1. *If S is prefactorised in G and N is a normal subgroup of G, then SN/N is prefactorised in $G/N = (G_1N/N)(G_2N/N) \cdots (G_rN/N)$.*
2. *If N is a normal subgroup of G, a subgroup S/N of the factorised quotient group $G/N = (G_1N/N)(G_2N/N) \cdots (G_rN/N)$ is factorised if and only if S is a factorised subgroup of $G = G_1 G_2 \cdots G_r$.*
3. *If U and V are factorised subgroups of G, then $U \cap V$ is a factorised subgroup of G.*

Proof. 1. Clearly, $SN/N = ((S \cap G_1)N/N)((S \cap G_2)N/N) \cdots ((S \cap G_r)N/N) \leq (SN/N \cap G_1N/N)(SN/N \cap G_2N/N) \cdots (SN/N \cap G_rN/N)$, which is contained in SN/N. This shows that SN/N is prefactorised.

2. Let S be a factorised subgroup of G containing N. If $xN = a_{i_1} a_{i_2} \cdots a_{i_r} N$ is an element of S/N with $x \in S$ and $a_{i_j} \in G_{i_j}$, then $x = a_{i_1} a_{i_2} \cdots a_{i_r} y$, where y belongs to $N \leq S$. Hence $a_{i_1} a_{i_2} \cdots a_{i_r} = xy^{-1}$ belongs to S, and so a_{i_j} belongs to S. Therefore S/N is a factorised subgroup of G/N.

Conversely, suppose that the subgroup S/N is factorised in G/N. Consider an element $x = a_{i_1}a_{i_2}\cdots a_{i_r}$ of S, with $a_{i_j} \in G_{i_j}$. Since $xN = a_{i_1}a_{i_2}\cdots a_{i_r}N$, it follows that $a_{i_j}N$ belongs to S/N. Hence a_{i_j} belongs to S, and so S is factorised.

3. If $a_{i_1}a_{i_2}\cdots a_{i_r} \in U \cap V$ with $a_{i_j} \in G_{i_j}$, where $\{i_1, i_2, \ldots, i_r\} = \{1, 2, \ldots, r\}$, then using the fact that U and V are factorised it is clear that a_{i_j} belongs to $U \cap V$ for all $i_j \in \{1, 2, \ldots, r\}$. Therefore $U \cap V$ is a factorised subgroup of G. □

The following example due to Höfling [152] shows that the intersection of prefactorised subgroups might not be prefactorised, even if both subgroups are normal.

Example 1.1.18. Let V be a vector space of dimension 3 over a field F with base $\{x_1, x_2, x_3\}$. We use additive notation. Let $A = \langle x_1, x_2\rangle$ and $B = \langle x_2, x_3\rangle$. Then the subgroups $H_1 = \langle x_1, x_2 + x_3\rangle = \langle x_1\rangle \oplus \langle x_2 + x_3\rangle$ and $H_2 = \langle x_1 + x_2, x_3\rangle = \langle x_1 + x_2\rangle \oplus \langle x_3\rangle$ are prefactorised, but their intersection $H = \langle x_1 + x_2 + x_3\rangle$ intersects A and B trivially. Hence H cannot be prefactorised.

Now we consider Hall subgroups, in particular Sylow subgroups, of factorised groups. The reader might guess that a factorised group has a prefactorised Hall subgroup, and indeed it has provided that it is a product of two subgroups and satisfies a Sylow-type theorem. We say that a group G *satisfies the π-Sylow theorem* for a set of primes π, or, following Hall [142], that G *satisfies D_π*, when the following property holds: If U is a π-subgroup of G and S is a π-subgroup of G such that S is not properly contained in any π-subgroup of G, then there exists an element g of G such that $U^g \leq S$ (a soluble group automatically satisfies the D_π-property for all π by [119, I, 3.3]).

The following result, whose proof can be found in Huppert [157, IV, 4.6], allows us to prove a number of useful facts.

Theorem 1.1.19. *Let π be a set of primes and let $G = AB$ be the product of the subgroups A and B. Assume also that G satisfies the π-Sylow theorem. There exist Hall π-subgroups G_π, A_π, and B_π of G, A, and B, respectively, such that $G_\pi = A_\pi B_\pi$. In particular, G_π is a prefactorised Hall π-subgroup of G.*

Proposition 4.1.45 shows that the above theorem is true for soluble pairwise mutually permutable products of more than two factors. Moreover, Example 4.1.46 demonstrates that the Sylow subgroups are not factorised in general, even for mutually permutable products of two factors.

The next two results, proved by Pennington [210], are important in the study of products of nilpotent groups in Chapter 3.

Lemma 1.1.20. *Let π be a set of primes and let $G = AB$ be a factorised group satisfying D_π. If the subgroups A and B of G possess normal Hall π-subgroups A_π and B_π, respectively, then $A_\pi B_\pi = B_\pi A_\pi$ is a Hall π-subgroup of G and $[A_\pi, B_\pi] \leq \mathrm{O}_\pi(G)$. In particular, if $\mathrm{O}_\pi(G) = 1$, then $[\langle A_\pi^G\rangle, \langle B_\pi^G\rangle] = 1$.*

Proof. Let g be an element of G. Then $G = AB^g$ and $A_\pi B_\pi^g$ is a Hall π-subgroup of G by Theorem 1.1.19. Therefore A_π permutes with every conjugate of B_π^g. We apply Theorem 1.1.9 (4) to deduce that $[A_\pi, B_\pi^g]$ is a subnormal π-subgroup of G, and so it is contained in $O_\pi(G)$. Assume $O_\pi(G) = 1$. Then A_π centralises every conjugate of B_π. Consequently, $[\langle A_\pi^G\rangle, \langle B_\pi^G\rangle] = 1$. □

A nice result about factorised groups with the D_π-property is that $O_\pi(G)$ is prefactorised provided that the factors have normal Hall π-subgroups. This fact will determine the prefactorisation of the Fitting subgroup of a product of two nilpotent groups (see Theorem 3.2.7).

Proposition 1.1.21. *Let $G = AB$ be the product of two subgroups satisfying D_π. If A and B have normal Hall π-subgroups, then*

$$O_\pi(G) = (A \cap O_\pi(G))(B \cap O_\pi(G)).$$

Proof. Let us denote with bars the image under the natural homomorphism $G \to G/O_\pi(G)$. Since G satisfies D_π, by Theorem 1.1.19 there exists a Hall π-subgroup of G of the form $H = A_\pi B_\pi$, where A_π and B_π denote the Hall π-subgroups of A and B, respectively. Let $ab \in O_\pi(G) \le H = A_\pi B_\pi$ with $a \in A_\pi$ and $b \in B_\pi$. Then $\bar{a} = \bar{b}^{-1}$. Thus, since $O_\pi(\bar{G}) = 1$, we have by Lemma 1.1.20 that

$$1 = [\bar{a}, \langle \bar{B}_\pi^{\bar{G}}\rangle] = [\bar{b}^{-1}, \langle \bar{B}_\pi^{\bar{G}}\rangle],$$

and so $\bar{b}$ lies in the centre of $\langle \bar{B}_\pi^{\bar{G}}\rangle$. But then $\bar{b} \in O_\pi(Z(\langle \bar{B}_\pi^{\bar{G}}\rangle) \le O_\pi(\bar{G}) = 1$, which means that $b \in O_\pi(G)$. A similar argument shows that $a \in O_\pi(G)$. By Lemma 1.1.14, we conclude that

$$O_\pi(G) = (A \cap O_\pi(G))(B \cap O_\pi(G)).$$

This completes the proof of the proposition. □

Remark 1.1.22. Clearly $O_\pi(G)$ is not factorised in general. For example, let G be a nilpotent group of order divisible by two different primes p and q and that $A = B = G$. Then $A \cap B = G$ is not contained in $O_p(G)$ and, by Corollary 1.1.16, $O_p(G)$ cannot be factorised.

The following lemma, due to Kazarin [176], was first proved by Johnson [167] for soluble groups and it is an extension of Corollary 1.1.10 to π-soluble products of two factors.

Lemma 1.1.23. *Let π be a set of prime numbers. Let the π-soluble group $G = AB$ be the product of its subgroups A and B. Then $O_\pi(A) \cap O_\pi(B) \le O_\pi(G)$.*

Proof. Let $x \in O_\pi(A) \cap O_\pi(B)$ and let $g \in G$. Then $g = ab$ for some $a \in A, b \in B$. Hence

$$\langle x, x^g \rangle = \langle x^{b^{-1}}, x^a \rangle^b \leq \langle O_\pi(A), O_\pi(B) \rangle^b.$$

Since G is a D_π-group, by Theorem 1.1.19 we have that some Hall π-subgroup K of G is a product of a Hall π-subgroup S of A and a Hall π-subgroup T of B. Since $O_\pi(A) \leq S$ and $O_\pi(B) \leq T$, the subgroup $\langle O_\pi(A), O_\pi(B) \rangle$ is contained in $ST = K$ and is also a π-group. Hence $\langle x, x^g \rangle$ is a π-group for every $g \in G$. Assume now that $g \in O_{\pi,\pi'}(G)$. Then $[x, g]$ is a π-element in $O_{\pi,\pi'}(G)$ and so $[x, g] \in O_\pi(G)$. Consequently,

$$x\, O_\pi(G) \in C_G(O_{\pi,\pi'}(G)/O_\pi(G)) \leq O_{\pi,\pi'}(G)/O_\pi(G)$$

by [158, IX, 1.3] and thus $x \in O_{\pi,\pi'}(G)$. Since x is a π-element, we conclude that $x \in O_\pi(G)$, as desired. □

The π-solubility of G is necessary in Lemma 1.1.23.

Example 1.1.24 (Johnson [167]). Let $G = PSL_2(2^n)$ with $n \geq 2$. Then G is a simple group of order $2^n(2^{2n} - 1)$ (see [157, III, 6.2]). By [157, III, 8.27], G possesses a subgroup A which is the semidirect product of an elementary abelian group of order 2^n with a cyclic group of order $2^n - 1$ and a subgroup B which is dihedral of order $2(2^n + 1)$. Then $|A \cap B| = 2$ and $AB = G$ by order considerations. Let $\pi = \pi(B)$ be the set of primes dividing $|B|$. We have that $|O_\pi(A) \cap O_\pi(B)| = 2$, but $|O_\pi(G)| = 1$. Hence $O_\pi(A) \cap O_\pi(B)$ cannot be contained in $O_\pi(G)$.

We end this section with a lemma of Amberg and Höfling [11], which will be used in Chapter 3.

Lemma 1.1.25. *If the group $G = AB = AC = BC$ is the product of three subgroups A, B, and C and a conjugate C^g with $g \in G$ is factorised with respect to the factorisation $G = AB$, then $G = C$.*

Proof. We clearly have $G = AC^g = BC^g$. Hence we may assume without loss of generality that C itself is factorised. Then $G = A(B \cap C) = (A \cap C)B$ and so

$$\begin{aligned} G &= A(B \cap C) \cap (A \cap C)B \\ &= (A \cap C)(A(B \cap C) \cap B) \\ &= (A \cap C)(A \cap B)(B \cap C), \end{aligned}$$

which is contained in C because C is factorised. □

1.2 Permutability, Sylow permutability, and related properties

One of the central concepts in this book is that of a *permutable* subgroup, that is, a subgroup which permutes with every subgroup of the group. Permutable subgroups have been subject of many papers since their introduction by Ore [207]. There they are called *quasinormal*. More recently the substitute term permutable has gained currency and so we prefer to use it for this and related concepts.

Sometimes the requirement for a subgroup is not to permute with all subgroups, but only with the members of some interesting families of subgroups. Therefore several subgroup embedding properties related to permutability appear. In this section we shall establish a number of fundamental results concerned with some of these embeddings. We begin by introducing them and thereafter we proceed to develop their properties.

Definition 1.2.1. Let Θ be a family of subgroups of a group G. A subgroup H of a group G is called Θ*-permutable* if H permutes with all subgroups $A \in \Theta$.

In case $\Theta = \mathrm{s}(G)$ is the family of all subgroups of a group G, the Θ-permutable subgroups of G are just the permutable subgroups of G. If $\Theta = \mathrm{Syl}(G)$ is the family of all Sylow subgroups of G, the Θ-permutable subgroups of G are called S-*permutable* subgroups. As in the case of permutable subgroups, S-permutable subgroups have received other names in the mathematical literature, like S-*quasinormal* or π-*quasinormal subgroups* (cf. Agrawal [2]). If $\Theta = \mathrm{Max}(G)$ is the family of all maximal subgroups of G, the Θ-permutable subgroups of G are called M-*permutable* subgroups (cf. Maier [193]).

Of course all normal subgroups are permutable and there are numerous examples of non-normal permutable subgroups. For example, let p be an odd prime, and let G be an extraspecial group of order p^3 and exponent p^2; G has all subgroups permutable, but G has non-normal subgroups. Permutable subgroups are S-permutable. However, the dihedral group of order 8 has subgroups which are not permutable, but all its subgroups are obviously S-permutable. This example also shows that an M-permutable subgroup is not permutable in general.

Theorem 1.2.2. *If a subgroup H of a group G permutes with the subgroups X and Y of G, then it also permutes with their join $\langle X, Y\rangle$. Hence, if Θ is a family of subgroups of G, the set of all Θ-permutable subgroups of G is closed with respect to joins.*

Proof. Let $g \in \langle X, Y\rangle$ and $h \in H$. Then $g = z_1 z_2 \cdots z_r$, where $z_i \in X$ or $z_i \in Y$ for $1 \le i \le r$. Set $h_0 = h$. For $1 \le i \le r$, by induction there exist $h_i \in H$ and $w_i \in X$ (if $z_i \in X$) or $w_i \in Y$ (if $z_i \in Y$) such that $h_{i-1} z_i = w_i h_i$. Hence $hg = (w_1 \cdots w_r) h_r$. This proves that $H\langle X, Y\rangle \subseteq \langle X, Y\rangle H$. An analogous argument proves the other inclusion. □

Remark 1.2.3. Permutability and S-permutability, like normality, are not in general transitive relations (consider, for instance, the alternating group of degree 4 and its subgroups of order 2). This observation is the point of departure of the study of some relevant classes of groups defined in terms of permutability properties of subnormal subgroups in Chapter 2.

Definition 1.2.4.

1. A *subgroup embedding functor* is a function f which associates to each group G a family of subgroups $f(G)$ of G such that for every group G and every isomorphism $\mu: G \to G^{\mu}$ we have that $X \in f(G)$ if and only if $X^{\mu} \in f(G^{\mu})$.
2. A subgroup H of a group G is said to be f*-permutable* if H is $f(G)$-permutable.

Remark 1.2.5. For every subgroup embedding functor f and every group $G, f(G)$ is a family of subgroups invariant by automorphisms.

Sometimes we will write $X f G$ instead of $X \in f(G)$.

Definition 1.2.6. Given two subgroup embedding functors f_1 and f_2, we write $f_1 \subseteq f_2$ if $f_1(G) \subseteq f_2(G)$ for every group G.

Examples of such functions include Syl, Max, sn, sfn, and s, defined in such a way that for every finite group G, $\mathrm{Syl}(G)$ is the set of Sylow subgroups, $\mathrm{Max}(G)$ is the set of all maximal subgroups, $\mathrm{sn}(G)$ is the set of all subnormal subgroups, $\mathrm{sfn}(G)$ is the set of all self-normalising subgroups, and $\mathrm{s}(G)$ is the set of all subgroups of G.

Note that, by Theorem 1.2.2, f-permutable subgroups are closed with respect to joins.

Lemma 1.2.7. *Let f be a subgroup embedding functor. If f is either* s *or* Syl, *then it satisfies the following two conditions:*

1. *If H is an f-permutable subgroup of a group G and H is contained in L, then H is f-permutable in L.*
2. *If N is a normal subgroup of G and H is an f-permutable subgroup of G, then HN/N is f-permutable in G/N.*

Proof. 1. It is clear that Statement 1 holds for $f = \mathrm{s}$. Assume that $f = \mathrm{Syl}$ and let H be an S-permutable subgroup of a group G. Let L be a subgroup of G containing H. If P_0 is a Sylow subgroup of L, there exists a Sylow subgroup P of G such that $P \cap L = P_0$. Hence $HP_0 = H(P \cap L) = HP \cap L = P_0H$ and so H is S-permutable in L.

2. Assume that N is normal in G and that H is a permutable subgroup of G. Consider a subgroup T/N of G/N. Then H permutes with T and so $(HN)T = H(NT) = HT$ is a subgroup of G. It follows that HN/N permutes with T/N.

Suppose now that $f = \text{Syl}$. If T/N is a Sylow p-subgroup of G/N, there exists Sylow p-subgroup T_p of G such that $T = NT_p$. Therefore H permutes with T_p and so $HT_p = (HN)T_p = H(NT_p) = HT$ is a subgroup of G. Consequently H permutes with T and so HN/N permutes with T/N. □

Since there are groups with subgroups which are not sn-permutable and every subgroup of non-abelian simple group is sn-permutable, it follows that the functor sn does not satisfy Assertion 1 of Lemma 1.2.7. Nevertheless, Max, sn, and sfn satisfy Assertion 2 of Lemma 1.2.7 (the proof runs along the same lines to the above one).

The next result shows that the Assertion 1 of Lemma 1.2.7 is also satisfied by sfn (cf. Maier [193, 4.4]).

Lemma 1.2.8. *If X is an* sfn*-permutable subgroup of a group G and $X \leq U \leq G$, then X is* sfn*-permutable in U.*

Proof. Let Y be a subgroup of U such that $\mathrm{N}_U(Y) = Y$. Set $Y_0 = Y$ and $Y_i = \mathrm{N}_G(Y_{i-1})$ for $i = 1, 2, \dots$. There exists an n such that $Y_n = Y_{n+1}$. Denote $Y_n = Z$. Hence we have that Y is a subnormal subgroup of Z and so Y is subnormal in $Z \cap U$. Since $\mathrm{N}_U(Y) = Y$, we conclude that $Y = Z \cap U$. Since X is sfn-permutable in G, it follows that X permutes with Z. Hence

$$XY = X(Z \cap U) = XZ \cap U = ZX \cap U = (Z \cap U)X = YX.$$

Therefore X permutes with all self-normalising subgroups of U. □

Lemma 1.2.8 enables us to prove that every sfn-permutable subgroup is subnormal ([193, 4.5]). We need a preliminary result ([193, 2.6]).

Lemma 1.2.9. *Let X be a subgroup of a group G and U a maximal subgroup of G such that $X \leq U$. If X permutes with all conjugates U^g of U in G, then $\langle X^G \rangle$ is contained in the core U_G of U in G.*

As a consequence, if X is an M*-permutable proper subgroup of G, then $\langle X^G \rangle \neq G$.*

Proof. Suppose that X is not contained in U^g for a $g \in G$. Then $U^g X = XU^g = G$ by hypothesis. It follows that $g = xg^{-1}ug$ for some $u \in U$ and $x \in X$. Hence $g = ux \in U$. Consequently $X \leq U^g = U$, against our assumption. Therefore we have that $X \leq U_G$ and $\langle X^G \rangle \leq U_G$. □

Theorem 1.2.10. *Let G be a group and let X be a subgroup of G. If X is an* sfn*-permutable subgroup of G, then X is a subnormal subgroup of G.*

Proof. We will argue by induction on $|G|$. We may assume that X is a proper subgroup of G. By Lemma 1.2.9, it follows that $Z = \langle X^G \rangle$ is a proper subgroup of G since X is M-permutable, and by Lemma 1.2.8, X is sfn-permutable in Z. By induction, X is subnormal in Z. It follows that X is subnormal in G. The induction argument is therefore complete. □

We now present an example of a group having a non-subnormal M-permutable subgroup. This example also shows that functor Max does not satisfy Assertion 1 of Lemma 1.2.7.

Example 1.2.11. Let $S = \mathrm{Alt}(5)$, the alternating group of degree 5, and let P denote the principal indecomposable module over GF(2), the finite field of 2-elements. Let $G = [P]S$ be the corresponding semidirect product. It is known that $P/\mathrm{Rad}(P)$, where $\mathrm{Rad}(P)$ is the radical of P regarded as a module for S (see [119, B, 3.7]), is isomorphic to the trivial irreducible $\mathrm{GF}(2)S$-module and $\mathrm{C}_S(P) = \mathrm{O}_{2'}(S) = 1$ (see [119, B, 4.23]). Applying [119, B, 3.14], $\mathrm{Rad}(P)$ is contained in the Frattini subgroup of G. Since $\mathrm{Rad}(P)S$ is a normal subgroup of G, it follows that S is an M-permutable subgroup of G. If S were subnormal in G, then S would centralise every chief factor of G below P by [119, A, 14.3]. Applying [119, IV, Theorem 6.9], S would be nilpotent, which is not the case. Therefore S is not subnormal in G.

Now we turn our attention to permutable and S-permutable subgroups. Knowledge of properties of these subgroups will be applied to give information about some important classes of groups in the next chapters.

Recall the following theorem of Wielandt ([258], see also [119, A, 14.10]).

Theorem 1.2.12. *For a subgroup H of a group G, the following statements are equivalent:*

1. *H is subnormal in G.*
2. *H is subnormal in $\langle H, H^g\rangle$ for all $g \in G$.*
3. *H is subnormal in $\langle H, g\rangle$ for all $g \in G$.*

It is of course an immediate corollary of the above result that the normal closure $\langle H^G\rangle$ is nilpotent if and only if $\langle H, H^g\rangle$ is nilpotent for all $g \in G$. This is a special case of a result of Baer [26, Satz L].

The equivalence between 1 and 2 does not hold neither for normality, nor permutability, nor S-permutability: Consider the symmetric group $G = \mathrm{Sym}(4)$ of degree 4 and $H = \langle (1,2)(3,4)\rangle$. For every $g \in G$, $\langle H, H^g\rangle \leq \mathrm{Soc}(G)$. In fact, if $g \notin \mathrm{N}_G(H)$, $\langle H, H^g\rangle = \mathrm{Soc}(G)$, which is a normal subgroup of G, and hence H is S-permutable in $\langle H, H^g\rangle$. Let $g = (1,2,3)$. Then H is not S-permutable in $\langle H, g\rangle = \mathrm{Alt}(4)$. In particular, H is not S-permutable in G.

Nevertheless, the equivalence between 1 and 3, which holds for normality and permutability, also holds for S-permutability, as it was shown by Ballester-Bolinches and Esteban-Romero in [46].

Theorem 1.2.13. *A subgroup H of a group G is S-permutable in G if and only if H is S-permutable in $\langle H, g\rangle$ for every $g \in G$.*

Before proving the theorem just cited, we digress in the following to discuss some interesting facts about S-permutability (and permutability). They are contained in the next "portmanteau" theorem, whose proof is due to Schmid [221].

Theorem 1.2.14. *Let G be a group. Then:*

1. *If N is a normal subgroup of G and H/N is a permutable (respectively, S-permutable) subgroup of G/N, then H is a permutable (respectively, S-permutable) subgroup of G.*
2. *(Deskins* [116], *Kegel* [181]*) If H is an S-permutable subgroup of G, then the quotient $\langle H^G\rangle/H_G$ is nilpotent. In particular, H/H_G is contained in the Fitting subgroup $\mathrm{F}(G/H_G)$ of G/H_G.*
3. *(Kegel* [181]*) If H is an S-permutable subgroup of G, then H is subnormal in G.*
4. *If X is a subgroup of G and H is a permutable (respectively S-permutable) subgroup of G, then $H \cap X$ is a permutable (respectively, S-permutable) subgroup of X.*

Proof. 1. Let T be a (Sylow) subgroup of G. Then NT/N is a (Sylow) subgroup of G/N and, by hypothesis, H/N permutes with NT/N. Hence H permutes with NT and so $H(NT) = (HN)T = HT$ is a subgroup of G. We conclude that H permutes with T.

2. Assume that (G, H) is a counterexample with $|G| + |G : H|$ minimum. Then $H_G = 1$. Write

$$D = \bigcap_{x \in G \setminus \mathrm{N}_G(H)} \langle H, H^x\rangle.$$

Clearly, $H \leq D$ and every term $\langle H, H^x\rangle$ is S-permutable in G by Theorem 1.2.2. It is clear that the normal closure of D is $\langle D^G\rangle = \langle H^G\rangle$ and that

$$D_G = \bigcap_{x \in G \setminus \mathrm{N}_G(H)} \langle H, H^x\rangle_G.$$

Suppose that $D = H$. Then for every $x \in G \setminus \mathrm{N}_G(H)$, $\langle\langle H, H^x\rangle^G\rangle = \langle H^G\rangle$ and $\langle H^G\rangle/\langle H, H^x\rangle_G$ is nilpotent by the choice of H. Hence $\langle H^G\rangle/H_G$ is nilpotent, against the hypothesis. Hence H is a proper subgroup of D. In particular, for some prime number p, there exists a p-element $x \in G \setminus \mathrm{N}_G(H)$. Consider a Sylow p-subgroup P of G such that $x \in P$. We have that H permutes with P and so

$$H < D \leq \langle H, H^x\rangle \leq \langle H, x\rangle \leq HP = PH.$$

We conclude that the index $|D : H|$ is a power of p. Suppose that a q-element $y \in G$ for a prime $q \neq p$ does not normalise H. Then the same argument shows that $|D : H|$

is a power of q, but then $D = H$. Therefore, all p'-elements of G normalise H. It follows that $O^p(G)$ normalises H and so H is normal in $HO^p(G)$. Since $HO^p(G)$ is subnormal in G, H is subnormal in $G = O^p(G)P$. Moreover, $\langle H^G \rangle = \langle H^P \rangle \leq HP$. By Lemma 1.1.11, $O^p(\langle H^G \rangle)$ is normal subgroup of G contained in H. Hence $O^p(\langle H^G \rangle) \leq H_G = 1$ and $\langle H^G \rangle$ is a p-group. This contradiction proves the result.

3. By Statement 2, we have that $\langle H^G \rangle / H_G$ is nilpotent. Therefore H/H_G is subnormal in $\langle H^G \rangle / H_G$, which is normal in G/H_G. Thus H/H_G is subnormal in G/H_G. We conclude that H is subnormal in G.

4. Assume that H is permutable in G. Let T be a subgroup of X. Then $(H \cap X)T = HT \cap X$ and HT is a subgroup of G. Therefore $H \cap X$ permutes with T and $H \cap X$ turns out to be permutable in X.

Suppose that H is S-permutable in G. Let Q be a Sylow subgroup of X. There exists a Sylow subgroup P of G such that $Q = P \cap X$. Since H is S-permutable in G, it follows that HP is a subgroup of G. We prove that $X \cap HP = (H \cap X)Q$. Since H is a subnormal subgroup of HP by Statement 3 and $|HP : H|$ is a power of p, it follows that $O^p(H) = O^p(HP)$ by Lemma 1.1.11. Hence every Sylow q-subgroup of HP is contained in H for every prime $q \neq p$. In particular, $|(X \cap HP) : (H \cap X)|$ is a power of p. Therefore $X \cap HP = (H \cap X)Q$ and $H \cap X$ is an S-permutable subgroup of X. □

By Theorem 1.2.14 (3), every permutable subgroup is subnormal. This is one of the earliest results about permutable subgroups and was proved by Ore [207]. The "permutable" version of the Deskins and Kegel's result was proved by Itô and Szép [164].

In infinite groups, permutable subgroups do not have to be subnormal as Stonehewer shows in [229]. They are, however, always ascendant. Celentani, Leone, and Robinson [94] show that the normal closure of a core-free permutable subgroup of a locally finite group is locally nilpotent and its Sylow p-subgroups are finite if the locally finite group satisfies min-p for all primes p.

The subnormality of an S-permutable subgroup is crucial in the proof of Theorem 1.2.13.

Proof of Theorem 1.2.13. Suppose that G is a group of least order with a subgroup H such that H is S-permutable in $\langle H, g \rangle$ for every $g \in G$, but H is not S-permutable in G. Since H is a subnormal subgroup of $\langle H, g \rangle$ for every $g \in G$ by Theorem 1.2.14 (3), it follows from Theorem 1.2.12 that H is a subnormal subgroup of G. Let M be a maximal normal subgroup of G containing H. Since H is not S-permutable in G, there exists a prime p and a Sylow p-subgroup P of G such that P does not permute with H.

Suppose that there exists a maximal subgroup M_1 of G such that $H \leq M_1$ and M is not contained in M_1. Then $MM_1 = G$. From the minimality of G, it follows that H is S-permutable in M and M_1. Moreover, there exists a Sylow p-subgroup Q of

M and a Sylow p-subgroup Q_1 of M_1 such that their product $QQ_1 = P_0$ is a Sylow p-subgroup of G by Theorem 1.1.19. Then H permutes with both Q and Q_1, hence H permutes with P_0. Consider a minimal normal subgroup N of G contained in M. The minimal choice of G implies that HN/N permutes with PN/N. Hence HN permutes with P and $P(HN)$ is a subgroup of G. If $P(HN)$ is a proper subgroup of G, then H permutes with P. This contradicts the choice of H. Consequently we have that $P(HN) = G$. There exists an element $x \in G$ such that $P_0 = P^x$, and x can be expressed as $x = x_1x_2$, with $x_1 \in P$ and $x_2 \in HN$. Therefore $P_0 = P^x = P^{x_2}$. Hence H permutes with P^{x_2}, or, equivalently, $H^{x_2^{-1}}$ permutes with P. Since H is a subnormal subgroup of G, by a theorem of Wielandt [119, A, 14.3], we have that $\mathrm{Soc}(G)$ normalises H. In particular, H is a normal subgroup of HN, and since $x_2 \in HN$, we have that $H = H^{x_2^{-1}}$. This implies that H permutes with P, against the choice of H. Consequently, if M_1 is a maximal subgroup of G containing H, then $M \leq M_1$. Since $P(HN) = G$ and $HN \leq M$, it follows that $|G : M|$ is a power of p. Hence all maximal subgroups of G/M are normal. Thus M is actually a maximal subgroup, and it is the unique maximal subgroup of G containing H. Therefore, if $x \in G \setminus M$, we have that $\langle H, x\rangle = G$: otherwise there would exist another maximal subgroup of G containing H. From the hypothesis, H is S-permutable in $\langle H, x\rangle = G$. This final contradiction proves the theorem.

The converse holds by Lemma 1.2.7 (1). □

There are examples of permutable subgroups H of groups G in which H/H_G is not abelian, as a construction of Thompson [240] shows. Since he incorrectly claims that the permutable subgroup has trivial core, we modify slightly his example to ensure it is a core-free subgroup.

Example 1.2.15. Let p be an odd prime. Let E be an elementary abelian p-group of order p^{p+1} with basis $\{e_1, e_2, \dots, e_{p+1}\}$. Let

$$G = \langle E, x, y : x^{p^2} = e_{p+1},\ e_i^x = e_ie_{i+1}\ (1 \leq i \leq p),$$
$$y^p = e_p^{-1},\ x^y = x^{1+p},\ e_1^y = e_1e_p^{-1},\ e_i^y = e_i\ (2 \leq i \leq p+1)\rangle.$$

Two applications of a result on cyclic group extensions ([223, 9.7.1]) lead to $|G| = p^{p+4}$ and so E is normal in G and G/E is an extraspecial group of order p^3 and exponent p^2.

Let $H = \langle e_1, \dots, e_p, y\rangle$. Then $|H| = p^{p+1}$, $H' = \langle e_p\rangle \neq 1$ and $e_{p+1} \notin H$. Note that $G = H\langle x\rangle$. Let S be a subgroup of G containing H. Then $S = S \cap H\langle x\rangle = H(S \cap \langle x\rangle)$. It follows that the unique possibilities for S are H, $H\langle e_{p+1}\rangle$, $H\langle x^p\rangle$ and G.

Let z be an element of $G \setminus H\langle x^p\rangle$. We show that $z^p \in H\langle x^p\rangle \setminus H\langle e_{p+1}\rangle$. Assume that $z = ey^mx^n$, where $e \in H \cap E$ and m and n are natural numbers. Denote with bars the images of elements of G in G/E. One can verify by direct calculation that

$\bar{z}^p = (\overline{ey^m x^n})^p = \bar{x}^{np}$. Since $z \in G \setminus H\langle x^p \rangle$, p does not divide n. Therefore x^{np} is not an element of E, and hence $z^p \notin E$. In particular, $z^p \notin H\langle e_{p+1} \rangle$.

Now let z be an element of $H\langle x^p \rangle \setminus H\langle e_{p+1} \rangle$. We show that $z^p \in H\langle e_{p+1} \rangle \setminus H$. Assume that $z = ey^m x^{pn}$ with $e \in H \cap E$ and m and n natural numbers. Since $e_1^y = e_1 e_p^{-1}$, $e_i^y = e_i$ for $2 \leq i \leq p+1$ and $e_1^{x^p} = e_1 e_{p+1}$, $e_i^{x^p} = e_i$ for $2 \leq i \leq p+1$, and y^m and x^{pn} commute, we obtain that $(ey^m x^{pn})^p = e_p^{-m} e_{p+1}^n$. If p does not divide n, hence $z \in H\langle e_{p+1} \rangle \setminus H$.

Finally, if $z \in H\langle e_{p+1} \rangle \setminus H$, then $z^p \in H$.

We prove that H is permutable in G, that is, if J is a subgroup of G, then HJ is a subgroup of G. We can suppose that $J = \langle j \rangle$ is cyclic and J is not contained in H.

Assume that $j \in G \setminus H\langle x^p \rangle$. The previous arguments show that $|J : H \cap J| = p^3$ and so $|HJ| = |G|$. It follows that H and J permute.

Suppose that $j \in H\langle x^p \rangle \setminus H\langle e_{p+1} \rangle$. Then $|J : H \cap J| = p^2$ and $\langle H, J \rangle = \langle H, x^p \rangle = H\langle x^p \rangle$. Hence $|HJ| = |\langle H, J \rangle|$. We conclude that H and J permute.

Assume that $j \in H\langle e_{p+1} \rangle \setminus H$. Then $|J : H \cap J| = p$ and $\langle H, J \rangle = \langle H, e_{p+1} \rangle = H\langle e_{p+1} \rangle$. Thus $|HJ| = |\langle H, J \rangle|$. Thus H and J permute.

Therefore H is permutable. It only remains to show that $H_G = 1$. This follows because $\mathrm{Z}(G) = \langle e_{p+1} \rangle$ has order p and $\langle e_{p+1} \rangle \cap H = 1$.

Bradway, Gross, and Scott [76] have shown that there is no bound for the nilpotency class of a core-free permutable subgroup. Let p be a prime. Gross and Berger [140] give a universal construction of a p-group $G = H\langle x \rangle$, such that H is a core-free permutable subgroup of G and x has order p^n and if $G^* = H^*\langle x^* \rangle$ is a p-group, H^* is a core-free permutable subgroup of G^* and x^* has order p^n, then there exists a monomorphism of G^* into G mapping x^* to x and H^* to H.

Our next goal is to show that the set of all S-permutable subgroups is a sublattice of the subnormal subgroup lattice. We first prove a criterion for a subgroup to be S-permutable. For this we need a useful preliminary result, which also appears in [221].

Lemma 1.2.16. *Let H be a p-subgroup of a group G, and let p be a prime. Then H is S-permutable in G if and only if $\mathrm{O}^p(G) \leq \mathrm{N}_G(H)$.*

Proof. Assume that H is S-permutable in G. Then H is subnormal in G by Theorem 1.2.14 (3) and so it is contained in each Sylow p-subgroup P of G. If Q is a Sylow q-subgroup of G for some prime $q \neq p$, then H is a subnormal (even normal) subgroup of $HQ = QH$. Hence $\mathrm{O}^p(G) \leq \mathrm{N}_G(H)$.

Conversely, if $\mathrm{O}^p(G) \leq \mathrm{N}_G(H)$, then H is a normal subgroup of $H\mathrm{O}^p(G)$, which is a subnormal subgroup of G. Therefore H is subnormal in G and so $H \leq \mathrm{O}_p(G)$. Hence H is contained in every Sylow p-subgroup of G and is normalised by all Sylow q-subgroups of G for all primes $q \neq p$. We can conclude that H is S-permutable in G. □

Theorem 1.2.17 (Schmid [221]). *Let H be a nilpotent subgroup of a group G. Then the following statements are equivalent:*

1. *H is S-permutable in G.*

2. *Each Sylow subgroup of H is S-permutable in G.*

3. *Each characteristic subgroup of H is S-permutable in G.*

Proof. Let p be a prime and let H_p denote the Sylow p-subgroup of H. Of course, H_p is a characteristic subgroup of H. Assume that $q \neq p$ is a prime and Q is a Sylow q-subgroup of G.

If H is S-permutable in G, then H is subnormal in HQ by Theorem 1.2.14 (3). If $p \neq q$, then H_p is a subnormal Sylow subgroup of $HQ = QH$. It follows that H_p is normalised by Q. Applying Lemma 1.2.16, H_p is S-permutable in G. Hence Statement 1 implies Statement 2.

It is obvious that Statement 3 implies Statement 1. To complete the proof we now show that Statement 2 implies Statement 3. Let X be a characteristic subgroup of H_p. By hypothesis, $XQ = QX$ and X is a subnormal Sylow subgroup of XQ by Theorem 1.2.14 (3). It follows that Q normalises X. Therefore $\mathrm{O}^p(G)$ normalises H_p and hence $\mathrm{O}^p(G)$ normalises X. Applying Lemma 1.2.16, we conclude that X is S-permutable in G.

If X is a characteristic subgroup of H, then for every prime p, the Sylow p-subgroup X_p of X is characteristic H_p. Consequently X_p is S-permutable in G. By Theorem 1.2.2, X is S-permutable in G. □

We are now ready to prove the necessary and sufficient condition for S-permutability.

Proposition 1.2.18 (Schmid [221, Proposition C]). *The subgroup H of a group G is S-permutable in G if and only if for each prime p there exists a Sylow p-subgroup H_p/H_G of H/H_G such that $\mathrm{O}^p(G) \leq \mathrm{N}_G(H_p)$.*

Proof. Assume that H is S-permutable in G. Then H/H_G is nilpotent by Theorem 1.2.14 (2). It follows from Lemma 1.2.16 and Theorem 1.2.17 that

$$\mathrm{O}^p(G/H_G) = (\mathrm{O}^p(G))H_G/H_G$$

normalises each Sylow subgroup H_p/H_G of H/H_G. This confirms that $\mathrm{O}^p(G) \leq \mathrm{N}_G(H_p)$ and the necessity of the condition is proved. To prove the sufficiency, assume that there exists a Sylow p-subgroup H_p/H_G of H/H_G such that $\mathrm{O}^p(G) \leq \mathrm{N}_G(H_p)$. Then $\mathrm{O}^p(G/H_G) = (\mathrm{O}^p(G))H_G/H_G$ normalises H_p/H_G and H_p/H_G is S-permutable in G/H_G. Applying Theorem 1.2.14 (1), H_p is S-permutable in G. The S-permutability of H in G now follows from Theorem 1.2.2. □

Our preparations are complete and now prove the lattice result.

Theorem 1.2.19 (Kegel [181, Satz 2]). *If H_1 and H_2 are two S-permutable subgroups of the group G, then $H_1 \cap H_2$ is an S-permutable subgroup of G. Consequently, the set of all S-permutable subgroups is a sublattice of the subnormal subgroup lattice.*

Proof. It is clear that only the intersection property is in doubt. Suppose that H_1 and H_2 are S-permutable subgroups of G. By Theorem 1.2.14 (3), we know that H_1 and H_2 are subnormal in G. Set $C_i = (H_i)_G$, $i \in \{1, 2\}$. Applying Theorem 1.2.14 (2), H_i/C_i is nilpotent for $i \in \{1, 2\}$. We may assume that $(H_1 \cap H_2)_G = C_1 \cap C_2 = 1$ by Theorem 1.2.14 (1) and Lemma 1.2.7 (2). Consider the Sylow p-subgroup P_i/C_i for H_i/C_i for $i \in \{1, 2\}$ and a prime p. Let us see that $P_1 \cap P_2$ is a p-group. Let P be a Sylow p-subgroup of G. By subnormality of H_i, $P \cap H_i$ is a Sylow p-subgroup of H_i. Hence $P_i = (P \cap H_i)C_i = PC_i \cap H_i$ for $i \in \{1, 2\}$. Therefore $P_1 \cap P_2 = PC_1 \cap PC_2 \cap H_1 \cap H_2$. Since $PC_2 \cap PC_1 = P(PC_2 \cap C_1)$, it is enough to show that $PC_2 \cap C_1$ is a p-group. Take a Sylow q-subgroup Q of $PC_2 \cap C_1$ for a prime $q \neq p$. We have that $Q \leq C_2$ since C_2 is normal in G. But then $Q \leq C_1 \cap C_2 = 1$. This shows that $PC_2 \cap C_1$ is a p-group. In particular, $P_1 \cap P_2$ is a p-group. Moreover, $P_1 \cap P_2$ is subnormal because it is the intersection of two subnormal subgroups. We have then that $P_1 \cap P_2 \leq P$. Hence $P_1 \cap P_2 = P \cap PC_1 \cap PC_2 \cap H_1 \cap H_2 = P \cap H_1 \cap H_2$ is a Sylow p-subgroup of the subnormal subgroup $H_1 \cap H_2$. On the other hand, by Proposition 1.2.18, $O^p(G)$ normalises P_i, $i = 1, 2$. We conclude that the Sylow p-subgroup of the nilpotent group $H_1 \cap H_2$ is normalised by $O^p(G)$. By Proposition 1.2.18, $H_1 \cap H_2$ is S-permutable in G. □

On the other hand, the intersection of two permutable subgroups is not permutable in general, as the following example demonstrates.

Example 1.2.20. Assume that p is an odd prime,

$$A = \langle a, b : a^{p^2} = b^p = 1, a^b = a^{1+p} \rangle$$

is an extraspecial group of order p^3 and exponent p^2 and $Z = \langle z \rangle$ is a cyclic group of order p^2. Consider $G = A \times Z$. We have that A is a normal subgroup of G and that $B = \langle b \rangle \times \langle z \rangle$ is a permutable subgroup of G because $\langle b \rangle$ is a permutable subgroup of A, as it can be easily checked. Let us see that $A \cap B = \langle b \rangle$ is not a permutable subgroup of G. For this, it will be enough to check that $\langle b \rangle$ does not permute with $\langle az \rangle$. Observe that $(az)^b(az)^{-1} = a^b z(az)^{-1} = a^{1+p}zz^{-1}a^{-1} = a^p \in \langle az, b \rangle$. This implies that $H = \langle a^p, az \rangle$ is a proper subgroup of $\langle az, b \rangle$ which is contained in $\langle a, z \rangle$. In addition, $|H| = p^3$. Moreover $\langle b \rangle \cap \langle a, z \rangle = 1$. But if $\langle b \rangle$ permutes with $\langle az \rangle$, then $|\langle b, az \rangle| = |\langle b \rangle \langle az \rangle| = |\langle b \rangle| \cdot |\langle az \rangle| / |\langle b \rangle \cap \langle az \rangle| = p^3$. This contradiction shows that $\langle b \rangle$ is not permutable in G.

An interesting consequence of the structural properties and embedding of the S-permutable subgroups is the following.

Corollary 1.2.21 (Schmid [221]). *If H is an S-permutable subgroup of a group G, then $\mathrm{N}_G(H)$ is also an S-permutable subgroup of G.*

Proof. With a contradiction in mind, assume that H is an S-permutable subgroup of a group G such that $\mathrm{N}_G(H)$ is not S-permutable in G. Let us choose H of minimal order. Applying Lemma 1.2.7 (2) and Theorem 1.2.14 (1), we may assume $H_G = 1$. In this case, H is a subnormal nilpotent subgroup of G by Theorem 1.2.14 (2 and 3). Assume that every Sylow subgroup of H is a proper subgroup of H. Then, since the Sylow subgroups of H are S-permutable in G by Theorem 1.2.17, we have that their normalisers are S-permutable in G by the choice of H. Since the intersections of these normalisers is $\mathrm{N}_G(H)$, it follows that $\mathrm{N}_G(H)$ is S-permutable by Theorem 1.2.19. This contradiction shows that H is a p-group for some prime p. Applying Proposition 1.2.18, $\mathrm{O}^p(G)$ normalises H. Hence $\mathrm{N}_G(H)$ contains every Sylow q-subgroup of G for all primes $q \neq p$ and $G = \mathrm{N}_G(H)P$ for every Sylow p-subgroup of G. This final contradiction proves that $\mathrm{N}_G(H)$ is S-permutable in G. □

Recall that a normal subgroup N of a group G is said to be *hypercentral* in G if every chief factor of G below N is central in G (see [119, IV, 6.2]). By [119, IV, 6.4] the product of two normal hypercentral subgroups of G is hypercentral in G. Therefore G has a unique maximal hypercentral normal subgroup, which is called the *hypercentre* of G and it is denoted by $\mathrm{Z}_\infty(G)$. We remark that $\mathrm{Z}_\infty(G)$ is just the last member of the ascending central series of G and also the intersection of the normalisers of all Sylow subgroups of G (see Section 6 of Appendix C of [253]).

Lemma 1.2.22. *Let p be a prime and let H be an S-permutable p-subgroup of a group G. The following statements are equivalent in pairs:*

1. *H normalises each Sylow subgroup of G.*
2. $\mathrm{O}^p(G) \leq \mathrm{C}_G(H)$.
3. *H is contained in the hypercentre $\mathrm{Z}_\infty(G)$ of G.*

Proof. If H normalises each Sylow subgroup of G, then H centralises all Sylow q-subgroups of G for $q \neq p$ by Lemma 1.2.16.

Assume now that $\mathrm{O}^p(G) \leq \mathrm{C}_G(H)$. Then $\mathrm{O}^p(G) \leq \mathrm{C}_G(\langle H^G \rangle)$ and so the group of automorphisms induced by G on $\langle H^G \rangle$ is a p-group. In particular, every chief factor of G below $\langle H^G \rangle$ is central and so $H \leq \mathrm{Z}_\infty(G)$.

If H is contained in $\mathrm{Z}_\infty(G)$, then H clearly normalises each Sylow subgroup of G. □

Following Carocca and Maier [88], a subgroup H of a group G such that $H/H_G \leq \mathrm{Z}_\infty(G/H_G)$ is called *hypercentrally embedded* in G. Of course, every hypercentrally embedded subgroup is S-permutable, but the converse is not true in general (see

Example 1.5.7 below). Our immediate objective is to describe and justify a necessary and sufficient condition for an S-permutable subgroup of a soluble group to be hypercentrally embedded. The essential ingredients are the system normalisers.

A *Hall system* Σ of a soluble group G is a set of Hall subgroups of G, one for every subset of prime numbers, such that if H_π denotes the Hall π-subgroup of G in Σ, H_{π_1} and H_{π_2} permute for every pair of sets π_1 and π_2 of prime numbers. The main properties of Hall systems are described in [119, I, 4]. In particular, it is proved there that the group acts transitively by conjugation on the set of its Hall systems. The stabilisers of this action are called *system normalisers*. They form a characteristic class of nilpotent subgroups covering the central chief factors and avoiding the eccentric ones. The main properties of these subgroups are nicely presented in [119, I, 5].

Theorem 1.2.23 (Schmid [221, Theorem C]). *Let G be a soluble group and let H be an S-permutable subgroup of G. Then H/H_G is hypercentrally embedded in G if and only if H permutes with some system normaliser of G.*

Proof. Let Σ be a Hall system of G. Assume that $H/H_G \leq \mathrm{Z}_\infty(G/H_G)$. It is known that $\mathrm{N}_G(\Sigma)H_G/H_G$ is a system normaliser of G/H_G whose core is $\mathrm{Z}_\infty(G/H_G)$ (see [119, VI, 11.3 and 11.11]). Thus H permutes with $\mathrm{N}_G(\Sigma)$ and the "only if" part is clear. For the proof of the converse, we suppose that the result is false and that (G, H) is a counterexample with H a non-hypercentrally embedded S-permutable subgroup of G such that H permutes with $\mathrm{N}_G(\Sigma)$ and $|G|+|H|$ minimal. We proceed in several steps:

1. *$H_G = 1$, $H \leq \mathrm{O}_p(G)$ for some prime p, and $\mathrm{O}^p(G) \leq \mathrm{N}_G(H)$.*

Since $\mathrm{N}_G(\Sigma)H_G/H_G$ is a system normaliser of G/H_G, by minimality $H_G = 1$. Hence H is nilpotent by Theorem 1.2.14 (2). Assume that H is not a p-group for a prime p. By Theorem 1.2.17, if H_p is a p-Sylow subgroup of H, then H_p is S-permutable in G. Moreover, by Theorem 1.1.19, H_p permutes with the Sylow p-subgroup of $\mathrm{N}_G(\Sigma)$. By Lemma 1.2.16, $\mathrm{O}^p(G) \leq \mathrm{N}_G(H_p)$. In particular, all Sylow q-subgroups of $\mathrm{N}_G(\Sigma)$ for all primes $q \neq p$ normalise H_p. It follows that H_p permutes with $\mathrm{N}_G(\Sigma)$. By minimality of H, H_p is contained in $\mathrm{Z}_\infty(G)$, contrary to assumption. Hence H is a p-group for a prime p. By Lemma 1.2.16, $\mathrm{O}^p(G) \leq \mathrm{N}_G(H)$.

Let P be the Sylow p-subgroup in Σ and let R be the corresponding Hall p'-subgroup of G in Σ. Then $N = P \cap \mathrm{N}_G(R)$ is the (unique) Sylow p-subgroup of $\mathrm{N}_G(\Sigma)$ (see [119, I, 5.4 (a)]).

2. *$G = \mathrm{O}^p(G)N$ and $H_N = 1$.*

Since all chief factors of G between $\mathrm{O}^p(G)$ and G are central, they are covered by $\mathrm{N}_G(\Sigma)$ by [119, I, 5.6] and so $G = \mathrm{O}^p(G)\mathrm{N}_G(\Sigma) = \mathrm{O}^p(G)N$. Since $\mathrm{O}^p(G) \leq \mathrm{N}_G(H)$, $H_N = H_G = 1$.

3. $[H, R] \neq 1$.

There is a Sylow subgroup of G not normalising H by [119, I, 5.9]. Then by Statement 1, Lemma 1.2.22, and the choice of (G, H), H is not centralised by $O^p(G)$. If $[H, R] = 1$, then we would have that $[H, R^x] = 1$ for all $x \in O^p(G)$ because $O^p(G) \leq N_G(H)$. Now $\langle R^G \rangle = O^p(G) = O^p(O^p(G))$ is generated by all $O^p(G)$-conjugates of R and we have $[H, O^p(G)] = 1$. This is a contradiction.

4. $P = HN$.

By hypothesis, H permutes with $N_G(\Sigma)$. Hence by Statement 1 and Theorem 1.1.19, $P_0 = HN$ is a subgroup of P. Let $G_0 = P_0R$; this is a subgroup of G because H is normalised by R and R is normalised by N. Suppose that P_0 is a proper subgroup of P. Then G_0 is a proper subgroup of G. Consider the Hall system Σ_0 of G_0 obtained from Σ by replacing P by P_0. The corresponding system normaliser $N_{G_0}(\Sigma_0)$ of G_0 contains $N_G(\Sigma)$ and $N = P_0 \cap N_{G_0}(R)$ is the Sylow p-subgroup of $N_{G_0}(\Sigma_0)$ as well. Since the Hall p'-subgroup of $N_{G_0}(\Sigma_0)$ normalises H by Statement 1, H permutes with $N_{G_0}(\Sigma_0)$. Moreover, by Statement 2, we have that $H_{G_0} = H_N = 1$. Since the theorem holds for G_0, H, $N_{G_0}(\Sigma_0)$, we have that H is centralised by $O^p(G_0)$ by Lemma 1.2.22. This contradicts Statement 3 because $R \leq O^p(G_0)$. Therefore $P = HN$.

5. $O^p(G) \leq HR = RH$.

We have that $O^p(G) = \langle R^G \rangle = \langle R^P \rangle = \langle R^H \rangle$, because $G = PR$, $P = NH$ by Statement 4, and N normalises R. Clearly, $HR = RH$ and $\langle R^H \rangle$ is contained in HR.

6. $[O_p(G), O^p(G)] \leq H$.

This is immediate from Statement 1 and 5, because the commutator group is contained in $O_p(G) \cap O^p(G) \leq P \cap HR = H(P \cap R) = H$.

7. *Final contradiction.*

It is clear that $[O_p(G), O^p(G)]$ is a normal subgroup of G contained in H by Statement 6. By Statement 1, it must be trivial. It follows that $[H, R] = 1$, which contradicts Statement 3. This completes the proof. □

As a consequence of the above theorem, every permutable subgroup of a soluble group is hypercentrally embedded. We shall prove in Section 1.5 that this property holds in the category of all finite groups.

1.3 Power automorphisms

Assume that a group $G = AB$ is the product of two permutable subgroups A and B such that $A \cap B = 1$. If X is a normal subgroup of G contained in B, then for each subgroup Y of X we have that AY is a subgroup of G and $Y = X \cap AY$. Thus

A normalises every subgroup of X. In this case, we say that A acts as a group of power automorphisms on X. Power automorphisms arise naturally in our work and in fact they play a major role in the study of factorised groups and permutability. The basic facts concerning power automorphisms were elucidated by Cooper [104] (see Section 1.5 of [222]). Our intention in this section is to establish results on power automorphisms which are used in our proofs. We begin with a brief review of the concept.

Definition 1.3.1. Let G be a group and let α be an automorphism of G (respectively, an endomorphism of G). We say that α is a *power automorphism* (respectively, a *power endomorphism*) of G if for every $g \in G$ there exists an integer $n(g)$ such that $g^\alpha = g^{n(g)}$. In other words, α is a power automorphism (respectively, a power endomorphism) of G if α fixes all the subgroups of G (respectively, if the image of every subgroup H of G by α is contained in H).

The set of all power automorphisms of a group G is a subgroup of $\mathrm{Aut}(G)$ which is denoted by $\mathrm{PAut}(G)$.

Power automorphisms have been also referred to in the literature as *uniform automorphisms*.

The following result was proved by Robinson [212] and appears in [214, 13.4.3].

Theorem 1.3.2. *Let α be a power automorphism of the abelian p-group G, where p is a prime number. Then α is universal, that is, there exists a fixed integer n such that $g^\alpha = g^n$ for all elements $g \in G$. Moreover, if p does not divide the order of α, then α has no fixed points unless α is the trivial automorphism.*

The next theorem is due to Cooper [104, Theorem 2.2.1] (see [222, Theorem 1.5.2] for a proof).

Theorem 1.3.3. *Power automorphisms are central, that is, if G is a group and α is a power automorphism of G, then $g^{-1}g^\alpha \in \mathrm{Z}(G)$ for all $g \in G$.*

Suppose that every subgroup of a p-subgroup H of a group G is S-permutable in G. Then, by Lemma 1.2.16, $\mathrm{O}^p(G)$ acts as a group of power automorphisms on H. Hence the information given in the following lemma comes in useful in this situation.

Lemma 1.3.4 (Huppert [156, Hilfssatz 5]). *Let p be a prime. If G is a non-abelian p-group, then the group of power automorphisms of G is an abelian p-group.*

Proof. Let α be a power automorphism of G. Suppose that α does not fix an element g of order p of G. Then p must be odd. Assume that $g^\alpha = g^r$ for some integer r. Since g is not fixed by α, $r \not\equiv 1 \pmod p$. Hence $\gcd(p, r-1) = 1$. By Theorem 1.3.3, $g \in \mathrm{Z}(G)$. Let h be an element of order p of G. Since $\langle g, h\rangle$ is abelian, $h^\alpha = h^r$ by Theorem 1.3.2. It follows that α does not fix any element of order p. If $\theta = -1 + \alpha$

is the endomorphism of G defined as $x^\theta = x^{-1}x^\alpha$, then θ is a power endomorphism of G such that $\operatorname{Ker}\theta = 1$ and the image of θ is contained in $Z(G)$. It follows that G is abelian, contrary to hypothesis.

Therefore α fixes all elements of order p of G. Let $1 \neq x$ be an element of G of order p^t and assume that $x^\alpha = x^r$. Since $x^{p^{t-1}}$ has order p, then $(x^{p^{t-1}})^\alpha = x^{rp^{t-1}} = x^{p^{t-1}}$ and so $r \equiv 1 \pmod p$, that is, $r = 1 + kp$ with k integer. Let p^e be the exponent of G. Of course, $p^e > 1$. We have then that $(1 + kp)^{p^{e-1}} \equiv 1 \pmod{p^e}$ and so $\alpha^{p^{e-1}}$ fixes all elements of G. Therefore α has order dividing p^{e-1}. It follows that $\operatorname{PAut}(G)$ is a p-group. □

1.4 Dedekind and Iwasawa groups

The first example of groups with many permutable subgroups are of course groups with all subgroups normal. Such groups are called *Dedekind groups* and the major initial impulse to their study was given precisely by Dedekind in 1896 [115]. He characterised them in the finite case, a result later extended by Baer [25] to infinite groups. The motivation for Dedekind's work was algebraic number theory: he wished to determine the algebraic number fields with the property that every subfield is normal. Dedekind and Baer's result can be found in [217, 5.3.7].

Theorem 1.4.1. *A group G is Dedekind if and only if G is abelian or G is a direct product of the quaternion group Q_8 of order* 8*, an elementary abelian* 2*-group, and an abelian group with all elements of odd order.*

Non-abelian Dedekind groups have been also called Hamiltonian groups.

According to [222, 5.1.1], a subgroup is permutable if and only if it is modular (in the lattice of all subgroups) and subnormal. In particular, a nilpotent group has a modular subgroup lattice (that is, it has all its subgroups modular) if and only if every subgroup is permutable. Iwasawa [165] studied the groups whose subgroups are modular.

Definition 1.4.2. A group G is called an *Iwasawa group* when every subgroup of G is permutable in G.

Clearly, the subgroup lattice of an Iwasawa group is modular and all nilpotent groups with a modular subgroup lattice are Iwasawa. Hence Iwasawa groups are often referred to as *modular groups*.

The following result characterises Iwasawa p-groups for a prime p. It was proved by Iwasawa [165] and its proof also appears in [222, 2.3.1].

Theorem 1.4.3. *Let p be a prime. A p-group G is an Iwasawa group if and only if G is a Dedekind group, or G contains an abelian normal subgroup N such that G/N*

is cyclic (and so $G = \langle x\rangle N$ for an element x of G) and $a^x = a^{1+p^s}$ for all $a \in N$, where $s \geq 1$ and $s \geq 2$ if $p = 2$.

Having identified the Iwasawa p-groups, we are led naturally to inquire about conditions for a p-group to be Iwasawa. In this context, the following two results (see [222, 2.3.3] and [222, 2.3.4]) are useful.

Lemma 1.4.4 (Napolitani [205, Teorema A]). *Let G be a p-group. Then G is an Iwasawa group if and only if each of its sections of order p^3 is an Iwasawa group. Therefore if G is not an Iwasawa group, then there exist subgroups H and K of G with K normal in H such that H/K is dihedral of order 8 or non-abelian of order p^3 and exponent p for $p > 2$.*

Lemma 1.4.5. *Let G be a p-group with an abelian subgroup A such that every subgroup of A is normal in G and G/A is cyclic. For $p = 2$ and A of exponent greater than 2, assume further that there exist subgroups $A_2 \leq A_1 \leq A$ with A_1/A_2 cyclic of order 4 and $[A_1, G] \leq A_2$. Then G is an Iwasawa group.*

Busetto conjectured that the normal closure of a cyclic permutable subgroup of a p-group, p an odd prime, is an Iwasawa group. In an attempt to give an answer to this conjecture, Cossey and Stonehewer [109] studied the structure of the normal closure of a cyclic permutable subgroup. They proved that if A is a cyclic permutable subgroup of odd order of a group G, then $[A, G]$ is abelian, A acts by conjugation on $[A, G]$ as a group of power automorphisms; and $\langle A^G\rangle$ is abelian by cyclic. It follows from this result and the previous lemma that the normal closure of a cyclic permutable subgroup of a p-group, p an odd prime, is an Iwasawa group, giving a positive answer to Busetto's conjecture as special case. The general case was also studied by the same authors in [110].

We bring the section to a close by proving a useful result of Longobardi. It appears as Proposizione 1.6 of [190].

Theorem 1.4.6. *Let p a prime and let G be a non-Iwasawa p-group all of whose proper factor groups are Iwasawa. Then G has a unique minimal normal subgroup.*

Proof. We prove the theorem by way of contradiction. Suppose that the p-group G possesses two different minimal normal subgroups M and N such that G/M and G/N are Iwasawa, but G is not Iwasawa. Since G is not Iwasawa, there exist two subgroups A and B of G with B normal in A such that $|A/B| = p^3$ and A/B not Iwasawa by Lemma 1.4.4. If M is not contained in A, then $A \cong AM/M \leq G/M$, which implies that A is Iwasawa. Therefore M is contained in A and, analogously, N is contained in A. On the other hand, if $M \leq B$, then $A/B \cong (A/M)/(B/M)$ is a quotient of the Iwasawa group $A/M \leq G/M$, against the choice of A/B. It follows that M is not contained in B and, by the same argument, N is not contained in B. Consequently, MB/B and NB/B are minimal normal subgroups of A/B and so they coincide.

Since $MB = NB$, there exists $b \in B \setminus \{1\}$ such that $n = mb$ for certain $m \in M$ and $n \in N$. Thus $b = m^{-1}n \in \mathrm{Z}(G)$ because M and N are minimal normal subgroups of G and $A/B \cong (A/\langle b \rangle)/(B/\langle b \rangle)$ is a quotient of $A/\langle b \rangle$, which is a subgroup of the Iwasawa group $G/\langle b \rangle$. This implies that A/B is Iwasawa, against the choice of A/B. This proves the theorem. □

1.5 Pronormality, weak normality, and the subnormaliser condition

In this section, we investigate some embedding properties of subgroups which have special relevance to Chapter 2, where the classes of T-, PT-, and PST-groups are studied.

We begin with the concept of pronormality which was first introduced by P. Hall.

Definition 1.5.1. A subgroup H of G is said to be *pronormal* in G if for every $g \in G$, H and H^g are conjugate in their join $\langle H, H^g \rangle$.

Maximal subgroups, Hall subgroups of soluble normal subgroups, and Sylow subgroups of normal subgroups are examples of pronormal subgroups.

The main properties of pronormal subgroups appear in [119, Section I, 6]. For the reader's convenience, we summarise them in the following.

Theorem 1.5.2 ([119, I, 6.3]). *Let H be a pronormal subgroup of a group G.*

1. *If $H \leq L \leq G$, then H is pronormal in L.*
2. *If H is subnormal in G, then H is normal in G.*
3. $\mathrm{N}_G(H)$ *is both pronormal and self-normalising in G.*

Another interesting notion in this context is the one of abnormal subgroup, which is closely related to pronormality.

Definition 1.5.3. A subgroup H of a group G is said to be *abnormal* in G if $g \in \langle H, H^g \rangle$ for every $g \in G$.

Examples of abnormal subgroups include the maximal subgroups which are not normal and the normalisers of pronormal subgroups.

Lemma 1.5.4 ([119, I, 6.20 and 6.21]). *Let H be a subgroup of G. Then:*

1. *If H is pronormal in G, then* $\mathrm{N}_G(H)$ *is abnormal in G.*
2. *H is abnormal in G if and only if H is pronormal and* $\mathrm{N}_G(H) = H$.
3. *Assume that $H \leq L \leq G$. If H is abnormal in G, then H is abnormal in L and L is abnormal in G.*

Let X be a pronormal subgroup of a group G. Then X is pronormal in $X\,Z_\infty(G)$ by Theorem 1.5.2 (1). Since X is subnormal in $X\,Z_\infty(G)$, it follows that X is normal in $X\,Z_\infty(G)$ by Theorem 1.5.2 (2). Consequently, $Z_\infty(G)$ normalises every pronormal subgroup of G.

Consider now the subgroup embedding functor f which assigns to each group G the set $f(G) = \mathrm{Pr}(G)$ of all pronormal subgroups of G. It clear that $f(G)$ contains the normalisers of the Sylow subgroups of G. Carocca and Maier [88] studied the f-permutable subgroups and proved that they are exactly the hypercentrally embedded ones. Schmid [221] gave a clever explanation of Carocca and Maier's result by proving the following.

Proposition 1.5.5. *Let G be a group and let H be a core-free* S*-permutable subgroup of G. If H permutes with the normaliser N of a Sylow subgroup of G, then H is contained in N.*

Proof. We know, by Theorem 1.2.14 (2), that H is nilpotent. By Theorem 1.2.17, every Sylow of H is S-permutable and core-free in G. Let p be a prime dividing $|G|$ and let A be a Sylow p-subgroup of H. Applying Lemma 1.2.16, $O^p(G)$ normalises A. In particular, A is normalised by every Sylow q-subgroup of N for every prime $q \neq p$. Moreover, since HN is a subgroup, we can apply Theorem 1.1.19 to conclude that A permutes with a Sylow p-subgroup of N. This means that AN is a subgroup of G. Consequently, we may assume without loss of generality that $A = H$. Since $H \leq O_p(G)$, we may clearly suppose that N is the normaliser of a Sylow q-subgroup S of G for a prime $q \neq p$. Set $K = HN$. Applying Lemma 1.2.7 (1), H is S-permutable in K and, since $G = O^p(G)N$, it follows that $H_G = H_K = 1$. Proceeding by induction on the order of G, we may assume that $K = G$. Let P be a Sylow p-subgroup of G. Then H is contained in P. On the other hand, $\langle S^G\rangle = \langle S^H\rangle$ is a subgroup of HS and $\langle S^G\rangle \cap P$ is a Sylow p-subgroup of $\langle S^G\rangle$. Since H is the normal Sylow p-subgroup of HS, it follows that $\langle S^G\rangle \cap P$ is contained in H. Note that $\langle S^G\rangle \cap P$ is a normal subgroup of P and so $\langle S^G\rangle \cap P$ is contained in H_P. Moreover, because $G = O^p(G)P$, it follows that $H_P = H_G = 1$ and $\langle S^G\rangle$ is a p'-group. Hence $\langle S^G\rangle$ is contained in S and $N = G$. The induction argument is therefore complete. □

The following important corollary follows directly from the previous result.

Corollary 1.5.6 (Maier and Schmid [195]). *If H is a permutable subgroup of a group G, then $H/H_G \leq Z_\infty(G/H_G)$.*

Example 1.5.7. The extraspecial group $E = \langle a, b\rangle$ of order 27 and exponent 3 has an automorphism c of order 2 such that $a^c = a$, $b^c = b^2$. The corresponding semidirect product $G = [E]\langle c\rangle$ has trivial centre and so the core-free subgroup $\langle b\rangle$ is not hypercentrally embedded. However, $\langle b\rangle$ is S-permutable in G.

Other properties related to pronormality are weak normality and the subnormaliser condition.

Definition 1.5.8 (Müller, [202]). A subgroup H of G is said to be *weakly normal* in G if $H^g \le \mathrm{N}_G(H)$ implies that $g \in \mathrm{N}_G(H)$.

Proposition 1.5.9. *If H is a pronormal subgroup of a group G, then H is weakly normal in G.*

Proof. Suppose that $H^g \le \mathrm{N}_G(H)$. Since $H \le \mathrm{N}_G(H)$, we have that $\langle H, H^g\rangle \le \mathrm{N}_G(H)$ and there exists $x \in \langle H, H^g\rangle$ such that $H^g = H^x$. Since $x \in \mathrm{N}_G(H)$, $H^x = H$ and $H^g = H$. □

The converse of Proposition 1.5.9 is not true.

Example 1.5.10. Consider an irreducible and faithful Sym(3)-module V_7 over GF(7), the finite field of 7 elements such that the restriction of V_7 to Alt(3), the alternating group of degree 3, is a direct sum of two irreducible and faithful Alt(3)-submodules W_1 and W_2 of dimension 1. Let $H = [W_1](\mathrm{Alt}(3))$ be the corresponding semidirect product. Since no elements of order 2 normalise H, it follows that $H \le \mathrm{N}_G(H) \le [V_7](\mathrm{Alt}(3))$. Moreover, W_2 does not centralise H. Consequently $H = \mathrm{N}_G(H)$ and so H is weakly normal in G.

Assume that H is pronormal in G. Let a be an element of order 2 of Sym(3). Since $[V_7](\mathrm{Alt}(3))$ is normal in G, we have that $\langle H, H^a\rangle \le [V_7](\mathrm{Alt}(3))$. There exists an element $x \in \langle H, H^a\rangle$ such that $H^a = H^x$. Therefore $a \in [V_7](\mathrm{Alt}(3))$ because $ax^{-1} \in \mathrm{N}_G(H) = H$. This contradiction shows that H is a weakly normal subgroup of G which is not pronormal in G.

The join of two pronormal subgroups is not pronormal in general.

Example 1.5.11 ([119, Exercise I, 6.2]). Consider the group G of Example 1.5.10. Let $W_1 = \langle w\rangle$ and let U be a Sylow 3-subgroup of H. Then $H = \langle U, U^w\rangle$ is the join of two pronormal subgroups, but is not pronormal as we have shown in Example 1.5.10.

However, we have:

Theorem 1.5.12 (Sementovskiĭ [224]). *If A and B are two pronormal subgroups of a group G, then $J = \langle A, B\rangle$ is a weakly normal subgroup of G.*

Proof. Suppose that A and B are pronormal subgroups of G, $J = \langle A, B\rangle$, and $J^g \le \mathrm{N}_G(J)$. Since $A \le J \le \mathrm{N}_G(J)$ and $A^g \le J^g \le \mathrm{N}_G(J)$, it follows that $\langle A, A^g\rangle \le \mathrm{N}_G(J)$. A similar argument shows that $\langle B, B^g\rangle \le \mathrm{N}_G(J)$. From the pronormality of A, there exists $x \in \langle A, A^g\rangle \le \mathrm{N}_G(J)$ such that $A^x = A^g$, and so $A^g \le J^x = J$. From the pronormality of B, there exists $y \in \langle B, B^g\rangle \le \mathrm{N}_G(J)$ such that $B^g = B^y$, and so $B^g \le J^y = J$. Hence $J^g = \langle A^g, B^g\rangle \le \langle J, J\rangle = J$. It follows that $g \in \mathrm{N}_G(J)$. Therefore J is weakly normal in G. □

Assume that H is weakly normal in G and that H is normal in a subgroup K of G. Then $N_G(K)$ is contained in $N_G(H)$. This statement is actually an embedding property of subgroups introduced by Peng in [209]. He showed the equivalence between this property and pronormality in nilpotent subgroups of groups with nilpotent derived subgroup. Mysovskikh gave it the name of *subnormaliser condition* in [203]. Kurdachenko and Subbotin called the subgroups with this property transitively normal in [187].

Definition 1.5.13. A subgroup H of G is said to *satisfy the subnormaliser condition* or to be *transitively normal* in G if for every subgroup K of G such that H is a normal subgroup of K, it follows that $N_G(K) \leq N_G(H)$.

The condition just stated is equivalent to affirming that if H is a normal subgroup of K and K is a normal subgroup of L, then H is a normal subgroup of L or, as Kurdachenko and Subbotin point out in [187], to affirming that H is normal in every subgroup L containing H such that H is subnormal in L (see Proposition 1.5.15).

The following lemma is elementary.

Lemma 1.5.14. *Let H be a subgroup of a group G and N a normal subgroup G contained in H. If H/N satisfies the subnormaliser condition in G, then H satisfies the subnormaliser condition in G. The same applies to weak normality.*

Proposition 1.5.15.

1. *If G is a group and H is a weakly normal subgroup of G, then H satisfies the subnormaliser condition in G.*
2. *If H is a subnormal subgroup of $K \leq G$ and H satisfies the subnormaliser condition in G, then H is a normal subgroup of K (compare with Theorem 1.5.2 (2)).*

Proof. 1. Suppose that H is a weakly normal subgroup of G and let K be a subgroup of G containing H and contained in $N_G(H)$. Consider an element $g \in N_G(K)$. Then $K^g = K$. Hence $H^g \leq K^g = K \leq N_G(H)$. Consequently $g \in N_G(H)$ because H is a weakly normal subgroup of G.

2. Suppose that H satisfies the subnormaliser condition in G and that H is a subnormal subgroup of $K \leq G$. By induction on the length of a series from H to K, we can suppose that $H \trianglelefteq T \trianglelefteq K$. Therefore $K \leq N_G(T) \leq N_G(H)$ by the subnormaliser condition. Hence H is a normal subgroup of K. □

Example 1.5.16. There exist subgroups satisfying the subnormaliser condition which are not weakly normal, as the following example, due to Mysovskikh [204], shows:

Consider the natural wreath product of the cyclic group C_3 or order 3 by the alternating group Alt(4) of degree 4. Let B be the base group, $B = \langle w_1, w_2, w_3, w_4 \rangle$, where Alt(4) acts naturally on the indices. Let $W = \langle w_4 w_1^{-1}, w_4 w_2^{-1}, w_4 w_3^{-1} \rangle$, then Alt(4) acts faithfully on W. Let $G = [W](\mathrm{Alt}(4))$ be the corresponding semidirect product. The subgroup W, as a module over the minimal normal subgroup V_4 of

Alt(4), can be decomposed as a direct sum of the V_4-submodules $W_1 = \langle d \rangle$, with $d = w_1^{-1} w_2^{-1} w_3 w_4$, $W_2 = \langle e \rangle$, with $e = w_1 w_2^{-1} w_3^{-1} w_4$, and $W_3 = \langle f \rangle$, with $f = w_1^{-1} w_2 w_3^{-1} w_4$. Denote $a = (1,2,3)$, $b = (1,2)(3,4)$, $c = (1,3)(2,4)$. We have that $d^b = d$, $e^b = e^2$, $f^b = f^2$, $d^c = d^2$, $e^c = e^2$, $f^c = f$, $d^a = e$, $e^a = f$, $f^a = d$, $b^a = bc$, and $c^a = b$. Consider $D = [W_3]\langle b \rangle$, then $D \cong \mathrm{Sym}(3)$, the symmetric group on three letters, and since $b^d = b$, $b^e = be^2$, and $b^f = bf^2$, it follows that $N = \mathrm{N}_G(D) = \langle d, f, b, c \rangle = [W_1 W_3]\langle b, c \rangle$, a self-normalising subgroup. Since $D^a = \langle bc, d \rangle \leq N$ and $D^a \neq D$, we have that D is not weakly normal in G. But D satisfies the subnormaliser condition, because the intermediate subgroups between D and N are $\langle b, c, f \rangle$, $\langle b, cd, f \rangle$, $\langle b, cd^2, f \rangle$, self-normalising subgroups of order 12, and $\langle b, d, f \rangle$, a subgroup of order 18 whose normaliser is N. Consequently D satisfies the subnormaliser condition, but is not a weakly normal subgroup.

The following lemma (see [119, Exercise I, 6.8]) shows that for p-subgroups, pronormality, weak normality, and the subnormaliser condition are equivalent properties.

Lemma 1.5.17. *Let p be a prime. If one of the following conditions about a p-subgroup H of a group G is true, then they are all true.*

1. *H is a pronormal subgroup of G.*
2. *H is a weakly normal subgroup of G.*
3. *H satisfies the subnormaliser condition in G.*
4. *H is normal in $\mathrm{N}_G(X)$ for every p-subgroup X such that $H \leq X$.*
5. *H is normal in $\mathrm{N}_G(S)$ for every Sylow p-subgroup S of G such that $H \leq S$.*

Proof. Proposition 1.5.9 shows that Statement 1 implies Statement 2, and by Proposition 1.5.15 we have that Statement 2 implies Statement 3.

Suppose now that H satisfies the subnormaliser condition in G and let X be a p-subgroup of G containing H. Then H satisfies the subnormaliser condition in $\mathrm{N}_G(X)$ and H is subnormal in $\mathrm{N}_G(X)$. By Proposition 1.5.15, H is a normal subgroup of $\mathrm{N}_G(X)$. Hence Statement 3 implies Statement 4.

It is clear that Statement 4 implies Statement 5.

To complete the proof we now show that Statement 5 implies Statement 1. Let $g \in G$ and $J = \langle H, H^g \rangle$. Since H is a p-group, there exists a Sylow p-subgroup P of J such that $H \leq P$ and there exists a Sylow p-subgroup S of G such that $H \leq P \leq S$. Let $x \in J$ such that H^g is contained in P^x. Then $H^{gx^{-1}} \leq P \leq S$. We write $t = gx^{-1}$. It is clear that $t^{-1} \in \langle \mathrm{N}_G(S^{t^{-1}}), \mathrm{N}_G(S) \rangle$ because $\mathrm{N}_G(S)$ is abnormal in G. Since $H \leq S^{t^{-1}}$, it follows that $\langle \mathrm{N}_G(S), \mathrm{N}_G(S^{t^{-1}}) \rangle \leq \mathrm{N}_G(H)$ by Statement 5. Hence $t^{-1} \in \mathrm{N}_G(H)$ and $H^x = H^g$. Consequently, H is pronormal in G. □

1.6 SC-groups

The theme of this section is a class of all groups characterised by the existence of a simple chief series, that is, a chief series whose chief factors are simple. Such groups are called SC-groups and the corresponding class is a natural extension of the class $\mathfrak{U}$ of all supersoluble groups. The initial impulse to its study was given by Robinson [217].

Definition 1.6.1. A group is said to be an SC-*group* if all its chief factors are simple.

Note that in Section VI, 13 of [157], the expression "SC-group" is used with a different meaning.

Remark 1.6.2. The class of all supersoluble groups coincides with the class of all soluble SC-groups.

Recall that a *formation* $\mathfrak{F}$ is a class of groups satisfying:

1. if G is a group in $\mathfrak{F}$ and N is a normal subgroup of G, then G/N belongs to $\mathfrak{F}$, and
2. if N and M are normal subgroups of G such that G/N and G/M belong to $\mathfrak{F}$, then $G/(N \cap M)$ belong to $\mathfrak{F}$.

If, moreover, G belongs to $\mathfrak{F}$ when $G/\Phi(G)$ belongs to $\mathfrak{F}$, where $\Phi(G)$ denotes the Frattini subgroup of G, then we say that the formation $\mathfrak{F}$ is *saturated.*

Given a formation $\mathfrak{F}$, the intersection of all normal subgroups of G whose quotients belong to $\mathfrak{F}$ is the smallest normal subgroup of G with quotient in $\mathfrak{F}$. Such a subgroup is called $\mathfrak{F}$-*residual* of G and is denoted by $G^{\mathfrak{F}}$. For each normal subgroup N of G, we have $G^{\mathfrak{F}}N/N = (G/N)^{\mathfrak{F}}$.

A class of groups $\mathfrak{G}$ is called a *Fitting class* if it satisfies:

1. if G belongs to $\mathfrak{G}$ and N is a normal subgroup of G, then N belongs to $\mathfrak{G}$, and
2. if $G = N_1N_2$ with N_1, N_2 normal subgroups of G belonging to $\mathfrak{G}$, then G belongs to $\mathfrak{G}$.

Given a Fitting class $\mathfrak{G}$, the product of all normal $\mathfrak{G}$-subgroups of G is the largest normal $\mathfrak{G}$-subgroup of G. This subgroup is called $\mathfrak{G}$-*radical* of G and is denoted by $G_{\mathfrak{G}}$. For each normal subgroup N of G, we have $N \cap G_{\mathfrak{G}} = N_{\mathfrak{G}}$.

The reader is referred to [47], [119] for further information about formations and Fitting classes.

Theorem 1.6.3. *The class of all* SC-*groups is a formation which is neither closed under taking subgroups nor saturated. It is not a Fitting class either.*

Proof. Let G be an SC-group and let N be a normal subgroup of G. Every chief factor of G/N is isomorphic to a chief factor of G. Hence G/N is an SC-group.

Suppose that N and M are normal subgroups of G such that G/N and G/M are SC-groups. Let

$$1 \cong H_0/N \trianglelefteq N_1/N \trianglelefteq \cdots \trianglelefteq H_{r-1}/N \trianglelefteq H_r/N = G/N$$

and

$$1 \cong K_0/M \trianglelefteq K_1/M \trianglelefteq \cdots \trianglelefteq K_{t-1}/M \trianglelefteq K_t/M = G/M$$

be chief series of the SC-groups G/N and G/M, respectively. Note that the chief factors of both series are simple. As a consequence, the factors of the series

$$N = H_0 \trianglelefteq H_1 \trianglelefteq \cdots \trianglelefteq H_{r-1} \trianglelefteq H_r = G$$

and

$$M = K_0 \trianglelefteq K_1 \trianglelefteq \cdots \trianglelefteq K_{t-1} \trianglelefteq K_t = G$$

are simple, too, because

$$H_{i+1}/H_i \cong (H_{i+1}/N)/(H_i/N) \quad \text{and} \quad K_{j+1}/K_j \cong (K_{j+1}/M)/(K_j/M).$$

Consider the series

$$\begin{aligned} 1 &\cong (N \cap M)/(N \cap M) \trianglelefteq (H_1 \cap M)/(N \cap M) \\ &\trianglelefteq \cdots \trianglelefteq (H_{r-1} \cap M)/(N \cap M) \trianglelefteq M/(N \cap M) \trianglelefteq K_1/(N \cap M) \\ &\trianglelefteq \cdots \trianglelefteq K_{t-1}/(N \cap M) \trianglelefteq K_t/(N \cap M) = G/(N \cap M). \end{aligned}$$

Since $((H_{i+1} \cap M)/(N \cap M))/((H_i \cap M)/(N \cap M)) \cong (H_{i+1} \cap M)/(H_i \cap M) \cong (H_{i+1} \cap M)H_i/H_i \trianglelefteq H_{i+1}/H_i$, we have that

$$((H_{i+1} \cap M)/(N \cap M))/((H_i \cap M)/(N \cap M))$$

is either trivial or simple. Moreover,

$$(K_{j+1}/(N \cap M))/(K_j/(N \cap M))$$

is simple.

Therefore, if we delete from the previous series all repetitions we obtain a chief series of $G/(N \cap M)$ with simple chief factors. Therefore, $G/(N \cap M)$ is an SC-group.

In order to see that the formation of all SC-groups is not closed under taking subgroups, it is enough to take the alternating group Alt(5) of degree 5, which is an SC-group, but has Alt(4) as a subgroup, which is not an SC-group.

Consider now the semidirect product G of a quaternion group of order 8 with a faithful and irreducible module V over the finite field of 5 elements. Clearly, V is not 1-dimensional. We have that $G = AB$, where A and B are two distinct subgroups of G containing V of index 2. Then A and B are normal supersoluble subgroups of G, but G is not supersoluble. As a consequence, the class of all SC-groups is not a Fitting class. □

The next proposition shows a useful property of SC-groups. As usual, we denote by $\mathfrak{S}$ the class of all soluble groups and so $G^{\mathfrak{S}}$ denotes the soluble residual of a group G and $G_{\mathfrak{S}}$ is the soluble radical of G. $\mathfrak{U}$ denotes the class of all supersoluble groups.

Proposition 1.6.4. *If G is an* SC*-group, then $G^{\mathfrak{S}} \leq \mathrm{C}_G(G_{\mathfrak{S}})$.*

Proof. Clearly all chief factors of G below $G_{\mathfrak{S}}$ are cyclic. As a consequence, $G_{\mathfrak{S}}$ is $\mathfrak{U}$-hypercentral and so $G_{\mathfrak{S}}$ is a subgroup of the supersoluble hypercentre $\mathrm{Z}_{\mathfrak{U}}(G)$ (see [119, IV, Section 6]). By virtue of [119, IV, 6.10], $\mathrm{Z}_{\mathfrak{U}}(G)$ is centralised by the supersoluble residual of G, which obviously contains the soluble residual of G. Therefore $G^{\mathfrak{S}} \leq \mathrm{C}_G(G_{\mathfrak{S}})$. □

It is an obvious remark that all simple groups are SC-groups. In fact what is needed to understand the structure of these groups is the positive solution of the Schreier Conjecture: of course the latter is now known to be true, on the basis of the classification of non-abelian simple groups.

The result that follows gives a good picture of an arbitrary SC-group.

Theorem 1.6.5 ([217]). *A group G is an* SC*-group if and only if G satisfies:*

1. *$G/G^{\mathfrak{S}}$ is supersoluble.*

2. *If $G^{\mathfrak{S}}$ is non-trivial, then $G^{\mathfrak{S}}/\mathrm{Z}(G^{\mathfrak{S}})$ is a direct product of G-invariant simple groups.*

3. *Every chief factor of G below $\mathrm{Z}(G^{\mathfrak{S}})$ is cyclic, that is, $\mathrm{Z}(G^{\mathfrak{S}})$ is hypercentral with respect to the saturated formation of all supersoluble groups.*

Proof. Let G be an SC-group. By convenience, D will denote the soluble residual $G^{\mathfrak{S}}$ of G. If G is soluble, then G is supersoluble by Remark 1.6.2 and so G satisfies the Statements 1, 2, and 3. Therefore we are left with the case G not soluble, that is, $D \neq 1$. Since G is an SC-group, G/D is an SC-group by Theorem 1.6.3. Since G/D is soluble, it follows from Remark 1.6.2 that G/D is supersoluble. We also note that all chief factors of G below $\mathrm{Z}(D)$ are cyclic. This means that $\mathrm{Z}(D)$ is $\mathfrak{U}$-hypercentral in G and Statement 3 holds.

Next we show that $D/\mathrm{Z}(D)$ is a direct product of G-invariant simple groups. Denote by E the normal subgroup of D such that $E/\mathrm{Z}(D) = \mathrm{Z}(D/\mathrm{Z}(D))$. Then E is a normal soluble subgroup of G, and so E is contained in $\mathrm{Z}(D)$ by Proposition 1.6.4. Therefore $\mathrm{Z}(D/\mathrm{Z}(D)) = 1$ and we can assume that $\mathrm{Z}(D)$ is trivial. Let N be a minimal normal subgroup of G contained in D. Since G is an SC-group, N is simple. If N were abelian, by Proposition 1.6.4, N would be central in D. Consequently, N cannot be abelian. Note that $D/\mathrm{C}_D(N) = \mathrm{N}_D(N)/\mathrm{C}_D(N)$ is isomorphic to a subgroup of $\mathrm{Aut}(N)$ and $N\mathrm{C}_D(N)/\mathrm{C}_D(N)$ is isomorphic to $\mathrm{Inn}(N)$.

Hence $D/N\mathrm{C}_D(N) \cong (D/\mathrm{C}_D(N))/(N\mathrm{C}_D(N)/\mathrm{C}_D(N))$ is isomorphic to a subgroup of $\mathrm{Aut}(N)/\mathrm{Inn}(N)$. From the validity of the Schreier conjecture, it follows that $D/N\mathrm{C}_D(N)$ is soluble. Since D is a perfect subgroup of G, it follows that $D = N\mathrm{C}_D(N)$. Clearly we can assume that D is not simple. Therefore, $\mathrm{C}_D(N)$ is not trivial. Assume that $N = N_1$ and that N_2 is a minimal normal subgroup of G contained in $\mathrm{C}_D(N_1)$. Since G is an SC-group, N_2 is simple. By replacing D in the previous argument by $\mathrm{C}_D(N_1)$ and N by N_2, it follows that $D = N_1N_2\mathrm{C}_X(N_2)$, where $X = \mathrm{C}_D(N_1)$. By repeating the argument as many times as needed, we can conclude that $D = N_1 \times N_2 \times \cdots \times N_t$ is the direct product of minimal normal subgroups of G, each of them simple, as desired.

Conversely, assume that a group G satisfies the Statements 1, 2, and 3. Let us see that G is an SC-group. By virtue of Jordan–Hölder's theorem for groups with operators ([119, A, 3.2]), it is enough to construct a chief series for G with simple chief factors. This can be easily obtained by considering a chief series of G passing through D and observing that every chief factor of G below D is simple by the Statement 3 and that $D/\mathrm{Z}(D)$ is a direct product of minimal normal subgroups of $G/\mathrm{Z}(D)$, which are simple by the Statement 2. □

Corollary 1.6.6. *If G is an* SC*-group, $D = G^{\mathfrak{S}}$ is the soluble residual of G and $R = D/\mathrm{Z}(D)$, then $\mathrm{C}_G(D) = \mathrm{C}_G(R)$ is the soluble radical of G.*

Proof. Let S be the soluble radical of G. We know that $S \leq \mathrm{C}_G(D)$ by Proposition 1.6.4. Set $C = \mathrm{C}_G(R)$. Then C is a normal subgroup of G, $\mathrm{C}_G(D) \leq C$ and $[D, C]$ is contained in $\mathrm{Z}(D)$. Hence $[D, C, D] = 1 = [C, D, D]$. By the three subgroups lemma [157, III, 1.10], $[D, D, C] = 1$ so that C centralises D because D is perfect. Therefore $C = \mathrm{C}_G(D)$. Suppose that $C^{\mathfrak{S}} \neq 1$. Since $C^{\mathfrak{S}}$ is perfect, it is not contained in $\mathrm{Z}(D)$. This means that $C^{\mathfrak{S}}\mathrm{Z}(D)/\mathrm{Z}(D)$ is a non-trivial normal subgroup of R, which is a direct product of non-abelian minimal normal subgroups of $G/\mathrm{Z}(D)$ by Theorem 1.6.5 (2). Thus $C^{\mathfrak{S}}\mathrm{Z}(D)/\mathrm{Z}(D)$ contains a non-abelian minimal normal subgroup of $G/\mathrm{Z}(D)$. This is impossible because C centralises R. Consequently $C \leq S$, as desired. □

The following example due to Robinson [217] shows that the supersoluble embedding property cannot be omitted from the Statement 3 of Theorem 1.6.5.

Example 1.6.7. Let $K = \mathrm{Sz}(8)$ be the smallest simple Suzuki group of Lie type. We know that the Schur multiplier $\mathrm{M}(K)$ of K is an elementary abelian group of order 4 (see [261]). Let K^* be the universal cover of K, so that $\mathrm{Z}(K^*) = \mathrm{M}(K)$. Now K has an automorphism of order 3 which induces an automorphism α of K^* and acts non-trivially on $\mathrm{M}(K)$. Let $G = [K^*]\langle\alpha\rangle$ be the corresponding semidirect product. Then G is not an SC-group since $\mathrm{M}(K)$ is a minimal normal subgroup of G.

Final remark 1.6.8. Let G be an SC-group and let N be a normal subgroup of G. According to Theorem 1.6.3, G/N is an SC-group and so it has the structure described in Theorem 1.6.5. This remark concerns the traffic of information between the structure of G and that of G/N in two particular important cases.

Denote by D the soluble residual of G. Assume that $D \neq 1$. By Theorem 1.6.5, $D/\mathrm{Z}(D) = U_1/\mathrm{Z}(D) \times \cdots \times U_k/\mathrm{Z}(D)$, where $U_i/\mathrm{Z}(D)$ is non-abelian simple and U_i is normal in G. In particular, $D = U_1 \cdots U_r$. Denote $Z = \mathrm{Z}(D)$. Applying [47, 2.2.8], DN/N is the soluble residual of G/N and $(DN/N)/\mathrm{Z}(DN/N)$ is a direct product of G/N-invariant simple groups. Suppose that $N = U_i'$ for some i. Note that, since U_i/Z is non-abelian simple, $U_i'Z = U_i$. By reordering the indices if needed, we can assume that $i = 1$. Hence $D/U_1' = (U_1/U_1')(U_2U_1'/U_1') \cdots (U_kU_1'/U_1')$. Moreover, $U_1/U_1' = ZU_1'/U_1' \leq \mathrm{Z}(D/U_1')$. Then

$$(D/U_1')/(U_1/U_1') = ((U_2U_1'/U_1')/(U_1/U_1')) \cdots ((U_kU_1'/U_1')/(U_1/U_1')).$$

Note that

$$\begin{aligned}(U_jU_1'/U_1')/(U_1/U_1') &\cong_G U_jU_1'/U_1 = U_jU_1/U_1 \cong_G U_j/(U_1 \cap U_j) \\ &= U_j/(ZU_1' \cap U_j) = U_j/Z(U_1' \cap U_j) = U_j/Z.\end{aligned}$$

Consequently if $D \neq U_1$, then $(D/U_1')/(U_1/U_1')$ is a direct product of the non-abelian simple G/U_1'-invariant subgroups $(U_jU_1'/U_1')/(U_1/U_1')$, $j \neq 1$. This implies that $U_1/U_1' = ZU_1'/U_1' = \mathrm{Z}(D/U_1')$ and, by Proposition 1.6.4, $\mathrm{Z}(D/U_1') = (D/U_1')_{\mathfrak{S}} = U_1/U_1'$. Note that every chief factor of G/U_1' below U_1/U_1' is cyclic.

Suppose now that N is a normal nilpotent subgroup of G. Then $DN/N \neq 1$ and $NZ/N \leq \mathrm{Z}(DN/N)$. Furthermore, since $D \cap N$ is contained in Z by Proposition 1.6.4, we have

$$\begin{aligned}(DN/N)/(NZ/N) &\cong_G DN/NZ \cong_G D/(D \cap NZ) = D/(D \cap N)Z \\ &\cong_G D/Z.\end{aligned}$$

Note that

$$(U_iN/N)/(NZ/N) \cong_G U_i/Z(N \cap U_i) = U_i/Z$$

for all i and $(DN/N)/(NZ/N)$ is the direct product of the non-abelian simple G/N-invariant subgroups $(U_iN/N)/(NZ/N)$. Therefore $ZN/N = \mathrm{Z}(DN/N)$ is the soluble radical of DN/N by Proposition 1.6.4 and every chief factor of G/N below NZ/N is cyclic.

1.7 Numerical invariants of π-soluble groups

In this section we focus on some invariants of π-soluble groups, π a set of primes, which are a rich source of information about the complexity of the group and its Hall π-subgroups.

Our point of departure is the following theorem proved in [158, IX, 5.4]:

Theorem 1.7.1. *If p is an odd prime, then the p-length of a p-soluble group is less or equal than the derived length of a Sylow p-subgroup.*

The 2-version of the above theorem was proved by Bryukhanova in [80]. Her treatment is based on earlier results of Hall and Higman [143], and Berger and Gross [69].

Theorem 1.7.2. *The 2-length of a soluble group is less or equal than the derived length of a Sylow 2-subgroup.*

As a full account of the proof of the theorem just cited, together with the background material, is too long to include in this book, we have to be content with treating fully interesting related results due to Kazarin [176]. They are extensions of Theorems 1.7.1 and 1.7.2 which will be applied in Chapter 3 to obtain bounds for some invariants of soluble products of soluble groups and nilpotent groups.

It is impossible to discuss invariants of π-soluble groups at any depth without introducing early some of them associated with nilpotent and soluble groups. For a nilpotent group G, $\mathrm{c}(G)$ will denote the nilpotency class of G, that is, the common length of the upper and lower central series of G. For a soluble group G, $\mathrm{d}(G)$ will denote the derived length of G, that is, the length of the derived series of G. The relationship between the terms of the derived series and the lower central series is enshrined in [157, III, 2.12].

Theorem 1.7.3. *For every group G, $G^{(i)} \leq \gamma_{2^i}(G)$.*

This leads directly to an estimate for the derived length of a nilpotent group.

Theorem 1.7.4. *If G is a non-trivial nilpotent group of nilpotency class c and derived length d, then*

$$d \leq \lceil \log_2(c+1) \rceil \leq \log_2 c + 1.$$

Here $\lceil x \rceil$ denotes the least integer number n such that $x \leq n$.

The proofs of Theorem 3.3.11 and Corollary 3.3.14 are greatly facilitated by use the following elementary result, which is an application of Theorem 1.7.4.

Lemma 1.7.5. *Let K be a p-group such that K' has order p^k. Then the nilpotency class of K are at most $k+1$. In particular, the derived length of K is bounded by $\lceil \log_2(k+2) \rceil \leq \log_2(k+1) + 1 \leq k+1$.*

Definition 1.7.6. Let p be a prime and let S be a Sylow p-subgroup of a group G. The Sylow p-invariants of G are:

1. $b_p = \mathrm{b}_p(G)$, where p^{b_p} is the order of S,
2. $c_p = \mathrm{c}_p(G) = \mathrm{c}(S)$, the nilpotency class of S,

3. $d_p = \mathrm{d}_p(G) = \mathrm{d}(S)$, the derived length of S,

4. $e_p = \mathrm{e}_p(G)$, where p^{e_p} is the exponent of S, that is, the largest order of the elements of S.

Definition 1.7.7. Assume that π is a set of primes and that G is a π-soluble group. Let H be a Hall π-subgroup of G.

1. We call $d_\pi = \mathrm{d}_\pi(G) = \mathrm{d}(H)$.

2. We denote by $\mathrm{l}_\pi(G)$ the *π-length* of G, that is, the smallest number r such that the series
$$1 = P_0 \leq N_0 < P_1 < \cdots < P_r \leq N_{r+1},$$
defined by
$$N_i/P_i = \mathrm{O}_{\pi'}(G/P_i) \quad \text{and} \quad P_{i+1}/N_i = \mathrm{O}_\pi(G/N_i),$$
reaches G. In the language of classes of groups, $\mathrm{l}_\pi(G)$ is the smallest number r such that $G \in (\mathfrak{E}_{\pi'}\mathfrak{S}_\pi)^r\mathfrak{E}_{\pi'}$.

   When $\pi = \{p\}$ consists of a single prime p, we write $\mathrm{l}_\pi(G) = \mathrm{l}_p(G)$ and call this number the *p-length* of G.

The subgroup N_1 in Definition 1.7.7 is usually denoted by $\mathrm{O}_{\pi',\pi}(G)$. When $\pi = \{p\}$, with p a prime, it is denoted by $\mathrm{O}_{p',p}(G)$.

It is clear that the vanishing of any one of the invariants b_p, c_p, d_p, e_p, or l_p is equivalent to G being a p'-group, and that the vanishing of the invariants d_π or l_π means that the π-soluble group G is a π'-group. We will refer to all these invariants under the generic names of *numerical invariants* or *Hall invariants*.

One of the key results in the study of Hall invariants in soluble groups is the important "Theorem B" of Hall and Higman [143]. This result has had considerable influence on the study of groups. Its proof, which appears in [158, IX, 2.9], involves a large number of results and techniques.

Theorem 1.7.8. *Let K be a field of characteristic p and let V be a finite-dimensional vector space over K. Let G be a p-soluble group of non-singular linear transformations of V such that $\mathrm{O}_p(G) = 1$. If g is an element of G of order p^n, the minimum polynomial of g is $(t-1)^r$, where $r \leq p^n$. If $r < p^n$, then there exists an integer $n_0 \leq n$ for which $p^{n_0} - 1$ is a power of a prime q and the Sylow q-subgroups of G are non-abelian; further, if n_0 is the least such integer, $r \geq p^{n-n_0}(p^{n_0}-1)$.*

It is clear that, if p is odd, $p^{m_0} - 1$ can be a power of a prime q only if $q = 2$, and that then $p - 1$ is also a power of 2. Such primes are called *Fermat primes*. In the case $p = 2$, if $2^a = q^b + 1$, then $a > 1$, and so q^b is congruent to 3 modulo 4. Hence b is odd, $q + 1$ is a power of 2, and q is a *Mersenne prime*. Therefore we can precise the thesis of Theorem 1.7.8:

Corollary 1.7.9. *In the hypothesis of Theorem* 1.7.8*:*

1. *If p is neither* 2 *nor a Fermat prime, then* $r = p^n$.
2. *If p is an odd Fermat prime,* $p^{n-1}(p-1) \leq r$ *always, and* $r = p^n$ *if G has an abelian Sylow* 2*-subgroup.*
3. *If* $p = 2$*, then* $3 \cdot 2^{n-2} \leq r$ *always; if* $q = 2^{n_0} - 1$ *is the least Mersenne prime for which H has a non-abelian q-subgroup, then* $q \cdot 2^{n-n_0} \leq r$*; if G has abelian Sylow q-subgroups for all Mersenne primes q less than* 2^n*, then* $r = 2^n$.

We give the following interpretation of the thesis of Theorem 1.7.8 in the following terms.

Lemma 1.7.10. *Let S be a Sylow p-subgroup of a group G. Let*

$$1 = P_0 \leq N_0 < P_1 < \cdots < P_l \leq N_l = G$$

be the series of G defined by $N_i/P_i = \mathrm{O}_{p'}(G/P_i)$ *and* $P_{i+1}/N_i = \mathrm{O}_p(G/N_i)$*. Assume that* $N_0 = 1$*, that is,* $\mathrm{O}_{p'}(G) = 1$ *and that P_1 is elementary abelian. Then P_1 can be seen as a module for* $H = G/P_1$ *over a field of characteristic p. If S contains an element g such that gP_1 is represented by a transformation whose minimal polynomial is* $(x-1)^k$*, then there is an element* $h \in P_1$ *such that*

$$[h, \overbrace{g, g, \ldots, g}^{(k-1 \text{ times})}] \neq 1.$$

In particular, we obtain the following consequence of Theorem 1.7.8.

Corollary 1.7.11. *Under the hypothesis of Theorem* 1.7.8*, the semidirect product* $H = [V]G$ *satisfies:*

1. *If p is neither* 2 *nor a Fermat prime, then* $\mathrm{c}_p(H) \geq p^n$.
2. *If p is an odd Fermat prime,* $\mathrm{c}_p(H) \geq p^{n-1}(p-1)$*, and* $\mathrm{c}_p(H) \geq p^n$ *if H has an abelian Sylow* 2*-subgroup.*
3. *If* $p = 2$*, then* $\mathrm{c}_p(H) \geq 3 \cdot 2^{n-2}$ *always; if* $q = 2^{n_0} - 1$ *is the least Mersenne prime for which G has a non-abelian q-subgroup, then* $\mathrm{c}_p(H) \geq q \cdot 2^{n-n_0}$*; if G has abelian Sylow q-subgroups for all Mersenne primes less than* 2^n*, then* $\mathrm{c}_p(H) \geq 2^n$.

We now prove a couple of lemmas which are very useful when working with Hall invariants. The first one was proved by Hall and Higman [143, Lemma 1.2.1]. It is an indispensable tool in establishing the structure of a minimal counterexample in proofs of theorems involving Hall invariants.

Lemma 1.7.12. *Let a and b two numerical invariants defined for a class of groups $\mathfrak{V}$ which is closed under taking subgroups, quotients, and direct products. Assume that the trivial group 1 satisfies $a(1) \leq b(1)$ and that*

1. *if H is a subgroup of $G \in \mathfrak{V}$, then $a(H) \leq a(G)$;*
2. *$a(G \times H) = \max\{a(G), a(H)\}$ for every G, $H \in \mathfrak{V}$;*
3. *if N is a normal subgroup of $G \in \mathfrak{V}$, then $b(G/N) \leq b(G)$.*

Then if G is a counterexample of least order to the inequality $a(G) \leq b(G)$, then G possesses a unique minimal normal subgroup.

Proof. Assume that G has two different minimal normal subgroups M and N. Then $M \cap N = 1$ and so G is isomorphic to a subgroup of the direct product $(G/M) \times (G/N)$. Since a is an invariant, Statements 1 and 2 give

$$a(G) \leq \max\{a(G/M), a(G/N)\}.$$

By minimality of G, we have that $a(G/M) \leq b(G/M)$ and by Statement 3, we get $b(G/M) \leq b(G)$. Thus $a(G/M) \leq b(G)$. A similar argument shows that $a(G/N) \leq b(G)$ and so $a(G) \leq b(G)$, as required. □

Remark 1.7.13. We note that if a is an invariant defined in a class of groups $\mathfrak{V}$ closed under taking subgroups, quotients, and direct products and satisfying

1. if H is a subgroup of a group G in $\mathfrak{V}$, then $a(H) \leq a(G)$,
2. if $a(G \times H) = \max\{a(G), a(H)\}$ if G and H are groups in $\mathfrak{V}$, and
3. if N is a normal subgroup of a group G in $\mathfrak{V}$, then $b(G/N) \leq b(G)$,

then the class of all groups G in $\mathfrak{V}$ satisfying $a(G) \leq r$ for a natural number r is a formation. Note that these conditions are satisfied by all Hall π-invariants introduced in Definitions 1.7.6 and 1.7.7 except for $\mathrm{b}_p(G)$.

As a consequence, the class of all π-soluble groups with π-length at most k is a saturated formation, for all k. This can be applied to deduce that every π-soluble group G with an abelian Hall π-subgroup, then G has π-length at most 1.

As an example of how Lemma 1.7.12 can be used, we offer the following. Suppose that G is a p-soluble group of minimal order with $\mathrm{c}_p(G) < \mathrm{l}_p(G)$, p a prime. Then G has a unique minimal normal subgroup and so it satisfies that $\mathrm{O}_{p'}(G) = 1$. Therefore $N = \mathrm{O}_p(G)$ contains the centre of a Sylow p-subgroup of G because $\mathrm{C}_G(\mathrm{O}_p(G)) \leq \mathrm{O}_p(G)$ by [158, IX, 1.3]. Thus $\mathrm{c}_p(G/N) \leq \mathrm{c}_p(G) - 1$. Since $\mathrm{l}_p(G/N) = \mathrm{l}_p(G) - 1$, the minimal choice of G implies that $\mathrm{l}_p(G) - 1 \leq \mathrm{c}_p(G) - 1$. This contradiction proves the following.

Theorem 1.7.14. *Let G be a p-soluble group. Then* $\mathrm{c}_p(G) \geq \mathrm{l}_p(G)$.

Let now π be a set of primes. If a group G is π-soluble, one would expect its chief factors whose orders are divisible by primes in π to play a more important role than they do in groups in general. Our object in the sequel is to confirm this expectation by proving an extension of Theorems 1.7.1 and 1.7.2 which finds applications later in Chapter 3.

The next observation will be useful in our approach (see Chernikov and Petravchuk [101]).

Lemma 1.7.15. *Let G be a π-soluble group and let $\pi_1, \pi_2, \dots, \pi_r$ be sets of primes whose union is π. Then* $\mathrm{l}_\pi(G) \leq \sum_{i=1}^{r} \mathrm{l}_{\pi_i}(G)$.

Proof. Suppose that the lemma is not true and derive a contradiction. Let G be a counterexample of least order. Since the class of all π-soluble groups with π-length at most k is a saturated formation, it follows that G has a unique minimal normal subgroup. Moreover, $\mathrm{O}_{\pi'}(G) = 1$ and $\mathrm{O}_{p'}(G) = 1$ for some prime $p \in \pi$. We may assume that $p \in \pi_1$. Then $\mathrm{O}_{\pi_1'}(G) = 1$ and, if $\bar{G} = G/\mathrm{O}_{\pi_1}(G)$, it follows that $\mathrm{l}_{\pi_1}(\bar{G}) = \mathrm{l}_{\pi_1}(G) - 1$. We will also show that $\mathrm{l}_\pi(\bar{G}) = \mathrm{l}_\pi(G) - 1$. Indeed, $\mathrm{l}_\pi(G/\mathrm{O}_\pi(G)) \leq \mathrm{l}_\pi(\bar{G})$ because of the inclusion $\mathrm{O}_{\pi_1}(G) \leq \mathrm{O}_\pi(G)$. Moreover, since $\mathrm{O}_{\pi'}(G) = 1$, $\mathrm{l}_\pi(G/\mathrm{O}_\pi(G)) = \mathrm{l}_\pi(G) - 1$. Thus $\mathrm{l}_\pi(G) - 1 = \mathrm{l}_\pi(G/\mathrm{O}_\pi(G)) \leq \mathrm{l}_\pi(\bar{G}) < \mathrm{l}_\pi(G)$, so $\mathrm{l}_\pi(\bar{G}) = \mathrm{l}_\pi(G) - 1$. Therefore $\mathrm{l}_\pi(G) - 1 \leq \sum_{i=1}^{n} \mathrm{l}_{\pi_i}(\bar{G}) = (\mathrm{l}_{\pi_1}(G) - 1) + \sum_{i=2}^{r} \mathrm{l}_{\pi_i}(\bar{G})$ and thus $\mathrm{l}_\pi(G) \leq \mathrm{l}_{\pi_1}(G) + \sum_{i=2}^{r} \mathrm{l}_{\pi_i}(\bar{G})$. This is impossible because $\mathrm{l}_{\pi_i}(\bar{G}) \leq \mathrm{l}_{\pi_i}(G)$, $i \in \{2, \dots, r\}$. The proof of the lemma is now complete. □

Suppose that we want to prove the following statement: *If G is a π-soluble group, then* $\mathrm{l}_\pi(G) \leq \mathrm{d}_\pi(G)$. Let G be a counterexample with $|G|$ as small as possible. Then $\mathrm{l}_\pi(G) > \mathrm{d}_\pi(G)$ but for each proper factor group $\bar{G}$ of G the inequalities $\mathrm{l}_\pi(\bar{G}) \leq \mathrm{d}_\pi(\bar{G}) \leq \mathrm{d}_\pi(G) < \mathrm{l}_\pi(G)$ hold. We then conclude that G has exactly one minimal normal subgroup N such that $N = \mathrm{C}_G(N)$ and $N = \mathrm{O}_{p',p}(G) = \mathrm{O}_p(G)$ for some $p \in \pi$. Let H be a Hall π-subgroup of the group G, let K be the last non-trivial term in the derived series of H, and let $K_{p'}$ be a Hall p'-subgroup of K. Then $K_{p'} \leq \mathrm{C}_G(N) = N$. Therefore $K_{p'} = 1$ and, thus, K is a p-group.

Assume that K is contained in N, then $\mathrm{d}(H/N) \leq \mathrm{d}(H) - 1$. Thus $\mathrm{d}_\pi(G/N) \leq \mathrm{d}_\pi(G) - 1$ and, thus, $\mathrm{l}_\pi(G/N) \leq \mathrm{d}_\pi(G/N) \leq \mathrm{d}_\pi(G) - 1$. Since N is a π-group, we obviously have $\mathrm{l}_\pi(G/N) \geq \mathrm{l}_\pi(G) - 1$. This would be the desired contradiction.

It turns out that finding sufficient conditions to ensure that an abelian normal p-subgroup of a Hall π-subgroup of a π-soluble group G is contained in $\mathrm{O}_p(G)$ is important in this context. For instance, applying one of the consequences of the Theorem 1.7.8 appeared in [158, IX, 5.1], p must be 2 or 3 in our counterexample. Moreover, if $p = 3$, the Sylow 2-subgroups of G cannot be abelian. Consequently, if either

$\{2,3\} \cap \pi = \emptyset$ or K is a subgroup of 3-power order and the Sylow 2-subgroups are abelian, no counterexample exists.

Assume that $\{2,3\} \cap \pi$ is non-empty. Then the following lemma, due to Kazarin [176, Lemma 2], applies.

Lemma 1.7.16. *Let A be a Hall π-subgroup of a π-soluble group G, K an abelian normal p-subgroup of A and assume that $p > 2$. If $\{2\} \cap \pi(G) \subseteq \pi$, then $K \leq \mathrm{O}_{p',p}(G)$.*

Proof. Suppose, by way a contradiction, that the lemma is false and choose for G a counterexample of least order. If $\mathrm{O}_{p'}(G) \neq 1$, then

$$K\mathrm{O}_{p'}(G)/\mathrm{O}_{p'}(G) \leq \mathrm{O}_{p',p}(G/\mathrm{O}_{p'}(G)) = \mathrm{O}_{p',p}(G)/\mathrm{O}_{p'}(G)$$

and so $K \leq \mathrm{O}_{p',p}(G)$, contrary to supposition. Therefore $\mathrm{O}_{p'}(G) = 1$. Let N be a normal p'-subgroup of A. Let $G_{p'}$ be a Hall p'-subgroup of G containing N. Note that $G = AG_{p'}$. Hence $\langle N^G \rangle = \langle N^{AG_{p'}} \rangle = \langle N^{G_{p'}} \rangle \leq G_{p'}$ and so $N = 1$ since $\mathrm{O}_{p'}(G) = 1$. Let $\sigma = (\pi(G) \setminus \pi) \cup \{p\}$. Assume that $\sigma \neq \pi$. Let X be a Hall σ-subgroup of G containing some Sylow p-subgroup A_p of A. Then $G = AX$. The same argument shows that $\mathrm{O}_{p'}(X) \leq \mathrm{O}_{p'}(G) = 1$. By minimality of G, we have that $K \leq \mathrm{O}_{p',p}(X) = \mathrm{O}_p(X)$. Then $K \leq \mathrm{O}_p(A) \cap \mathrm{O}_p(X)$. By Lemma 1.1.23 we conclude that $K \leq \mathrm{O}_p(G)$. This contradiction shows that $\sigma = \pi$ and so $\pi = \{p\}$. In this case G has odd order. By [158, IX, 5.1], we conclude that $K \leq \mathrm{O}_{p',p}(G)$. This final contradiction proves the lemma. □

Suppose that $|\{2,3\} \cap \pi| = 1$. Let $p \in \{2,3\} \cap \pi$. Then $(\pi \setminus \{p\}) \cap \{2,3\} = \emptyset$ and so $\mathrm{l}_{\pi \setminus \{p\}}(G) \leq \mathrm{d}_{\pi \setminus \{p\}}(G)$. Now we can apply Lemma 1.7.15 to obtain that $\mathrm{l}_\pi(G) \leq \mathrm{l}_p(G) + \mathrm{l}_{\pi \setminus \{p\}}(G)$. By Theorems 1.7.1 and 1.7.2, we have $\mathrm{l}_p(G) \leq \mathrm{d}_p(G) \leq \mathrm{d}_\pi(G)$. In this case, we have $\mathrm{l}_\pi(G) \leq 2\mathrm{d}_\pi(G)$.

Assume now that $\{2,3\}$ contained in π. Then, if $p = 3$, we can use Lemma 1.7.16 to conclude that K is contained in N and so G is not a counterexample. Suppose that $p = 2$. To deal with this case, we need some additional results, which were also proved by Kazarin in [176].

Lemma 1.7.17. *Let G be a π-soluble group, $p \in \pi$, and assume that $\mathrm{O}_{\pi'}(G) = 1$. Then there exist groups H, X, and Y and a homomorphism $\varphi\colon G \to H = X \times Y$ such that $G^\varphi X = G^\varphi Y = H$, $\mathrm{Ker}\,\varphi = 1$, $\mathrm{O}_{\pi'}(X) = \mathrm{O}_{\pi'}(Y) = 1 = \mathrm{O}_p(Y) = \mathrm{O}_\sigma(Y)$ and $\mathrm{O}_p(X) = \mathrm{F}(X)$, where $\sigma = \pi' \cup \{p\}$. Furthermore, X can be taken as $G/\mathrm{O}_{p'}(G)$ and Y as $G/\mathrm{O}_\sigma(G)$.*

Proof. Let $S = \mathrm{O}_{p'}(G)$, $\sigma = \pi' \cup \{p\}$, and $T = \mathrm{O}_\sigma(G)$. Clearly $S \cap T \leq \mathrm{O}_{\pi'}(G) = 1$. Let $X = G/S$, $Y = G/T$, and $H = X \times Y$, and let $\varphi\colon G \to H$ the natural homomorphism given by the rule $g^\varphi = (gS, gT) \in H$ for every $g \in G$. Then $\mathrm{Ker}\,\varphi = 1$, and G^φ is subdirect in H. Moreover $\mathrm{F}(X) = \mathrm{O}_p(X)$ and $\mathrm{O}_{\pi'}(X) = \mathrm{O}_{\pi'}(Y) = \mathrm{O}_\sigma(Y) = \mathrm{O}_p(Y) = 1$. □

Lemma 1.7.18. *Let G be a π-soluble group, let A be a Hall π-subgroup of G for some set π of primes, let K be an abelian normal subgroup of A, and suppose that* $\mathrm{O}_{\pi'}(G) = 1$. *Then K is contained in the Fitting subgroup* $\mathrm{F}(G)$ *of G when at least one of the following conditions holds:*

1. $\{2,3\} \cap \pi = \emptyset$;
2. $\{2,3\} \cap \pi(G) \subseteq \pi$ *and* $|K| \equiv 1 \pmod 2$;
3. *the Sylow* 2*-subgroups of G are abelian.*

Proof. Let $p \in \pi(K)$ and put $R = \mathrm{O}_p(K)$. Consider a group $H = X \times Y$ and a homomorphism φ as in Lemma 1.7.17. We can identify the group G with its image G^{φ} in H. It is enough to prove that $R \leq \mathrm{O}_p(X) \cap G \leq \mathrm{O}_p(G)$. It is clear that RX/X is contained in the Hall π-subgroup AX/X of the group $H/X \cong Y$. Since $\mathrm{O}_{\pi'}(Y) = \mathrm{O}_p(Y) = 1$, $\mathrm{F}(Y)$ is a π-group and $\mathrm{F}(Y)X/X \leq AX/X$. Since R is normal in A, $[R, A] \leq R$ and so RX/X centralises $\mathrm{F}(Y)X/X$. Now $\mathrm{F}(Y)X/X$ contains its centraliser by [119, A, 10.6] and so $RX/X \leq \mathrm{F}(Y)X/X$. Since $\mathrm{O}_p(Y) = 1$, we conclude that $RX/X = 1$ and so $R \leq X$. Then R is an abelian normal p-subgroup of a Hall π-subgroup of X. Applying [158, IX, 5.1] we conclude that $R \leq \mathrm{O}_p(X)$ in each of the cases 1–3 with the exception of the case $p = 3$ and $\{2,3\} \subseteq \pi$. In this case Lemma 1.7.16 also allows us to deduce that $R \leq \mathrm{O}_p(X)$. This completes the proof. □

It is now possible, following the pattern established above (first the minimal counterexample argument, then an application of Lemma 1.7.18), to deduce that if $\{2,3\} \subseteq \pi$ and $\rho = \pi \setminus \{2\}$, then $\mathrm{l}_{\rho}(X) \leq \mathrm{d}_{\pi}(X)$ for every π-soluble group X.

Consequently, our counterexample G satisfies

$$\mathrm{l}_{\pi}(G) \leq \mathrm{l}_2(G) + \mathrm{l}_{\pi\setminus\{2\}}(G) \leq \mathrm{d}_2(G) + \mathrm{d}_{\pi}(G) \leq 2\,\mathrm{d}_{\pi}(G).$$

These conclusions are summarised in:

Theorem 1.7.19 (Kazarin [176]). *Let π be a set of primes, and let G be a π-soluble group. Then $\mathrm{l}_{\pi}(G) \leq 2\,\mathrm{d}_{\pi}(G)$. If $\{2,3\} \cap \pi = \emptyset$, then $\mathrm{l}_{\pi}(G) \leq \mathrm{d}_{\pi}(G)$.*

We now come to a theorem of fundamental importance, confirming that the prime 2 has an anomalous position in this theory.

Theorem 1.7.20 (Kazarin [176]). *Let π be a set of primes, and let G be a soluble group. If $2 \notin \pi$, then $\mathrm{l}_{\pi}(G) \leq \mathrm{d}_{\pi}(G)$.*

The proof of this theorem will be our main preoccupation in the rest of the section.

The following result is part of [158, IX, 5.4].

Lemma 1.7.21. *Let P be a Sylow 3-subgroup of the 3-soluble group G. Then the derived length of $P\mathrm{O}_{3',3}(G)/\mathrm{O}_{3',3}(G)$ is smaller than the derived length of P.*

Lemma 1.7.22. *Let π be a set of odd primes. Let G be a soluble group, let A be a Hall π-subgroup of G and that R is a normal subgroup of A such that $R = KL$, $K = \mathrm{O}_3(R)$ is abelian, L is a $3'$-group, and $K = [K, L] = R'$. Then K is contained in $\mathrm{O}_{3',3}(G)$.*

Proof. Assume that the result is false and let G be a counterexample of minimal order. We will divide the proof into a series of steps from which a contradiction will follow.

1. $\mathrm{O}_{3'}(G) = 1$. *In particular,* $\mathrm{O}_{3',3}(G) = \mathrm{O}_3(G)$.

Suppose that N is a minimal normal subgroup of G and that N is a $3'$-group. Then AN/N is a Hall π-subgroup of G/N, RN/N is a normal subgroup of AN/N, and KN/N is the normal Sylow 3-subgroup of RN/N. Moreover, KN/N is abelian, LN/N is a $3'$-group, and $KN/N = [K, L]N/N = [KN/N, LN/N] = R'N/N = (RN/N)'$. Therefore G/N satisfies the hypotheses of the lemma and we conclude that $KN/N \leq \mathrm{O}_{3',3}(G/N) = \mathrm{O}_{3',3}(G)/N$.

2. *G possesses a unique minimal normal subgroup, N say.*

Suppose that N_1 and N_2 are two different minimal normal subgroups of G. For $i \in \{1, 2\}$, G/N_i satisfies the hypotheses of the lemma and so $KN_i/N_i \leq \mathrm{O}_{3',3}(G/N_i)$. Since $\mathrm{O}_{3',3}(G/N_i)$ centralises all 3-chief factors of G/N_i by [119, A, 13.8], we obtain that $KN_i/N_i \leq \mathrm{C}_G(N_{3-i}N_i/N_i)$. But then $[K, N_{3-i}] \leq N_i \cap N_{3-i} = 1$. Consequently $K \leq \mathrm{C}_G(N_{3-i})$ for $i \in \{1, 2\}$. Since K centralises all 3-chief factors of G/N_i and K centralises N_i, we conclude that $K \leq \mathrm{O}_{3',3}(G)$. This contradicts our choice of G. Hence G has a unique minimal normal subgroup, N say.

3. $\Phi(G) = 1$.

By [119, A, 13.4 (f)], $\mathrm{O}_{3',3}(G/\Phi(G)) = \mathrm{O}_{3',3}(G)/\Phi(G)$. The statement follows by our minimal choice of G.

As a consequence of the previous statements we conclude:

4. *G is a primitive group, $N = \mathrm{F}(G) = \mathrm{C}_G(N)$, and there exists a maximal subgroup M of G such that $G = NM$ and $N \cap M = 1$.*

5. *$A_0 = A \cap M$ is a Hall π-subgroup of M, $A = NA_0$, and $F = \mathrm{F}(M)$ is a $3'$-group.*

It is obvious that $3 \in \pi$, because otherwise $K = 1$ and G would not be a counterexample. Hence $N \leq A$ and so $A = A \cap NM = N(A \cap M)$. Moreover A_0 is a Hall π-subgroup of M. Applying [119, A, 13.6], $F = \mathrm{F}(M)$ is a $3'$-group.

6. $\pi(A) = \pi(G) \setminus \{2\}$. *In particular, if S is a Sylow 2-subgroup of G, then $G = AS$.*

Suppose that q is a prime in $\pi(G) \setminus (\pi(A) \cup \{2\})$. Suppose that $2 \notin \pi(G)$. In this case, the Sylow 2-subgroups of G are trivial and so they are abelian. By [158, IX, 5.1], this implies that $K \leq \mathrm{F}(G)$, against the choice of G. Therefore $2 \in \pi(G)$ and so $\{2, 3\} \subseteq \pi(G)$. Let X be a Hall $2'$-subgroup of G containing A and let T be a Hall q'-subgroup of G containing A. Note that X and T are proper subgroups of G of coprime indices and so $G = XT$. Now $\mathrm{O}_{3'}(X)$ is normalised by a Sylow 3-subgroup A_3 of A and, in particular, by $\mathrm{O}_3(G)$. Since $\mathrm{O}_3(G)$ is a normal subgroup of G, we get $[\mathrm{O}_{3'}(X), \mathrm{O}_3(G)] = 1$. But $N = \mathrm{O}_3(G)$ is self-centralising and so $\mathrm{O}_{3'}(X) = 1$. Note that X satisfies the hypotheses and, by minimality of G, $K \leq \mathrm{O}_{3',3}(X) = \mathrm{O}_3(X)$. A similar argument shows that $K \leq \mathrm{O}_3(T)$. Since $G = XT$ and K is subnormal in both X and T, we conclude that K is subnormal in G and so $K \leq \mathrm{O}_3(G)$. This contradicts our choice of G. Hence $\pi(A) = \pi(G) \setminus \{2\}$.

7. *Let $Q = \mathrm{O}_2(M)$. Then $\mathrm{F}(M) = Q$, $G = NQR$ and $A = NR$. Moreover, if $U = \mathrm{O}_3(A_0)$, then $A_0 = UL$, $U = [U, L]$ and the Sylow 3-subgroup NU of A has nilpotency class 2.*

Let $F = \mathrm{F}(M)$. We can assume without loss of generality that $L \leq M$. Set $U = NK \cap M$, then $UN = (NK \cap M)N = NK \cap NM = NK$. Let $Z = QY$, where $Y = NR$. Now N is a normal subgroup of Z and $Z = NQR = NQKL = NQUL$, where N and U are 3-groups, $Q = \mathrm{O}_2(M)$ is a 2-group, and L is a $3'$-group. Assume that Z is a proper subgroup of G. By minimality of G, we obtain that $K \leq \mathrm{O}_{3',3}(Z)$, which coincides with $\mathrm{O}_3(Z)$ because N is self-centralising. Now $NK \leq \mathrm{O}_3(Z)$ and NK is a Sylow 3-subgroup of Z, which implies that Q normalises NK. Hence Q normalises $NK \cap M = U$. Let $q \neq 2$ be a prime dividing $|\mathrm{F}(M)|$ and let F_q be a Sylow q-subgroup of A_0. We have that K is normalised by A and so F_q normalises K and N. This implies that F_q normalises $NK \cap M = U$. It follows that U is normalised by $\mathrm{F}(M)$. Since $U \leq M$ and U and $\mathrm{F}(M)$ have coprime orders, $[U, \mathrm{F}(M)] = 1$. Since $\mathrm{F}(M)$ is self-centralising in M by [119, A, 10.6], we conclude that $U = 1$. Consequently $NK = UN = N$ and so $K \leq N$. This contradicts the minimal choice of G. Therefore $G = QY$ and so $A = NR$.

Since $UN = KN$ and $K = [K, L]$, we obtain that

$$\begin{aligned} UN/N &= KN/N = [K, L]N/N \\ &= [KN/N, LN/N] = [UN/N, LN/N] = [U, L]N/N. \end{aligned}$$

Since U and $[U, L]$ are both subgroups of M, it follows that $UN \cap M = U(N \cap M) = U = [U, L]N \cap M = [U, L](N \cap M) = [U, L]$.

Now $U \cong U/(N \cap U) \cong UN/N = KN/N$ is an abelian group. This implies that NU is metabelian and $(NU)' = [N, U]$. Moreover $\mathrm{O}_3(A) = NK$ and so $(\mathrm{O}_3(A))' = (NK)' = [N, K] \leq K \cap N$, which is contained in $\mathrm{Z}(\mathrm{O}_3(A))$ because K and N are abelian. It follows that $\mathrm{O}_3(A)$ has nilpotency class 2. Since $A = NKL = NUL$ and

$NK = NU$ is a normal 3-subgroup of A, we have that $NK = O_3(A)$ is the Sylow 3-subgroup of A.

8. *A' is abelian and $[N, L]$ centralises U.*

We have that $NK = O_3(A)$ has class 2 and $O_3(A)' \leq Z(O_3(A))$. On the other hand, $[N, L] \leq [N, R] \leq N \cap R \leq O_3(R) \cap N = K \cap N$. Hence $[O_3(A), L] = [NK, L] = [N, L][K, L] \leq K$, which is abelian. It follows that $A' = (O_3(A)L)' = (O_3(A))'[O_3(A), L]$ is abelian. Since $[N, L] \leq A'$ and $U = [U, L] \leq A'$, we conclude that $[N, L]$ centralises U.

9. *The subgroup $Q = O_2(M)$ is a non-abelian 2-group, $\Phi(Q) = Z(Q)$ is elementary abelian, and A_0 acts irreducibly on $Q/\Phi(Q)$. Moreover $[U, \Phi(Q)] = 1$.*

Assume that A_0 does not act irreducibly on $Q/\Phi(Q)$. By Maschke's theorem ([119, A, 11.4]), there exist two proper A_0-admissible subgroups Q_1 and Q_2 of Q such that $Q_1Q_2 = Q$ and $Q_1 \neq Q \neq Q_2$. It follows that $X = NQ_1A_0$ and $Y = NQ_2A_0$ are proper subgroups of G and $G = XY$. By the choice of G, we have that $U \leq O_3(X) \cap O_3(Y)$. By Lemma 1.1.23, we have that $U \leq O_3(G) \leq N$, which contradicts the choice of G. Therefore A_0 acts as an irreducible group of automorphisms on $Q/\Phi(Q)$.

Assume now that Q is abelian. By Lemma 1.7.18 (3), $U \leq A' \leq N$ and once again this contradicts the choice of G. Thus Q is non-abelian.

Note that $[Q, U] \neq 1$ because otherwise $NU = O_3(G) = N$, against the choice of G. Since A_0 acts irreducibly on $Q/\Phi(Q)$, we obtain that $[Q, U] = Q$ by [119, A, 12.5]. Now $X = N\Phi(Q)A_0$ is a proper subgroup of G. By minimality of G, we have that $U \leq O_3(X)$ and so $[U, \Phi(Q)] = 1$. Since $[U, \Phi(Q), Q] = 1 = [Q, \Phi(Q), U]$ (note that $[Q, \Phi(Q)] \leq \Phi(Q)$ and $[\Phi(Q), U] = 1$), we obtain by the three subgroups lemma [119, A, 7.6] that $[U, Q, \Phi(Q)] = 1$. Since $[U, Q] = Q$, it follows that $[Q, \Phi(Q)] = 1$ and so $\Phi(Q) \leq Z(Q)$. Since Q is non-abelian and A_0 acts irreducibly on $Q/\Phi(Q)$, we have that $Z(Q) = \Phi(Q)$. Since $Q' \leq \Phi(Q) = Z(Q)$, Q has nilpotency class 2.

Now we prove that $\Phi(Q)$ has exponent 2. Let $x, y \in Q$. Then $[x, y]^2 = [x^2, y] = 1$ because $x^2 \in \Phi(Q) \leq Z(Q)$. This implies that all commutators have order at most 2 and, since they are central, G' has exponent 2. It will be enough to check that Q has exponent 4. Note that since Q has class 2, the map ψ given by $x \mapsto x^4$ is a group homomorphism between Q and $\Phi(Q)$, since $(xy)^4 = x^4y^4[y, x]^6$ by [157, III, 1.3] and $[y, x]^6 = 1$. Obviously this is an M-homomorphism. Therefore $Q/\operatorname{Ker}\psi \cong \operatorname{Im}\psi = Q^4$. Assume that $\operatorname{Ker}\psi \neq Q$. In this case, $(\operatorname{Ker}\psi)\Phi(Q)$ is also a proper normal subgroup of M contained in Q and so $\operatorname{Ker}\psi \leq \Phi(Q)$ because A_0 acts irreducibly on $Q/\Phi(Q)$. But $\operatorname{Im}\psi \leq \Phi(Q)$ is centralised by U, while $Q/\Phi(Q)$ is not centralised by U. This contradiction shows that $\operatorname{Ker}\psi = Q$ and so Q has exponent 4. Consequently $\Phi(Q)$ has exponent 2.

10. $L = \langle a \rangle$ *is a cyclic* p*-group for some prime* $p \in \pi$ *and* $a^p \in C_L(U)$.

Note that $[U, L] = U$. Therefore, there exist a prime p and an element of p-power order of L, a say, such that $[U, \langle a \rangle]$ is not contained in N. Moreover, by [119, A, 12.5], $U = C_U(\langle a \rangle) \times [U, \langle a \rangle]$ and $[U, \langle a \rangle] = [U, \langle a \rangle, \langle a \rangle]$ by [119, A, 12.4]. In that case, the subgroup $NQ[U, \langle a \rangle]\langle a \rangle$ satisfies the hypotheses of the statement we want to prove. Suppose that $NQ[U, \langle a \rangle]\langle a \rangle$ is a proper subgroup of G. We deduce then that $[U, \langle a \rangle]$ is contained in $O_{3',3}(NQ[U, \langle a \rangle]\langle a \rangle)$, which coincides with $O_3(NQ[U, \langle a \rangle]\langle a \rangle)$ because N is self-centralising. But now $[U, \langle a \rangle] \leq O_3(NQ[U, \langle a \rangle]\langle a \rangle) \cap Q[U, \langle a \rangle]\langle a \rangle$, and so $[U, \langle a \rangle] \leq O_3(Q[U, \langle a \rangle]\langle a \rangle) = 1$ because Q is self-centralising in M. This contradiction implies that $G = NQU\langle a \rangle$, and so $L = \langle a \rangle$ for a p-element a of L.

Let $K_0 = NQ[U, \langle a^p \rangle]\langle a^p \rangle$. We note that K_0 satisfies the hypotheses of the statement we want to prove. We can argue as above to deduce that $[U, \langle a^p \rangle] \leq N$. It follows then that $[U, \langle a^p \rangle] = 1$, that is, a^p centralises U.

11. U *is an elementary abelian* 3*-group.*

Note that $[U, L] = U$. Therefore $C_U(L) = 1$ and so $C_{\Phi(U)}(L) = 1$. In particular, $[\Phi(U), L] = \Phi(U)$ by [119, A, 12.5]. Hence $X = NQ\Phi(U)L$ satisfies the hypothesis of our statement and so $\Phi(U) \leq F(X)$ by the minimality of G. Thus $\Phi(U)$ would be contained in $C_U(Q) = 1$ and so U is elementary abelian.

12. $\Phi(Q)$ *has order* 2.

We know by Statement 9 that $\Phi(Q)$ is an elementary abelian group. Let z be an element of order 2 of $\Phi(Q)$. Then N, regarded as a $\mathrm{GF}(3)M$-module, can be decomposed into a direct sum of homogeneous components under the action of $\langle z \rangle$. The only possibilities for such homogeneous components are $N = V_1 \times V_2$ with $v^z = v^j$ for each $v \in V_i$, $i \in \{1, 2\}$. Note that if $v \in V_i$ and $y \in QU$, then $yz = zy$ and so $(v^y)^z = v^{yz} = v^{zy} = (v^z)^y = (v^j)^y = (v^y)^j$, which implies that $v^y \in V_i$ for $i \in \{1, 2\}$. It follows that V_1 and V_2 are invariant under the action of QU. Since N is a minimal normal subgroup of G, it follows that either a permutes both components, or one of them must be trivial. Suppose that $V_1^a = V_2$. Let $u \in U$ and $v \in V_1$. Then $[[v, a], u] = 1$, that is, $[v^{-1}v^a, u] = 1$, and since N is abelian, $v^{-1}v^u = (v^a)^{-1}(v^a)^u \in V_1 \cap V_2 = 1$. Hence u centralises V_1 and $V_1^a = V_2$. This means that u centralises N, so $u \in C_M(N)$. Therefore $U = 1$. In this case, G cannot be a counterexample to Lemma 1.7.22.

Hence N is homogeneous under the action of $\langle z \rangle$. Obviously, $N \neq V_1$ because in this case z would centralise N and so $z \in C_M(N) = 1$. Consequently $N = V_2$ and $v^z = v^{-1}$ for all $v \in N$. Applying [119, B, 9.4], $\Phi(Q)$ is cyclic.

13. $Q = Q_1 \curlyvee Q_2 \curlyvee \cdots \curlyvee Q_p$, *where the* Q_i *are quaternion groups,* $i \in \{1, 2, \ldots, p\}$, *and* QU *can be embedded into a central product of* p *copies of groups isomorphic to* $\mathrm{SL}_2(3)$. *In addition, the element* a *permutes the* Q_i *cyclically.*

Since $p > 3$, U is not a cyclic group. Moreover, the action of UL on $\bar{Q} = Q/\Phi(Q)$ is irreducible. It follows that under the action of U,

$$\bar{Q} = \bar{Q}_1 \times \bar{Q}_2 \times \cdots \times \bar{Q}_s,$$

where the $\bar{Q}_i$ are minimal U-invariant subgroups of $\bar{Q}$. Moreover, the $\bar{Q}_i$ are irreducible under the action of $U/\mathrm{C}_U(\bar{Q}_i)$ and the kernels $\mathrm{C}_U(\bar{Q}_i)$ of the actions are conjugate. Since U does not act trivially on Q and $U/\mathrm{C}_U(\bar{Q}_i)$ is cyclic of order 3, it follows that $|\bar{Q}_i| = 4$.

As $\langle a \rangle$ permutes the $\bar{Q}_i$ cyclically, it follows that $s = p$. Finally, the inverse images Q_i of the $\bar{Q}_i$ in Q are groups of order 8 which admit a cyclic group of order 3 as an automorphism group for each $i \in \{1, 2, \dots, s\}$. Observe that Q_i is normal in QU. Suppose that Q_i is an abelian group. Then by Corollary 1.7.11 applied to NQ_iU, NU has nilpotency class at least 3. This contradicts Statement 7. Since the dihedral group of order 8 has no automorphisms of order 3, it follows that the Q_i are quaternion groups of order 8. The rest of the assertions are now clear.

14. $|L| = p$.

The subgroup $L = \langle a \rangle$, whose order is a power of p, acts transitively on the set $\{\bar{Q}_1, \bar{Q}_2, \dots, \bar{Q}_p\}$. Hence the index of the stabiliser in L of $\bar{Q}_i$ is p. This implies that the unique maximal $\langle a^p \rangle$ of L belongs to the stabilisers of all copies. In particular, all $\bar{Q}_i$ are $\mathrm{GF}(2)\langle a^p \rangle$-modules of dimension 2. Since the group of automorphisms induced by $\langle a^p \rangle$ on $\bar{Q}_i$ must be a subgroup of $\mathrm{GL}_2(2) \cong \mathrm{Sym}(3)$, it turns out that a^p must centralise all copies $\bar{Q}$. Since $\Phi(Q)$ has order 2, by [119, A, 12.3] it turns out that a^p centralises Q. But $Q = \mathrm{F}(M)$ is self-centralising in M. Therefore $a^p = 1$, as desired.

15. *G is not a counterexample to Lemma* 1.7.22.

The normal subgroup N, regarded as a module for QU over the field $F = \mathrm{GF}(3)$ of three elements, is decomposed into a direct sum of irreducible QU-modules which are permuted by L. Note that all kernels are conjugate. Suppose first that N as a QU-module is irreducible. By [119, B, 9.17], Q possesses a unique irreducible and faithful module of dimension 2^p over F and 2^{2p} modules of dimension 1, and all of them are absolutely irreducible. On the other hand, N, regarded as a module for Q, is a direct sum of its homogeneous components and all of them are conjugate by Clifford's theorem [119, B, 7.3]. If the simple FQ-module in any of these components is one-dimensional, then $\Phi(Q)$ is contained in the kernel of all components and so $\Phi(Q) \leq \mathrm{C}_G(N) = N$. Therefore there is a unique homogeneous component corresponding to the module of dimension 2^p. By [119, B, 8.3], N is also irreducible as an FQ-module. In particular, N is irreducible as an FQU-module.

We can now identify $M = QUL$ with a subgroup of $\mathrm{GL}(N)$. We know that N has F-dimension 2^p. Since $Q = Q_1 \curlyvee Q_2 \curlyvee \cdots \curlyvee Q_p$ is a central product of copies of a quaternion group of order 8, in an appropriate basis for N the elements of Q

can be expressed as Kronecker matrices $h_1 \otimes h_2 \otimes \cdots \otimes h_p$, where $h_i \in \mathrm{SL}_2(3)'$, $1 \le i \le p$, and that the elements of QU can be expressed in the same basis as $h_1 \otimes h_2 \otimes \cdots \otimes h_p$ with $h_i \in \mathrm{SL}_2(3)$, $1 \le i \le p$. Thus N can be regarded as a tensor product $V \otimes V \otimes \cdots \otimes V$ of p copies of a two-dimensional subspace over F (cf. [119, B, 1.12] and the remark following [119, B, 9.19]). The action of the elements of QU on $x_1 \otimes x_2 \otimes \cdots \otimes x_p \in N$ ($x_i \in V$, $1 \le i \le p$) can be described as

$$(x_1 \otimes x_2 \otimes \cdots \otimes x_p)(h_1 \otimes h_2 \otimes \cdots \otimes h_p) = x_1h_1 \otimes x_2h_2 \otimes \cdots \otimes x_ph_p.$$

Let a be a generator of L which permutes the subgroups Q_i ($1 \le i \le p$) cyclically so that $Q_i^a = Q_{i+1}$ if $1 \le i < p$ and $Q_p^a = Q_1$. Let θ be the permutation of the set $\{1, 2, \ldots, p\}$ corresponding to this action (that is, $i\theta = i + 1$ for $1 \le i < p$ and $p\theta = 1$). We define an element a^* by the rule

$$(x_1 \otimes x_2 \otimes \cdots \otimes x_p)a^* = x_{1\theta} \otimes x_{2\theta} \otimes \cdots \otimes x_{p\theta},$$

then

$$\begin{aligned}(x_1 \otimes x_2 \otimes \cdots \otimes x_p)&(a^*)^{-1}(h_1 \otimes \cdots \otimes h_p)a^* \\ &= (x_{1\theta^{-1}} \otimes x_{2\theta^{-1}} \otimes \cdots \otimes x_{p\theta^{-1}})(h_1 \otimes h_2 \otimes \cdots \otimes h_p)a^* \\ &= (x_{1\theta^{-1}}h_1 \otimes x_{2\theta^{-1}}h_2 \otimes \cdots \otimes x_{p\theta^{-1}}h_p)a^* \\ &= x_1h_{1\theta} \otimes x_2h_{2\theta} \otimes \cdots \otimes x_ph_{p\theta} \\ &= (x_1 \otimes x_2 \otimes \cdots \otimes x_p)(h_{1\theta} \otimes h_{2\theta} \otimes \cdots \otimes h_{p\theta}),\end{aligned}$$

and so

$$(a^*)^{-1}(h_1 \otimes h_2 \otimes \cdots \otimes h_p)a^* = h_{1\theta} \otimes h_{2\theta} \otimes \cdots \otimes h_{p\theta}.$$

Now a^*, regarded as an element of $\mathrm{GL}(N)$, satisfies that a^*a^{-1} centralises Q. Therefore $a^*a^{-1} \in \mathrm{C}_G(Q) = \mathrm{Z}(Q)$, and since the generator of $\mathrm{Z}(Q)$ inverts all elements of N, $a^* = \lambda a$ with $\lambda \in F^* = \{1, -1\}$. Since a and a^* have both order p, we obtain that $\lambda = 1$.

Now we calculate the dimension of a fixed-point subspace of a on N as a vector space over F. It is clear that

$$x_1 \otimes x_2 \otimes \cdots \otimes x_p = (x_1 \otimes x_2 \otimes \cdots \otimes x_p)a = x_2 \otimes x_3 \otimes \cdots \otimes x_p \otimes x_1$$

if and only if $x_1 \otimes x_2 \otimes \cdots \otimes x_p$ is proportional to $x_1 \otimes x_1 \otimes \cdots \otimes x_1$ where $x_1 \in V$. Hence $\mathrm{C}_N(a)$ has F-dimension 2. The map $-1 + a$ defined by $n \mapsto [n, a]$ is a group homomorphism from N to N because $[n_1n_2, a] = [n_1, a]^{n_2}[n_2, a] = [n_1, a][n_2, a]$. Hence $\mathrm{C}_N(a) = \mathrm{Ker}(-1 + a)$. Consequently $\mathrm{Im}(-1 + a) = [N, L] \le \mathrm{C}_N(U)$ by Statement 8 and so

$$\dim_F \mathrm{C}_N(U) \ge 2^p - 2.$$

Note that if $u = h_1 \otimes h_2 \otimes \cdots \otimes h_p$ and $u_1 = h'_1 \otimes h'_2 \otimes \cdots \otimes h'_p$ are contained in U, then it follows from the commutativity of U and the properties of a Kronecker product of matrices that

$$\begin{aligned} uu_1 &= (h_1 \otimes h_2 \otimes \cdots \otimes h_p)(h'_1 \otimes h'_2 \otimes \cdots \otimes h'_p) \\ &= h_1h'_1 \otimes h_2h'_2 \otimes \cdots \otimes h_ph'_p \\ &= h'_1h_1 \otimes h'_2h_2 \otimes \cdots \otimes h'_ph_p \\ &= u_1u. \end{aligned}$$

Since $h_ih'_i = h'_ih_i$, it follows that h'_i is a power of h_i if both have order 3 because the subgroups of order 3 of $SL_2(3)$ are self-centralising. Now we can identify the set U with a subspace S of $T = F^p$ which is invariant under cyclic shifts. Define the weight $w = w(y)$ of an element $y = (f_1, f_2 \dots, f_p) \in T$ as the number of non-zero f_i. Suppose that $u = h_1 \otimes h_2 \otimes \cdots \otimes h_p \in U$ has weight $r \geq 3$ (in S). To simplify the notation, assume that the first r components of $h_1 \otimes h_2 \otimes \cdots \otimes h_p$ correspond to elements of order 3. Assume that $r \geq 3$, x_i, y_i in V satisfy $x_ih_i = x_i + y_i$ (in additive notation) and that y_i is fixed by h_i, in other words, that x_i, y_i form a basis for which h_i is expressed as the Jordan canonical form. Hence the elements $x_1 \otimes y_2 \otimes y_3 \otimes \cdots \otimes y_p$, $y_1 \otimes x_2 \otimes y_3 \otimes \cdots \otimes y_p$, $y_1 \otimes y_2 \otimes x_3 \otimes \cdots \otimes y_p$ are linearly independent elements of N which generate a subspace W of dimension 3 such that $W \cap C_N(U)$ is trivial. This is impossible since $\dim_F C_N(U) \geq \dim_F N - 2$. Consequently the weight of all elements of U cannot exceed 2.

We can now calculate the average weight of an element in S. Note that a subgroup U_i $(1 \leq i \leq p)$ of U consisting of those elements $u \in U$ such that the corresponding h_i operates as the identity in the expression $u = h_1 \otimes h_2 \otimes \cdots \otimes h_p$ is conjugate to U_j for each $j \in \{1, 2, \dots, p\}$. If we consider subspaces of S, we find that a subspace $S_i \leq S$ consisting of all $y = (f_1, f_2, \dots, f_p)$ with the property that $f_i = 0$ is a shift of the analogous subspace S_1. Since $U \neq 1$, $|S| > 1$. We can arrange all elements of S in an $m \times p$-matrix whose rows consist of elements of S and $m = |S|$. Since S_i is a subgroup of the additive group S of index 3, it turns out that $2/3$ of each column in this table consists of non-zero elements. Therefore the number of non-zero components in this matrix is $\frac{2}{3}mp$. On the other hand, this number is a sum of weights of all elements in S. Hence the average weight of an element of S is $\frac{2}{3}p$. Since all elements have weight at most 2, we conclude that $\frac{2}{3}p \leq 2$. Therefore $p \leq 3$, which contradicts the definition of p.

Therefore there cannot be a counterexample to Lemma 1.7.22. This completes the proof. □

Corollary 1.7.23. *Let A be a Hall π-subgroup of the soluble group G. If $2 \notin \pi$ and the smallest non-trivial term K of the derived series of A is p-group, p a prime, and $O_{\pi'}(G) = 1$, then K is contained in the Fitting subgroup of G.*

Proof. As in the proof of Lemma 1.7.18, it will be enough to show that K is contained in $\mathrm{O}_p(X) \cap G \leq \mathrm{O}_p(G)$, where X and Y are the groups given by Lemma 1.7.17. As in the proof of Lemma 1.7.18, K can be identified with an abelian normal subgroup of a Hall π-subgroup of X. Hence we may assume that $A \leq X$.

Suppose that $p \neq 3$. Applying [158, IX, 5.1] we conclude that $K \leq \mathrm{O}_{p',p}(X)$. Since $\mathrm{O}_p(X) = \mathrm{F}(X)$ and X is soluble, $\mathrm{O}_{p'}(X) = 1$. Hence $K \leq \mathrm{O}_p(X)$ and so $K \leq \mathrm{O}_p(X) \cap G \leq \mathrm{O}_p(G)$. Therefore we can suppose that $p = 3$. Hence $\mathrm{F}(X) = \mathrm{O}_3(X)$ and $\mathrm{O}_{3'}(X) = 1$.

Let $R = A^{(d-2)}$, where $d = \mathrm{d}(A)$; then $K = R'$. Since $\mathrm{F}(X) \leq A$, it follows that $\mathrm{O}_{3'}(A)$ centralises $\mathrm{F}(X)$. Therefore $\mathrm{O}_{3'}(A) = 1$. Hence $\mathrm{F}(A) = \mathrm{O}_3(A)$ and K is contained in $\mathrm{O}_3(A)$. Since $R' = K$ is a 3-group, it follows $R = [\mathrm{O}_3(R)]L$, where L is an abelian Hall $3'$-subgroup of R. Moreover, $[R, \mathrm{O}_3(R)] \leq \mathrm{O}_3(R)$ and $\mathrm{O}_3(R)' \leq K$, which is abelian. We conclude that

$$[R, \mathrm{O}_3(R), \mathrm{O}_3(R), \mathrm{O}_3(R)'] \leq [\mathrm{O}_3(R), \mathrm{O}_3(R), \mathrm{O}_3(R)'] \leq \mathrm{O}_3(R)'' \leq K' = 1.$$

By [158, IX, 5.3], it follows that $\mathrm{O}_3(R)' \leq \mathrm{O}_{3',3}(X) = \mathrm{F}(X)$. Consequently $K = [R, R] = [\mathrm{O}_3(R)L, \mathrm{O}_3(R)L] = [\mathrm{O}_3(R), L]\mathrm{O}_3(R)'$. Since $\mathrm{O}_3(R)' \leq \mathrm{F}(X)$, we will reach a contradiction if we show that $[\mathrm{O}_3(R), L] \leq \mathrm{F}(X)$. By virtue of [119, A, 12.5], we have that $\mathrm{O}_3(R) = [\mathrm{O}_3(R), L]\mathrm{C}_{\mathrm{O}_3(R)}(L)$. Furthermore $[\mathrm{O}_3(R), L] \leq R' = K$. Hence

$$[K, L] \leq [\mathrm{O}_3(R), L] = [\mathrm{O}_3(R), L, L] \leq [K, L].$$

It follows that $[\mathrm{O}_3(R), L] = [K, L]$. Now consider $C = [K, L]L$. Since $R = \mathrm{O}_3(R)L$, the nilpotent residual $R^{\mathfrak{N}}$ of R is contained in $[\mathrm{O}_3(R), L] = [K, L]$. Since $R^{\mathfrak{N}} \leq R' = K$ and $KL \leq R$, it turns out that $KL/R^{\mathfrak{N}} \leq R/R^{\mathfrak{N}}$ is also a nilpotent group and so $KL/R^{\mathfrak{N}} = (K/R^{\mathfrak{N}}) \times (LR^{\mathfrak{N}}/R^{\mathfrak{N}})$. Consequently $[K, L] \leq R^{\mathfrak{N}}$. It follows that $[K, L] = R^{\mathfrak{N}}$. Since $LR^{\mathfrak{N}}/R^{\mathfrak{N}}$ is a normal Hall subgroup of $R/R^{\mathfrak{N}}$ and $R/R^{\mathfrak{N}}$ is normal in $A/R^{\mathfrak{N}}$, it follows that $LR^{\mathfrak{N}} = L[K, L] = C$ is a normal subgroup of A. Then $[K, L] = \mathrm{O}_3(C)$ is an abelian normal subgroup of C, L is a $3'$-group, and $C' = [K, L, L] = [K, L]$ by [119, A, 12.4]. By Lemma 1.7.22 applied to C, $[K, L]$ is contained in $\mathrm{O}_{3',3}(X) = \mathrm{O}_3(X)$. The result is therefore proved. □

We are now ready to prove our last significant result of the section.

Proof of Theorem 1.7.20. If $3 \notin \pi$, the result follows from Theorem 1.7.19. Hence we assume that $3 \in \pi$ and we shall now derive a contradiction by supposing that the theorem is false. Let G be a group of minimal order with $\mathrm{l}_\pi(G) > \mathrm{d}_\pi(G)$. Then $\mathrm{O}_{\pi'}(G) = 1$ and G has a unique minimal normal subgroup, N say. Since G is soluble, N is an elementary abelian p-group for some prime $p \in \pi$. In this case, $\mathrm{F}(G) = \mathrm{O}_p(G)$. Let $t = \mathrm{d}_\pi(G)$. If A is a Hall π-subgroup of G and $K = A^{(t-1)}$, then K is an abelian normal subgroup of A. Obviously we may assume that A is a proper subgroup of G and so $\pi \cap \pi(G) \neq \pi(G)$. Let

$$\pi(G) = (\pi \cap \pi(G)) \cup \{r_1, r_2, \dots, r_m\},$$

where the r_i are the primes in $\pi(G) \setminus \pi$. Set $\sigma_i = \pi \cup \{r_i\}$. Applying [119, I, 4.16], there exists a Hall system Σ such that $A \in \Sigma$. Let S_i be the Hall σ_i-subgroup in Σ. Then S_i contains A. As $\mathrm{F}(G) \leq S_i$ and $\mathrm{C}_G(\mathrm{F}(G)) \leq \mathrm{F}(G)$ by [158, IX, 1.3], we have that $\mathrm{O}_{\pi'}(S_i) = 1$ and $\mathrm{F}(S_i) = \mathrm{O}_p(S_i)$. Applying Corollary 1.7.23, we have that $K \leq \mathrm{O}_p(S_i)$ for $1 \leq i \leq m$. Since $S_i S_j$ is a subgroup of G, it follows that $K \leq \mathrm{O}_p(S_i S_j)$ for all i, j by Corollary 1.1.8. Since $G = S_1 \cdots S_m$, we conclude that $K \leq \mathrm{O}_p(G) \leq \mathrm{O}_\pi(G)$. Consequently $\mathrm{d}_\pi(G/\mathrm{O}_\pi(G)) \leq \mathrm{d}_\pi(G) - 1$. The minimal choice of G implies that $\mathrm{l}_\pi(G/\mathrm{O}_\pi(G)) \leq \mathrm{d}_\pi(G) - 1$. Therefore we are forced to the conclusion that $\mathrm{l}_\pi(G) \leq \mathrm{d}_\pi(G)$. Hence our initial supposition is not true and the proof of the theorem is complete. □

Chapter 2

Groups whose subnormal subgroups are normal, permutable, or Sylow-permutable

Having studied in some detail permutability and S-permutability in Section 1.2 of Chapter 1 and knowing that they are not transitive in general, the structure of the groups in which these properties are transitive is crying out to be studied. In fact, there has been a widespread interest in this question and there has been considerable activity in this area in recent years. Many structural features of groups in these classes have been discovered, as well as characterisations of various sorts. Our central objective in this chapter is to give a full account of the main accomplishments in the area.

We begin by introducing the various classes of groups studied in the chapter. The first class we broach concerns the groups in which normality is a transitive relation.

Definition 2.0.1. A group G is a T-*group* if every subnormal subgroup if G is normal in G.

It is rather clear that the T-groups are exactly the groups G in which normality is a transitive relation, that is, if H is a normal subgroup of K and K is a normal subgroup of G, then H is a normal subgroup of G.

A relevant subclass of the class of T-groups is the class of all Dedekind groups (see Section 1.4). Note that the nilpotent T-groups are exactly the Dedekind ones.

The first explicit mention of T-groups in the literature is in a paper of Best and Taussky [71]. They show that any group with cyclic Sylow subgroups is a T-group. Subsequently Zacher [263] characterised the soluble T-groups by means of Sylow tower properties. The decisive structure theorem for soluble T-groups was proved by Gaschütz [137]. They are exactly the soluble groups G with an abelian normal Hall subgroup L of odd order such that G/L is a Dedekind group and L is acted upon by conjugation as a group of power automorphisms by G.

Another natural class of groups, but one which is much wider than the class of all T-groups, is the class of all PT-groups, or groups in which permutability is transitive.

Definition 2.0.2. A group G is said to be a PT-*group* when if H is a permutable subgroup of K and K is a permutable subgroup of G, then H is a permutable subgroup of G.

By Theorem 1.2.14 (3), every permutable subgroup is subnormal. Therefore, PT-groups are exactly the groups in which every subnormal subgroup is permutable. As examples of PT-groups we mention Iwasawa groups, or groups in which every subgroup is permutable. These groups are precisely the nilpotent PT-groups. A non-Dedekind Iwasawa group shows that the class of all T-groups is a proper subclass of the class of all PT-groups. Zacher [264] gave a precise structural description of the soluble PT-groups: one has to replace the word "Dedekind" by "Iwasawa" in Gaschütz's result.

We will now broaden the investigation by considering the effect of imposing transitivity on the S-permutable subgroups. We incorporate this property in a formal definition:

Definition 2.0.3. A group G is said to be a PST-*group* when if H is an S-permutable subgroup of K and K is an S-permutable subgroup of G, then H is an S-permutable subgroup of G.

Again by Theorem 1.2.14 (3), S-permutable subgroups are necessarily subnormal. Hence the PST-groups are exactly the groups in which every subnormal subgroup is S-permutable. Furthermore, the nilpotent groups are exactly the groups in which every subgroup is S-permutable. Since there are nilpotent groups which are not modular, it follows that the class of all PST-groups contains properly the class of all PT-groups. The work of Agrawal [2] has increased our understanding of the structure of soluble PST-groups. He showed that a group G is a soluble PST-group if and only if the nilpotent residual is a Hall subgroup of odd order acted upon by conjugation as a group of power automorphisms by G.

Theorems of Gaschütz, Zacher, and Agrawal make it transparent that *the classes of all soluble* T-, PT-, *and* PST-*groups are closed under taking subgroups* (see Corollaries 2.1.9 and 2.1.13). While every non-abelian simple group is T-group, it will emerge that solubility imposes considerable restrictions on the structure of a group belonging to one of these classes.

There are in essence three different ways of approaching the question of characterising T-, PT-, and PST-groups: normal structure, local criteria, and characterisations based on the embedding of some relevant subgroups. Before we move on to their description, we pause briefly to show a result which is very useful in induction arguments.

Lemma 2.0.4. *Assume that $\mathfrak{X}$ is one of the classes* T, PT, *or* PST. *If N and H are normal subgroups of G such that $N \leq H$ and G/N is an $\mathfrak{X}$-group, then G/H is an $\mathfrak{X}$-group. In particular, $\mathfrak{X}$ is closed under taking epimorphic images.*

Moreover, the class $\mathfrak{X}$ is closed under taking normal subgroups, but $\mathfrak{X}$ is not closed under taking subgroups nor direct products. In particular, $\mathfrak{X}$ is neither a formation nor a Fitting class.

Proof. Let A/H be a subnormal subgroup of G/H. Then A is a subnormal subgroup of G and so A/N is a subnormal subgroup of G/N. Assume first that G/N is a PST-group. Then A/N is an S-permutable subgroup of G/N. But this implies that A is an S-permutable subgroup of G by Theorem 1.2.14 (1) and so A/H is an S-permutable subgroup of G/H by Lemma 1.2.7 (2). It follows that G/H is a PST-group.

If we replace S-permutable by permutable or normal, we obtain the proof for PT and T, respectively.

Let K be a normal subgroup of an $\mathfrak{X}$-group G. Let U be a subnormal subgroup of K. Then U is a subnormal subgroup of G. It follows that U is S-permutable (respectively, permutable, normal) in G if G is a PST-group (respectively, PT-group, T-group) and so U is S-permutable (respectively, permutable, normal) in K by Lemma 1.2.7 (1). Consequently, K is a PST-group (respectively, PT-group, T-group).

Consider now the alternating group Alt(5) of degree 5. This group is a T-group because it is simple, but it contains Alt(4) as a subgroup, which is not a PST-group. This shows that all three classes are not closed under taking subgroups. The symmetric group Sym(3) of degree 3 is a T-group, but Sym(3) × Sym(3) is not a PST-group, because the diagonal of its Sylow 3-subgroup is a subnormal subgroup which is not S-permutable. It follows that PST, PT, and T are neither formations nor Fitting classes. □

2.1 Characterisations based on the normal structure

In this section we shall establish a number of fundamental theorems which, when taken together, give a good picture of the groups in the abovementioned classes. We begin with an extremely useful result.

Proposition 2.1.1. *Let G be a* PST*-group. Then:*

1. *G is an* SC*-group, and*
2. *p'-elements of G induce power automorphisms in every normal p-subgroup of G for every prime p.*

Proof. 1. Proceeding by induction on the order of G, bearing in mind Lemma 2.0.4, it is enough to see that every minimal normal subgroup of G is simple. Let N be a minimal normal subgroup of G. Suppose that N is not abelian. Then $N = S_1 \times \cdots \times S_k$, where S_i is isomorphic to a non-abelian simple group S for $1 \leq i \leq k$. Note that each S_i is a subnormal subgroup of G. Since G is a PST-group, every S_i is S-permutable in G.

Let P be a Sylow p-subgroup of G for a prime p. Then $PS_i = S_iP$ for $1 \leq i \leq k$. Hence

$$\langle S_i^P \rangle = \langle S_i^P \rangle \cap (S_i P) = S_i(\langle S_i^P \rangle \cap P).$$

We also have that S_i is a normal subgroup of $\langle S_i^P \rangle$, which is contained in N, and that $\langle S_i^P \rangle / S_i$ is a p-group. But since $\langle S_i^P \rangle$ is a direct product of non-abelian simple groups, we can conclude that $\langle S_i^P \rangle = S_i$. Consequently, S_i is a normal subgroup of G. Since N is a minimal normal subgroup of G, we obtain that $N = S_1$ is simple.

Assume now that N is an abelian group and so p-elementary abelian for some prime p. Let Q be a Sylow q-subgroup of G for a prime $q \neq p$. If a is a non-trivial element of N, then $\langle a \rangle$ is a subnormal subgroup of G. Thus $\langle a \rangle Q = Q \langle a \rangle$. Hence

$$\langle \langle a \rangle^Q \rangle = \langle \langle a \rangle^Q \rangle \cap \langle a \rangle Q = \langle a \rangle$$

because $\langle \langle a \rangle^Q \rangle$ is contained in N and so is a p-group. We conclude then that the elements of Q induce power automorphisms in N. If we choose a of order p such that a belongs to $N \cap \mathrm{Z}(P)$, we conclude that $|N| = p$, because $\langle a \rangle$ is a normal subgroup of G. Therefore N is simple.

2. It follows by arguments essentially identical to those given in the previous paragraph. □

Corollary 2.1.2. *Every soluble* PST-*group is supersoluble.*

We have seen that Sym(3) × Sym(3) is a supersoluble group which is not a PST-group. Therefore the converse of Corollary 2.1.2 is not valid.

From now on we prepare the way for the classical characterisations of Agrawal, Zacher, and Gaschütz. A major step in our proof is to give necessary and sufficient conditions for the subgroups of a normal p-subgroup to be S-permutable.

Lemma 2.1.3. *Let p be a prime and let N be a normal p-subgroup of G. The following statements are equivalent:*

1. *p'-elements of G induce power automorphisms in N,*
2. *all chief factors of G below N are cyclic and G-isomorphic when regarded as modules over G, and*
3. *all subgroups of N are* S-*permutable in G.*

Proof. 1 implies 2. Assume first that all p'-elements of G induce power automorphisms in N. Arguing like in the proof of Proposition 2.1.1, every chief factor of G below N is cyclic and so is centralised by all p-elements of G. If a p'-element g of G does not centralise N, we have by Lemma 1.3.4 that N is abelian. Therefore, by Theorem 1.3.2, the power automorphism induced by g is universal. It follows that all chief factors of G below N are G-isomorphic. If all p'-elements of G centralise N, then all chief factors of G below N are central.

2 implies 1. Suppose now that all chief factors of G below N are cyclic and G-isomorphic. We will show that p'-elements of G induce power automorphisms in N by induction on $|G|$. Assume first that $\Phi(N) \neq 1$. Then $G/\Phi(N)$ satisfies the hypothesis

and hence p'-elements of $G/\Phi(N)$ induce power automorphisms on $N/\Phi(N)$. Let H be a subgroup of N. In order to see that H is normalised by the p'-elements of G we can assume that $H < N$. Let M be a maximal subgroup of N such that $H \leq M$. Since $M/\Phi(N)$ is normalised by all p'-elements of $G/\Phi(N)$, we have that M is normalised by all p'-elements of G. Let g be a p'-element of G and set $X = \langle g\rangle M$. Then X is a proper subgroup of G. The hypotheses of the lemma clearly hold in X. By induction, we have that g induces a power automorphism in M. We conclude that H is normalised by g, as desired. Assume now that $\Phi(N) = 1$. Then N is a G-module over the finite field of p elements and, regarded as $\langle g\rangle$-module, is a direct product of $\langle g\rangle$-irreducible submodules $N = N_1 \times \cdots \times N_r$ by the theorem of Maschke [119, B, 4.5]. Moreover, all N_i are cyclic, $N_i = \langle n_i\rangle$ say. It follows that $n_i^g = n_i^m$ for $1 \leq i \leq r$, where m is an integer which does not depend on i because all N_i are G-isomorphic. Hence we conclude that for each element $n = n_1^{a_1} \cdots n_r^{a_r}$ of N, $n^g = n_i^m$ and so g induces a power automorphism in N.

1 implies 3. Assume that p'-elements of G induce power automorphisms in N. Let S be a subgroup of N. It is clear that S is normalised by all Sylow q-subgroups of G for all primes $q \neq p$. On the other hand, S is contained in all Sylow p-subgroups of G. It follows that S is S-permutable in G.

3 implies 1. Let S be a subgroup of N and let g be a q-element of G for a prime $q \neq p$. Let Q be a Sylow q-subgroup of G such that g belongs to Q. Then S permutes with Q. Moreover $N \cap SQ = S(N \cap Q) = S$, which is normal in SQ. This implies that Q normalises S. In particular, g normalises S. It follows that all p'-elements of G normalise S. Hence p'-elements of G act on N as a group of power automorphisms. □

Remark 2.1.4. We now present some counterexamples which establish limits to the validity of the above lemma.

In order to prove that p'-elements induce power automorphisms in a normal p-subgroup N, the hypothesis that every p-chief factor of G is cyclic is essential. Let $N = \langle a\rangle \times \langle b\rangle \times \langle c\rangle \times \langle d\rangle$ be an elementary abelian group of order 16 and let x be an automorphism of N defined by $a^x = b$, $b^x = ab$, $c^x = d$, $d^x = cd$. Let G be the semidirect product $G = [N]\langle x\rangle$. It is clear that the 2-chief factors of G are the minimal normal subgroups $N_1 = \langle a, b\rangle$ and $N_2 = \langle c, d\rangle$, and that $a \mapsto c$, $b \mapsto d$ defines a G-isomorphism between N_1 and N_2. However, it is clear that $2'$-elements of G do not induce power automorphisms in N.

On the other hand, the G-isomorphism between the p-chief factors of G below N is essential, too. Let $G = \mathrm{Sym}(3) \times C_3$ be the direct product of a symmetric group of degree 3 and a cyclic group of degree 3. In G, C_3 is a central 3-chief factor, but $\mathrm{Alt}(3)$ is a non-central 3-chief factor. Therefore 3-chief factors of G are not G-isomorphic. Moreover, it is clear that the diagonal subgroup of the normal Sylow 3-subgroup $N = \mathrm{Alt}(3) \times C_3$ is not normalised by any $3'$-element of G. Therefore $3'$-elements of G do not induce power automorphisms in N.

Definition 2.1.5. Let $\mathfrak{X}$ be a class of groups. We say that $\mathfrak{X}$ *admits a local characterisation*, or $\mathfrak{X}$ is a *local class*, if for every prime p there exists a class of groups $\mathfrak{X}_p$ (usually related to the behaviour of p-subgroups or p-chief factors) such that $\mathfrak{X} = \bigcap_{p \in \mathbb{P}} \mathfrak{X}_p$. In this case, we say that $\mathfrak{X}_p$ is a *local definition* of $\mathfrak{X}$.

Local characterisations have been a useful tool in the study of finite groups. Some known classes admit such a treatment. This is the case of solubility (with p-solubility), supersolubility (with p-supersolubility) or nilpotency (with p-nilpotency). This notion is behind the idea of local formation (cf. [119, IV, Section 3], [47, Section 3.1]).

Recall that if p is a prime, a group G is *p-nilpotent* if it has a normal Hall p'-subgroup and G is said to be *p-supersoluble* if all chief factors of G having order divisible by p are exactly of order p.

Lemma 2.1.6. *Let p be a prime, and let G be a p-supersoluble group. Then the derived subgroup G' of G is p-nilpotent. In particular, if $\mathrm{O}_{p'}(G) = 1$, G has a unique Sylow p-subgroup.*

Proof. Let H/K be a p-chief factor of G. Clearly, we have that H/K has dimension one as a $\mathrm{GF}(p)G$-module, since G is a p-supersoluble group. Hence the factor group $G/\mathrm{C}_G(H/K)$ is isomorphic to a subgroup of $\mathrm{Aut}(C_p)$, which is isomorphic to C_{p-1}. Therefore $G/\mathrm{C}_G(H/K)$ is an abelian group of exponent dividing $p-1$, and thus the derived subgroup G' of G is contained in $\mathrm{C}_G(H/K)$. This argument holds for every p-chief factor of G, and hence G' is in fact contained in $\mathrm{O}_{p',p}(G)$ and so G' is p-nilpotent. If $\mathrm{O}_{p'}(G) = 1$, then $\mathrm{O}_{p',p}(G) = \mathrm{O}_p(G) = \mathrm{F}(G)$. Consequently, G' is contained in a Sylow p-subgroup P of G, and thus P is a normal subgroup of G. □

The following class, introduced by Alejandre, Ballester-Bolinches, and Pedraza-Aguilera in [5], plays a central role in our approach.

Definition 2.1.7. Let p be a prime number. We say that a group G *belongs to* $\mathcal{U}_p^*$ when all chief factors of G of order divisible by p are cyclic (that is, G is p-supersoluble) and G-isomorphic when regarded as G-modules by conjugation.

It is clear that the class $\mathcal{U}_p^*$ is closed under taking subgroups and quotients.

We are now ready to state and prove some characterisations of soluble PST-groups.

Theorem 2.1.8. *Let G be a group with nilpotent residual L. Any two of the following statements are equivalent:*

1. *G belongs to $\mathcal{U}_p^*$ for all primes p.*
2. *L is an abelian Hall subgroup of odd order acted on by conjugation by G as a group of power automorphisms.*
3. *G is a soluble* PST-*group.*

Proof. 1 implies 2. Assume that G belongs to $\mathcal{U}_p^*$ for all p. It is clear that G is supersoluble. Since $L \leq G'$, we have that L is nilpotent by Lemma 2.1.6. Assume that L has even order. Since L is nilpotent, there exists a Hall $2'$-subgroup $L_{2'}$ of L and a Sylow 2-subgroup L_2 of L such that $L = L_{2'} \times L_2$ and $L_2 \neq 1$. In this case G has a normal subgroup A such that L/A is a 2-chief factor of G. Since G is supersoluble, $|L/A| = 2$ and so L/A is contained in $\mathrm{Z}(G/A)$. As a consequence $(G/A)/(L/A) \cong G/L$, which is nilpotent, and so G/A is nilpotent. This implies that $L \leq A$, which is a contradiction. Therefore L is a nilpotent subgroup of odd order.

Let us see now that L is a Hall subgroup of G by induction on the order of G. Assume first that $L \neq 1$. Let N be a minimal normal subgroup of G such that $N \leq L$. Since G/N is a PST-group and L/N is the nilpotent residual of G/N, we obtain by induction that $\gcd(|L/N|, |G/N : L/N|) = 1$. Let $\pi = \pi(L/N)$ be the set of prime divisors of the order of L/N and p be the divisor of the order of N. If p belongs to π, then L is a Hall π-subgroup of G. We can assume that p does not belong to π, and so N is a Sylow p-subgroup of L. Then p divides $|G/L|$ and G has a chief factor C/D such that $L \leq D < C \leq G$ and is central in G. Since G is a $\mathcal{U}_p^*$-group, C/D is G-isomorphic to N, and so $N \leq \mathrm{Z}(G)$. On the other hand, $L = N \times L_{p'}$ for a Hall p'-subgroup $L_{p'}$ of L. Hence $L/L_{p'}$ is central in G and so $G/L_{p'}$ is nilpotent, which implies that $L \leq L_{p'}$, a contradiction. Consequently L is a Hall π-subgroup of G, where $\pi = \pi(L)$. In particular, there exists a Hall π'-subgroup $G_{\pi'}$ of G such that $G = LG_{\pi'} = L\mathrm{O}^{\pi}(G)$.

For every p in π, let L_p be the Sylow p-subgroup of L. Let g be a p'-element of G. By Lemma 2.1.3, g induces a power automorphism in L_p. Thus all π'-elements of G induce power automorphisms in L_p for all $p \in \pi$. This implies $G_{\pi'}$ normalises each subgroup of L. Suppose that $L \neq 1$. Then every chief factor of G of the form L_p/T is non-central in G. Therefore, there exists a p'-element g of G which does not centralise L_p. Applying Lemma 1.3.4, L_p is abelian. Consequently L is abelian and so every element of G normalises each subgroup of L.

2 implies 3. Suppose now that L is an abelian normal Hall subgroup of odd order acted on by conjugation by G as a group of power automorphisms. The group G is soluble because G/L is nilpotent and L is abelian. Let H be a subnormal subgroup of G. Assume that $H \cap L \neq 1$. It is clear that $H \cap L$ is a normal subgroup of G because $G/(H \cap L)$ satisfies the hypotheses of the theorem. By induction on $|G|$, we have that $G/(H \cap L)$ is a PST-group. Hence H is an S-permutable subgroup of G. Therefore we can suppose that $H \cap L = 1$. In this case, H is contained in every complement C of L in G. Since C is nilpotent, we have that H permutes with all Sylow subgroups whose order divides the order of C. Now if P is a Sylow p-subgroup of G, with p in $\pi(L)$, then P is a normal subgroup of G. Hence $HP = PH$. We conclude that H permutes with all Sylow subgroups of G and so G is a PST-group.

3 implies 1. Assume that G is a soluble PST-group. By Corollary 2.1.2, G is supersoluble and all chief factors of G are cyclic. Let p be a prime dividing $|G|$. Applying [119, VII, 2.1], G has a normal Hall π-subgroup N, where π is the set of all

primes greater than p ($N = 1$ if $\pi = \emptyset$). Hence G/N is supersoluble and has a normal Sylow p-subgroup P/N. By Lemma 2.1.3, all chief factors of G/N below P/N are (G/N)-isomorphic. It follows that all p-chief factors of G are G-isomorphic. This proves that G satisfies $\mathcal{U}_p^*$ for all primes p. □

The equivalence between 3 and 2 was proved by Agrawal in [2]. The equivalence with 1 was proved by Alejandre, Ballester-Bolinches, and Pedraza-Aguilera in [5].

Corollary 2.1.9. *The class of soluble* PST-*groups is a local class which is closed under taking subgroups.*

The real key to help us to understand the difference between soluble T-, PT-, and PST-groups is the following result.

Theorem 2.1.10 ([44]). *Let G be a group.*

1. *Suppose that p is a prime number and H is an* S-*permutable p-subgroup of G. If the Sylow p-subgroups of G are Iwasawa (Dedekind, respectively), then H is permutable (normal, respectively) in G.*
2. *Assume that H is an* S-*permutable subgroup of G. If the Sylow subgroups of G are Iwasawa (Dedekind, respectively), then H is permutable (normal, respectively) in G.*

Proof. 1. Let A be a subgroup of G. Set $T = \langle A, H\rangle$. Since H is S-permutable in T by Lemma 1.2.7 (1), we have that H is subnormal in T and H is contained in $\mathrm{O}_p(T)$, which is contained in every Sylow p-subgroup P of T. Hence $T = \langle H, A\rangle \leq \langle \mathrm{O}_p(T), A\rangle = \mathrm{O}_p(T)A \leq T$. Let A_q be a Sylow q-subgroup of A for a prime $q \neq p$, and let G_q be a Sylow q-subgroup of G containing A_q. Then A_q is a Sylow q-subgroup of T and $A_q = G_q \cap T$ because $A_q \leq G_q \cap T$. From this, $HA_q = H(G_q \cap T) = HG_q \cap T$ is a subgroup of T. Moreover, $\mathrm{O}_p(T) \cap HA_q = H$. Hence H is normalised by A_q. On the other hand, since P is Iwasawa (Dedekind, respectively), we have that H permutes with (is normalised by, respectively) a Sylow p-subgroup A_p of A. Hence H permutes with (is normalised by, respectively) all Sylow subgroups of A. In particular, H permutes with A (H is normalised by A, respectively). This implies that H is a permutable (normal, respectively) subgroup of G.

2. We argue by contradiction. Assume that G is a counterexample of least order to the result. Then there exists an S-permutable subgroup H of G which is not permutable (normal, respectively) in G. Take such an H of least order. Assume that $H_G \neq 1$. Let N be a minimal normal subgroup of G contained in H. Since H/N is S-permutable in G/N by Lemma 1.2.7 (2), we have that H/N is permutable (normal, respectively) in G/N and so H is permutable (normal, respectively) in G by Theorem 1.2.14. Then $H_G = 1$. By Theorem 1.2.14 (2) and Theorem 1.2.17, H is a nilpotent subgroup. By minimality of H, we can assume that H is a p-group for some prime p. Then H is permutable (normal, respectively) by the first part. This is the desired contradiction. □

Next we derive some results from Theorem 2.1.10, the first being the Gaschütz and Zacher characterisations, the second showing that in the soluble universe the difference between PST-, PT-, and T-groups is just the Sylow structure and the third stating that the classes of soluble PT- and T-groups are subgroup-closed local classes.

Theorem 2.1.11 (Zacher [264] and Gaschütz [137]). *A group G is a soluble* PT*-group (respectively, a soluble* T*-group) if and only if the following conditions are satisfied:*

1. *the nilpotent residual L of G is an abelian Hall subgroup of odd order,*
2. *G acts by conjugation on L as a group of power automorphisms, and*
3. *G/L is an Iwasawa group (respectively, a Dedekind group).*

Corollary 2.1.12. *Let G be a group.*

1. *G is a soluble* PT*-group if and only if G is a soluble* PST*-group whose Sylow subgroups are Iwasawa.*
2. *G is a soluble* T*-group if and only if G is a soluble* PST*-group whose Sylow subgroups are Dedekind.*

Since subgroups of Iwasawa groups are Iwasawa groups and subgroups of Dedekind groups are Dedekind groups, we obtain:

Corollary 2.1.13. *The class of all soluble* PT*-groups (respectively, soluble* T*-groups) is a local class which is closed under taking subgroups.*

Some results of Agrawal, Zacher, and Gaschütz type in some classes of infinite groups are presented in [212], [49], and [50].

Given the above satisfying characterisations of PST-, PT-, and T-groups in the soluble universe, it is natural to ask for characterisations in the general finite one. We note that every simple group is obviously a PST-group. Therefore before the classification of finite simple groups there was no hope of describing those groups in the insoluble case. In fact, an essential role here it is played by the fact that every PST-group is an SC-group by Proposition 2.1.1 and so it has the structure described in Theorem 1.6.5, whose proof in the insoluble case depend on the positive solution of the Schreier Conjecture. Unfortunately, the simplicity of all chief factors is not enough to ensure the S-permutability of the subnormal subgroups. This was already noted for soluble groups at the beginning of the section. To show that there are insoluble SC-groups which are not PST-groups is the purpose of the following example.

Example 2.1.14. Let A be a group such that $\mathrm{Z}(A) = \Phi(A) \cong C_3$ and $A/\Phi(A) \cong \mathrm{Alt}(6)$, the alternating group on 6 letters (such a group exists because the Schur multiplier of $\mathrm{Alt}(6)$ is isomorphic to C_6). Let $G = A \times \mathrm{Sym}(3)$. Clearly G is an

SC-group which is not an PST-group: the diagonal subgroup of the elementary abelian 3-subgroup $O_3(G)$ is a subnormal subgroup which does not permute with a cyclic subgroup of order 2 of Sym(3).

As in the soluble universe, the above example confirms that in an insoluble PST-group there must be additional restrictions on normal p-subgroups, p a prime, beyond what is implied by Theorem 1.6.5. Our strategy in this case is to find out a "finite" analogue of the class $\mathfrak{U}_p^*$. In order to carry out our task the following lemma is crucial.

Lemma 2.1.15. *Let p be a prime and let G be a p-soluble group. The following statements are equivalent:*

1. *G is a $\mathfrak{U}_p^*$-group;*
2. *(Property n_p) Every subgroup of $O_p(G/N)$ is S-permutable in G/N for each normal subgroup N of G.*

Proof. Assume that G is a $\mathfrak{U}_p^*$-group. Then G/N also belongs $\mathfrak{U}_p^*$ and so all chief factors of G/N below $O_p(G/N)$, are cyclic and G/N-isomorphic. It follows that all subgroups of $O_p(G/N)$ are S-permutable in G/N by Lemma 2.1.3.

Conversely, suppose that every subgroup of $O_p(G/N)$ is S-permutable in G/N for all normal subgroups N of G. We prove that G belongs to $\mathfrak{U}_p^*$ by induction on the order of G. Suppose that $O_{p'}(G) \neq 1$ and let B a minimal normal subgroup of G contained in $O_{p'}(G)$. Clearly the condition remains true when G is replaced by G/B. Applying induction, it follows that G/B is a $\mathfrak{U}_p^*$-group. Hence G belongs to $\mathfrak{U}_p^*$, as required. Thus we may assume that $O_{p'}(G) = 1$. Then $O_p(G) \neq 1$ and, by Lemma 2.1.3, all chief factors of G below $O_p(G)$ are cyclic and G-isomorphic. Let A be a minimal normal subgroup of G. Then A is a p-group. It is clear that the hypothesis of the lemma holds in G/A, so we apply the induction hypothesis to conclude that G/A is a $\mathfrak{U}_p^*$-group. Since A is cyclic, it follows that G is p-supersoluble. This implies that G' is p-nilpotent and so it is a p-group because $O_{p'}(G) = 1$. In particular, a Sylow p-subgroup is normal and so all p-chief factors are G-isomorphic. This yields $G \in \mathfrak{U}_p^*$. □

The S-permutability of every subgroup of $O_p(G/N)$ in G/N is a "p-soluble free" property which was first introduced in the soluble universe by Beidleman and Heineken in [62]. There they prove that a soluble group G is a PST-group if and only if G satisfies n_p for all p. Note that this result follows from the above lemma and Theorem 2.1.8. The situation for insoluble groups is rather different. For instance, the regular wreath product of the alternating group of degree 5 with a cyclic group of order 2 is a non-PST-group satisfying n_p for all p. Since every group satisfying n_p for all p has all the abelian chief factors cyclic by Lemma 2.1.3, one would expect that the properties of the non-abelian chief factors have a noticeable influence of the structure of a group satisfying n_p for all p. This is already seen in the observation that the simplicity of the non-abelian chief factors of a group satisfying n_p for all p implies the S-permutability of its subnormal subgroups as it is proved in [33, Theorem A].

The next lemma describes a minimal configuration often encountered in the study of PST- and PT-groups. The information given in it comes in useful when analysing SC-groups which are not PST-groups.

Lemma 2.1.16. *Let G be an* SC*-group and let D be the soluble residual of G. By Theorem* 1.6.5, $D/Z(D) = D_1/Z(D) \times \cdots \times D_r/Z(D)$, *where $D_i/Z(D)$ is simple and G-invariant for all i. Let f be a subgroup embedding functor and assume that f is either* s *or* Syl. *Let U be a non-f-permutable subnormal subgroup of G of minimal order. Then:*

1. *If UD_i'/D_i' is f-permutable in G/D_i' for all i, then U is soluble.*
2. *If U is soluble and UL/L is f-permutable in G/L for all non-trivial nilpotent normal subgroups L of G, then U is a cyclic q-group for a prime q.*

Proof. Assume that U be a non-f-permutable subnormal subgroup of G of minimal order. Then there exist a prime p and a Sylow p-subgroup S of G if $f = \mathrm{Syl}$ or a p-subgroup S of G if $f = \mathrm{s}$ such that U does not permute with S.

Suppose that UD_i'/D_i' is f-permutable in G/D_i' for all i, but U is non-soluble. Then $(U \cap D)Z(D)/Z(D)$ is a non-trivial subnormal subgroup of $D/Z(D)$ and so it contains some $D_i/Z(D)$ for some i. Therefore D_i' is a normal subgroup of G contained in $((U \cap D)Z(D))'$, which is a subgroup of $U \cap D$. By hypothesis, U/D_i' permutes with SD_i'/D_i'. In particular, U permutes with S. This contradiction means that U is soluble and Assertion 1 holds.

We prove now Assertion 2. Assume that U is soluble. Clearly U is an SC-group and so it is supersoluble. Thus the nilpotent residual $U^{\mathfrak{N}}$ of U is nilpotent. All proper subnormal subgroups of U permute with S and hence, by minimality of U, U possesses just one maximal normal subgroup. Now supersolubility of U yields $U/U^{\mathfrak{N}}$ is a cyclic q-group for some prime q.

Suppose that UL/L is f-permutable in G/L for every normal nilpotent subgroup L of G. Since $U^{\mathfrak{N}}$ is nilpotent and supplements every Sylow q-subgroup of U, it follows that $U^{\mathfrak{N}}$ is a q'-group. Denote by A the normal closure of $U^{\mathfrak{N}}$ in G. Evidently A is a nilpotent q'-subgroup of G.

Suppose that $p = q$. The solubility of U implies that $U^{\mathfrak{N}}$ is a proper normal subgroup of U. Hence $U^{\mathfrak{N}}$ is f-permutable in G by minimality of U. Furthermore $U^{\mathfrak{N}}S \cap A = U^{\mathfrak{N}}(S \cap A) = U^{\mathfrak{N}}$. Thus $U^{\mathfrak{N}}$ is a normal subgroup of $\langle U, S\rangle$. If $f = \mathrm{Syl}$, then S is a Sylow p-subgroup of G and so $\langle U, S\rangle/U^{\mathfrak{N}}$ is generated by the subnormal p-group $U/U^{\mathfrak{N}}$ and the Sylow p-subgroup $SU^{\mathfrak{N}}/U^{\mathfrak{N}}$. It follows that U is contained in $SU^{\mathfrak{N}}$ and $U = U^{\mathfrak{N}}(S \cap U)$. Since both factors permute with U, we have that U permutes with S, contrary to construction. Assume now that $f = \mathrm{s}$. If $A \neq 1$, there exists a minimal normal subgroup L of G which is an r-group for some prime $r \neq p$. By hypothesis, UL/L permutes with SL/L. In particular, $(UL)S$ is a subgroup of G. By Theorem 1.1.19, there exists a Sylow p-subgroup V of UL such that VS is a Sylow p-subgroup of $(UL)S$. Clearly V is contained in U because U is subnormal in G and L is a p'-group. Since S permutes with $U^{\mathfrak{N}}$ and $U = (U^{\mathfrak{N}})V$, it follows that U permutes with S. This contradicts the choice of the pair (U, S). Consequently, $A = 1$ and so U is a cyclic q-group in this case.

Suppose that q is different from p. Assume further that $A \neq 1$. Then, by hypothesis, UA/A permutes with SA/A. Since UA/A is a q-group, we have that S normalises UA. Let π be the set of primes dividing $|UA : U|$. Applying Lemma 1.1.11, $\mathrm{O}^{\pi}(UA) = \mathrm{O}^{\pi}(U) = U$ because $q \notin \pi$. In particular, S normalises U, contrary to assumption. Hence $A = 1$ and U is a cyclic q-group. This proves Assertion 2. □

Remark 2.1.17. Note that the above lemma holds if "f-permutable" is replaced by "normal".

The following property, introduced by Robinson in [217], plays a significant role in the characterisation of the PST-groups in the general case.

Definition 2.1.18. Let p be a prime number. We say that a group G *satisfies* $\mathbf{N}_p$ if whenever N is a soluble normal subgroup of G, p'-elements of G induce power automorphisms in $\mathrm{O}_p(G/N)$.

Applying Lemma 2.1.3, a group G has the property $\mathbf{N}_p$ if and only if every subgroup of $\mathrm{O}_p(G/N)$ is S-permutable in G/N for each soluble normal subgroup N of G. Therefore, in the soluble universe, property $\mathbf{N}_p$ is just the property n_p studied above.

The importance of the property $\mathbf{N}_p$ is underscored by the next result, which is a very satisfying characterisation of PST-groups.

Theorem 2.1.19 (Robinson [217]). *A group G is a* PST-*group if and only if G has a perfect normal subgroup D such that:*

1. *G/D is a soluble* PST-*group,*

2. *if D is non-trivial, then $D/\mathrm{Z}(D) = U_1/\mathrm{Z}(D) \times \cdots \times U_k/\mathrm{Z}(D)$, where $U_i/\mathrm{Z}(D)$ is simple and U_i is normal in G, and*

3. *if $\{i_1, i_2, \dots, i_r\} \subseteq \{1, 2, \dots, k\}$, where $1 \leq r < k$, then G and $G/U'_{i_1}U'_{i_2} \cdots U'_{i_r}$ satisfy $\mathbf{N}_p$ for all $p \in \pi(\mathrm{Z}(D))$.*

Proof. Since PST-groups are SC-groups by Proposition 2.1.1, the Statement 2 holds. Moreover, since all quotients of PST-groups are PST-groups, the Statement 1 holds. By Proposition 2.1.1, every PST-group satisfies condition $\mathbf{N}_p$ for all p. Hence Statement 3 holds. Therefore it is enough to check the sufficiency of all three statements of the theorem.

Assume that G satisfies all three conditions but is not a PST-group. Among all these groups, we choose one of least order. Let U be a non-S-permutable subgroup of G of minimal order. Clearly, $1 \neq D$ is the soluble residual of G and G is an SC-group. Applying Remark 1.6.8, G/U_i' satisfies the statements of the theorem for all i. The minimal choice of G implies that UU_i'/U_i' is S-permutable in G/U_i' for all i. By Lemma 2.1.16, U is soluble. Let L be a nilpotent subnormal subgroup of G. Then, G/L and $G/LU_{i_1}'U_{i_2}' \cdots U_{i_r}'$, $\{i_1, i_2, \dots, i_r\} \subseteq \{1, 2, \dots, k\}$, satisfy $\mathbf{N}_p$ for all $p \in \pi(\mathrm{Z}(DL/L))$. Applying Remark 1.6.8, G/L satisfies the three statements of the theorem. Therefore UL/L is S-permutable in G/L by minimality of G. By Lemma 2.1.16, U is a q-group for some prime q. If $q \in \pi(\mathrm{Z}(D))$, since $U \leq \mathrm{O}_q(G)$, the condition $\mathbf{N}_q$ implies that U is S-permutable in G, which is not the case. Thus $q \notin \pi(\mathrm{Z}(D))$. By Proposition 1.6.4, $\mathrm{O}_q(G) \cap D = 1$ and $\mathrm{O}_q(G)$ is G-isomorphic to $\mathrm{O}_q(G)D/D$. Since G/D is a PST-group, it follows that q'-elements of G/D induce power automorphisms in $\mathrm{O}_q(G/D)$. Therefore q'-elements of G induce power automorphisms in $\mathrm{O}_q(G)$. By Lemma 2.1.3, U is S-permutable in G. This final contradiction completes the proof. □

As a consequence we obtain the following result of Ballester-Bolinches, Beidleman, and Heineken [33, Theorem A (i)].

Corollary 2.1.20. *A group G is a* PST*-group if and only if G satisfies n_p for all primes p, and every non-abelian chief factor of G is simple.*

The following property is needed to force a PST-group to be a PT-group. It was also introduced by Robinson in [217].

Definition 2.1.21. Let p be a prime number. We say that a group G *satisfies* $\mathbf{P}_p$ if whenever N is a soluble normal subgroup of G, every subgroup of $\mathrm{O}_p(G/N)$ is permutable in a Sylow p-subgroup of G/N.

It is clear that PT-groups have property $\mathbf{P}_p$ for all primes p. They also satisfies the following stronger property introduced in the soluble universe by Beidleman and Heineken in [62].

Definition 2.1.22. Let p be a prime number. We say that a group G *satisfies* m_p if whenever N is a normal subgroup of G, every subgroup of $\mathrm{O}_p(G/N)$ is permutable in G/N.

We are now in position to state and prove two characterisations of PT-groups. The first one is:

Theorem 2.1.23 (Robinson [217, Theorem 3.1]). *A group G is a* PT*-group if and only if G has a perfect normal subgroup D such that:*

1. *G/D is a soluble* PT*-group,*
2. *if D is non-trivial, then $D/\mathrm{Z}(D) = U_1/\mathrm{Z}(D) \times \cdots \times U_k/\mathrm{Z}(D)$, where $U_i/\mathrm{Z}(D)$ is simple and U_i is normal in G, and*
3. *if $\{i_1, i_2, \dots, i_r\} \subseteq \{1, 2, \dots, k\}$, where $1 \le r < k$, then G and $G/U'_{i_1}U'_{i_2} \cdots U'_{i_r}$ satisfy $\mathbf{N}_p$ for all $p \in \pi(\mathrm{Z}(D))$ and $\mathbf{P}_p$ for all $p \in \pi(D)$.*

Proof. Since PT-groups are PST-groups and quotients of PT-groups are PT-groups, the Conditions 1, 2, 3 are satisfied by a PT-group. Assume, arguing by contradiction, that G is non-PT-group of minimal order satisfying all three conditions. We consider a subnormal non-permutable subgroup U of G of minimal order. Note that G is an insoluble PST-group by Theorem 2.1.19. Applying Remark 1.6.8, G/U'_i satisfies the conditions of the theorem for all i. The minimal choice of G implies that UU'_i/U'_i is permutable in G/U'_i for all i. By Lemma 2.1.16, U is soluble. Let L be a nilpotent subnormal subgroup of G. Then, G/L and $G/LU'_{i_1}U'_{i_2} \cdots U'_{i_r}$, $\{i_1, i_2, \dots, i_r\} \subseteq \{1, 2, \dots, k\}$, satisfy $\mathbf{N}_p$ for all $p \in \pi(\mathrm{Z}(DL/L))$ and $\mathbf{P}_p$ for all $p \in \pi(DL/L)$. Applying Remark 1.6.8, G/L satisfies the three conditions of the theorem. Therefore UL/L is permutable in G/L by minimality of G. By Lemma 2.1.16, U is a q-group for some prime q. Since G is PST-group, q'-elements of G normalise U. Let Q be a Sylow q-subgroup of G. Then U is contained in Q. If $q \in \pi(D)$, we have that U is permutable in Q by condition $\mathbf{P}_q$ and if $q \notin \pi(D)$, then $Q \cap D = 1$ and so $Q \cong QD/D$, which shows that Q is an Iwasawa group because QD/D is a Sylow subgroup of the soluble PT-group G/D (see Corollary 2.1.12). In this case, U is also permutable in Q. This implies that U is a permutable subgroup of G, contrary to assumption. □

The second one is due to Ballester-Bolinches, Beidleman, and Heineken [33, Theorem A (ii)].

Corollary 2.1.24. *A group G is a* PT*-group if and only if G satisfies m_p for all primes p and every non-abelian chief factor of G is simple.*

We now break off the general discussion to describe an example.

Example 2.1.25 (Robinson [217]). An SC-group with $\mathbf{N}_p$ and $\mathbf{P}_p$ for all primes p need not be a PT-group.

Let $D_1 = \mathrm{SL}_3(4)$ and $D_2 = \mathrm{PSL}_2(8)$. Then D_1 and D_2 have field automorphisms α_1 and α_2 of orders 2 and 3, respectively. Let $G_i = [D]\langle \alpha_i \rangle$ be the corresponding semidirect product and put $G = G_1 \times G_2$. Clearly G is an SC-group. Note that $G'' = D = D_1 \times D_2$ and $\mathrm{Z}(D) = \mathrm{Z}(D_1)$ has order 3, while $G/D \cong C_6$. Hence G satisfies $\mathbf{N}_p$ and $\mathbf{P}_p$ for all p. However G/D_2 is not a PT-group, because it does not satisfy $\mathbf{N}_3$.

Our next objective is to establish analogues of Theorem 2.1.23 and Corollary 2.1.24 for T-groups. We begin with the following property on soluble normal subgroups introduced by Robinson in [217].

Definition 2.1.26. Let p be a prime number and let G be a group. We say that G *satisfies* $\mathbf{T}_p$ if whenever N is a soluble normal subgroup of G, every element of G induces a power automorphism in each G-invariant p-factor X/N of nilpotency class at most 2.

T-groups satisfy $\mathbf{T}_p$ and the following property introduced in the soluble universe by Beidleman and Heineken in [62].

Definition 2.1.27. Let p be a prime number. We say that a group G *satisfies* t_p if whenever N is a normal subgroup of G, every subgroup of $\mathrm{O}_p(G/N)$ is normal in G/N.

Theorem 2.1.28 ([217, Theorem 4.1]). *A group G is a* T*-group if and only if G has a perfect normal subgroup D such that:*

1. *G/D is a soluble* T*-group,*
2. *if D is non-trivial, then $D/\mathrm{Z}(D) = U_1/\mathrm{Z}(D) \times \cdots \times U_k/\mathrm{Z}(D)$, where $U_i/\mathrm{Z}(D)$ is simple and U_i is normal in G, and*
3. *if $\{i_1, i_2, \dots, i_r\} \subseteq \{1, 2, \dots, k\}$, where $1 \le r < k$, then G and $G/U'_{i_1}U'_{i_2} \cdots U'_{i_r}$ satisfy $\mathbf{T}_p$ for all $p \in \pi(\mathrm{Z}(D))$.*

Proof. Only the sufficiency requires a proof. Let G be a counterexample of least order and U a non-normal subnormal subgroup of G of minimal order. Note that G is an insoluble SC-group. By Remark 1.6.8, G/U'_i satisfies the conditions of the theorem for all i. The minimal choice of G implies that UU'_i/U'_i is normal in G/U'_i for all i. By Lemma 2.1.16 and Remark 2.1.17, U is soluble. Let L be a nilpotent normal subgroup of G. Then, G/L and $G/LU'_{i_1}U'_{i_2} \cdots U'_{i_r}$, $\{i_1, i_2, \dots, i_r\} \subseteq \{1, 2, \dots, k\}$, satisfy $\mathbf{T}_p$ for all $p \in \pi(\mathrm{Z}(DL/L))$. Applying Remark 1.6.8, G/L satisfies the three conditions of the theorem. Therefore UL/L is normal in G/L by minimality of G. By Lemma 2.1.16 and Remark 2.1.17, U is a q-group for some prime q, and U is contained S, the soluble radical of G. Let $K = \gamma_3(S)$, the third term of the lower central series of S. Since $K \le \gamma_3(G)$, it follows that $KD/D \le \gamma_3(G/D) = \gamma_3(G)D/D$. On the other hand, G/D is a soluble T-group and so, by Theorem 2.1.11, $(G/D)/(G/D)^{\mathfrak{N}}$ is a Dedekind group, which has nilpotency class at most 2. It follows that $\gamma_3(G/D)$ is contained in $(G/D)^{\mathfrak{N}}$. Consequently KD/D is abelian by Theorem 2.1.11. Since $K \cap D$ is central in K by Proposition 1.6.4, we have that K is nilpotent of class at most 2. Assume that $U \cap K \neq 1$. If $U \cap K$ is a proper subgroup of U, then it is normal in G by the minimal choice of U. This implies that U is normal in G because $G/U \cap K$ is a T-group, against supposition. Therefore U is contained in K and U has nilpotent class at most 2. Suppose that $U \cap K = 1$. Then $U \cong UK/K$ is nilpotent of class at most 2, because $\gamma_3(UK) \leq \gamma_3(S) = K$. In both cases, we have that U is a subnormal q-group of nilpotent class at most 2. If q does not belong to $\pi(\mathrm{Z}(D))$, then $\langle U^G \rangle \cap D = 1$ by Proposition 1.6.4 and $\langle U^G \rangle$ is G-isomorphic to $\langle U^G \rangle D/D$; this implies that U is a normal subgroup of G, because G/D is a T-group. Hence q belongs to $\pi(\mathrm{Z}(D))$. Since UD is a normal subgroup of G, $\langle U^G \rangle \leq UD$ and so $\langle U^G \rangle \leq UD \cap S = U\mathrm{Z}(D)$. Then $\gamma_3(\langle U^G \rangle) \leq \gamma_3(U) = 1$. By the condition $\mathbf{T}_q$, elements of G induce power automorphisms in $\langle U^G \rangle$. Thus U is a normal subgroup of G, and we have a contradiction. □

As a consequence we obtain the next result of Ballester-Bolinches, Beidleman, and Heineken [33, Theorem A (iii)]:

Corollary 2.1.29. *A group G is a* T*-group if and only if G satisfies t_p for all primes p, and every non-abelian chief factor of G is simple.*

Remark 2.1.30. The statements of Theorems 2.1.23 and 2.1.28 were incorrectly written in Theorems 3.1 and 4.1 of [217]. In this paper, the conditions about $\mathbf{N}_p$ and $\mathbf{P}_p$ or $\mathbf{T}_p$ for G were missing. Unfortunately, with the weaker hypothesis the results become false, as the following example shows.

Example 2.1.31. Consider the group G as in Example 2.1.14. We see that G satisfies the conditions of [217, Theorem 3.1] and [217, Theorem 4.1], but G is not a PST-group. Consider $D = A$. Then D is a perfect normal subgroup of G. Moreover:

1. $G/D \cong \mathrm{Sym}(3)$ is a soluble T-group.

2. $D/\mathrm{Z}(D) = U_1/\mathrm{Z}(D) \cong \mathrm{Alt}(6)$ is simple and $U_1 = D$ is normal in G.

3. If $\{i_1, i_2, \dots, i_r\} \subseteq \{1\}$, where $1 \leq r < 1$, then $G/U'_{i_1}U'_{i_2}\cdots U'_{i_r}$ satisfies $\mathbf{N}_p$ and $\mathbf{T}_p$ for all $p \in \pi(\mathrm{Z}(D))$ and $\mathbf{P}_p$ for all $p \in \pi(D)$.

   This condition is trivially satisfied because there does not exist such a subset.

2.2 Local characterisations

A natural outcome of the study of soluble T-, PT-, and PST-groups is the fact that the difference between these three local classes is simply the Sylow structure. Hence it seems reasonable to think about local characterisations of these classes involving Sylow subgroups and their normalisers. Our intention in this section is to establish local results which can be applied to shed light on the behaviour of the Sylow normalisers of soluble T-, PT-, and PST-groups. The following definitions are the main keys which allows us to describe these three classes in a natural way. They have been introduced by Robinson (T-groups), Beidleman, Brewster, and Robinson (PT-groups), and Ballester-Bolinches and Esteban-Romero (PST-groups).

Definition 2.2.1. Let G be a group and let p be a prime. We say that G:

1. enjoys property $\mathcal{C}_p$ (see [213]) if each subgroup of a Sylow p-subgroup P of G is normal in the normaliser $\mathrm{N}_G(P)$,
2. satisfies property $\mathcal{X}_p$ (as in [56]) if each subgroup of a Sylow p-subgroup P of G is permutable in the normaliser $\mathrm{N}_G(P)$, and
3. enjoys property $\mathcal{Y}_p$ (see [44]) if whenever $H \leq K$ are two p-subgroups of G, H is S-permutable in $\mathrm{N}_G(K)$.

It is clear, by its definition, that property $\mathcal{Y}_p$, that is, the class of all $\mathcal{Y}_p$-groups, is inherited by subgroups. Applying Lemma 1.5.17 property $\mathcal{C}_p$ is equivalent to the pronormality, weak normality, and subnormaliser condition of all p-subgroups. Therefore the class of all $\mathcal{C}_p$-groups is also closed under formation of subgroups.

The fact that the property $\mathcal{C}_p$ is subgroup-closed yields the following local characterisation of soluble T-groups.

Theorem 2.2.2 (Peng [208], Robinson [213]). *A group G is a soluble* T*-group if and only if G satisfies $\mathcal{C}_p$ for all primes p.*

Thirty-one years later a similar characterisation of soluble PT-groups was found by Beidleman, Brewster, and Robinson [56].

Theorem 2.2.3. *A group G is a soluble* PT*-group if and only if G satisfies $\mathcal{X}_p$ for all primes p.*

As in the T-case, the above result would follow easily if we could have the subgroup-closed character of the property $\mathcal{X}_p$ at our disposal. In the Beidleman, Brewster, and Robinson approach, this property follows after an intensive study of $\mathcal{X}_p$ and its consequences for the group structure (see [56, Corollary 3]).

The following theorem, proved by Ballester-Bolinches and Esteban-Romero in [44, Theorem 3], summarises the relationship between the three properties listed above and is crucial to having a global knowledge of their behaviour.

Theorem 2.2.4. *A group G satisfies $\mathcal{X}_p$ (respectively, $\mathcal{C}_p$) if and only if G satisfies $\mathcal{Y}_p$ and the Sylow p-subgroups of G are Iwasawa (respectively, Dedekind).*

This theorem is a consequence of the following p-nilpotency criterion for groups with Iwasawa Sylow p-subgroups ([44, Theorem 1]).

Theorem 2.2.5. *Let p be a prime and let G be a group with an Iwasawa Sylow p-subgroup P. Then G is p-nilpotent if and only if $\mathrm{N}_G(P)$ is p-nilpotent.*

Since $\mathcal{Y}_p$ is subgroup-closed, Theorem 2.2.4 has the virtue of showing that the subgroup-closed character of $\mathcal{X}_p$ depends exclusively on the fact that the Sylow p-subgroups are Iwasawa groups. It also shows that, in order to get a global characterisation of the soluble PST-groups, it is necessary to impose the subgroup-closed character in the definition of $\mathcal{Y}_p$, because in the PST-case there are no restrictions on the Sylow p-subgroups.

The proofs of Theorem 2.2.4 and Theorem 2.2.5 will occupy us the first part of this section. Before we move on to these proofs, we pause briefly to take a look at the local characterisation of the soluble PST-groups, which may help to illuminate the theory so far. In what follows p will denote a prime.

Assume that G belongs to $\mathcal{U}_p^*$. Let $H \leq K$ be p-subgroups of G. Since the class of all $\mathcal{U}_p^*$-groups is closed under the formation of subgroups, $\mathrm{N}_G(K)$ is a $\mathcal{U}_p^*$-group. By Lemma 2.1.3, H is S-permutable $\mathrm{N}_G(K)$. Therefore G belongs to $\mathcal{Y}_p$. In fact, our next result shows that the classes $\mathcal{U}_p^*$ and $\mathcal{Y}_p$ coincide in the p-soluble universe.

Theorem 2.2.6. *Let G be a p-soluble group. Then G satisfies $\mathcal{U}_p^*$ if and only if G satisfies $\mathcal{Y}_p$.*

Since p-nilpotent groups clearly satisfy $\mathcal{U}_p^*$, we obtain:

Corollary 2.2.7. *If G is p-nilpotent, then G has $\mathcal{Y}_p$.*

The proof of Theorem 2.2.6 depend on the following lemma, which analyses the behaviour of the property $\mathcal{Y}_p$ in epimorphic images.

Lemma 2.2.8 ([44, Lemma 2]). *Let G be a group.*

1. *If G has property $\mathcal{Y}_p$ and A is a normal p-subgroup of G, then G/A has property $\mathcal{Y}_p$.*
2. *If G has property $\mathcal{Y}_p$ and N is a normal p'-subgroup of G, then G/N has property $\mathcal{Y}_p$.*

Proof. 1. This follows immediately from the definition since A is contained in every Sylow p-subgroup of G and $\mathrm{N}_G(P/A) = \mathrm{N}_G(P)/A$ if P is a Sylow p-subgroup of G.

2. Assume that G has property $\mathcal{Y}_p$ and let $H/N \leq S/N$ be p-subgroups of G/N. Let H_1 and S_1 be Sylow p-subgroups of H and S, respectively, such that H_1 is contained in S_1 and $H = H_1N$ and $S = S_1N$. Since G has $\mathcal{Y}_p$, it follows that H_1 is S-permutable in $\mathrm{N}_G(S_1)$. Therefore $H/N = H_1N/N$ is S-permutable in $\mathrm{N}_G(S_1)N/N = \mathrm{N}_G(S_1N/N)$ by Lemma 1.2.7 (2). This implies that G/N has $\mathcal{Y}_p$. □

Proof of Theorem 2.2.6. Only the inclusion $\mathcal{Y}_p \subseteq \mathcal{U}_p^*$ for p-soluble groups is in doubt. Let G be a p-soluble $\mathcal{Y}_p$-group. We show that G belongs to $\mathcal{U}_p^*$ by induction on the order of G. Let $A = O_{p'}(G)$. By Lemma 2.2.8 (2), G/A belongs to $\mathcal{Y}_p$, so that if $A \neq 1$, we can conclude by induction that G/A belongs to $\mathcal{U}_p^*$. In this case we have that G belongs to $\mathcal{U}_p^*$. Therefore we can assume that $A = 1$ and, in particular, every minimal normal subgroup of G is a p-group. Now let N be an arbitrary minimal normal p-subgroup of G. If N_0 is a subgroup of N, then N_0 is S-permutable in $G = N_G(N)$ by the $\mathcal{Y}_p$-property. By Lemma 2.1.3, N is cyclic. Applying Lemma 2.2.8 (1), G/N is a $\mathcal{Y}_p$-group. Therefore by induction G/N belongs to $\mathcal{U}_p^*$. In particular, G/N is p-supersoluble and so G is p-supersoluble. By Lemma 2.1.6, G has a normal Sylow p-subgroup P and by Lemma 2.1.3, all chief factors of G below P are cyclic and G-isomorphic. We conclude that G is a $\mathcal{U}_p^*$-group, as desired. □

Property $\mathcal{Y}_p$ does not imply property $\mathcal{U}_p^*$ in general. The alternating group $G = \mathrm{Alt}(5)$ of degree 5 has Sylow 3-subgroups of order 3 and Sylow 5-subgroups of order 5. Hence it satisfies $\mathcal{C}_3$ and $\mathcal{C}_5$. By Theorem 2.2.4, G satisfies $\mathcal{Y}_3$ and $\mathcal{Y}_5$. But G is neither 3-soluble nor 5-soluble. Hence it is clear that G is neither a $\mathcal{U}_3^*$-group nor a $\mathcal{U}_5^*$-group.

The following result, proved by Ballester-Bolinches and Esteban-Romero in [44, Theorem 4] shows that the classes $\mathcal{Y}_p$ are local definitions of the local class of all soluble PST-groups.

Theorem 2.2.9. *A group G is a soluble* PST*-group if and only if G satisfies $\mathcal{Y}_p$ for all primes p.*

Proof. Assume that G is a soluble PST-group. Then G satisfies $\mathcal{U}_p^*$ for all primes p and so G satisfies $\mathcal{Y}_p$ for all primes p by Theorem 2.2.6.

Conversely, suppose that G satisfies $\mathcal{Y}_p$ for all primes p. Then every subgroup of G has the same property. Therefore, if G is a group with least order subject to not being a soluble PST-group, then every proper subgroup of G is a soluble PST-group. According to Theorem 2.1.8, every soluble PST-group is supersoluble. Therefore either G is supersoluble, or G is a minimal non-supersoluble group. In both cases, we have that G is soluble (in the second case, the solubility of G follows from [157, VI, 9.6]). Since $\mathcal{Y}_p$ coincides with $\mathcal{U}_p^*$ in the p-soluble universe by Theorem 2.2.6, it follows that G satisfies $\mathcal{U}_p^*$ for all p. Then G is a soluble PST-group by Theorem 2.1.8. □

Assume now that a group G has $\mathcal{X}_p$ for all primes p. Then, G satisfies $\mathcal{Y}_p$ for all primes p by Theorem 2.2.4, and so G is soluble PST-group by Theorem 2.2.9. Since Sylow subgroups are Iwasawa, it follows that G is a PT-group by Corollary 2.1.12. The same arguments could be apply to groups satisfying $\mathcal{C}_p$. Therefore Theorem 2.2.2 and Theorem 2.2.3 are important consequences of Theorem 2.2.4 and Theorem 2.2.9.

There are two main ingredients in the proof of Theorem 2.2.5. The first is its p-soluble version.

Lemma 2.2.10. *Assume that G is a p-soluble group with Iwasawa Sylow p-subgroups. If P is a Sylow subgroup of G such that $\mathrm{N}_G(P)$ is p-nilpotent, then G is p-nilpotent.*

Proof. Assume, by way a contradiction, that the result is false and let G be a counterexample of least order. Then, for each non-trivial normal subgroup N of G, we have that PN/N is a Sylow p-subgroup of G/N and $\mathrm{N}_G(P)N/N = \mathrm{N}_G(PN/N)$ is p-nilpotent. It follows from the minimal choice of G that G/N is p-nilpotent. Therefore, since the class of all p-nilpotent groups is a saturated formation, we have that G has a unique minimal normal subgroup N, N is an elementary abelian p-group, $\mathrm{C}_G(N) = N$, and N is complemented in G by a core-free maximal subgroup M. It is clear that N is contained in P. Suppose that N is a proper subgroup of P. Then, since $G = NM$, we have that $P = N(P \cap M)$ and $P \cap M \neq 1$. Let $x \in P \cap M$ be an element of order p. If $n \in N$, then $\langle n, x\rangle = \langle n\rangle\langle x\rangle$ is an elementary abelian p-group because P is Iwasawa. Therefore $x \in \mathrm{C}_G(n)$. This implies that $1 \neq P \cap M \cap \mathrm{C}_G(N) = P \cap M \cap N$, which is a contradiction. Hence $P = N$ and so $G = \mathrm{N}_G(P)$ is p-nilpotent. This is the desired contradiction. □

The second result needed is a powerful theorem of Yoshida [262] strengthening earlier results of Hall and Wielandt. It guarantees that the Sylow normaliser controls p-transfer if the Sylow p-subgroup is not too far from being abelian. A full account of its proof, with essential background material, is too long to include in this book. The reader is referred to [233, 5.5.2].

Theorem 2.2.11. *Let p be a prime and let P be a Sylow p-subgroup of a group G. If P has no epimorphic image isomorphic to $C_p \wr C_p$, then $P \cap G' = P \cap (\mathrm{N}_G(P))'$.*

We are now in position to undertake the proof of Theorem 2.2.5.

Proof of Theorem 2.2.5. Let G be a group with Iwasawa Sylow p-subgroups and p-nilpotent Sylow normalisers with least order subject to not being p-nilpotent. Let P be a Sylow p-subgroup of G. Then $1 \neq P$. Since $C_p \wr C_p$ is not an Iwasawa group, it cannot be an epimorphic image of P. Therefore $P \cap G' = P \cap (\mathrm{N}_G(P))'$ by Theorem 2.2.11. By hypothesis, $\mathrm{N}_G(P)$ is p-nilpotent and so it has a normal subgroup of index p. It follows that P cannot be contained in $(\mathrm{N}_G(P))'$. Therefore $P \cap (\mathrm{N}_G(P))' \neq P$ and so $P \cap G' \neq P$. This implies that P is not contained in G' and therefore G/G' has a subgroup of index p. Consequently, $N = \mathrm{O}^p(G) \neq G$. It is clear that $\mathrm{N}_G(P) \leq \mathrm{N}_G(P \cap N)$ and $P \cap N$ is an Iwasawa Sylow p-subgroup of N. Suppose that $P \cap N = 1$. Then N is a normal Hall p'-subgroup of G because $G = NP$. Hence G is p-nilpotent, contrary to assumption. Therefore $P \cap N \neq 1$. Suppose that $\mathrm{N}_G(P \cap N) = G$. Let A be a minimal normal subgroup of G such that $A \leq P \cap N$. Clearly the group G/A satisfies the hypotheses of the theorem. The minimal choice of G implies that G/A is p-nilpotent. In particular, G is p-soluble. Applying Lemma 2.2.10, we have that G is p-nilpotent. This contradiction means that

$N_G(P \cap N)$ is a proper subgroup of G. But then, since $N_G(P \cap N)$ inherits the properties of G, $N_G(P \cap N)$ is p-nilpotent by minimality of G. Hence $N_N(P \cap N)$ is also p-nilpotent and therefore N satisfies the hypotheses of the theorem. Since $N \neq G$, it follows that N is p-nilpotent. Consequently G is p-nilpotent, and we have reached a final contradiction. □

It is now time to present a proof of Theorem 2.2.4. Its proof depends on a relevant property of $\mathcal{Y}_p$-groups contained in the next lemma.

Lemma 2.2.12. *If G satisfies $\mathcal{Y}_p$ and has a non-abelian Sylow p-subgroup P, then $N_G(P)$ is p-nilpotent.*

Proof. By Lemma 2.1.3 every p'-element of $N_G(P)$ normalises every subgroup of P. Since P is non-abelian, we can apply Lemma 1.3.4 to conclude that every p'-element of $N_G(P)$ actually centralises P. Consequently, $N_G(P)$ is p-nilpotent. □

Proof of Theorem 2.2.4. We only give a proof for property $\mathcal{X}_p$. The one for $\mathcal{C}_p$ is analogous.

Suppose that G satisfies $\mathcal{Y}_p$ and a Sylow p-subgroup P of G is Iwasawa. By Theorem 2.1.10, we have that every subgroup of P is permutable in $N_G(P)$.

Conversely, suppose that G satisfies $\mathcal{X}_p$. Then it is clear that every Sylow p-subgroup P of G is Iwasawa and p'-elements of $N_G(P)$ induce power automorphisms in P. Therefore, if P is abelian, every subgroup of P is normal in $N_G(P)$ by Theorem 2.1.10 and then G satisfies property $\mathcal{C}_p$. Since $\mathcal{C}_p$ is subgroup-closed, G satisfies $\mathcal{Y}_p$. If P is non-abelian, then $N_G(P)$ is p-nilpotent by Lemma 2.2.12. By Theorem 2.2.5, we have that G itself is p-nilpotent. Hence G satisfies $\mathcal{U}_p^*$ and so G satisfies $\mathcal{Y}_p$ by Theorem 2.2.6. □

The above results show that in any investigation of soluble PST-groups the property $\mathcal{Y}_p$ is likely to be a useful tool. Therefore it could be interesting to have interesting properties of $\mathcal{Y}_p$-groups at our disposal. The following result of Ballester-Bolinches and Esteban-Romero [44, Theorem 5] provides valuable information about a group in this class.

Theorem 2.2.13. *A group G satisfies $\mathcal{Y}_p$ if and only if G is p-nilpotent or G has an abelian Sylow p-subgroup and G satisfies $\mathcal{C}_p$.*

This result may be compared to the following result of Beidleman, Brewster, and Robinson [56].

Theorem 2.2.14. *A group G satisfies $\mathcal{X}_p$ if and only if G is p-nilpotent and has Iwasawa Sylow p-subgroups or G has abelian Sylow p-subgroups and satisfies $\mathcal{C}_p$.*

This theorem follows at once from Theorem 2.1.10 and Theorem 2.2.13.

$\mathcal{Y}_p$-groups are also p-normal in the following sense:

Definition 2.2.15. A group G is said to be *p-normal* if G satisfies the following property: If P is a Sylow p-subgroup of G and $\mathrm{Z}(P)$ is contained in P^g for some $g \in G$, then $\mathrm{Z}(P) = \mathrm{Z}(P^g)$.

Lemma 2.2.16. *If G satisfies $\mathcal{Y}_p$, then G is p-normal.*

Proof. Suppose that G satisfies $\mathcal{Y}_p$. Let P be a Sylow p-subgroup of G and let g be an element of G such that $Z = \mathrm{Z}(P) \leq P^g$. Suppose that Z is not a normal subgroup of P^g. Then, by [157, IV, 5.1], there exists an element $g \in G$ of order q^b for a prime $q \neq p$ such that $g \notin \mathrm{N}_G(Z)$, $J = ZZ^g \cdots Z^{g^{q^b-1}}$ is a p-group, and $g \in \mathrm{N}_G(J) \setminus \mathrm{C}_G(J)$. But g is a p'-element of $\mathrm{N}_G(J)$ and G is a $\mathcal{Y}_p$-group. Consequently g induces a power automorphism in J. In particular, we get the contradiction $g \in \mathrm{N}_G(Z)$. Therefore $\mathrm{Z}(P)$ is a normal subgroup of P^g. Then $\mathrm{Z}(P^{g^{-1}}) = (\mathrm{Z}(P))^{g^{-1}}$ is a normal subgroup of P. By [157, IV, 5.2], since $\mathrm{Z}(P)$ is a characteristic subgroup of P, we have that $\mathrm{Z}(P) = \mathrm{Z}(P^{g^{-1}})$ and $\mathrm{Z}(P) = \mathrm{Z}(P^g)$. This proves that G is p-normal. □

Theorem 2.2.17 ([220, Exercise 594]). *Let G be a group. The following conditions are equivalent:*

1. *G is p-nilpotent.*

2. *G is p-normal and $\mathrm{N}_G(\mathrm{Z}(P))$ is p-nilpotent.*

Proof. Assume that G is p-nilpotent. Then $\mathrm{N}_G(\mathrm{Z}(P))$ is p-nilpotent. Let P and P^* be two Sylow p-subgroups of G such that $\mathrm{Z}(P) \leq P^*$. Let g be an element of G such that $P^* = P^g$. Since G is p-nilpotent, G has a normal Hall p'-subgroup N. Moreover $N\mathrm{Z}(P)$ is a normal subgroup of G. Hence $N\mathrm{Z}(P) = N\mathrm{Z}(P^g)$. Therefore

$$\mathrm{Z}(P) = (N \cap P)\mathrm{Z}(P) = N\,\mathrm{Z}(P) \cap P = N\,\mathrm{Z}(P^*) \cap P = (N \cap P^*)\mathrm{Z}(P^*) = \mathrm{Z}(P^*),$$

which implies that G is p-normal.

Conversely, suppose that G is p-normal and $W = \mathrm{N}_G(\mathrm{Z}(P))$ is p-nilpotent. By a theorem of Grün ([141, Satz 5], see [220, Theorem 10.45]), we have that

$$G/\mathrm{O}^p(G) \cong W/\mathrm{O}^p(W).$$

Since W is p-nilpotent and $P \leq W$, $W/\mathrm{O}^p(W) \cong P$ and so G has a normal Hall p'-subgroup $\mathrm{O}^p(G)$. □

Proof of Theorem 2.2.13. Suppose that G is p-nilpotent. Then G satisfies $\mathcal{Y}_p$ by Corollary 2.2.7. Assume now that G has abelian Sylow p-subgroups and that G satisfies $\mathfrak{C}_p$. It follows from Theorem 2.2.4 that G satisfies $\mathcal{Y}_p$.

Assume that the converse is not true and let G be a counterexample of minimal order. If G had an abelian Sylow p-subgroup, then G would satisfy $\mathcal{C}_p$ by Theorem 2.2.4. Therefore G would have a non-abelian Sylow p-subgroup P and G would not be p-nilpotent. Suppose that $P_G = 1$. Therefore $\mathrm{N}_G(\mathrm{Z}(P))$ is a proper subgroup of G. Hence $\mathrm{N}_G(\mathrm{Z}(P))$ is p-nilpotent by the minimal choice of G. Applying Lemma 2.2.16 and Theorem 2.2.17, we have that G is p-nilpotent. This contradiction yields $P_G \neq 1$. Let N be a minimal normal subgroup of G contained in P. Since G has minimal order and G/N is a $\mathcal{Y}_p$-group, it follows that G/N is p-nilpotent or P/N is abelian.

Suppose that P/N is abelian. Since P is non-abelian, then $\mathrm{N}_G(P)$ is p-nilpotent by Lemma 2.2.12, and so $\mathrm{N}_G(P)/N = P/N \times \mathrm{O}_{p'}(\mathrm{N}_G(P))N/N$ and P/N lies in the centre of $\mathrm{N}_{G/N}(P/N)$. By [157, Hauptsatz IV, 2.6]), G/N is p-nilpotent. But if G/N is p-nilpotent, bearing in mind that G is a $\mathcal{Y}_p$-group and hence a $\mathcal{U}_p^*$-group by Theorem 2.2.6, we have that $|N| = p$ and p divides $|G/N|$ (otherwise, G would have an abelian Sylow p-subgroup $N = P$). It follows that G is p-nilpotent, because G acts centrally on the chief p-factors of G/N and hence G must act centrally on N. This contradiction proves the theorem. □

Corollary 2.2.18 (Ballester-Bolinches and Esteban-Romero [44, Corollary 1]). *If p is the smallest prime divisor of the order of G, then G is a $\mathcal{Y}_p$-group if and only if G is p-nilpotent.*

Proof. Suppose that G is a non-p-nilpotent $\mathcal{Y}_p$-group of minimal order. Since all proper subgroups of G satisfy $\mathcal{Y}_p$, it follows that all the proper subgroups of G are p-nilpotent. By [157, Satz IV, 5.4], we have that G has a normal Sylow p-subgroup P. Applying Theorem 2.2.13, we have that every subgroup of P is normal in $G = \mathrm{N}_G(P)$. Since p is the smallest prime dividing the order of G, it follows that G is p-nilpotent, contrary to assumption. □

As another consequence we recover the following result of Beidleman, Brewster, and Robinson [56, Theorem C]:

Corollary 2.2.19. *If p is the smallest prime divisor of the order of G, then G satisfies $\mathcal{X}_p$ if and only if G is p-nilpotent and has Iwasawa Sylow p-subgroups.*

As it is mentioned at the beginning of the section, property $\mathcal{C}_p$ can be easily characterised by the pronormality of all p-subgroups. It follows at once from this result that $\mathcal{C}_p$ is a subgroup-closed class and local definition of the local class of all soluble T-groups. There are also pronormal characterisations of the classes $\mathcal{Y}_p$ and $\mathcal{X}_p$. In order to formulate them, we first need a definition which was introduced by Beidleman, Brewster, and Robinson in [56].

Definition 2.2.20. A group G *satisfies* $\mathbf{H}_p$ if every normal subgroup of a Sylow p-subgroup of G is pronormal in G.

Beidleman, Brewster, and Robinson proved that a group G is a soluble PT-group if and only if every subgroup Sylow p-subgroup of G is Iwasawa and G satisfies $\mathbf{H}_p$ for every prime p (see Theorem 2.2.25). It seems natural to wonder whether a soluble group G is a PST-group if and only if G satisfies $\mathbf{H}_p$ for every prime p. Unfortunately, the answer is negative in general, as the following example shows.

Example 2.2.21. Let $\langle a\rangle$ be a cyclic group of order 2, and let V be a one-dimensional irreducible and faithful $\langle a\rangle$-module over the field of seven elements. Then V is one-dimensional. We write $V = \langle v\rangle$. Clearly, $v^a = -v$. We consider now the tensor product $V \otimes V$, whose basis consists of the single element $v \otimes v$. Since $(v \otimes v)^a = v^a \otimes v^a = (-v) \otimes (-v) = v \otimes v$, we have that $V \otimes V$ is the trivial $\langle a\rangle$-module.

We take now the associated Hartley group $P = \mathrm{H}(V, V)$. Thus P is a 7-group such that $P/\Phi(P) \cong V \oplus V$ and $\mathrm{Z}(P) = \Phi(P) = P' \cong V \otimes V$. Finally, let G be the corresponding semidirect product $G = [P]\langle a\rangle$, which is a 2-nilpotent group. It is known (see [119, B, 12.11]) that there exist two $\langle a\rangle$-invariant subgroups P_1 and P_2 of P such that $P_1 \cap \Phi(P) = P_2 \cap \Phi(P) = 1$, $[P_1, P_2] = \Phi(P)$, and $P_i\Phi(G)/\Phi(G)$ is $\langle a\rangle$-isomorphic to V for $i = 1, 2$. Hence $|P_i| = 7$ for $i = 1, 2$. Assume $P_i = \langle x_i\rangle$. Using the above-mentioned properties of the subgroups P_1 and P_2, it is not difficult to prove that the subgroup $H = \langle x_1x_2\rangle$ of P is not $\langle a\rangle$-invariant. Therefore H does not permute with the subgroup $\langle a\rangle$. However, H is a subnormal subgroup of G. Thus G cannot be a PST-group.

But note that G satisfies the property $\mathbf{H}_2$. Furthermore, G satisfies the property $\mathbf{H}_7$ as well: if P_0 is a non-trivial normal subgroup of P, then $P_0 \cap \mathrm{Z}(P) \neq 1$ and hence $P' = \mathrm{Z}(P) \leq P_0$. Since P_0 is subnormal in G, we have that P_0/P' is subnormal in G/P'. Clearly G/P' satisfies $\mathcal{Y}_2$ and $\mathcal{Y}_7$ and its Sylow subgroups are abelian. Applying Theorem 2.2.9 and Corollary 2.1.12, G/P' is a T-group, so that P_0/P' is a normal subgroup of G/P'. We conclude that P_0 is a normal subgroup of G, and therefore G satisfies $\mathbf{H}_7$.

Assume that G is an $\mathcal{Y}_p$-group. Then, by Theorem 2.2.13, G is p-nilpotent or G has an abelian Sylow subgroup and G satisfies $\mathcal{C}_p$. If G is p-nilpotent and A is the normal Hall p'-subgroup of G, then $G = AP$ for each Sylow p-subgroup P of G. Let B be a normal subgroup of P. Then AB is normal in G and B is a Sylow p-subgroup of AB. Therefore B is pronormal in G. Now, if G has $\mathcal{C}_p$, then every p-subgroup of G is pronormal in G by Lemma 1.5.17. We therefore conclude that G is an $\mathbf{H}_p$-group. Since $\mathcal{Y}_p$ is a subgroup-closed property, we have that every subgroup of a $\mathcal{Y}_p$-group satisfies $\mathbf{H}_p$. Our next objective is to show that the converse is also true. To prove this result the following lemma is needed.

Lemma 2.2.22. *Let G be a p-soluble group. If G satisfies $\mathbf{H}_p$ and N is a normal subgroup of G, then G/N satisfies $\mathbf{H}_p$.*

Proof. Arguing by induction, we may clearly assume that N is a minimal normal subgroup of G and so N is either a p-group or a p'-group. Assume that N is a p-subgroup of G. Then N is contained in every Sylow p-subgroup P of G. If H/N is a normal subgroup of P/N, then H is a normal subgroup of P and so H is pronormal in G because G has $\mathbf{H}_p$. Therefore H/N is pronormal in G/N. Suppose now that N is a normal p'-subgroup of G. Let R/N be a Sylow p-subgroup of G/N and let H/N be a normal subgroup of R/N. Let H_p be a Sylow p-subgroup of H and let Q be a Sylow p-subgroup of G containing H_p such that $R = QN$. Then $\langle H_p^Q \rangle N/N = \langle (H_pN/N)^{QN/N} \rangle = \langle (H/N)^{R/N} \rangle = H/N = H_pN/N$. Hence $\langle H_p^Q \rangle N = H_pN$ and so $\langle H_p^Q \rangle = H_p$ by order considerations. Therefore H_p is a normal subgroup of Q and so H_p is pronormal in G because of the $\mathbf{H}_p$-property. We therefore conclude that $H/N = H_pN/N$ is pronormal in G/N and that G/N is therefore an $\mathbf{H}_p$-group. □

The second result we need is the p-soluble version proved by Alejandre, Ballester-Bolinches, and Pedraza-Aguilera in [5, Theorem 5].

Theorem 2.2.23. *Let G be a p-soluble group. Then G satisfies $\mathcal{U}_p^*$ if and only if every subgroup of G satisfies $\mathbf{H}_p$.*

Proof. Assume that G is a $\mathcal{U}_p^*$-group. By Theorem 2.2.6, G satisfies $\mathcal{Y}_p$. Hence every subgroup of G satisfies $\mathbf{H}_p$ by the argument just given above.

Conversely, assume now that every subgroup of G satisfies $\mathbf{H}_p$. Assume first that $\mathrm{O}_{p'}(G) = 1$. Let N be a normal p-subgroup of G. Let N_0 be a subgroup of N and let g be a p'-element of G. Then there exists a chain

$$N_0 \trianglelefteq N_1 \trianglelefteq \cdots \trianglelefteq N_{r-1} \trianglelefteq N_r = N.$$

Hence N_{r-1} is pronormal in G. Thus N_{r-1} is normal in G. Hence $N_{r-1}\langle g \rangle$ is a subgroup of G. By hypothesis, $N_{r-1}\langle g \rangle$ satisfies $\mathbf{H}_p$. An inductive argument shows that N_0 is normalised by g. Therefore p'-elements of G induce power automorphisms in N. By Lemma 2.1.3, all chief factors of G below N are cyclic and G-isomorphic.

First we show that G is p-supersoluble by induction on $|G|$. Assume that N is a minimal normal subgroup of G. Then G/N is p-supersoluble by induction and N is cyclic by the previous argument. It follows that G is p-supersoluble. By Lemma 2.1.6, G has a normal Sylow p-subgroup P and, by the previous argument, G satisfies $\mathcal{U}_p^*$.

Assume now that $\mathrm{O}_{p'}(G) \neq 1$. By Lemma 2.2.22, $G/\mathrm{O}_{p'}(G)$ and all its subgroups satisfy $\mathbf{H}_p$. Therefore $G/\mathrm{O}_{p'}(G)$ satisfies $\mathcal{U}_p^*$ and so does G. □

Corollary 2.2.24. *A group G satisfies $\mathcal{Y}_p$ if and only if every subgroup of G is an $\mathbf{H}_p$-group.*

Proof. Only the sufficiency of the condition is in doubt. We proceed by induction on the order of G. Let H and K be p-subgroups of G such that H is contained in K. Assume that $N_G(K)$ is a proper subgroup of G. Then, by induction, $N_G(K)$ is a $\mathcal{Y}_p$-group and so H is S-permutable in $N_G(K)$. Thus we can suppose that K is a normal subgroup of G. Let g be a p'-element of G. If $K\langle g \rangle$ is a proper subgroup of G, we apply induction and conclude that every subgroup of K is S-permutable in $K\langle g\rangle$. By Lemma 2.1.3, g induces a power automorphism on K. In particular, g normalises H. Suppose that $G = K\langle g\rangle$. Then G is p-soluble. Applying Theorem 2.2.23 and Theorem 2.2.6, G satisfies $\mathcal{Y}_p$. Therefore, H is S-permutable in G and, by Lemma 2.1.3, g normalises H. We conclude that p'-elements of G induce power automorphisms in K. By Lemma 2.1.3, H is S-permutable in G. Consequently, G satisfies $\mathcal{Y}_p$. □

Theorem 2.2.25 (Beidleman, Brewster, and Robinson [56, Theorem D]). *Let G be a group. Then G satisfies $\mathcal{X}_p$ if and only if G satisfies $\mathbf{H}_p$ and G has Iwasawa Sylow p-subgroups.*

Proof. Suppose that G satisfies $\mathcal{X}_p$. By Theorem 2.2.4, G satisfies $\mathcal{Y}_p$ and has Iwasawa Sylow p-subgroups. Applying the previous corollary, G satisfies $\mathbf{H}_p$ and G has Iwasawa Sylow p-subgroups.

Conversely, assume that G satisfies $\mathbf{H}_p$ and G has Iwasawa Sylow p-subgroups. Let P be a Sylow p-subgroup of G. Assume that P is Dedekind. Then given a subgroup H of P, H is normal in P and by the property $\mathbf{H}_p$, H is pronormal in G. It follows that H is normal in $\mathrm{N}_G(P)$. Therefore G satisfies $\mathcal{C}_p$; in particular, G satisfies $\mathcal{X}_p$. Assume now that P is not Dedekind. By Theorem 1.4.3, P has an abelian normal subgroup A and an element x such that $P = A\langle x\rangle$ and $a^x = a^{1+p^n}$ for every $a \in A$, where n is a fixed natural number such that $n \geq 2$ if $p = 2$. Every subgroup of P below A is normal in P, as well as every subgroup between P' and P. Hence every subgroup of P below A is pronormal by property $\mathbf{H}_p$, as well as every subgroup between P' and P. Therefore all subgroups of P below A and all subgroups between P' and P are normal in $\mathrm{N}_G(P)$. It follows that p'-elements of $\mathrm{N}_G(P)$ induce power automorphisms in A and in P/P'. By Lemma 2.1.3, all chief factors of $\mathrm{N}_G(P)$ below A are $\mathrm{N}_G(P)$-isomorphic, as well as all chief factors of $\mathrm{N}_G(P)$ between P' and P. Since P/A is cyclic, $P' \leq A$. If $P' = A$, then P/P' is cyclic and so P is cyclic, a contradiction. Hence P' is a proper subgroup of A. This implies that all chief factors of $\mathrm{N}_G(P)$ are $\mathrm{N}_G(P)$-isomorphic. Then $\mathrm{N}_G(P)$ satisfies $\mathcal{U}_p^*$, and by Theorem 2.2.6, $\mathrm{N}_G(P)$ satisfies $\mathcal{Y}_p$. Since P is an Iwasawa group, $\mathrm{N}_G(P)$ satisfies $\mathcal{X}_p$. Therefore G satisfies $\mathcal{X}_p$, as desired. □

Remarks 2.2.26. 1. One might think that the most natural candidate for the "PST-version" of properties $\mathcal{C}_p$ and $\mathcal{X}_p$ could be:

> A group G satisfies $\mathcal{Y}'_p$ if every subgroup of a Sylow p-subgroup is S-permutable in $\mathrm{N}_G(P)$.

In $G = \mathrm{Sym}(4)$, the symmetric group of degree 4, the Sylow 2-subgroups are self-normalising. Hence every subgroup of a Sylow 2-subgroup P of G is S-permutable in $\mathrm{N}_G(P)$, and the subgroups of a Sylow 3-subgroup Q of G are S-permutable in $\mathrm{N}_G(Q)$. Consequently, G satisfies $\mathcal{Y}'_p$ for every p, but G is not a PST-group, because the cyclic subgroups of the Klein 4-group are not permutable with the Sylow 3-subgroups of G.

Note that the above example shows that $\mathcal{Y}'_p$ is not subgroup-closed.

2. Any hope of defining subgroup-closed local definitions for PST, PT, and T outside of the soluble universe is soon dispelled. As soon as we have a local property $\mathcal{J}_p$ which is subgroup-closed and such that a group G is a PST-group if and only if G satisfies $\mathcal{J}_p$ for every prime p, then $\bigcap_{p\in\mathbb{P}} \mathcal{J}_p$ is contained in the class of all soluble groups. Therefore a group satisfying $\mathcal{J}_p$ for all primes p should be soluble.

We mention that other local versions of T, PT, and PST-groups have been discussed by some authors. Thus, in a seminal paper, Bryce and Cossey [78] have examined soluble groups with the property T_p: in these groups each subnormal p'-perfect subgroup is normal. It is shown that a soluble group is a T-group if and only if it satisfies T_p for all primes p. The class T'_p of all soluble groups for which every p'-perfect subnormal subgroup of G is permutable and the class T''_p of all soluble groups for which every p'-perfect subnormal subgroup is S-permutable were introduced by Beidleman and Heineken in [62]. They proved that a soluble group G is a PT-group if and only if it is a T'_p-group for all primes p, and G is a PST-group if and only if it is a T''_p-group for all primes p. Alejandre, Ballester-Bolinches, and Pedraza-Aguilera [5] examined soluble groups with the following property PST_p: in these groups every subnormal p'-perfect subgroup of G permutes with every Hall p'-subgroup of G. They proved in [5, Theorem 6] that properties $\mathcal{U}^*_p$ and PST_p define the same classes in the soluble universe. In fact, they coincide in the p-soluble universe. In particular, PST_p is a local definition of the local class of all soluble PST-groups.

Ballester-Bolinches and Esteban-Romero in [43, Theorem A] gave a characterisation of the property $\mathcal{U}^*_p$ which can be regarded as a local version of the Agrawal theorem. Let $\mathfrak{E}_{p'}\mathfrak{S}_p$ be the saturated formation of all p-nilpotent groups. For every group G, $G(p)$ will denote the $\mathfrak{E}_{p'}\mathfrak{S}_p$-residual of G.

Theorem 2.2.27. *A group G satisfies $\mathcal{U}^*_p$ if and only if*

1. *G is p-nilpotent, or*
2. *$G(p)/\mathrm{O}_{p'}(G(p))$ is an abelian normal Sylow p-subgroup of $G/\mathrm{O}_{p'}(G(p))$ such that the elements of $G/\mathrm{O}_{p'}(G(p))$ induce power automorphisms in $G(p)/\mathrm{O}_{p'}(G(p))$.*

Proof. Let G be a $\mathcal{U}^*_p$-group and assume that G is not p-nilpotent. Then $G(p) \neq 1$ and p divides $|G(p)|$. By Theorems 2.2.6 and 2.2.13, G has an abelian Sylow p-subgroup P such that every subgroup of P is normal in $\mathrm{N}_G(P)$.

Since $G/\mathrm{O}_{p'}(G(p))$ inherits the hypotheses of G, we can assume without loss of generality that $\mathrm{O}_{p'}(G(p)) = 1$. Hence $G(p)$ is contained in the commutator subgroup G' of G, which is p-nilpotent because G is p-supersoluble. Hence $G(p)$ is p-nilpotent and so $G(p)$ is a p-group contained in P. In particular, $G(p)$ is abelian.

Assume now that $G(p)$ is a proper subgroup of P. Then all G-chief factors between $G(p)$ and P are central. Since all p-chief factors of G are cyclic, there exists a normal subgroup M of G such that $|G(p)/M| = p$. Then $G(p)/M$ is a chief factor of G which is not central. This contradicts the fact that G satisfies $\mathcal{U}_p^*$. Therefore $G(p) = P$ and so $\mathrm{N}_G(G(p)) = G$. This implies that every subgroup of $G(p)$ is normal in G, that is, the elements of G induce power automorphisms in $G(p)$.

Conversely, suppose that G is either p-nilpotent or $G(p)/\mathrm{O}_{p'}(G)$ is an abelian normal Sylow p-subgroup of $G/\mathrm{O}_{p'}(G(p))$ such that the elements of $G/\mathrm{O}_{p'}(G)$ induce power automorphisms in $G(p)/\mathrm{O}_{p'}(G(p))$.

If G is p-nilpotent, then G is p-supersoluble and all p-chief factors of G are central. Hence G is a $\mathcal{U}_p^*$-group.

Assume now that $G(p)/\mathrm{O}_{p'}(G(p))$ is an abelian Sylow p-subgroup of the group $G/\mathrm{O}_{p'}(G(p))$ and every subgroup of $G(p)/\mathrm{O}_{p'}(G(p))$ is normal in $G/\mathrm{O}_{p'}(G(p))$. Since $G/\mathrm{O}_{p'}(G(p))$ satisfies the hypothesis and G satisfies $\mathcal{U}_p^*$, we can assume that $\mathrm{O}_{p'}(G(p)) = 1$. Then $G(p)$ is a p-subgroup acted on by G as a group of power automorphisms. By Lemma 2.1.3, all p-chief factors of G are cyclic and G-isomorphic. Hence G satisfies $\mathcal{U}_p^*$. □

2.3 Characterisations based on subgroup embedding properties

In this section we study characterisations of the classes of all soluble PT- and PST-groups, concentrating on results which depend specifically on some subgroup embedding properties, naturally arising in the attempt to give permutable and S-permutable versions of the weak normality and subnormaliser condition introduced in Definitions 1.5.8 and 1.5.13. These embedding properties provides criteria for a group to be a soluble T-group.

According to Proposition 1.5.9, pronormality implies weak normality and, by virtue of Proposition 1.5.15, every weakly normal subgroup satisfies the subnormaliser condition. They are, however, three different subgroup embedding properties (Examples 1.5.10 and 1.5.16). Fortunately, they coincide in p-subgroups and characterise property $\mathcal{C}_p$ by Lemma 1.5.17. This fact enables us to prove the following result (see [45, Theorem A]).

Theorem 2.3.1. *Let G be a group. Any two of the following conditions are equivalent:*

1. *G is a soluble* T*-group.*
2. *Every subgroup of G is weakly normal in G.*
3. *For every prime number p, every p-subgroup of G is weakly normal in G.*
4. *Every subgroup of G satisfies the subnormaliser condition in G.*
5. *For every prime number p, every p-subgroup of G satisfies the subnormaliser condition in G.*

Proof. Clearly 2 implies 3 and 4 implies 5. If Statement 5 holds, it follows from Lemma 1.5.17 that G satisfies $\mathscr{C}_p$ for all p. Therefore by Theorem 2.2.2 Statement 1 is true.

To complete the proof we now show that 1 implies 2. Suppose that there exists a soluble T-group G with a non-weakly normal subgroup H. We choose for G a counterexample of least order. Then there exists an element $g \in G$ such that $H^g \leq \mathrm{N}_G(H)$ but $g \notin \mathrm{N}_G(H)$. Let $S = \langle H, g \rangle$. Then S is a T-group by Corollary 2.1.13. If $S \neq G$, then, by the minimal choice of G, it follows that H is a weakly normal subgroup of S. Hence $g \in \mathrm{N}_S(H) \leq \mathrm{N}_G(H)$, contrary to assumption. Therefore $G = \langle H, g \rangle$.

On the other hand, G is supersoluble by Corollary 2.1.2. Let p be the largest prime dividing the order of H. Then, if A is a Sylow p-subgroup of H, we have that A is normal in H. Moreover, A is weakly normal in G by Theorem 2.2.2 and Lemma 1.5.17. Since $A^g \leq H^g \leq \mathrm{N}_G(H) \leq \mathrm{N}_G(A)$, it follows that $g \in \mathrm{N}_G(A)$ and so A is a normal subgroup of G. Since G/A is a soluble T-group by Lemma 2.0.4, we can conclude, by the choice of G, that H/A is weakly normal in G/A. By Lemma 1.5.14, H is weakly normal in G. □

Bianchi, Gillio Berta Mauri, Herzog, and Verardi [72, Theorem 10] proved a characterisation of soluble T-groups by means of $\mathscr{H}$-subgroups: a subgroup H of a group G is said to be an $\mathscr{H}$-subgroup of G if $\mathrm{N}_G(H) \cap H^g$ is contained in H for all $g \in G$. This result, as it is showed in [45], is a consequence of the above theorem.

Our next major objective is to characterise soluble PST-groups. As a point of departure, we note that condition $H^g \leq \mathrm{N}_G(H)$ can be seen as H is a normal subgroup of $\langle H, H^g \rangle$, and the fact that g normalises H can be seen as H is a normal subgroup of $\langle H, g \rangle$. Therefore H is weakly normal if and only if whenever H is normal in $\langle H, H^g \rangle$, then H is normal in $\langle H, g \rangle$. On the other hand, a subgroup G satisfies the subnormaliser condition when if H is normal in K and x is an element of G such that K is normal in $\langle K, x \rangle$ then H is normal in $\langle H, x \rangle$. This motivates the following generalisations of both conditions, introduced by Ballester-Bolinches and Esteban-Romero in [46]:

Definition 2.3.2. We say that a subgroup H of a group G is *weakly* S*-permutable* in G when the following condition holds: if $g \in G$ and H is S-permutable in $\langle H, H^g \rangle$, then H is S-permutable in $\langle H, g \rangle$.

Definition 2.3.3. We say that a subgroup H of a group G *satisfies the* S*-subpermutiser condition* in G or that H is *transitively* S*-permutable* in G when the following condition holds: if H is S-permutable in K and x is an element of G such that K is S-permutable in $\langle K, x \rangle$, then H is S-permutable in $\langle H, x \rangle$.

Clearly, S-permutable subgroups are weakly S-permutable. Maximal subgroups, Sylow subgroups, and self-normalising subgroups are weakly S-permutable, too.

The following proposition shows the relation between the above properties and the corresponding T-versions.

Proposition 2.3.4. *Let H be a subgroup of a group G. Then:*

1. *If H is weakly normal in G, then H is weakly* S*-permutable in G.*
2. *If H satisfies the subnormaliser condition in G, then H satisfies the* S*-subpermutiser condition in G.*

Proof. 1. Suppose that H is a weakly normal subgroup of G. Let g be an element of G such that H is S-permutable in $\langle H, H^g\rangle$. By Kegel's theorem (Theorem 1.2.14 (3)), we know that H is subnormal in $\langle H, H^g\rangle$. Now we can apply Proposition 1.5.15 to conclude that H is normal in $\langle H, H^g\rangle$. The weak normality of H in G implies that H is normal in $\langle H, g\rangle$ and, in particular, H is S-permutable in $\langle H, g\rangle$. Consequently, H is weakly S-permutable in G.

2. The same arguments used in the proof of Statement 1 show that every subgroup satisfying the subnormaliser condition also satisfies the S-subpermutiser condition in G. □

The next results analyse the relation between weak S-permutability and the S-subpermutiser condition. There exist p-subgroups satisfying the S-subpermutiser condition which are not weakly S-permutable (see Example 2.3.9 below). However, weak S-permutability implies the S-subpermutiser condition.

By Theorem 1.2.13, a subgroup H of a group G satisfies the S-subpermutiser condition in G if and only if H satisfies the following property: if H is S-permutable in K and K is S-permutable in L, then H is S-permutable in L.

Corollary 2.3.5. *If H satisfies the* S*-subpermutiser condition in a group G and H is a subnormal subgroup of a subgroup K of G, then H is* S*-permutable in K.*

Proof. Suppose that H satisfies the S-subpermutiser condition in G and that H is subnormal in a subgroup K of G. An inductive argument allows us to suppose, without loss of generality, that H is S-permutable in a proper normal subgroup L of K. Consider $g \in K$. Since H is S-permutable in L and L is S-permutable in $\langle L, g\rangle$, from the S-subpermutiser condition we have that H is S-permutable in $\langle H, g\rangle$. Since this happens for every $g \in K$, from Theorem 1.2.13 we obtain that H is an S-permutable subgroup of K. □

Corollary 2.3.6. *If H is weakly* S*-permutable in G, then H satisfies the* S*-subpermutiser condition in G.*

Proof. Assume that H is a weakly S-permutable subgroup of G. Let K be a subgroup of G such that H is S-permutable in K. Suppose, in addition, that x is an element of G such that K is S-permutable in $\langle K, x\rangle$. By Kegel's theorem (Theorem 1.2.14 (3)), we have that H is subnormal in $\langle K, x\rangle$. By Corollary 2.3.5, we obtain that H is S-permutable in $\langle K, x\rangle$, as desired. □

It is natural to ask for the relation between the class $\mathcal{Y}_p$ and weak S-permutability and S-subpermutiser condition. There exist groups in the class $\mathcal{Y}_p$ with p'-perfect subnormal subgroups which neither are S-permutable nor satisfy the S-subpermutiser condition (see Example 2.3.9 below). This example also shows that the PST-version of Lemma 1.5.17 does not hold, and so the PST-results are not just an empty exercise in generalisation.

The following theorem is a partial extension of Lemma 1.5.17 given by Ballester-Bolinches and Esteban-Romero in [46, Theorem B].

Theorem 2.3.7. *Let G be a group. The following statements are equivalent:*

1. *G is a $\mathcal{Y}_p$-group.*
2. *Every p-subgroup of G satisfies the* S*-subpermutiser condition in G.*

Proof. Suppose that every p-subgroup of G satisfies the S-subpermutiser condition in G. Suppose that $H \leq L \leq P$, where P is a Sylow p-subgroup of G. Since H is a subnormal subgroup of $\mathrm{N}_G(L)$ and H satisfies the S-subpermutiser condition in G, we have that H is S-permutable in $\mathrm{N}_G(L)$ by Corollary 2.3.5. Therefore G is in the class $\mathcal{Y}_p$.

Conversely, suppose now that G is in the class $\mathcal{Y}_p$. Assume that the p-subgroup H is S-permutable in K and K is S-permutable in L. Note that H is subnormal in L by Theorem 1.2.14 (3). By Theorem 2.2.13, G is p-nilpotent or G has an abelian Sylow p-subgroup and G satisfies $\mathcal{C}_p$. Assume first that G is p-nilpotent. In this case, H is contained in $\mathrm{O}_p(L)$ and so in all Sylow p-subgroups of L. Moreover, L is p-nilpotent and so $\mathrm{O}_p(L)$ is centralised by $\mathrm{O}_{p'}(L)$; in particular, all Sylow q-subgroups of L for primes $q \neq p$ centralise H. This implies that H is S-permutable in L. Assume now that G has an abelian Sylow p-subgroup and that G satisfies $\mathcal{C}_p$. By Lemma 1.5.17, H is pronormal is G and so is in L. By Theorem 1.5.2, H is normal in L. In particular, H is S-permutable in L. □

We have arrived now at our promised objective, namely a characterisation of soluble PST-groups (Ballester-Bolinches and Esteban-Romero [46]).

Theorem 2.3.8. *Let G be a group. The following statements are equivalent:*

1. *G is a soluble* PST*-group.*
2. *Every subgroup of G is weakly* S*-permutable in G.*
3. *For every prime number p, every p-subgroup of G is weakly* S*-permutable in G.*
4. *Every subgroup of G satisfies the* S*-subpermutiser condition in G.*
5. *For every prime number p, every p-subgroup of G satisfies the* S*-subpermutiser condition in G.*

Proof. Let us see that 1 implies 2. Suppose that G is a soluble PST-group. By Theorem 2.1.8, $G = AB$, where A is the nilpotent residual of G, A is abelian of odd order, $|A|$ and $|B|$ are coprime and every subgroup of A is normal in G. Let $g \in G$ and $H \leq G$ such that H is S-permutable in $\langle H, H^g \rangle$. We prove that H is S-permutable in $\langle H, g \rangle$ by induction on the order of G. We can suppose that G is not nilpotent and so $A \neq 1$. Let N be a minimal normal subgroup of G contained in A. By induction, HN/N is weakly S-permutable in G/N. Hence HN/N is S-permutable in $\langle H, g \rangle N/N$. Consequently HN is S-permutable in $\langle H, g \rangle N$ by Theorem 1.2.14 (1). Assume that $\langle H, g \rangle$ is a proper subgroup of G. Since, by Corollary 2.1.9, $\langle H, g \rangle$ is a PST-group, we can apply the induction hypothesis to $\langle H, g \rangle$ to conclude that H is S-permutable in $\langle H, g \rangle$. Therefore we may assume $G = \langle H, g \rangle$ and then HN is S-permutable in G. This implies that HN is a subnormal subgroup of G by Theorem 1.2.14 (3).

Suppose that H is not weakly S-permutable in G and derive a contradiction. Let p be a prime dividing $|G|$ and let P be a Sylow p-subgroup of G such that H does not permute with P. If $(HN)P$ were a proper subgroup of G, then $(HN)P$ would be a PST-group by Corollary 2.1.9 and H would permute with P by induction. Consequently, $G = (HN)P$. Suppose that p divides $|A|$. Then $P \leq A$ and P is a normal subgroup of G. Hence H permutes with P, a contradiction. Therefore $|P|$ and $|A|$ are coprime. Moreover, we may also assume that $H_G = 1$. Thus $H \cap A = 1$ and $|H|$ and $|A|$ are coprime. As a consequence, if π is the set of primes dividing $|A|$ and n_π is the π-part of the number n, then

$$|G|_\pi = \frac{|HN|_\pi |H|_\pi}{|HN \cap P|_\pi} = |HN|_\pi = |N|_\pi$$

and hence $A = N$.

Let us denote $T = \langle H, H^g \rangle$ and let q be the prime dividing $|N|$. If $|T|_q \neq 1$, then $N \cap T$ is a non-trivial normal subgroup of G. Hence $N \leq T$. Since H is S-permutable in T, we have that H is a subnormal subgroup of T and so H is subnormal in HN. Therefore H is a subnormal subgroup of G. Since G is a PST-group, we have that H is S-permutable in G, contrary to supposition. Therefore $|T|_q = 1$. We may assume that $T \leq B$. The element g can be expressed as $g = bn$, with $b \in B$ and $1 \neq n \in N = \langle x \rangle$, where x has order p (recall that G is supersoluble by Corollary 2.1.2). Hence $N = \langle n \rangle$, and $H^g \leq B^g = B^n$. Therefore $H^g \leq B \cap B^n = \mathrm{C}_B(n)$ (see [119, A, 16.3]). Consequently $H^g \leq \mathrm{C}_G(n)$. This implies that $H^b \leq \mathrm{C}_G(N)$ and so $H^b N$ is a nilpotent group. But in this case H^b is a subnormal subgroup of G, because $H^b N$ is a subnormal subgroup of G. Therefore H is a subnormal subgroup of G. Since G is a PST-group, we have that H is S-permutable in G. This contradiction proves that H is S-permutable in G and 1 implies 2.

It is obvious that 2 implies 3 and 4 implies 5. From Corollary 2.3.6, it follows that 2 implies 4 and that 3 implies 5. From Theorems 2.2.9 and 2.3.7, it follows that 5 implies 1. This completes the proof. □

Example 2.3.9. Consider the Iwasawa group $P = \langle x, y : x^2 = y^8 = 1, y^x = y^5 \rangle$ of order 16. This group has an irreducible and faithful module over the field of 17 elements, $V = \langle w_1, w_2 \rangle$, such that the action of P is described by $w_1^x = w_2$, $w_2^x = w_1$, $w_1^y = w_1^9$, $w_2^y = w_2^8$. We construct the semidirect product $G = [V]P$. We observe that x centralises the element $w_1 w_2$. Let $g = w_1 w_2 y$. Let $H = \langle x \rangle$. We have that $H^g = \langle x^y \rangle = \langle x y^4 \rangle \leq P$. Consequently the subgroup $H = \langle x \rangle$ is S-permutable in $\langle H, H^g \rangle$. But H is not S-permutable in $\langle H, g \rangle = G$: it suffices to see that H does not permute with P^{w_1}.

It is clear that G is a 2-nilpotent group, and so G belongs to the class $\mathcal{Y}_2$ by Theorem 2.2.13. Applying Theorem 2.3.7, all 2-subgroups of G, in particular H, satisfy the S-subpermutiser condition in G (the reader is invited to prove directly that H satisfies the S-subpermutiser condition in G).

Consider now the subgroup $L = \langle x, w_1 w_2^{-1} \rangle$. Then L is a $2'$-perfect subnormal subgroup of G which does not permute with P. However, L is S-permutable in $M = \langle x, y^2, w_1, w_2 \rangle$, which is a normal subgroup of G, and M is S-permutable in $G = \langle M, g \rangle$, but L is not S-permutable in $G = \langle L, g \rangle$. It follows that L does not satisfy the S-subpermutiser condition in G.

The definitions of weak S-permutability and the S-subpermutiser condition can be modified in order to obtain characterisations of soluble PT-groups.

Definition 2.3.10. We say that a subgroup H of a group G is *weakly permutable* when the following condition holds: if H is permutable in $\langle H, H^g \rangle$, then H is permutable in $\langle H, g \rangle$.

Definition 2.3.11. We say that a subgroup H of a group G satisfies the *subpermutiser condition* or that H is *transitively permutable* in G when the following condition holds: if H is permutable in K and x is an element of G such that K is permutable in $\langle K, x \rangle$, then H is permutable in $\langle H, x \rangle$.

Weak permutability and the subpermutiser condition extend weak normality and the subnormaliser condition, respectively, to permutability. The following result holds.

Theorem 2.3.12.
1. *If H is a weakly normal subgroup of G, then H is a weakly permutable subgroup of G.*
2. *If H is a weakly permutable subgroup of G, then H is a weakly* S*-permutable subgroup of G.*
3. *If H satisfies the subnormaliser condition in G, then H satisfies the subpermutiser condition in G.*
4. *If H satisfies the subpermutiser condition in G, then H satisfies the* S*-subpermutiser condition in G.*

5. *If H is a weakly permutable subgroup of G, then H satisfies the subpermutiser condition in G.*

6. *If H is weakly permutable in G and H is a subnormal subgroup of a subgroup K of G, then H is permutable in K.*

7. *If H satisfies the subpermutiser condition in G and H is a subnormal subgroup of a subgroup K of G, then H is permutable in K.*

We can give now PT-versions of Theorem 2.3.7 and 2.3.8 (see Ballester-Bolinches and Esteban-Romero [46]).

Theorem 2.3.13. *Let G be a group. The following statements are equivalent:*

1. *G belongs to $\mathcal{X}_p$.*

2. *Every p-subgroup of G satisfies the subpermutiser condition.*

Theorem 2.3.14. *Let G be a group. The following statements are equivalent:*

1. *G is a soluble* PT*-group.*

2. *Every subgroup of G is weakly permutable in G.*

3. *For every prime p, every p-subgroup of G is weakly permutable in G.*

4. *Every subgroup of G satisfies the subpermutiser condition in G.*

5. *For every prime p, every p-subgroup of G satisfies the subpermutiser condition in G.*

2.4 Chief factors of PST-groups

According to Theorem 2.1.8, soluble PST-groups G are characterised as supersoluble groups having all its p-chief factors G-isomorphic when regarded as G-modules by conjugation for every prime p. In the insoluble case, things can be very different. Let G be a non-soluble PST-group with non-trivial soluble residual D. Then, by Theorem 2.1.19, all p-chief factors covered by D are G-isomorphic and all p-chief factors avoided by D are G-isomorphic. Hence G has at most two G-isomorphism classes of p-chief factors. The following example, due to Thompson and presented by Cossey in [105], shows that this number can be two.

Example 2.4.1. Let p be an odd prime and let q be a prime such that $q \equiv 1 \pmod{p}$. Put $D = \mathrm{SL}_p(q^p)$ and define two automorphisms τ and φ of D as follows: $A^\tau = (A^{-1})^T$, while φ arises from the field automorphism $a \mapsto a^q$. Then $X = \langle \tau, \varphi \rangle$ is cyclic of order $2p$ and the semidirect product $[D]X$ is a T-group by Theorem 2.1.28. Nevertheless, $\mathrm{Z}(D)$ and $\langle \varphi D \rangle$ are non-G-isomorphic p-chief factors.

Ballester-Bolinches, Beidleman, and Heineken [33, Theorem 2] took this study further and proved the following stronger result.

Theorem 2.4.2. *Assume that G is a* PST-*group. If H_1/K_1 and H_2/K_2 are isomorphic abelian chief factors of G and $H_1H_2 \le G'$, then they are G-isomorphic.*

The proof of this result will occupy us in much of the current section.

During the proof we shall require the following lemma about simple groups. Although it can be given in a more general form for a wider class of simple groups, we state and prove only the exact result needed for our purposes.

Lemma 2.4.3. *Assume that G is one of the simple groups*

$$\begin{aligned} &A_l(q) = \mathrm{PSL}_{l+1}(q) && \text{where } \gcd(l+1, q-1) \neq 1, \\ &{}^2A_l(q) = \mathrm{PSU}_{l+1}(q) && \text{where } \gcd(l+1, q+1) \neq 1, \\ &E_6(q) && \text{where 3 divides } q-1, \text{ and} \\ &{}^2E_6(q) && \text{where 3 divides } q+1. \end{aligned}$$

Let $\bar{G}$ be the corresponding universal Chevalley group such that $\bar{G}/\mathrm{Z}(\bar{G})$ is isomorphic to G. The natural extension of the action of the group $A = \mathrm{Out}(G)$ of outer automorphisms of G on $\bar{G}$ makes the isomorphism between $\bar{G}/\mathrm{Z}(\bar{G})$ and G to be an A-isomorphism and the group of diagonal outer automorphisms of G and $\mathrm{Z}(\bar{G})$ are A-isomorphic.

Proof. In the proof of this lemma we will follow all notational conventions about simple groups of Lie type given in the book of Carter [90]. We will prove the result for the untwisted groups. For the twisted groups, most of these arguments are valid by replacing the roots by the orbits of the set of roots under the graph symmetry (in this case, only field automorphisms must be considered).

Let $P = \mathbb{Z}\Phi$ be the set of all linear combinations of elements of the set of roots of the corresponding Lie algebra $\mathfrak{L}$ with rational integer coefficients, then P is a free abelian group with a basis consisting of the set of fundamental roots

$$\Pi = \{p_1, p_2, \dots, p_l\}$$

(see Section 7.1 of [90]) and let $\{q_1, q_2, \dots, q_l\}$ be the fundamental weights of the Lie algebra $\mathfrak{L}$. Let Q be the set of all linear combinations of $\{q_1, \dots, q_l\}$ with rational integer coefficients. Then Q is also a free abelian group of rank l, P is a subgroup of Q, and $|Q : P|$ is the determinant of the Cartan matrix of $\mathfrak{L}$. Let $K = \mathrm{GF}(q)$ be the field of q elements.

Let χ be a K-character of P, that is, a homomorphism of G into the multiplicative group K^* of non-zero elements of K. The automorphisms of $\mathfrak{L}_K$ of the form $h(\chi)$ where $h(\chi)h_s = h_s$ and $h(\chi)e_s = \chi(s)e_s$ generate an abelian group $\hat{H}$ which is isomorphic to the group of K-characters of P (see Section 7.1 of [90]). By the note before Lemma 7.1.2 of [90], $\hat{H}$ normalises all roots subgroups H_r of G. According to Theorem 7.1.1 and Section 12.2 of [90], the diagonal automorphisms of G are the maps of the form $g \mapsto h(\chi)gh(\chi^{-1})$, where $h(\chi)$ is an element of $\hat{H}$ for which χ is a K-character of P which cannot be extended to a K-character of Q. It is not difficult to check that the action of these diagonal automorphisms on the generators sends $x_r(t)$ to $x_r(\chi(r)t)$.

Let f be an automorphism of the field K. Then the map $x_r(t) \mapsto x_r(f(t))$ can be extended to an automorphism of G. The field automorphisms of G are the automorphisms obtained in this way.

Let $r \mapsto \bar{r}$ be the map of Φ onto itself arising from a symmetry of the Dynkin diagram of the Lie algebra $\mathfrak{L}$. By Theorem 12.2.3 of [90], there exist numbers $\gamma_r = \pm 1$ such that the map $x_r(t) \mapsto x_{\bar{r}}(\gamma_r t)$ can be extended to an automorphism of G. These automorphisms are the graph automorphisms.

Theorem 12.1.1 of [90] gives a presentation of the universal Chevalley group $\bar{G}$. Call $\bar{Z} = \mathrm{Z}(\bar{G})$, then $\bar{Z}$ is composed of the elements $\prod_{i=1}^{l} \bar{h}_{p_i}(t_i)$ such that $(t_1, \dots, t_l)$ determines a K-character of Q/P.

First of all, we define the action of these automorphisms on the generators of $\bar{G}$ in the natural way: The diagonal automorphism of the form $h(\chi)$ applies $\bar{x}_r(t)$ onto $\bar{x}_r(\chi(r)t)$, the field automorphism defined by f applies $\bar{x}_r(t)$ onto $\bar{x}_r(f(t))$ and the graph automorphism defined by $\tau: r \to \bar{r}$ applies $x_r(t)$ onto $x_{\bar{r}}(\gamma_r t)$. Since these maps are compatible with the relations of Proposition 12.2.1 of [90], they define automorphisms of $\bar{G}$.

Let q_0 be the prime dividing q. The group of field automorphisms of K is cyclic generated by the Frobenius automorphism $\varphi: x \mapsto x^{q_0}$. Therefore the group of field automorphisms of G is cyclic and its generator is induced from this Frobenius automorphism. The third relation of Theorem 12.1.1 of [90] shows that the Frobenius automorphism sends $\bar{h}_r(t)$ onto $\bar{h}_r(t^{q_0}) = \bar{h}_r(t)^{q_0}$.

It is easy to check that the result of the conjugation of $h(\chi)$ by the field automorphism induced by a field automorphism f is the automorphism $h(f \circ \chi)$. Since f is a power of φ, we only have to consider $h(\varphi \circ \chi) = h(\chi)^{q_0}$.

It can be checked that conjugation by the graph automorphism induced by $\tau: r \mapsto \bar{r}$ sends $h(\chi)$ to $h(\chi \circ \tau^{-1})$. On the other hand, we can check that $\bar{h}_r(t)$ is transformed into $\bar{h}_{\bar{r}}(t)$ by the conjugation by the graph automorphism.

On the other hand, we can verify that $\bar{h}_r(t)$ is left invariant by the action of $h(\chi)$. In particular, $\bar{Z}$ is invariant by the action of $h(\chi)$ and, of course, by conjugation by elements of $\bar{G}$.

Suppose that $G = A_l(q)$. Let μ be a generator of K^*. The character χ of P defined by $\chi(p_1) = \mu$, $\chi(p_l) = \mu^{-1}$, and $\chi(p_i) = 1$ if $2 \leq i \leq l-1$ cannot be extended to a character of Q if $m = \gcd(l+1, q-1) \neq 1$. Since m is the least natural number n such that χ^n can be extended to a character of Q, we have that the outer diagonal automorphism induced by χ has order m and this is a generator of the group of outer diagonal automorphisms by [103, Table 5]. On the other hand, $\psi(\bar{q}_i) = \xi^i$ for $1 \leq i \leq l$, where ξ is a primitive mth root of unity, defines a character of Q/P (here $\bar{q}_i = q_i + P$) and so $\prod_{i=1}^{l} \bar{h}_{p_i}(\xi^i)$ defines an element of $\bar{Z}$. Since $\bar{Z}$ has order m, this element is a generator of $\bar{Z}$ (see Table 5 and Section 3.3 of [103]). Since $\chi \circ \tau^{-1} = \chi^{-1}$, we obtain that the graph automorphism induced by the non-trivial symmetry τ of the Dynkin diagram of $\mathfrak{L}$ inverts both $h(\chi)$ and the generator of $\bar{Z}$.

If $G = E_6(q)$ and 3 divides $q - 1$, then the character χ of P given by $\chi(p_1) = \mu$, $\chi(p_6) = \mu^{-1}$, $\chi(p_i) = 1$, $2 \leq i \leq 5$, cannot be extended to a character of Q. It follows that $h(\chi)$ generates the group of outer diagonal automorphisms of G. On the other hand, if ξ is a primitive cubic root of unity in K, then ψ given by $\psi(\bar{q}_1) = \psi(\bar{q}_5) = \xi$, $\psi(\bar{q}_2) = \psi(\bar{q}_6) = \xi^{-1}$, $\psi(\bar{q}_3) = \psi(\bar{q}_4) = 1$ is a character of Q/P and $\bar{z} = \bar{h}_{p_1}(\xi)\bar{h}_{p_2}(\xi)^{-1}\bar{h}_{p_5}(\xi)\bar{h}_{p_6}(\xi)^{-1}$ is a generator of $\bar{Z}$. Once again, we have that the graph automorphism induced by the non-trivial automorphism τ of $\mathfrak{L}$ inverts $h(\chi)$ and $\bar{z}$.

Consequently, in both cases we have that the group of diagonal outer automorphisms is A-isomorphic to $\bar{Z}$, as desired. □

Proof of Theorem 2.4.2. By Proposition 2.1.1, G is an SC-group. Assume that H_1/K_1 and H_2/K_2 are two isomorphic abelian chief factors of G of order p, p a prime, below G'. Since every chief factor of order two is central, we may assume that p is an odd prime. Let D be the soluble residual of G. Recall that, by Theorem 2.1.19, $\mathrm{Z}(D)$ is the soluble radical of D. Assume that D avoids these factors. Then they are G-isomorphic to chief factors of G/D. Since G/D is a soluble PST-group by Theorem 2.1.19, we have that H_1/K_1 and H_2/K_2 are G-isomorphic by Theorem 2.1.8. Now assume that D covers H_1/K_1 and H_2/K_2. Hence $\mathrm{Z}(D)$ covers both factors and they are G-isomorphic to chief factors of G below $\mathrm{O}_p(\mathrm{Z}(D))$. Since, by Theorem 2.1.19, the p'-elements of G induce power automorphisms on $\mathrm{O}_p(\mathrm{Z}(D))$, we can apply Lemma 2.1.3 to conclude that hence H_1/K_1 and H_2/K_2 are G-isomorphic. Therefore we can assume that $K_1 < H_1 \leq \mathrm{Z}(D)$ and $D \leq K_2 < H_2 \leq G'$. Let us choose a normal subgroup M of G maximal subject to $M \cap H_1 = K_1$.

Suppose first that p divides the order of MD/D. By the previous paragraph, all p-chief factors of G avoided by D are G-isomorphic. This allows us to assume that $H_2 \leq MD$. Now MD/D is a normal subgroup of the soluble group G/D, and $M/D \cong M/D \cap M = M/K_1$. This implies that M/K_1 is soluble. Since $\mathrm{Z}(D)/K_1$ is obviously soluble, we have that $M\mathrm{Z}(D)/K_1$ is soluble. By Lemma 2.0.4, $M\mathrm{Z}(D)/K_1$ is a PST-group. On the other hand, the image of H_2/K_2 under the isomorphism between MD/D and M/K_1 is $(H_2 \cap M)/(K_2 \cap M)$. Furthermore $H_1(K_2 \cap M)/(K_2 \cap M)$ is G-isomorphic to $H_1/(K_1 \cap M \cap H_1) = H_1/K_1$. It follows that $H_1(K_2 \cap M)(H_2 \cap M)/(K_2 \cap M) = ((H_2 \cap M)/(K_2 \cap M)) \times (H_1(K_2 \cap M)/(K_2 \cap M))$ is a normal p-subgroup of the PST-group $G/(K_2 \cap M)$. By Lemma 2.1.3, both direct factors are G-isomorphic. Therefore H_2/K_2 and H_1/K_1 are G-isomorphic.

Assume next that p does not divide the order of MD/D. Then we can assume that $MD \leq K_2 < H_2 \leq G$. Since M centralises the chief factors H_1/K_1 and H_2/K_2, we may replace G by G/M and so we may assume that $K_1 = 1$. This means that H_1 is the unique minimal normal subgroup of G. It follows that $O_{p'}(G) = 1$ and $Z = Z(D)$ is a p-group.

Let $C = C_G(D/Z)$. By Proposition 2.1.1 and Corollary 1.6.6, $C = C_G(D)$ is the soluble radical of G. Note that C is a PST-group by Lemma 2.0.4. Moreover $Z \leq C$ and $Z \leq C_G(C)$, because $[Z, C] \leq [D, C] = 1$. Hence by Theorem 2.1.8, the nilpotent residual $C^{\mathfrak{N}}$ of C is abelian and by [47, 4.2.5, 4.2.17] we have that $Z \cap C^{\mathfrak{N}} = 1$. Since G has a unique minimal normal subgroup, we obtain that $C^{\mathfrak{N}} = 1$ and so C is a p-group.

Suppose that H_2/K_2 is G-isomorphic with a chief factor of G between D and CD. Since $CD/D \cong C/Z$, we can assume that $Z \leq K_2 < H_2 \leq C$. Write $H_2 = \langle x_2, K_2\rangle$ and $H_1 = \langle x_1\rangle$. Since C centralises Z, we have that $A = \langle x_1, x_2\rangle$ is abelian and $A \leq C$. Since G is a PST-group, by Theorem 2.1.8 we obtain that p'-elements of G induce power automorphisms in C and hence in A. It follows that H_1 and H_2/K_2 are G-isomorphic.

Thus we can assume that $C = Z$ and that H_2/K_2 is G-isomorphic with a factor of $(G/D)'$. By Theorem 1.6.5, we have that $D/Z = (U_1/Z) \times \cdots \times (U_k/Z)$, where for $1 \leq i \leq k$, U_i/Z is a non-abelian simple normal subgroup of G/D. Now G/D is isomorphic to a subgroup of $\mathrm{Out}(U_1/Z)\times\cdots\times\mathrm{Out}(U_k/Z)$, where $\mathrm{Out}(X)$ denotes the outer automorphism group of the group X. Hence p divides the order of $(\mathrm{Out}(U_i/Z))'$ for some i. Moreover, $U_i' \cap Z \neq 1$ because U_i' is a non-trivial normal subgroup of G and G has a unique minimal normal subgroup contained in Z. Let $S_i = U_i/Z$, then we have that p divides the order of the Schur multiplier $M(S_i)$ (see [157, V, 23.5]).

Here we appeal to the classification of finite simple groups. From the Atlas [103], we see that S_i cannot be a sporadic group. Since p is odd, it follows that S_i is not an alternating group. Therefore we are left only with Chevalley and twisted Chevalley groups. According to [103, Section 3.3 and Table 5], the outer automorphism group is a semidirect product of a cyclic group of a diagonal automorphisms, a cyclic group of field automorphisms and a group of graph automorphisms modulo field automorphisms, except for the groups $B_2(2^f)$, $G_2(3^f)$, and $F_4(2^f)$, in which the extraordinary graph automorphism squares to the generating field automorphism and gives rise to the Suzuki and Ree groups. The group of graph automorphisms is always cyclic except for $D_4(q)$, in which it is isomorphic to the symmetric group of degree 3. Therefore, consulting [103], we see that we have to consider only

$$\begin{aligned}
A_l(q) &= \mathrm{PSL}_{l+1}(q) && \text{if } p \text{ divides } \gcd(l+1, q-1),\\
{}^2A_l(q) &= \mathrm{PSU}_{l+1}(q) && \text{if } p \text{ divides } \gcd(l+1, q+1),\\
E_6(q) & && \text{if } p = 3 \text{ and } 3 \text{ divides } q-1, \text{ and}\\
{}^2E_6(q) & && \text{if } p = 3 \text{ and } 3 \text{ divides } q+1.
\end{aligned}$$

Here, as it is usual when studying simple groups, q is a prime power. We adhere to the "smallest field convention" used in the Atlas [103]. By Lemma 2.4.3, we see that the group of diagonal outer automorphisms of S_i is A-isomorphic to the Schur multiplier, where A is the group of outer automorphisms of S_i. Since the field and graph automorphisms (modulo inner and diagonal automorphisms) induce abelian groups, we have that $(\mathrm{Out}(U_i/Z))'$ is contained in the group of diagonal outer automorphisms. In particular, the p-chief factors of $\mathrm{Z}(D)$ and $G'\mathrm{C}_G(U_i)/\mathrm{C}_G(U_i)U_i$ are G-isomorphic. This completes the proof of the theorem. □

If a subnormal subgroup S of a group G is hypercentrally embedded in G, then S/S_G is contained in the hypercentre of G and so it normalises all Sylow subgroups of G/S_G. In particular, S is S-permutable in G. Therefore the hypercentral embedding of the subnormal subgroups is a sufficient condition for their S-permutability. The next result shows that this condition is also necessary in soluble groups. It was proved by Ballester-Bolinches and Esteban-Romero in [43, Corollary 2].

Theorem 2.4.4. *A soluble group G is a* PST-*group if and only if every subnormal subgroup of G is hypercentrally embedded in G.*

Proof. Only the necessity of the condition is in doubt. Assume that G is a soluble PST-group with a subnormal subgroup H which is not hypercentrally embedded in G. Among all these groups, we choose one with least $|G| + |H|$. Since G/H_G is a PST-group, we can assume that $H_G = 1$. Hence H is nilpotent by Theorem 1.2.14 (2). If H is not of prime power order, then every Sylow subgroup is a core-free subnormal subgroup of G properly contained in H. Hence every Sylow subgroup of H is contained in the hypercentre of G by minimality of H and therefore H is hypercentrally embedded in G, against assumption. Thus H is a p-group for some prime p and so is $T = \langle H^G \rangle$. By Theorem 2.1.8, G belongs to $\mathcal{U}_p^*$. By Theorem 2.2.13, G is p-nilpotent, or G has an abelian Sylow p-subgroup and G satisfies $\mathcal{C}_p$. In the first case, all p-chief factors of G are central and so H is hypercentrally embedded in G. In the second case, by Theorem 2.1.10, H is a normal subgroup of G and so H is hypercentrally embedded in G. In both cases, we obtain a contradiction which completes the proof of the theorem. □

A local result formulated in terms of the subnormal p'-perfect subgroups and the p-nilpotent hypercentre gives a characterisation of the $\mathcal{U}_p^*$-groups (see [43, Theorem B]).

The following example, presented by Ballester-Bolinches, Beidleman, and Heineken in [32], shows that the above condition is not necessary in general.

Example 2.4.5. Let p be an odd prime and let q be a prime such that $q-1$ is divisible by p but not by p^2. The group $\mathrm{SL}_{p^2}(q^p)$ possesses a duality automorphism δ (which maps every matrix onto the inverse of its transpose) of order 2 and an automorphism σ of order p arising from applying the field automorphism to every matrix entry. Since $q^p - 1 = (1+np)^p - 1 = \sum_{k=1}^{p} \binom{p}{k} (np)^k$ is divisible by p^2, we obtain that

$\gcd(p^2, q^p - 1) = p^2$ and so the centre of $\mathrm{SL}_{p^2}(q^p)$ is cyclic of order p^2. Now let $H \cong \mathrm{SL}_{p^2}(q^p)$ and

$$K = \langle H, d, s : [d, s] = d^2 = s^p = 1; d^{-1}hd = \delta(h), s^{-1}hs = \sigma(h) \text{ for all } h \in H\rangle.$$

We choose the subgroup $L = \langle d_1 d_2, s_1 s_2^{-1}, H_1, H_2\rangle$ of the direct product $K_1 \times K_2$ of two copies of K. We have that $\mathrm{Z}(L) = 1$ because $d_1 d_2$ inverts by conjugation all elements of $\mathrm{Z}(H_1 H_2)$. On the other hand, if t_1, t_2 are generators of $\mathrm{Z}(H_1)$, $\mathrm{Z}(H_2)$, respectively, then $[t_1 t_2, s_1 s_2^{-1}] = t_1^{kp} t_2^{-kp}$ for some k prime to p such that $q = 1+kp$. It is easy to see that

$$\langle t_1 t_2, t_1^p t_2^{-p}\rangle = \langle \langle t_1 t_2\rangle^L\rangle \leq \mathrm{Z}(H_1 H_2),$$
$$\langle t_1^p t_2^p\rangle = \langle t_1 t_2\rangle_L,$$

and so $\langle t_1 t_2\rangle$ is subnormal and not hypercentrally embedded in L. On the other hand, L is a PST-group.

The subnormal subgroups of PST-groups G, in which the isomorphic abelian chief factors are G-isomorphic, are hypercentrally embedded in G as the following theorem, proved by Ballester-Bolinches, Beidleman, and Heineken [32, Theorem 3], shows.

Theorem 2.4.6. *Assume that G is a* PST-*group and that all abelian isomorphic chief factors of G are G-isomorphic. Then all subnormal subgroups of G are hypercentrally embedded in G.*

Proof. We suppose that the result is false and that (G, S) is a counterexample with S a non-hypercentrally embedded subnormal subgroup with $|G| + |S|$ is minimal. Then $1 \neq S$ is S-permutable in G. Since G/S_G satisfies the hypotheses of the theorem, we can assume that $S_G = 1$. In this case, S nilpotent by Theorem 1.2.14 (2). If S is not of prime power order, then every Sylow subgroup is a core-free subnormal subgroup of G properly contained in S. Hence every Sylow subgroup of S is contained in the hypercentre of G by minimality of S. Therefore, S is hypercentrally embedded in G, against assumption. Consequently, S is a p-group for some prime p and so is $T = \langle S^G\rangle$. Let X be the centraliser of a minimal normal subgroup N of G contained in T. Clearly the hypotheses of the theorem hold in X and so X is p-nilpotent. Since N is cyclic, G/X is abelian of exponent dividing $p - 1$. In particular, G is soluble. This contradicts the above theorem. □

As it was proved in Section 1.5, the hypercentrally embedded subgroups and subgroups permuting with all the pronormal subgroups are the same thing. Therefore, we have:

Corollary 2.4.7. *Assume that G is a* PST-*group and that all abelian isomorphic chief factors of G are G-isomorphic. Then all subnormal subgroups of G permute with all the pronormal subgroups of G.*

Chapter 3

Products of nilpotent groups

Undoubtedly two of the most important basic results in products of finite groups are the theorem of Burnside, which states that groups whose orders contain at most two prime divisors are soluble, and the Hall characterisation of soluble groups by the existence of Sylow complements. Hall's result, which calls for an application of Burnside's theorem, implies in particular that a group is soluble if and only if it is the product of pairwise permutable Sylow subgroups. Having this result in mind, the question of finding structural theorems for groups which are the pairwise permutable products of nilpotent groups naturally arose. The ground-breaking theorems of Hall and Higman [143] could be regarded as the prototypes of such results. One of them shows that the derived length of a soluble group which is the product of pairwise nilpotent subgroups of coprime orders is bounded by the sum of the nilpotency classes of the factors. However, the first question to be answered was the solubility of a group which is the product of pairwise permutable nilpotent subgroups. Wielandt in [254] began to study products of nilpotent groups with a series of results which he finished in [255]. His main theorem, together with the Kegel's main one [181], answered affirmatively to this question.

By the time of the appearance of Wielandt's papers an investigation of products of cyclic and abelian groups was under way. Huppert [154] showed that a group which is the pairwise permutable product of cyclic subgroups is supersoluble, and Itô [162] shows that any (not necessarily finite) group is metabelian whenever it is the product of two abelian subgroups. This result has provided the inspiration for many variations, which have extended and deepened our knowledge of the theory's scope and limitations. We shall give a full account many of these results in Section 3.1.

The structure of a group which is a product of two nilpotent groups has been extensively studied by a number of authors, with many interesting results available. This will be the main concern of Section 3.2.

In an attempt to extend the aforesaid theorem of Hall and Higman, one can wonder whether the derived length of a group which is the product of pairwise nilpotent subgroups should be bounded by a function of the nilpotency classes of the factors, and in particular that the bound should be that given by Hall and Higman, the sum of the nilpotency classes of the factors. The case of a product of two subgroups is especially important. The first occurrence of this conjecture we know is of Kegel [183]. It was in the late 1990s that Cossey and Stonehewer [107] produced examples to show that

Kegel's conjecture is false in general. Since the soluble groups with derived length at most d form a formation, a minimal counterexample to this conjecture must have a unique minimal normal subgroup and so in particular its Fitting subgroup must be a prime power order group. It seemed natural then to consider products of two nilpotent groups in which the Fitting subgroup is a p-group for some prime p. Cossey and Stonehewer [108] began to consider these groups in detail. In fact, the authors recognise that counterexamples to Kegel's conjecture came from the analysis of the structure of such groups. Another approach to this question was started by Kazarin [170], who showed that the derived length of a product of two nilpotent subgroups could be bounded by the orders of the derived subgroups of the factors. Several authors have improved on his bounds in some special cases. Another invariant of nilpotent groups which would seem relevant to the derived length of a product is the derived length of the factors and this was also considered by Kazarin in [176]. He showed that the derived length of a soluble product of two subgroups of coprime order can be bounded by the derived lengths of the factors and perhaps not surprisingly a better bound can be found if the product has odd order. However his bounds are not best possible. We shall take a close look at these results in Section 3.3.

3.1 Products of abelian groups

Groups which are factorised as a product of two abelian subgroups are the main theme of this section. We investigate some of their properties, concentrating on results which depend specifically on its Sylow and normal structure.

We will begin with the particular case of a product of two cyclic groups. These factorised groups will play a relevant role in the study of totally permutable products in Chapters 4 and 5.

Our first result is due to Douglas [121] and Itô [163].

Lemma 3.1.1. *Let the group $G = \langle a\rangle\langle b\rangle$ be the product of the cyclic subgroups $\langle a\rangle$ and $\langle b\rangle$. If $|G| > 1$ and $|\langle a\rangle| \geq |\langle b\rangle|$, then $\langle a\rangle$ contains a non-identity normal subgroup of G. Moreover, G is supersoluble.*

Proof. We derive a contradiction by supposing that $\langle a\rangle_G = 1$. Then $\langle a\rangle \cap \langle b\rangle = 1$ and $b \notin \langle a\rangle$. We know that a induces a permutation on the right coclasses of $\langle a\rangle$ in G of the form

$$\begin{pmatrix} \langle a\rangle & \langle a\rangle b & \langle a\rangle b^2 & \dots \\ \langle a\rangle & \langle a\rangle ba & \langle a\rangle b^2 a & \dots \end{pmatrix}.$$

We consider a cycle $(\langle a\rangle b, \langle a\rangle ba, \dots, \langle a\rangle ba^{r-1})$ of this permutation. Note also that $\langle a\rangle$ cannot appear in this cycle because this element is fixed by multiplication by a. Moreover $|G : \langle a\rangle| = |\langle b\rangle : \langle a\rangle \cap \langle b\rangle| = |\langle b\rangle|$. Hence $r \leq |\langle b\rangle| - 1 < |a|$. Since $\langle a\rangle ba^r = \langle a\rangle b$, it follows that $ba^r b^{-1} \in \langle a\rangle$. Consequently $a^r \neq 1$ and thus

$\langle a^r \rangle$ is a normal subgroup of $\langle a, b \rangle = G$ contained in $\langle a \rangle$, which yields the desired contradiction.

Since every homomorphic image of G is again a product of two cyclic groups, the supersolubility of follows from the previous argument by induction on $|G|$. □

The alternating group of degree 4 shows that the product of two abelian subgroups is not supersoluble in general.

Lemma 3.1.1 was extended by Huppert in [154] to products of pairwise permutable cyclic subgroups. The proof can be found in [157, VI, 10.3].

Theorem 3.1.2. *Let the group $G = G_1 G_2 \cdots G_r$ be the product of the cyclic groups $G_1, G_2, \ldots, G_r$. Then G is supersoluble.*

The following particular case was studied by the same author [154] (see [157, III, 11.5]).

Theorem 3.1.3. *Let $G = \langle a \rangle \langle b \rangle$ be a p-group with $p > 2$. Then G is metacyclic.*

There are 2-groups which are product of two cyclic groups in which the derived subgroup is not cyclic as the following example shows.

Example 3.1.4. Consider the group

$$G = \langle a, b, c : a^8 = 1, b^2 = a^4, c^2 = 1, b^a = bc, c^a = ca^4, c^b = c \rangle$$

satisfies that $G = \langle a \rangle \langle ab \rangle$ is a product of two cyclic subgroups of order 8, but its derived subgroup is $\langle a^4, c \rangle$, which is elementary abelian of order 4.

Our next observation shows that the derived subgroup of a product of two abelian groups is just the commutator subgroup of the factors.

Lemma 3.1.5. *Let the group $G = AB$ be the product of two abelian groups A and B. Then $G' = [A, B]$.*

Proof. Note that $[A, B]$ is a normal subgroup of G by [157, I, 1.6] and is contained in the derived subgroup of G. In order to see that $G' = [A, B]$, we note that

$$G/[A, B] = (A[A, B]/[A, B])(B[A, B]/[A, B])$$

is a central product of $A[A, B]/[A, B]$ and $B[A, B]/[A, B]$. Since both subgroups are abelian, $G/[A, B]$ is abelian and so $G' \leq [A, B]$. □

We can put the above lemma to immediate use in the following result of Mann [197], which is concerned with a generalisation of the celebrated theorem of Itô. In fact, its proof is the same as the one given by Itô in [162].

Theorem 3.1.6. *Let A, B, V, and W be subgroups of a group G satisfying $[A, V] = [B, W] = 1$, $VB = BV$, and $WA = AW$. Then $[[V, W], [A, B]] = 1$.*

Proof. Let $a \in A$, $b \in B$, $a_1 \in V$, $b_1 \in W$, and put $b^{a_1} = a_2b_2$, $a^{b_1} = b_3a_3$, where $a_2 \in V$, $a_3 \in A$, $b_2 \in B$, and $b_3 \in W$. Then

$$[a,b]^{a_1b_1} = [a,b^{a_1}]^{b_1} = [a,a_2b_2]^{b_1} = [a,b_2]^{b_1} = [a^{b_1},b_2] = [b_3a_3,b_2] = [a_3,b_2],$$

$$[a,b]^{b_1a_1} = [a^{b_1},b]^{a_1} = [b_3a_3,b]^{a_1} = [a_3,b]^{a_1} = [a_3,b^{a_1}] = [a_3,a_2b_2] = [a_3,b_2].$$

The statement follows immediately. □

The celebrated theorem of Itô [162] about product of two abelian groups follows as special case of this theorem.

Theorem 3.1.7. *Let the group $G = AB$ be the product of two abelian subgroups A and B. Then G is metabelian.*

Proof. Take $V = A$ and $W = B$ in Theorem 3.1.6. □

At the time of writing a satisfactory generalisation of the previous theorem to pairwise permutable products of more than two factors seems to be unresolved.

Conder and Isaacs [102] proved, as a consequence of a more general result, that if either of the subgroups A or B in the previous theorem is cyclic, $G'/(G' \cap A)$ is isomorphic to a subgroup of B.

We now state and prove a more detailed description of the structure of a product of two cyclic subgroups ([95, Lemma 3.7]).

Lemma 3.1.8. *Let the non-trivial group $G = AB$ be the product of two cyclic subgroups, and let p be the largest prime dividing $|G|$. Then the following statements hold:*

1. *Either A or B contains some normal subgroup of G of order p.*
2. *Any q-subgroup of A permutes with any q-subgroup of B for all primes q.*
3. *If p does not divide $|B|$, then a Sylow p-subgroup of A is normal in G.*

Proof. 1. Assume that the result is false and let G be a counterexample of least order. Since $G \neq 1$, we can appeal to Lemma 3.1.1 to conclude that there exists a minimal normal subgroup K of G contained in either A or B. We may assume without loss of generality that $K \leq A$. The fact that A is cyclic implies that K has prime order, and minimality of G as a counterexample yields $|K| < p$. Clearly the hypothesis of the lemma is satisfied by $G/K = (A/K)(BK/K)$. Since $|G/K| < |G|$, G/K has a normal subgroup H/K of order p such that H/K is contained in either A/K or BK/K. Since $|K| < p$, it follows that K is contained in $Z(H)$. Then H is cyclic.

Let P be a subgroup of H of order p. Then P is normal in H and so P is normal in G. Since G is a counterexample, H is not contained in A. Consequently $H/K \leq BK/K$ and so $H \leq KB$. Therefore $H = K(H \cap B)$ and $P \leq H \cap B$. This contradiction proves Statement 1.

2. Assume, by way of contradiction, that the result fails, and let G be a counterexample of least order. Let A_1 and B_1 be q-subgroups of A and B, respectively. Clearly, $A_1 \neq 1$ and $B_1 \neq 1$. Since $G \neq 1$, by Lemma 3.1.1, there is a normal subgroup R of G of prime order contained in either A or B. We can assume without loss of generality that $R \leq A$. Once again the hypothesis is inherited by G/R. The minimality of G implies that A_1R/R permutes with B_1R/R. Therefore $(A_1R)B_1$ is a subgroup of G. If $(A_1R)B_1 \neq G$, then the choice of G and the fact that $(A_1R)B_1$ is the product of the cyclic subgroups A_1R and B_1 ensures that $A_1B_1 = B_1A_1$, a contradiction. Therefore $G = (A_1R)B_1$. Suppose that $|R| = q$, then $R \leq A_1$ because A is cyclic. Hence $G = A_1B_1$ and we reach the same contradiction. Consequently, $|R| = r \neq q$, and thus R is a Sylow r-subgroup of G. If $G' \leq R$, then A_1R is normal in G. Therefore A_1 is normal in G and A_1 permutes with B_1. This contradiction shows that G' is not contained in R. By Theorem 3.1.7, G' is abelian. Then the Sylow q-subgroup Q of G' is normal in G. It is clear that $Q \neq 1$ because G' is not contained in R. Hence the hypothesis holds in the group $G/Q = (AQ/Q)(BQ/Q)$ and so A_1Q/Q permutes with B_1Q/Q by the minimal choice of G. Therefore A_1B_1Q is a subgroup of G. In particular, $\langle A_1, B_1\rangle$ is a q-group, and $\langle A_1, B_1\rangle \cap R = 1$. Since $G = A_1B_1R$, we have that $G = (A_1B_1)R = \langle A_1, B_1\rangle R$. It follows that $\langle A_1, B_1\rangle = A_1B_1$, that is, A_1 and B_1 permute, contrary to supposition. Therefore, Statement 2 holds.

3. Assume that, on the contrary, the result is false. Let G be a counterexample of least order. Since p does not divide $|B|$, A contains a normal subgroup R of G such that $|R| = p$ by Statement 1. Let P be a Sylow p-subgroup of A. Then $R \leq P$. It is clear that the hypothesis is inherited by G/R. Since $|G/R| < |G|$, a Sylow p-subgroup P/R of A/R is normal in G/R. Then P is normal in G, a contradiction which shows Statement 3. □

The following result appears as [95, Corollary 3.7] and enables us to give the first example of a totally permutable product.

Corollary 3.1.9. *Let the group $G = AB$ be the product of the cyclic subgroups of prime-power order A and B. Then every subgroup of A permutes with every subgroup of B.*

Proof. Suppose that A is a p-group and B is a q-group. Suppose that $p = q$, by Lemma 3.1.8 (2), all subgroups of A permute with all subgroups of B. Suppose now that $p \neq q$. There is no loss of generality in assuming $p > q$. By Lemma 3.1.8 (3), A is normal in G. Since all subgroups of A are characteristic in A, we obtain that all subgroups of A are normal in G. Consequently, all subgroups of A permute with all subgroups of B, as desired. □

The following lemma provides a sufficient condition for a product of two cyclic groups of prime power order to be abelian ([147, Lemma 2.3]).

Lemma 3.1.10. *Let the group $T = \langle x\rangle\langle y\rangle$ be the product of two cyclic q-groups, with q an odd prime. Assume that H is a q'-group of automorphisms of T such that $T = [H, T]$ and $\langle x\rangle$ and $\langle y\rangle$ are H-invariant. Then T is an abelian group.*

Proof. According to Theorem 3.1.3, T is metacyclic, that is, there exists a normal subgroup A of T such that A and T/A are cyclic. We now deduce that T is an Iwasawa group by Lemma 1.4.5. Moreover, since q is odd, T does not involve the quaternion group Q_8 of order 8.

Assume that T is non-abelian. Since T is an Iwasawa group and does not involve Q_8, by [222, 2.3.23] there exist characteristic subgroups R and S of T such that $\Phi(T) \leq S < R$ and $[R, \mathrm{Aut}(T)] \leq S$. Since $T = [H, T]$, it follows that R is a proper subgroup of T. On the other hand, $T/\Phi(T) \cong C_q \times C_q$ and so $|T : R| = q$ and $S = \Phi(T)$. Consequently $[R, H] \leq \Phi(T)$. Moreover, $R = [R, H]\mathrm{C}_R(H)$ by [119, A, 12.5]. We also know that any two of the subgroups of T permute. We may assume that $\langle x\rangle$ is not contained in R. Then

$$T = R\langle x\rangle = \Phi(T)\mathrm{C}_R(H)\langle x\rangle = \mathrm{C}_R(H)\langle x\rangle.$$

Hence

$$T = [H, T] = [H, \mathrm{C}_R(H)\langle x\rangle] = [H, \langle x\rangle] \leq \langle x\rangle,$$

a contradiction which proves the result. □

The following lemma will be needed in Theorem 5.1.16 to give information about totally permutable products of groups of prime power order.

Lemma 3.1.11. *Let p_1, p_2, and p_3 be primes, $p_2 > p_3$, and suppose that a group G is factorised as the product of pairwise permutable cyclic p_i-subgroups H_i for $i \in \{1, 2, 3\}$. If H_1H_2 and H_2H_3 are non-abelian, then the following statements hold:*

1. *$G = [H_2](H_1H_3)$.*
2. *$p_1 \leq p_2$.*
3. *If $p_1 = p_2$, then $H_1H_3 = H_1 \times H_3$.*

Proof. First of all, since $p_2 > p_3$, it follows that H_2 is normal in H_2H_3 because of Lemma 3.1.8 (3). Then $(H_2H_3)' \leq H_2$. Suppose that $(H_2H_3)' < H_2$. Since H_2 is cyclic, $(H_2H_3)' \leq \Phi(H_2)$, and since H_2 is normal in H_2H_3, we have that $(H_2H_3)' \leq \Phi(H_2H_3)$ by [119, A, 9.2 (d)]. Then [119, A, 9.3 (b)] implies that H_2H_3 is nilpotent and so H_2H_3 is abelian, a contradiction. Thus

$$(H_2H_3)' = H_2. \tag{3.1}$$

Now let P be a subgroup of H_2 with $|P| = p_2$. Since H_2 is normal in H_2H_3 and P is characteristic in H_2, we have that P is a normal subgroup of PH_3 and $(PH_3)' \leq P$. Suppose that PH_3 is abelian. Then since H_2 is abelian, $P \leq \mathrm{Z}(H_2H_3)$. By [119, A, 12.5], $H_2 = [H_2, H_3] \times (H_2 \cap \mathrm{Z}(H_2H_3))$. Since H_2 is cyclic, $H_2 \leq \mathrm{Z}(H_2H_3)$. But then H_2H_3 is abelian. Thus

$$(PH_3)' = P. \tag{3.2}$$

Analogously, if $p_1 > p_3$, then H_1 is a normal subgroup of H_1H_3, and, since H_1H_3 is non-abelian, we conclude that $(H_1H_3)' = H_1$.

Assume that Statement 1 fails to hold and derive a contradiction. Let G be a group of minimal order subject to the condition that H_2 is not a normal subgroup of G. Suppose that $p_1 < p_2$. Then H_2 is normal in H_2H_3 and in H_1H_2. Hence H_2 is a normal subgroup of G. This contradiction shows that $p_1 \geq p_2$. By Corollary 3.1.9, $PH_1 = H_1P$. Obviously $|H_1P : H_1| \leq p_2$. Since p_2 is the smallest prime dividing $|H_1P|$, it follows that H_1 is a normal subgroup of H_1P. On the other hand, since $p_1 \geq p_2 > p_3$, H_1 is normal in H_1H_3 by Lemma 3.1.8 (3). Therefore H_1 is a normal subgroup of $K = H_1PH_3$. Since H_1 is cyclic, $K/\mathrm{C}_K(H_1)$ is abelian. Then, by (3.2), we have that $P = (PH_3)' \leq K' \leq \mathrm{C}_K(H_1)$. Thus $H_1H_2 \leq \mathrm{C}_G(P)$ and P is a normal subgroup of $G = H_1H_2H_3$.

If $(H_1P/P)(H_2/P)$ and $(H_2/P)(H_3P/P)$ are non-abelian, then G/P inherits the hypothesis and since $|G/P| < |G|$, H_2/P is a normal subgroup of G/P by the minimal choice of G, contrary to assumption. If $(H_2/P)(H_3P/P) = H_2H_3/P$ is abelian, then in view of (3.1), $H_2 = (H_2H_3)' \leq P$ and H_2 is normal in G, another contradiction. Suppose now that $(H_1P/P)(H_2/P)$ is abelian. Since H_2 is normal in H_2H_3, we obtain that H_2/P is a normal in $(H_1/P)(H_2/P)(H_3P/P)$, and so H_2 is a normal subgroup of G. This is the desired contradiction. Hence H_2 is a normal subgroup of G and Statement 1 holds.

Since H_2 is a cyclic p_2-group, the divisors of $|\mathrm{Aut}(H_2)|$ are not greater than p_2. The non-abelian character of the subgroup H_1H_2 ensures that $p_2 \geq p_1$, and Statement 2 is proved.

Suppose now that $p_1 = p_2$. Since H_2 is cyclic, we have that $G/\mathrm{C}_G(H_2)$ is abelian. Hence $\mathrm{C}_G(H_2) \geq G' \geq (H_1H_3)'$. Since H_1 is not contained in $\mathrm{C}_G(H_2)$, $H_1 \neq (H_1H_3)'$. We now repeat the first argument of the proof and conclude that H_1H_3 is nilpotent. Hence $H_1H_3 = H_1 \times H_3$ and Statement 3 holds. □

We now state an interesting result of Itô [162, Satz 2 (II)] about products of abelian groups. It might be seen as an extension of Lemma 3.1.1. Its proof appears in [9, 2.1.4].

Theorem 3.1.12. *Let the group $G = AB$ be the product of two abelian subgroups A and B. If G is not trivial, then A or B contains a non-trivial normal subgroup of G.*

The next result is proved in [9, 2.1.5, 2.1.6, and 2.1.7]. The first and the second parts are due to Amberg [6], while the last part was first proved by Itô [162, Satz 2 (II)].

Theorem 3.1.13. *Let the group $G = AB$ be the product of two distinct abelian subgroups A and B.*

1. *If G does not have prime order, then it contains a factorised normal subgroup N with $1 \neq N \neq G$.*

2. *Every maximal proper factorised normal subgroup of G contains one of the factors A and B.*

3. *There exists a proper normal subgroup of G containing A or B.*

We note that the last part of Theorem 3.1.13 holds in a more general context, namely for products of two nilpotent groups (see Theorem 3.2.27).

We end the section with a result of Heineken [148] which gives information about the index of Fitting subgroup of a product of two abelian subgroups. It is even valid for infinite groups provided the subgroup B is finite.

Theorem 3.1.14. *Assume that the group $G = AB$ is the product of two abelian subgroups A and B. Then the index of the Fitting subgroup $\mathrm{F}(G)$ of G in G is not greater than $|B|$.*

Proof. It is enough to find a nilpotent subnormal subgroup of G whose index is not greater than $|B|$. We will argue by induction of $|B|$. The statement holds trivially in the case $B = 1$, so we assume that the result is proved for all pairs of nilpotent subgroups (U, V) such that $UV = VU$ and $|V| < |B|$.

Let $W = A_G$. Since $A \leq \mathrm{C}_G(W)$, we obtain that $\mathrm{C}_G(W) = AB \cap \mathrm{C}_G(W) = A(B \cap \mathrm{C}_G(W))$. We distinguish two cases:

Assume first that $\mathrm{C}_G(W) \neq W$. In this case, $B \cap \mathrm{C}_G(W) \neq B$ and we can use the induction hypothesis for $\mathrm{C}_G(W)$. There is a nilpotent subnormal subgroup S of $\mathrm{C}_G(W)$ such that $|\mathrm{C}_G(W) : S| \leq |B \cap \mathrm{C}_G(W)|$. We obtain

$$\begin{aligned} |G : S| &= |G : \mathrm{C}_G(W)|\,|\mathrm{C}_G(W) : S| = |B : B \cap \mathrm{C}_G(W)|\,|\mathrm{C}_G(W) : S| \\ &\leq |B : B \cap \mathrm{C}_G(W)|\,|B \cap \mathrm{C}_G(W)| = |B|. \end{aligned}$$

Now assume that $\mathrm{C}_G(W) = G$. By Theorem 3.1.12, one of the factors A/W and BW/W contains a non-trivial normal subgroup of G/W, V/W say. Since A/W is core-free in G/W, it follows that V/W is contained in BW/W. Then $V = WB \cap V = W(B \cap V)$. In particular, $B \cap V \neq 1$. Since $G/V = (AV/V)(BV/V)$ and $BV/V \cong B/(B \cap V)$, the inductive hypothesis is valid for G/V. Hence there is a nilpotent subnormal subgroup R/V such that $|G : R| = |(G/V) : (R/V)| \leq |B/(B \cap V)|$. Let $C/W = \mathrm{C}_G(V/W)$. Since BW/W centralises C/W, we have in particular that $B \leq C$ and $G/C = AC/C \cong A(A \cap C)$ is abelian. Since V/W is an abelian minimal normal subgroup of G/W and the group $(G/W)/(C/W) \cong$

G/C of automorphisms induced by G/W on V/W is also abelian, we conclude that $|G:C| < |V:W|$ by [119, B, 9.8]. Hence

$$\begin{aligned} |G : R \cap C| &\le |G : R|\,|G : C| \le |B : B \cap V|\,|V : W| \\ &= |B : B \cap V|\,|B \cap V : B \cap W| \le |B|. \end{aligned}$$

(note that $|B \cap V : B \cap W| = |V : W|$ by [119, A, 1.7]). The nilpotency of the subnormal subgroup $R \cap C$ now follows: $(R \cap C)/W \le R/V$ is nilpotent, $W \le \mathrm{Z}(G)$, and $[R \cap C, V] \le [C, V] \le W$. This completes the proof of the theorem. □

3.2 The subgroup structure of a dinilpotent group

In this section we focus on the subgroup structure of a group that can be written as the product of two nilpotent groups; following Stonehewer [231] we call such a group *dinilpotent group*. The milestone is the theorem that every dinilpotent group is soluble. This was proved by Wielandt [255] in the coprime case and by Kegel [180] in the general one. Both undertakings, especially the coprime one, are labour-intensive. We summarise their results here.

Theorem 3.2.1. *Every dinilpotent group is soluble.*

Proof. See [157, VI, 4.3] or [9, 2.4.3]. □

Corollary 3.2.2. *Let $G = G_1 G_2 \cdots G_r$ be the product of the pairwise permutable nilpotent subgroups $G_1, G_2, \dots, G_r$. Then G is soluble.*

Proof. See [9, 2.4.4]. □

An interesting extension of the Kegel–Wielandt theorem was proved by Kazarin in [168]. It confirms a conjecture of Scott and Shemetkov.

Theorem 3.2.3. *Let the group $G = AB$ be the product the subgroups A and B. If A and B have nilpotent subgroups of index at most two, then G is soluble.*

Suppose that a group G is dinilpotent. Then $G = AB$ is a product of two nilpotent subgroups A and B. Since G is soluble, it follows that G satisfies D_π for every set of primes π ([119, A, 3.3)]). Therefore, if A_π and B_π are the Hall π-subgroups of A and B respectively, we can apply Theorem 1.1.19 to deduce that A_π permutes with every conjugate of B_π in G. Hence, by Lemma 1.1.20, $[\langle A_\pi^G \rangle, \langle B_\pi^G \rangle] = 1$ provided that $\mathrm{O}_\pi(G) = 1$. Since $A \cap B$ is subnormal in A and B, by Theorem 1.1.7 it follows that $A \cap B$ is subnormal in G and so it is contained in $\mathrm{F}(G)$. These facts are turn out to be crucial for deriving information about dinilpotent groups and they will be applied in the sequel without further reference.

The behaviour of the minimal normal subgroups of a dinilpotent group is analysed in the following result, due to Stonehewer [231]. It will be useful for our immediate purposes and for later applications.

Theorem 3.2.4. *Let the group $G = AB$ be the product of the nilpotent subgroups A and B. If N is a minimal normal subgroup of G, then either AN or BN is nilpotent.*

Proof. Assume that the result is false and let G be a counterexample of least order. Suppose first that L is a minimal normal subgroup of G different from N. Since $G/L = (AL/L)(BL/L)$ is a dinilpotent group of smaller order than G, it follows from the minimal choice of G that either ANL/L or BNL/L is nilpotent. Suppose ANL/L nilpotent. Then $ANL/L \cong AN/(L \cap AN)$ is nilpotent and $AN/N \cong A/(A \cap N)$ is nilpotent. Therefore $AN/(L \cap AN \cap N) \cong AN$ is nilpotent, a conclusion which contradicts our supposition. We can then conclude that N is the unique minimal normal subgroup of G.

Since G is soluble, it follows that N is an elementary abelian p-group for some prime p. Consequently, $\mathrm{O}_{p'}(G) = 1$. If either A or B is a p-group, then either AN or BN is a p-group. This contradiction implies that $A_{p'} \neq 1$ and $B_{p'} \neq 1$. Moreover, $\mathrm{O}_{p'}(G) = 1$. It then follows that $[\langle A_{p'}^G \rangle, \langle B_{p'}^G \rangle] = 1$. Then $N \leq \langle A_{p'}^G \rangle \cap \langle B_{p'}^G \rangle$ is centralised by $A_{p'}$ and $B_{p'}$. Let A_p and B_p the Sylow p-subgroups of A and B respectively. Then $P = A_p B_p$ is a Sylow p-subgroup of G and N is a minimal normal subgroup of the P. In particular, N is central in P. Consequently $N \leq \mathrm{Z}(G)$. It follows that AN is nilpotent, and therefore the theorem is proved. □

As an application of Theorem 3.2.4, we offer a result which says that certain types of chief factors of dinilpotent groups ensure the existence of minimal normal subgroups of the same type. To this end we need the following observation.

Lemma 3.2.5. *Let p be a prime. Let N be an abelian normal p-group of a soluble group G. If N contains no central chief factors of G, then the Hall p'-subgroup $G_{p'}$ of G acts fixed point freely on N.*

Proof. By [119, A, 12.5], $N = \mathrm{C}_N(G_{p'}) \times [N, G_{p'}]$. Suppose that $\mathrm{C}_N(G_{p'}) \neq 1$. Let n be a non-trivial element of $\mathrm{C}_N(G_{p'})$. Let

$$1 = N_0 < N_1 < N_2 < \cdots < N_t = N$$

be a part of a chief series of G passing through N. Since n is a non-trivial element of N, there exists $r \leq t$ such that $n \in N_t \setminus N_{t-1}$. Then $\mathrm{C}_{N_t/N_{t-1}}(G_{p'}N_{t-1}/N_{t-1}) \neq 1$. By Dade's theorem [119, B, 11.4], it follows that N_t/N_{t-1} is centralised by $G_{p'}$. It follows that $G/\mathrm{C}_G(N_t/N_{t-1})$ is a p-group and so N_t/N_{t-1} is a central chief factor of G below N, a contradiction which shows that $N = [N, G_{p'}]$ and so $G_{p'}$ acts fixed point freely on N. □

Lemma 3.2.6. *Let $G = AB$ be a dinilpotent group and let p be a prime. Suppose that N is a normal p-subgroup of G which contains no central chief factors of G and that K/L is a chief factor of G with $K \leq N$ and AK/L nilpotent. Then there is a minimal normal subgroup M of G with $M \leq N$ and AM nilpotent.*

Proof. If B is a p-group, the result is vacuously true, since a chief factor K/L of G in N, with AK/L nilpotent, would have its centraliser supplemented in G by a Sylow p-subgroup of G and so it would be central in G. If A is a p-group, the result is true, because AM is nilpotent for every normal p-subgroup M of G. Thus we may assume that neither A nor B is a p-group.

We suppose that the result is not true and choose G satisfying the hypothesis, but not the conclusion of the lemma, with $|G|$ as small as possible. We suppose also that among such G we choose one with $|L|$ as small as possible. Then we have $|L| \neq 1$.

Let U be a minimal normal subgroup of G contained in L. Then G/U satisfies the hypotheses of the lemma, and so by minimality of G there is a chief factor V/U with AV/U nilpotent. It follows from the minimality of L that $L = U$. Note that AL is not nilpotent and hence, by Theorem 3.2.4, we must have BL nilpotent. If K is not elementary abelian, then $\Phi(K) \neq 1$ and $\Phi(K)$ is normal in G, giving $\Phi(K) = L$. But then the Hall p'-subgroup of A, $A_{p'}$, centralises K/L and so, by [119, A, 12.7], $A_{p'}$ centralises K. In particular, $A_{p'}$ centralises L, against the choice of L. Therefore K must be elementary abelian and we may regard K as a $\mathrm{GF}(p)G$-module.

As a module for the Hall p'-subgroup $G_{p'} = A_{p'}B_{p'}$ of G, L has a complement in K, V say, by Maschke's theorem ([119, A, 11.5]). Since BL is nilpotent, $L \leq \mathrm{C}_G(B_{p'})$. Let $X = \mathrm{C}_L(A_{p'}) \cap \mathrm{C}_L(B_{p'}) = \mathrm{C}_L(A_{p'})$. Suppose that $X \neq 1$. Then $X \leq \mathrm{C}_L(G_{p'})$ and so $\mathrm{C}_L(G_{p'}) \neq 1$. This contradicts Lemma 3.2.5. Hence $X = 1$. Now AK/L is nilpotent by hypothesis. Therefore $[K/L, A_{p'}L/L] = 1$. Consequently $[K/L, G_{p'}L/L] = [K/L, B_{p'}L/L]$, and $[K, B_{p'}]L = [K, G_p]L$. By [119, A, 12.5], we have $K = \mathrm{C}_K(A_{p'}) \times [K, A_{p'}]$. Since $[K/L, A_{p'}L/L] = 1$, we deduce that $[K, A_{p'}] \leq L$, and so $L = [K, A_{p'}](L \cap \mathrm{C}_K(A_{p'})) = [K, A_{p'}]\mathrm{C}_L(A_{p'}) = [K, A_{p'}]$. On the other hand, K/L is $G_{p'}$-isomorphic to V. Since $A_{p'} \leq \mathrm{C}_G(K/L)$, it follows that $A_{p'} \leq \mathrm{C}_G(V)$. Consequently $V \leq \mathrm{C}_K(A_{p'})$. Since $K = \mathrm{C}_K(A_{p'}) \times [K, A_{p'}] = \mathrm{C}_K(A_{p'}) \times L$, we conclude that $\mathrm{C}_K(A_{p'}) = V(\mathrm{C}_K(A_{p'}) \cap L) = V$, and V is A-invariant. Now we argue in a similar way with $B_{p'}$. We know that $K = [K, B_{p'}]L$. Applying [119, A, 12.5], $K = [K, B_{p'}] \times \mathrm{C}_K(B_{p'})$ and so $L = \mathrm{C}_K(B_{p'})$. Hence $\mathrm{C}_V(B_{p'}) = L \cap V = 1$. Consider now V regarded as a $B_{p'}$-module. Then $V = \mathrm{C}_V(B_{p'}) \times [V, B_{p'}]$ by [119, A, 12.5]. It follows that $V = [V, B_{p'}] \leq [K, B_{p'}]$. But now $[K, B_{p'}] = V(L \cap [K, B_{p'}]) = V$ because $L \cap [K, B_{p'}] = 1$. Hence $V = [K, B_{p'}]$ is B-invariant. Consequently, V is a normal subgroup of G. But this yields that V is a minimal normal subgroup of G and AV, which is isomorphic to AK/L, is nilpotent. This final contradiction completes the proof. □

The next theorem is also an application of Proposition 1.1.21. It shows that the Fitting subgroup of a dinilpotent group is always factorised.

Theorem 3.2.7. *If $G = AB$ is a dinilpotent group, then the Fitting subgroup $\mathrm{F}(G)$ of G is factorised.*

Proof. By Sylow's theorem, G satisfies D_p for all primes p. Therefore $\mathrm{O}_p(G) = (A \cap \mathrm{O}_p(G))(B \cap \mathrm{O}_p(G))$ for every prime p by Proposition 1.1.21. Since $\mathrm{F}(G) = \prod_{p \in \mathbb{P}} \mathrm{O}_p(G)$, it follows that

$$\mathrm{F}(G) = (A \cap \mathrm{F}(G))(B \cap \mathrm{F}(G)).$$

Since $A \cap B \leq \mathrm{F}(G)$, $\mathrm{F}(G)$ is factorised in G by Corollary 1.1.16. □

Many questions about dinilpotent groups can be resolved by examining primitive dinilpotent groups. The following theorem of Heineken [149], which strengthens some earlier results of Gross [139], gives information about these groups.

Theorem 3.2.8. *Let $G = AB$ be a dinilpotent group with a minimal normal subgroup N such that $N = \mathrm{C}_G(N) \neq G$. Then:*

1. $A \cap B = 1$,
2. *either $N \leq A$ or $N \leq B$,*
3. *if $N \leq A$, then A is a p-group for some prime p and B is a p'-group,*
4. *there is a unique complement W of N in G which is factorised.*

We prove first some preliminary results, due to Amberg and Höfling [11], which will be also useful in analysing prefactorised and factorised subgroups of dinilpotent groups.

If the soluble group $G = AB$ is the product of two subgroups with coprime orders, then every subgroup has a factorised conjugate, as the following lemma shows.

Lemma 3.2.9. *Let the soluble group $G = AB$ be the product of two subgroups A and B with coprime orders. Then every subgroup of G has a factorised conjugate, and every subnormal subgroup of G is factorised.*

Proof. By hypothesis, A is a Hall π-subgroup of G for some set of primes π and B is a Hall π'-subgroup of G. Let S be a subgroup of G. If S_π is a Hall π-subgroup and $S_{\pi'}$ is a Hall π'-subgroup of S, then $S = S_\pi S_{\pi'}$. Now S_π and $S_{\pi'}$ are contained in Hall subgroups A^x and B^y of G with x and y in G respectively. Write $xy^{-1} = ab$ with $a \in A$ and $b \in B$. Then $A^x = A^{by}$ and $B^y = B^{by}$. If $g^{-1} = by$, then $S_\pi^g \leq A$ and $S_{\pi'}^g \leq B$. Hence $S^g = S_\pi^g S_{\pi'}^g = (S^g \cap A)(S^g \cap B)$.

If S is subnormal in G, then $S \cap A$ is a Hall π-subgroup of S and $S \cap B$ is a Hall π'-subgroup of S by [119, I, 4.21]. Therefore $S = (S \cap A)(S \cap B)$, and S is factorised. □

Recall that if $G = AB$ is a dinilpotent group and π is a set of primes and A_π and B_π are the Hall π-subgroups of A and B, respectively, then $A_\pi B_\pi$ is a prefactorised Hall π-subgroup of G. We can assume without loss of generality that $\Sigma = \{A_\pi B_\pi : \pi \text{ is a set of primes}\}$ is a Hall system of G.

Lemma 3.2.10. *Let $G = AB$ be a dinilpotent group. Then the Hall system $\Sigma = \{A_\pi B_\pi : \pi$ is a set of primes$\}$ reduces into every prefactorised subgroup S of G, that is, $S \cap A_\pi B_\pi$ is a Hall π-subgroup of S for every set of primes π.*

Proof. If $(A \cap S)_\pi$ and $(B \cap S)_\pi$ are the Hall π-subgroups of $A \cap S$ and $B \cap S$, respectively, then the subgroup $(A \cap S)_\pi (B \cap S)_\pi$ is a Hall π-subgroup of $S = (S \cap A)(S \cap B)$. Since $(A \cap S)_\pi = A_\pi \cap S$ and $(B \cap S)_\pi = B_\pi \cap S$, the subgroup $(A \cap S)_\pi (B \cap S)_\pi = A_\pi B_\pi \cap S$ is a Hall π-subgroup of S. □

The following result will be used to prove the uniqueness of the factorised conjugates in Theorem 3.2.8, Corollaries 3.2.19 and 3.2.22 and Theorem 3.2.25.

Proposition 3.2.11. *Let $G = AB$ be a dinilpotent group and let P be a pronormal subgroup of G. Suppose that either P is prefactorised or $\mathrm{N}_G(P)$ is prefactorised. If some conjugate P^g with $g \in G$ is prefactorised, then $P = P^g$.*

Proof. By Lemma 3.2.10, the Hall system $\Sigma = \{A_\pi B_\pi : \pi \subseteq \mathbb{P}\}$ reduces into P^g and P in the first case and into $\mathrm{N}_G(P)$ in the second case. If Σ reduces into $\mathrm{N}_G(P)$, then it also reduces into P, since P is normal in $\mathrm{N}_G(P)$. Since P is a pronormal subgroup of G, it follows that $P = P^g$ by [119, I, 6.6]. □

Proof of Theorem 3.2.8. We know that N is an elementary abelian p-group for some prime p, and $N = \mathrm{F}(G)$ by [119, A, 15.6]. Applying Theorem 3.2.4, AN is nilpotent or BN is nilpotent. Suppose that AN and BN are both nilpotent. Then A and B are p-groups and so is G. In this case, G is cyclic of order p and the result holds trivially. Therefore we can assume that BN is not nilpotent, and so A is a p-group. By Theorem 3.2.7, $N = (N \cap A)(N \cap B)$. Hence $AN = A(B \cap N)$. Then $AN \cap B$ is subnormal in AN and B, so that $AN \cap B$ is a subnormal subgroup of G by Theorem 1.1.7. Thus $AN \cap B \leq N$. Write $C/N = \mathrm{F}(G/N)$. By [119, A, 13.6], C/N is a p'-group. Let $C_{p'}$ be a Hall p'-subgroup of C contained in B. Then $C = NC_{p'}$ and $AN \cap B \leq \mathrm{C}_G(C_{p'})$ because B is nilpotent. We conclude then that $AN \cap B \leq \mathrm{Z}(C)$. On the other hand, if $\mathrm{Z}(C) \neq 1$, then $N \leq \mathrm{Z}(C)$ and so $N = C$, a contradiction which shows that $\mathrm{Z}(C) = 1$. Consequently, $AN \cap B = 1$ and $A \cap B = N \cap B = 1$. Now $B \cap C = B \cap NC_{p'} = (B \cap N)C_{p'} = C_{p'}$. Hence $C = NC_{p'} = N(B \cap C)$. Let B_p be the Sylow p-subgroup of B. Since $B_p \leq \mathrm{C}_G(B \cap C)$ because B is nilpotent, it follows that

$$B_p N/N \leq \mathrm{C}_{G/N}((B \cap C)N/N) = \mathrm{C}_{G/N}(C/N) \leq C/N.$$

Hence B_p is contained in C and so $B_p \leq N$, and $B_p \leq B \cap N = 1$. This implies that B is a p'-group. Therefore Assertions 1, 2 and 3 are proved.

Finally, we prove that there exists a unique complement of N in G which is factorised. Clearly, the complements of N in G are just the core-free maximal subgroups of G. Let M be one of them. Then $G = NM$ and $N \cap M = 1$ and M is pronormal in G. The above arguments show that A is a Sylow p-subgroup of G and B is a Hall p'-subgroup of G. Hence $A = N(A \cap M)$, and we can assume without loss of generality that a Hall p'-subgroup $M_{p'}$ of M is contained in B. Since $C \cap B$ is a Hall p'-subgroup of C, by the Frattini argument [119, A, 6.3 (b)] we obtain that $G = C\mathrm{N}_G(C \cap B) = N\mathrm{N}_G(C \cap B)$. Moreover $C \cap B$ is not normal in G because it does not contain N. Hence $W = \mathrm{N}_G(C \cap B)$ is a core-free maximal subgroup of G containing B and $W = B(W \cap A)$. In particular, W is factorised by Corollary 1.1.16. Now Assertion 4 follows from Proposition 3.2.11 together with the well-known fact that a primitive soluble group has a single conjugacy class of core-free maximal subgroups of G. □

Armed with Theorem 3.2.8 and Lemma 3.2.9 it is not hard to prove the following

Corollary 3.2.12. *Let $G = AB$ be a primitive dinilpotent group with a unique minimal normal subgroup $N = \mathrm{C}_G(N) \neq G$. Then every subgroup of G has a factorised conjugate, and every subnormal subgroup of G is factorised.*

The following observation follows from Theorem 3.2.8.

Corollary 3.2.13. *Let $G = AB$ be a primitive dinilpotent group such that $|\mathrm{F}(G)|$ and $|G/\mathrm{F}(G)|$ are coprime. Then $\mathrm{F}(G)$ coincides with one of the factors A or B.*

We present now an application of the above corollary due to Detomi [117].

Proposition 3.2.14. *Let $G = AB$ be a dinilpotent group. If L and N are normal subgroups of G with $L \leq N$ such that G/N, N/L, and L are nilpotent groups, and $\gcd(|L|, |G/N|) = 1$, then G is metanilpotent and $[A, B] \leq \mathrm{F}(G)$. Moreover $G/\mathrm{F}(G)$ is the direct product of $A\mathrm{F}(G)/\mathrm{F}(G)$ and $B\mathrm{F}(G)/\mathrm{F}(G)$.*

Proof. First of all, observe that if $\Phi(G)$ is the Frattini subgroup of G, then $G/\Phi(G)$ is a subdirect product of primitive soluble groups G/M_G, where M is a maximal subgroup of G.

Let us show that the hypotheses of Corollary 3.2.13 hold for G/M_G, where M is a maximal subgroup of G. Without loss of generality, we can assume that $M_G = 1$. Let $H = \mathrm{F}(G)$. By [119, A, 15.6], H is the unique minimal normal subgroup of G so that $H = \mathrm{O}_p(G)$ for some prime p. Moreover $|\mathrm{F}_2(G)/H|$ is a p'-group. On one hand, if L is non-trivial, then $L = H$. Since $\mathrm{F}_2(G)$ contains N and $\gcd(|L|, |G/N|) = 1$, it follows that H is a Sylow p-subgroup of G. By Corollary 3.2.13, it follows that G/H is isomorphic to A or B.

Consequently, if $H_{(M)}/M_G$ denotes the Fitting subgroup of G/M_G, we have that $G/H_{(M)}$ is nilpotent for all maximal subgroups M of G. Therefore $G/\bigcap\{H_{(M)} : M \text{ is maximal in } G\}$ is nilpotent.

Since

$$\Big(\bigcap\{H_{(M)} : M \text{ is maximal in } G\}\Big)' \leq \Phi(G),$$

it follows that $\bigcap\{H_{(M)} : M \text{ is maximal in } G\}$ is nilpotent by [157, III, 3.5]. Thus G is metanilpotent and $\bigcap\{H_{(M)} : M \text{ is maximal in } G\} = \mathrm{F}(G)$.

Moreover, since $H_{(M)}/M_G$ coincides with AM_G/M_G or BM_G/M_G, for every maximal subgroup M of G, we conclude that $H_{(M)}$ contains A or B for every maximal subgroup M of G and

$$[A, B] \leq \bigcap\{H_{(M)} : M \text{ is maximal in } G\} = \mathrm{F}(G).$$

Now $A \cap B$ is contained in $\mathrm{F}(G)$, so that $G/\mathrm{F}(G)$ is the direct product of $A\mathrm{F}(G)/\mathrm{F}(G)$ and $B\mathrm{F}(G)/\mathrm{F}(G)$. This completes the proof of the proposition. □

Bearing in mind that a group which is the product of two abelian subgroups is metabelian, whereas a group which is the product of two nilpotent groups is not metanilpotent in general (consider the symmetric group of degree 4), Maier [191] studied conditions which could characterise the metanilpotency of a dinilpotent group. He proved that a dinilpotent group is metanilpotent if and only if it is of p-length at most one for all primes p. Note that this result is a consequence of Theorem 3.2.8 since the classes under consideration are saturated formations.

We now move on to an investigation of prefactorised and factorised subgroups of dinilpotent groups. We establish a number of results of Amberg and Höfling [11] that put a rich source of examples of such subgroups at our disposal.

Definition 3.2.15. Let $\mathfrak{X}$ be a class of groups. A maximal subgroup M of a group G is said to be *$\mathfrak{X}$-abnormal* in G if G/M_G does not belong to $\mathfrak{X}$. A subgroup S is called *sub-$\mathfrak{X}$-abnormal* in G if either $G = S$ or there exists a chain

$$S = S_0 \leq S_1 \leq \cdots \leq S_n = G$$

such that S_i is an $\mathfrak{X}$-abnormal maximal subgroup of S_{i+1} for all $i \in \{0, 1, \ldots, n-1\}$.

Definition 3.2.16. Let $\mathfrak{X}$ be a class of groups. A subgroup S of a group G is *$\mathfrak{X}$-maximal* in G if S belongs to $\mathfrak{X}$ and for every $\mathfrak{X}$-subgroup T containing S, the equality $S = T$ holds.

Recall that if $\mathfrak{X}$ is a class of groups, the *characteristic* $\operatorname{char}\mathfrak{X}$ of $\mathfrak{X}$ is the set of all primes p such that C_p belongs to $\mathfrak{X}$.

Theorem 3.2.17. *Let $G = AB$ be a dinilpotent group. If $\mathfrak{X}$ is a class of groups such that* $\operatorname{char}\mathfrak{X}$ *contains the common prime divisors of the orders of A and B, then every sub-$\mathfrak{X}$-abnormal subgroup of G has a factorised conjugate.*

Proof. Assume that this is false and let $G = AB$ be a counterexample of least order. Among the sub-$\mathfrak{X}$-abnormal subgroups X of G not having any factorised conjugate, we choose S of maximal order. Then S is an $\mathfrak{X}$-abnormal maximal subgroup of G. Assume that S is normal in G. Then G/S is not an $\mathfrak{X}$-group. Hence $AS/S = 1$ or $BS/S = 1$ by hypothesis. Applying Corollary 1.1.16, it follows that S is factorised. Suppose now that S is not normal in G. Then the primitive group G/S_G cannot be nilpotent. Hence by Corollary 3.2.12, there exists an element $g \in G$ such that S^g/S_G is factorised in G/S_G. Thus S^g is also a factorised subgroup of G by Lemma 1.1.17 (2). This contradiction proves the theorem. □

$\mathfrak{F}$-normalisers of soluble groups associated to saturated formations $\mathfrak{F}$ are typical examples of sub-$\mathfrak{F}$-abnormal subgroups ([47, Chapter 4]). Therefore the following result is an immediate consequence of the above theorem.

Corollary 3.2.18. *Let $G = AB$ be a dinilpotent group. If $\mathfrak{F}$ is a saturated formation such that* char $\mathfrak{F}$ *contains the common prime divisors of the orders of A and B, then G has a factorised $\mathfrak{F}$-normaliser.*

A special case of the above results concerns the situation in which $\mathfrak{F} = \mathfrak{N}$, the saturated formation of all nilpotent groups, for then it is clear that abnormality implies sub-$\mathfrak{N}$-abnormality ([119, I, 6.20]) and pronormality (Lemma 1.5.4).

Corollary 3.2.19. *Every abnormal subgroup of a dinilpotent group has a unique factorised conjugate. In particular, G always has a factorised Carter subgroup.*

Recall that a class $\mathfrak{H}$ of groups is a *Schunck class* if it contains all groups whose primitive factor groups are in $\mathfrak{H}$. In particular, every saturated formation is also a Schunck class.

A subgroup P of a group G is an *$\mathfrak{H}$-projector* of G if the factor group PN/N is $\mathfrak{H}$-maximal in G/N for every normal subgroup N of G. If $\mathfrak{H}$ is a Schunck class, every soluble group G has $\mathfrak{H}$-projectors and every $\mathfrak{H}$-projector is pronormal in G ([119, Chapter III] and [47, Section 2.3]).

Theorem 3.2.20. *Let $G = AB$ be a dinilpotent group.*

1. *If $\mathfrak{H}$ is a Schunck class such that* char $\mathfrak{H}$ *contains every prime dividing the orders of both A and B, then every $\mathfrak{H}$-maximal subgroup of G has a factorised conjugate.*

2. *If $\mathfrak{H}$ is a saturated formation, then every $\mathfrak{H}$-maximal subgroup of G has a prefactorised conjugate.*

Proof. 1. Assume, by way of contradiction, that Statement 1 fails, and choose a dinilpotent group G of least order having an $\mathfrak{H}$-maximal subgroup which has no factorised conjugate. Let N be a non-trivial normal subgroup of G. Then the $\mathfrak{H}$-subgroup

HN/N of G/N is contained in a maximal $\mathfrak{H}$-subgroup L/N of the dinilpotent group G/N. Since the order of G/N is less than the order of G, there exists a factorised conjugate L^g/N of L/N in $G/N = (AN/N)(BN/N)$. Then L^g is a factorised subgroup of G by Lemma 1.1.17 (2) and so L^g is dinilpotent. Clearly H^g is an $\mathfrak{H}$-maximal subgroup of L^g. Thus if L^g is a proper subgroup of G, then by the choice of G the subgroup H^g has a factorised conjugate in L^g which is clearly a factorised conjugate of H. This contradiction shows that every proper epimorphic image of G is an $\mathfrak{H}$-group.

Of course we may assume that G is not an $\mathfrak{H}$-group, and so it must be primitive by the definition of a Schunck class. Then, by [119, A, 15.6], G has a unique minimal normal subgroup N. If N is properly contained in G, then Corollary 3.2.12 yields that every subgroup of G has a factorised conjugate, against the hypothesis.

Thus $G = N$ is cyclic of prime order p. If $p \in \operatorname{char} \mathfrak{H}$, then $G \in \mathfrak{H}$ and $G = H$, a contradiction. Finally, if p does not belong to $\operatorname{char} \mathfrak{H}$, then we have $H = 1$. Since p is not a common prime divisor of the orders of A and B, we must have $A = 1$ or $B = 1$. Hence H is factorised. This contradiction proves Statement 1.

2. Let π be the characteristic of the saturated formation $\mathfrak{H}$. Then the $\mathfrak{H}$-subgroup H of G is a π-group by [119, IV, 4.3]. We know that if A_π and B_π are Hall π-subgroups of A and B respectively, then $A_\pi B_\pi$ is a Hall π-subgroup of G. By replacing H by a suitable conjugate, we may suppose that H is contained in $A_\pi B_\pi$. Then H is also $\mathfrak{H}$-maximal in the factorised group $A_\pi B_\pi$. By Statement 1, there exists a factorised conjugate of H in $A_\pi B_\pi$. This is a prefactorised conjugate of H in $G = AB$. □

An $\mathfrak{H}$-maximal subgroup of a dinilpotent group need not have a prefactorised conjugate if $\operatorname{char} \mathfrak{H}$ does not contain the common prime divisors of the orders of A and B.

Example 3.2.21 (Amberg and Höfling [11]). Let $G = C_p \wr C_q$ be the regular wreath product of a cyclic group of order p with a cyclic group of order q, where p and q are different primes. Let $A = \mathrm{C}_G(C_q)$ and suppose that B is a subgroup of the base group of G of the form

$$B = \overbrace{C_p \times \cdots \times C_p}^{(q-1)} \times 1.$$

Then $G = AB$ and

$$\mathrm{O}^p(G) = \langle C_q^G \rangle = \Big\{(x_1, \dots, x_q) : \prod_{i=1}^{q} x_i = 1\Big\} C_q \neq (A \cap \mathrm{O}^p(G))(B \cap \mathrm{O}^p(G)).$$

Here $\mathrm{O}^p(G)$ is an $\mathfrak{H}$-projector for the class $\mathfrak{H}$ of all p-perfect groups. Observe also that $\mathrm{O}^q(G)$ equals the base group of G and the nilpotent residual of G is

$$G^{\mathfrak{N}} = \mathrm{O}^p(G) \cap \mathrm{O}^q(G) \neq (A \cap G^{\mathfrak{N}})(B \cap G^{\mathfrak{N}}).$$

Since $G^{\mathfrak{N}} = G' = [A, B]$, this example also shows that neither G' nor $[A, B]$ needs to be prefactorised.

As a consequence of Theorem 3.2.20 and Proposition 3.2.11, we obtain the following generalisation of a result of Heineken [149] (see [9, 2.5.15]).

Corollary 3.2.22. *Let $G = AB$ be a dinilpotent group.*

1. *If $\mathfrak{H}$ is a Schunck class such that every common prime divisor of the orders of A and B belongs to* char $\mathfrak{H}$, *then G has a unique factorised $\mathfrak{H}$-projector.*

2. *If $\mathfrak{H}$ is a saturated formation, then G has a unique prefactorised $\mathfrak{H}$-projector.*

Corollary 3.2.23. *Let $\mathfrak{H}$ be a Schunck class. Let the group $G = AB = AC = BC$ be the product of two nilpotent subgroups A and B and an $\mathfrak{H}$-subgroup C. If every common prime divisor p of the orders of A and B belongs to* char $\mathfrak{H}$, *then G is a soluble $\mathfrak{H}$-group.*

Proof. Le D be an $\mathfrak{H}$-maximal subgroup of G containing C. Then also $G = AB = AD = BD$. Now by Theorem 3.2.20, D has a factorised conjugate D^g with $g \in G$. Therefore $G = D$ is an $\mathfrak{H}$-group by Lemma 1.1.25. □

Recall that if $\mathfrak{F}$ is a class of groups, a subgroup I of a group G is an *$\mathfrak{F}$-injector* of G if for every subnormal subgroup S of G, the intersection $I \cap S$ is $\mathfrak{F}$-maximal in S. It is well known that if $\mathfrak{F}$ is a Fitting class, then every soluble group has a unique conjugacy class of $\mathfrak{F}$-injectors ([47, 2.4.8], [119, IX, 1.4]). From this point of view, Fitting classes are dual to Schunck classes. However, for $\mathfrak{F}$-injectors of products of nilpotent groups, there is no result corresponding to Corollary 3.2.22, as we can see in the following example, in which a dinilpotent group need not contain prefactorised $\mathfrak{F}$-injectors for a Fitting class $\mathfrak{F}$.

Example 3.2.24 (Amberg and Höfling [11]). Let G be a soluble group and

$$1 = G_0 \trianglelefteq G_1 \trianglelefteq G_2 \trianglelefteq \cdots \trianglelefteq G_n = G$$

be a chief series of G. Every p-chief factor G_i/G_{i+1} can be regarded as a vector space over the field GF(p) of p elements. Let $\lambda_i(g)$ be the linear transformation of the p-chief factor G_i/G_{i+1} induced by g by conjugation. Let

$$\Delta(g) = \prod \det \lambda_i(g),$$

where the product is taken over all i such that G_{i+1}/G_i is a p-group. Then Δ is a homomorphism from G into the multiplicative group of GF(p). Then the class $\mathfrak{D}(p)$ of soluble groups G such that $\Delta(g) = 1$ for all $g \in G$ is a normal Fitting class, that is, a Fitting class such that every soluble group has a unique normal $\mathfrak{D}(p)$-injector by a result of Blessenohl and Gaschütz (see [119, IX, 2.14 (b)]).

Let S and S^* be symmetric groups of degree 3 acting on the sets $\{1, 2, 3\}$ and $\{1^*, 2^*, 3^*\}$, respectively. Consider the direct product $G = S \times S^*$. Then the normal subgroup $D = \langle(1,2,3),(1^*,2^*,3^*),(1,2)(1^*,2^*)\rangle$ of index 2 in G is the unique $\mathfrak{D}(3)$-injector as well as the $\mathfrak{D}(3)$-radical of G.

The subgroups $A = \langle(1,2,3),(1^*,2^*)\rangle$ and $B = \langle(1^*,2^*,3^*),(1,2)\rangle$ of G are cyclic of order 6 and $A \cap B = 1$ and $G = AB$. Finally, $A \cap D$ and $B \cap D$ are both cyclic of order 3. Hence the $\mathfrak{D}(3)$-injector D of G is not prefactorised.

Note that in this example every subgroup of A permutes with every subgroup of B.

The next result of Amberg and Höfling [11] provides a test for the factorisation of the injectors in terms of the radicals.

Theorem 3.2.25. *Let $\mathfrak{F}$ be a Fitting class of soluble groups. Then every dinilpotent group contains a unique prefactorised (respectively, factorised) $\mathfrak{F}$-injector if and only if the $\mathfrak{F}$-radical of every dinilpotent group is prefactorised (respectively, factorised).*

Proof. Assume that $G = AB$ is a dinilpotent group with a unique prefactorised $\mathfrak{F}$-injector, I say, and suppose inductively that in dinilpotent groups of order less than $|G|$ the $\mathfrak{F}$-radical is prefactorised. As A and B are nilpotent, the subgroup $X = AG_{\mathfrak{F}} \cap BG_{\mathfrak{F}}$ is a factorised subnormal subgroup of G and $I \cap X$ is a prefactorised $\mathfrak{F}$-injector of X by [9, 1.1.4, 6.3.11]. Observe that in this case $X_{\mathfrak{F}} = G_{\mathfrak{F}}$. Hence if X is properly contained in G, it follows by induction that $G_{\mathfrak{F}}$ is prefactorised in X. In particular, $G_{\mathfrak{F}}$ is prefactorised in G. Therefore suppose that $X = G$, and so $G = AG_{\mathfrak{F}}$. It implies that $G/G_{\mathfrak{F}}$ is nilpotent and $I/G_{\mathfrak{F}}$ is subnormal in $G/G_{\mathfrak{F}}$. Thus the subnormal subgroup I of G is contained in $G_{\mathfrak{F}}$ and $I = G_{\mathfrak{F}}$ is prefactorised in G, as desired.

Suppose that the $\mathfrak{F}$-radical of every dinilpotent group is factorised, and assume that G is a dinilpotent group. We proceed by induction on the order of G to show that G has a unique prefactorised $\mathfrak{F}$-injector. Let I be an $\mathfrak{F}$-injector of G. If I is normal in G, then clearly $I = G_{\mathfrak{F}}$ and the result follows. Therefore $\mathrm{N}_G(I)$ is a proper subgroup of G. Now, by [119, IX, 1.5], I is pronormal in G and so $\mathrm{N}_G(I)$ is abnormal in G by Lemma 1.5.4. Then it has a factorised conjugate $\mathrm{N}_G(I)^g = \mathrm{N}_G(I^g)$ by Corollary 3.2.19. Suppose that $\mathrm{N}_G(I)$ is a proper subgroup of G. Since I^g is the $\mathfrak{F}$-radical of $\mathrm{N}_G(I^g)$, it is prefactorised in $\mathrm{N}_G(I^g)$ by [119, IX, 1.5] and it follows at once that I^g is prefactorised in G. The uniqueness of I^g follows from Proposition 3.2.11 because I is pronormal in G.

If $A \cap B$ is contained in $G_{\mathfrak{F}}$, then it is also contained in I because $G_{\mathfrak{F}} \leq I$. If, on the other hand, $A \cap B$ is contained in I, then it is an $\mathfrak{F}$-group because it is subnormal in G, and hence in the $\mathfrak{F}$-group I. Therefore $A \cap B$ is contained in $G_{\mathfrak{F}}$. The proposition is proved. □

Let π be a set of primes. In a dinilpotent group $G = AB$, we have that

$$\mathrm{O}_\pi(A)\,\mathrm{O}_\pi(B) = \mathrm{O}_\pi(B)\,\mathrm{O}_\pi(A)$$

is a Hall π-subgroup of G. The following result of Kazarin, Martínez-Pastor, and Pérez-Ramos in [177] confirms that the same is true for products of π-decomposable groups if π does not contain the prime 2.

Recall that a group X is *π-decomposable* if $X = X_\pi \times X_{\pi'}$ is the direct product of a π-subgroup $X_\pi = \mathrm{O}_\pi(X)$ and a π'-subgroup $X_{\pi'} = \mathrm{O}_{\pi'}(X)$.

Theorem 3.2.26. *Let π be a set of odd primes. If the group $G = AB$ is the product of two π-decomposable subgroups $A = A_\pi \times A_{\pi'}$ and $B = B_\pi \times B_{\pi'}$, then $A_\pi B_\pi = B_\pi A_\pi$ and this is a Hall π-subgroup of G.*

In the rest of the section we shall concentrate on structure theorems which will play an important role to prove bounds on the derived length of dinilpotent groups.

The proof of the following result, which generalises Theorem 3.1.13 (3), appears in [223, 13.2.10]. It is also contained in the proof of [9, 7.1.6].

Theorem 3.2.27 (Kegel [180]). *Let $G = AB$ be a dinilpotent group for which at least one of the factors A or B is proper. Then there exists a proper normal subgroup H of G containing A or B.*

In the case of products of an abelian and a nilpotent subgroup, we obtain the following result of Franciosi, de Giovanni, Heineken, and Newell [129]. Its proof appears in [9, 2.5.9]. This result will be extended in the next section (see Theorem 3.3.44).

Theorem 3.2.28. *Let the group $G = AB$ be the product of an abelian subgroup A and a nilpotent subgroup B. Then $A\mathrm{F}(G)$ is a normal subgroup of G. In particular, G has Fitting length at most* 3.

As the symmetric group of degree 4 has Fitting length 3 and it is the product of an abelian subgroup and a nilpotent subgroup, the bound in the above theorem is best possible. Moreover, Theorem 3.2.28 does not hold for arbitrary dinilpotent groups: for each positive integer n and different primes p and q, there exists a dinilpotent group of order $p^a q^b$ of Fitting length n.

Our next goal is to prove the normality of the subgroup $(\mathrm{F}_2(G) \cap A)\mathrm{F}(G)$ of a dinilpotent group G. It was proved by Heineken [150]. We need one preparatory lemma.

Lemma 3.2.29. *Let G be a group. Assume that $\{X_i\}_{i \in I}$ is a family of normal subgroups of G and $\bigcap_{i\in I} X_i = 1$, and for every $i \in I$, define Y_i such that $Y_i/X_i = \mathrm{F}(G/X_i)$. Then $\bigcap_{i\in I} Y_i = \mathrm{F}(G)$.*

Proof. Since $\mathrm{F}(G)X_i/X_i \le \mathrm{F}(G/X_i) = Y_i/X_i$, we deduce that the intersection of all Y_i contains $\mathrm{F}(G)$. On the other hand, choose an element x which is contained in this intersection. There is an integer k such that $[g, \overbrace{x, \dots, x}^{(k \text{ times})}]$ is contained in X_i for all i and for all elements $g \in G$ by [157, III, 6.15]. Hence x is a left Engel element and is contained in $\mathrm{F}(G)$ by [157, III, 6.15]. This shows the other inclusion and completes the proof. □

Theorem 3.2.30. *Let $G = AB$ be a dinilpotent group. Then*

$$(\mathrm{F}_2(G) \cap A)\mathrm{F}(G)$$

is a normal subgroup of G.

Proof. Consider a minimal set of maximal subgroups M_i satisfying the following conditions:

1. $M_i\mathrm{F}(G) = G$, and
2. $\mathrm{F}(G) \cap \bigcap_i M_i = \Phi(G)$.

Fix an index j and let D_j denote the core of M_j in G. Put $R_j = (\bigcap_{i \neq j} M_i) \cap \mathrm{F}(G)$. Since $M_i \cap \mathrm{F}(G)$ is normal in G for all i, it follows that R_j is a normal subgroup of G. Moreover, G/D_j is a primitive soluble group such that $\mathrm{F}(G)D_j/D_j = \mathrm{Soc}(G/D_j)$ is complemented in G/D_j by M_j/D_j. Hence, by [119, A, 15.6] $R_jD_j = \mathrm{F}(G)D_j$ and R_jD_j/D_j is a self-centralising minimal normal subgroup of G/D_j which is G-isomorphic to $R_j/\Phi(G)$. In particular, $\mathrm{C}_{G/\Phi(G)}(R_j/\Phi(G)) = R_jD_j/\Phi(G)$. Furthermore, $\mathrm{F}(G)$ is the product of all R_j and $\mathrm{F}(G)/\Phi(G) = \mathrm{C}_{G/\Phi(G)}(\mathrm{F}(G)/\Phi(G))$. Consequently, $\mathrm{F}(G) = \bigcap_j R_jD_j$.

We will show that $(\mathrm{F}_2(G/D_j) \cap AD_j/D_j)\mathrm{F}(G/D_j)$ is a normal subgroup of G/D_j. In order to prove this, we will distinguish two cases, according to whether $\mathrm{F}(G/D_j) = R_jD_j/D_j$ is contained in AD_j/D_j or in BD_j/D_j by Theorem 3.2.8. In the first case, $\mathrm{F}_2(G/D_j) \cap AD_j/D_j = R_jD_j/D_j$. Assume that $\mathrm{F}(G/D_j)$ is contained in BD_j/D_j. We have that $\mathrm{F}_2(G/D_j)$ is contained in $\mathrm{F}(G/D_j) \cdot (AD_j/D_j)$, and

$$\mathrm{F}_2(G/D_j) = \mathrm{F}(G/D_j) \cdot (\mathrm{F}_2(G/D_j) \cap AD_j/D_j).$$

Let S_j be defined by $S_j/D_j = \mathrm{F}_2(G/D_j)$. Then $S_j = R_j(S_j \cap AD_j)$ is a normal subgroup of G, as desired.

Consequently, $K = \bigcap_j R_j(S_j \cap AD_j)$ is a normal subgroup of G, and

$$\mathrm{F}(G) = \bigcap_j R_jD_j \leq K \leq \bigcap_j S_j = \mathrm{F}_2(G),$$

where the last equality follows from Lemma 3.2.29. Furthermore, we have that

$$\begin{aligned}\bigcap_j (S_j \cap A) &= \Big(\bigcap_j S_j\Big) \cap A \\ &= \mathrm{F}_2(G) \cap A \leq K \leq \mathrm{F}_2(G) = (\mathrm{F}_2(G) \cap A)(\mathrm{F}_2(G) \cap B),\end{aligned}$$

where the last equality holds by Theorem 3.2.7. Now

$$K = (\mathrm{F}_2(G) \cap A)(K \cap \mathrm{F}_2(G) \cap B) = (\mathrm{F}_2(G) \cap A)(K \cap B),$$

and since $R_j(S_j \cap AD_j) \cap BD_j \leq R_j D_j$ because modulo D_j it is a nilpotent subnormal subgroup of G/D_j, we obtain that $K \cap B \leq \bigcap_j R_j D_j = \mathrm{F}(G)$ and so $K \cap B = \mathrm{F}(G) \cap B$. Hence

$$\begin{aligned} K &= (\mathrm{F}_2(G) \cap A)(\mathrm{F}(G) \cap B) \\ &= (\mathrm{F}_2(G) \cap A)(\mathrm{F}(G) \cap A)(\mathrm{F}(G) \cap B) \\ &= (\mathrm{F}_2(G) \cap A)\mathrm{F}(G), \end{aligned}$$

and the proof is complete. □

The next result of Heineken [149] provides useful information about the structure of the Fitting factor of dinilpotent groups.

Theorem 3.2.31. *Let $G = AB$ be a dinilpotent group. Then $G/\mathrm{F}(G)$ is a subdirect product of $\{p, q\}$-groups which are products of a p-group isomorphic to a quotient group of A with a q-group isomorphic to a quotient group of B, where p and q are different primes.*

Proof. Consider a complemented chief factor H/K of G such that $\mathrm{C}_{G/K}(H/K) = H/K$ (we can take for K the core of a maximal subgroup of G). Then G/K and H/K satisfy the hypotheses of Theorem 3.2.8. Hence H is contained in AK or BK, and if H/K is a p-group and $H \leq AK$, then G/H is the product of the p-group AH/H and the p'-group BH/H, and all primes dividing $|BH : H|$ divide $|Y : H|$, where Y/H is the Fitting subgroup of G/H, because that Y is contained in BH. Consider a non-trivial Sylow q-subgroup Q/H of Y/H for some prime q. Then $p \neq q$. Since G/H is soluble, $\mathrm{C}_{G/H}(Y/H)$ is contained in Y/H, thus the Sylow q-subgroup of $\mathrm{C}_{BH/H}(Q/H)$ coincides with $\mathrm{Z}(Q/H)$, while the Hall q'-subgroup of BH/H is contained in $\mathrm{C}_{BH/H}(Q/H)$. Now $\mathrm{C}_{G/H}(Q/H)$ is the direct product of a q'-group T/H and $\mathrm{Z}(Q/H)$, and G/T is the product of the p-group $AT/T \cong (AT/H)/(T/H)$ and the q-group $BT/T \cong (BT/H)(T/H)$. Continuing like this for all primes dividing $|Y : H|$, we obtain that G/H satisfies the thesis of the proposition given for G/F. Since F is the intersection of all H appearing in complemented self-centralising chief factors H/K, the statement follows for G/F itself, too. □

It is clear that a dinilpotent group can be written as a subdirect product of dinilpotent groups with Fitting subgroup of prime power order. Moreover, it is especially easy to find dinilpotent groups G for which $\mathrm{F}(G)$ is a p-group for some prime p, and this is the situation we study in the rest of the section. The structural results included here were proved by Cossey and Stonehewer [108].

If a dinilpotent group $G = AB$ has $\mathrm{F}(G)$ a p-group, for some prime p, and neither A nor B is a p-group, then G is close to being a direct product.

Theorem 3.2.32. *Let $G = AB$ be a dinilpotent group. Assume also that* $\mathrm{F}(G)$ *is a p-group for some prime p and neither A nor B is a p-group. Then* $\mathrm{O}^p(G) = XY$, *where X is the normal closure in G of $A_{p'}$ and Y is the normal closure in G of $B_{p'}$. Moreover* $[X, Y] = 1$.

Proof. Let X and Y denote the normal closures of $A_{p'}$ and $B_{p'}$, respectively, in G. Since $\mathrm{O}_{p'}(G) = 1$, it follows from Lemma 1.1.20 that $[X, Y] = 1$. Since $A_{p'}$ and $B_{p'}$ are contained in $\mathrm{O}^p(G)$, we have $XY \leq \mathrm{O}^p(G)$. On the other hand, we know that $A_{p'}B_{p'}$ is a Hall p'-subgroup of G and so $\mathrm{O}^p(G) = \langle (A_{p'}B_{p'})^G \rangle \leq XY$. This completes the proof. □

On the quest for dinilpotent groups with Fitting subgroup of prime power order, Cossey and Stonehewer made the following observation.

Theorem 3.2.33. *A soluble group G for which* $\mathrm{F}(G)$ *is a p-group for some prime p is dinilpotent if and only if it has a nilpotent Hall p'-subgroup.*

Proof. It is clear that if G has a nilpotent Hall p'-subgroup $G_{p'}$, then we can write $G = G_p G_{p'}$ for a Sylow p-subgroup G_p of G, with G_p and $G_{p'}$ nilpotent, so that G is dinilpotent.

Conversely, suppose that $G = AB$ with A and B nilpotent and $\mathrm{F}(G)$ a p-group. Since $G/\Phi(G)$ satisfies the same hypotheses and has Hall p'-subgroup isomorphic to those of G, we may assume that $\Phi(G) = 1$. Thus $\mathrm{F}(G)$ is the direct product of minimal normal subgroups of G. Let $\{M_i : i \in I\}$ be the set of minimal normal subgroups of G. Since $\Phi(G) = 1$, each minimal normal subgroup M_i is complemented, say by H_i. Let N_i be the centraliser of M_i in H_i. Note that N_i is normal in G. We then have that G/N_i is a primitive soluble group by [119, A, 15.5]. Since G/N_i is a nilpotent product, too, it now follows from Theorem 3.2.8 that one of AN_i/N_i and BN_i/N_i is a p-group and the other is a p'-group. In particular, a Hall p'-subgroup of G/N_i is nilpotent. Since G is a subdirect product of the G/N_i, we obtain immediately that a Hall p'-subgroup of G is nilpotent. □

As a consequence of the above result, if a dinilpotent group G has $\mathrm{F}(G)$ a p-group for some prime p, then G has a natural nilpotent factorisation $G = G_p G_{p'}$, for a Sylow p-subgroup G_p and and a Hall p'-subgroup $G_{p'}$ of G. For primitive dinilpotent groups, this is essentially the only factorisation by virtue of Theorem 3.2.8. In some other special cases, we can often obtain further information about the factorisation.

Lemma 3.2.34. *Suppose that $G = AB$ is a dinilpotent group,* $\mathrm{F}(G)$ *is a p-group for some prime p, and* $\mathrm{F}(G)/\Phi(\mathrm{F}(G))$ *contains no central chief factors of G. If A is a p-group, then A is a Sylow p-subgroup of G.*

Proof. We note first that $\mathrm{F}(G/\mathrm{F}(G)) \leq B_{p'}\mathrm{F}(G)/\mathrm{F}(G)$. Let P be the Sylow p-subgroup of B. We have

$$\begin{aligned} P\mathrm{F}(G)/\mathrm{F}(G) &\leq \mathrm{C}_{G/\mathrm{F}(G)}(B_{p'}\mathrm{F}(G)/\mathrm{F}(G)) \\ &\leq \mathrm{C}_{G/\mathrm{F}(G)}(\mathrm{F}(G/\mathrm{F}(G))) \leq \mathrm{F}(G/\mathrm{F}(G)) \leq B_{p'}\mathrm{F}(G)/\mathrm{F}(G). \end{aligned}$$

Hence we must have $P \leq \mathrm{F}(G)$.

Suppose that the result is not true and let G be chosen minimal satisfying that $G = AB$ is a dinilpotent group, $\mathrm{F}(G)$ is a p-group for some prime p, and $\mathrm{F}(G)/\Phi(\mathrm{F}(G))$ contains no chief factors of G such that A a p-group, but not a Sylow p-subgroup. Note that we cannot have $P = 1$, since then A would be a Sylow subgroup. Assume $\Phi(\mathrm{F}(G)) \neq 1$. Then $G/\Phi(\mathrm{F}(G))$ satisfies the hypotheses of the lemma. By our choice of G, $A\Phi(\mathrm{F}(G))/\Phi(\mathrm{F}(G))$ is a Sylow subgroup of $G/\Phi(\mathrm{F}(G))$ and so $A\Phi(\mathrm{F}(G))$ is a Sylow subgroup of G. Now $\mathrm{F}(G)$ is contained in every Sylow p-subgroup of G. Therefore $\Phi(\mathrm{F}(G))$ is contained in the Frattini subgroup of every Sylow p-subgroup of G and we have that A is a Sylow p-subgroup of G. This contradicts the choice of G. Thus $\Phi(\mathrm{F}(G)) = 1$. But then $\mathrm{F}(G)$ contains fixed points for the action of $B_{p'}$, since $P \neq 1$. Hence, by Lemma 3.2.5, $\mathrm{F}(G)$ contains a central chief factor of G. This contradicts the hypothesis. Therefore A must be a Sylow p-subgroup of G. □

Corollary 3.2.35. *With G and N as in Lemma* 3.2.6 *and N abelian, we obtain that $N = U \times V$, with U and V normal in G, U the maximal subgroup of N with UA nilpotent, and V the maximal subgroup of N with VB nilpotent.*

Proof. Choose U and V maximal in N and normal in G with UA and VB nilpotent. Since $U \cap V$ is a normal subgroup of G contained in N, $U \cap V$ is centralised by a Hall p'-subgroup of G, and N contains no central chief factors of G, it follows that $U \cap V = 1$ by Lemma 3.2.5. Note that if A is a p-group, then $U = N$ and $V = 1$, and if B is a p-group, then $U = 1$ and $V = N$.

Suppose that $UV \neq N$ and let T/UV be a chief factor of G with $T \leq N$. Then either AT/UV or BT/UV is nilpotent by Theorem 3.2.4; we may suppose that AT/UV is nilpotent. Now N/U satisfies the hypotheses of Lemma 3.2.6 and hence there is a minimal normal subgroup S/U of G/U with $S \leq N$ and AS/U nilpotent. But then $A_{p'}$ centralises S/U and U, and by [119, A, 12.3], it turns out that $A_{p'}$ centralises S and so AS is nilpotent. This contradicts the choice of U. Thus $N = U \times V$. Note moreover that $A_{p'}$ acts fixed point freely on N/U, because $B_{p'}$ centralises N/U and $G_{p'}$ acts fixed point freely on N/U by Lemma 3.2.5. Hence U is also the subgroup maximal with respect to AU being nilpotent. This completes the proof of the corollary. □

We can now assemble the pieces and prove the following.

Theorem 3.2.36. *Suppose that the group $G = AB$ is a dinilpotent group,* $\mathrm{F}(G)$ *is a p-group for some prime p, and* $\mathrm{F}(G)/\Phi(\mathrm{F}(G))$ *contains no central chief factors of G, neither A nor B is a p-group, and that $\Phi(G) = 1$. Then $G = U \times V$, with $U = A_p B_{p'}$ and $V = A_{p'} B_p$.*

Proof. Note that $\Phi(\mathrm{F}(G)) = 1$. Since $F = \mathrm{F}(G)$ contains no central chief factors, by Corollary 3.2.35 we obtain $F = X \times Y$, with AX and BY nilpotent and X and Y normal in G. Note that $X = \mathrm{C}_F(A_{p'})$ and $Y = \mathrm{C}_F(B_{p'})$ because $A\,\mathrm{C}_F(A_{p'})$ and $B\,\mathrm{C}_F(B_{p'})$ are nilpotent. Moreover we have $\mathrm{F}(G) = (A \cap F)(B \cap F)$ by Theorem 3.2.7 and hence

$$A \cap F \leq \mathrm{C}_F(A_{p'}) = X, \quad B \cap F \leq \mathrm{C}_F(B_{p'}) = Y.$$

If $A \cap F \neq X$, then we would have $X \cap (B \cap F) = X \cap B \neq 1$ and hence $X \cap B$ would be a non-trivial subgroup of fixed points for $A_{p'} B_{p'}$ and F would have central G-chief factors by Lemma 3.2.5.

It follows that $X = A \cap F$ and $Y = B \cap F$. Since $A \cap B \leq F$, we have

$$1 = X \cap Y = (A \cap F) \cap (B \cap F) = A \cap B.$$

Since $\Phi(G) = 1$, we have that F is a direct product of minimal normal subgroups of G by [119, A, 10.6], and complemented in G by [119, A, 9.2 (f)]. Let D be a complement. Set

$$U = (A \cap F)\mathrm{C}_D(B \cap F), \quad V = (B \cap F)\mathrm{C}_D(A \cap F).$$

We have that U is normal in G because $\mathrm{C}_D(B \cap F)$ is normalised by D and centralised by $B \cap F$. Similarly, V is normal in G. Assume that $U \cap V$ is non-trivial. Then it contains a minimal normal subgroup of G and so $U \cap V \cap F \neq 1$. Note that $U \cap V \cap F = (U \cap F) \cap (V \cap F) = (A \cap F) \cap (B \cap F) = 1$, a contradiction. Hence $U \cap V = 1$.

We can choose the complement D so that there is a Hall q'-subgroup Q of G with $Q = (Q \cap A)(Q \cap B) \leq D$ by Theorem 1.1.19. Then we have $Q \cap A \leq \mathrm{C}_D(A \cap F) \leq V$ and similarly $Q \cap B \leq U$. Hence $Q \leq UV$. In particular, G/UV is a p-group. Let M be a minimal normal subgroup of G contained in $B \cap F$ and let K be a complement for M in G with $U \leq K$. If $W = \mathrm{C}_K(M)$, then W is a normal subgroup of G containing U. Also MW/W is a self-centralising minimal normal subgroup of G/W and so, by Theorem 3.2.8, one of AW/W and BW/W is a p-group and the other is a p'-group. Since $MW/W \leq BW/W$, we have that BW/W is a p-group and AW/W is a p'-group. If $B \cap F = M_1 \times \cdots \times M_n$, with each M_i a minimal normal subgroup of G, and K_i is a complement for M_i in G as above, and W_i is the centraliser of M_i in K_i, then $W_1 \cap \cdots \cap W_n$ is normal in G, contains U, and intersects $B \cap F$ trivially, therefore it must equal U because $\mathrm{C}_G(B \cap F) = U(B \cap F)$. It then follows that BU/U is a p-group and AU/U is a p'-group. Hence AUV/UV

is a p'-group. Thus we have $A \leq UV$. Similarly we have $B \leq UV$ and hence $G = UV$. This yields $G = U \times V$.

By our choice of D, we have

$$Q \cap A \leq C_A(A \cap F) \cap D = C_D(A \cap F) \leq V.$$

Since $(Q \cap A)(Q \cap B) = Q$, it follows that $Q \cap A$ is a Hall p'-subgroup of V. Analogously, $Q \cap B$ is a Hall p'-subgroup of U. Moreover, we can choose a Sylow p-subgroup P of G with $P = (P \cap A)(P \cap B)$. Then, since $A/(A \cap U) \cong AU/U$ is a p'-group, we have $P \cap A \leq U$ and similarly $P \cap B \leq V$. Thus $P \cap A$ is a Sylow p-subgroup of U and $P \cap B$ is a Sylow p-subgroup of V. The theorem follows. □

If the condition $\Phi(G) = 1$ is removed from the hypotheses of Theorem 3.2.36, then the conclusion is no longer true.

Example 3.2.37 (Cossey and Stonehewer [108]). Let N be the extraspecial group of order 27 and exponent 3. Suppose that N is generated by two elements x and y and let θ denote the automorphism of N that inverts x and fixes y. Note that θ also inverts the commutator $[x, y]$. Let $H = [N]\langle\theta\rangle$ be the semidirect product of N and $\langle\theta\rangle$. Let M be a cyclic group of order 7. We can define a homomorphism of H into $\mathrm{Aut}(M)$ by letting x and θ centralise M and y act as the automorphism that squares elements of M. Now let K be the semidirect product of M and H. Note that $L = \langle x, [x, y], \theta\rangle$ is a normal subgroup of index 21 in K. Finally, let U be a faithful and irreducible module for K/L over the field of 3 elements and let G be the semidirect product of U and K.

Observe that $\mathrm{F}(G) = U \times \langle x, [x, y]\rangle$ is a 3-group containing no central chief factors and $\Phi(G) = \langle [x, y]\rangle$. Moreover $G = AB$, where $A = U\langle y, \theta\rangle$ and $B = M\langle x, [x, y]\rangle$ are both nilpotent and neither is a 3-group. If G were directly decomposable, say $G = S \times T$, then U would be contained in one of the factors, S say, since U is the unique non-cyclic chief factor of G. But then G/U would be isomorphic to the direct product of T and S/U. This would give that a Sylow 3-subgroup of G/U is the direct product of the Sylow 3-subgroups of T and S/U. Since the Sylow 3-subgroup of G/U is isomorphic to N, which is directly indecomposable, we would have either $|T|$ or $|S/U|$ prime to 3. If $|T|$ were prime to 3, we would have $T \leq C_G(U)$ and so $|T| = 2$. Since a Sylow 2-subgroup of G is non-central, this would be a contradiction. Now assume that $|S/U|$ is prime to 3. Hence U is the Sylow 3-subgroup of S. By Theorem 1.1.19, there exists a Sylow 3-subgroup T_3 of T such that UT_3 is a Sylow 3 subgroup of G. Note that UT_3 is conjugate to the Sylow 3-subgroup UN of G, say $UT_3 = (UN)^z$ with $z \in G$. Since $U \leq S$, U is central in UT_3. But then $U^{z^{-1}} = U$ is central in UN. It follows that N centralises U. This contradicts the fact that $C_G(U) = UL$ by the construction of G. Therefore G is directly indecomposable.

3.3 Numerical invariants of dinilpotent groups

In this section, we consider further contributions to the study of the relation of the structure of a dinilpotent group to the structure of its nilpotent factors. Our approach can be described in terms of the following general philosophy: Choose some numerical invariant of a dinilpotent group, for instance the derived length, the Fitting length, or the π-length. Try to bound it in terms of a given numerical invariant of the factors. These results can be applied to shed light on the structural complexity of a dinilpotent group.

A classical result of Hall and Higman establishes the standpoint for the rest of the section.

Theorem 3.3.1. *Let G be a group of order $n_1 n_2 \cdots n_r$, where $\gcd(n_i, n_j) = 1$ if $i \neq j$. Assume that, for every $i \in \{1, 2, \ldots, r\}$, G possesses a nilpotent subgroup of order n_i and class k_i. Then $d \leq k_1 + k_2 + \cdots + k_r$, where d is the derived length of G. In particular, for any soluble group G, the derived length of G is bounded by the sum of the nilpotency classes of its Sylow subgroups.*

A full account of the proof of this result can be found in [157, VI, 14.16]. Note that the group G in the above theorem must be soluble by Corollary 3.2.2.

Fix an positive integer n. Then the class $\mathfrak{F}$ of all groups X such that the nth term $X^{(n)}$ of the derived series is nilpotent is a saturated formation. Hence every group G not belonging to $\mathfrak{F}$ has an epimorphic image G/K in the boundary of $\mathfrak{F}$ and so G/K is a primitive group. This elementary observation, together with Theorems 3.2.8 and 3.3.1, allows us to establish the following results of Pennington [210] and Gross [139].

Theorem 3.3.2. *Let $G = AB$ be a dinilpotent group with A and B of nilpotency classes c and d, respectively. Then the $(c + d)$th term $G^{(c+d)}$ of the derived series of G is a nilpotent π-group, where $\pi = \pi(A) \cap \pi(B)$.*

Theorem 3.3.3. *Let $G = AB$ be a dinilpotent group with A and B of classes c and d, respectively. Then the Frattini factor group $G/\Phi(G)$ has derived length at most $c + d$. In particular, the Fitting length of G is at most $c + d$.*

Remark 3.3.4. For any dinilpotent group $G = AB$ with A and B of nilpotency classes c and d, respectively, we obtain with the help of Proposition 1.1.21 that

$$\begin{aligned} \mathrm{d}(G) &\leq \mathrm{d}(G/G^{(c+d)}) + \mathrm{d}(G^{(c+d)}) \\ &\leq c + d + \max_{p \in \pi}\{\mathrm{d}(\mathrm{O}_p(G))\} \\ &= c + d + \max_{p \in \pi}\{\mathrm{d}((A \cap \mathrm{O}_p(G))(B \cap \mathrm{O}_p(G)))\}. \end{aligned}$$

It is thus natural to conjecture that the derived length of a dinilpotent group should be bounded by a function of the nilpotency classes of the factors and in particular that the bound should be that given by Hall and Higman, the sum of the nilpotency classes of the factors.

The first explicit reference to this conjecture that we know of is in Kegel [182]. There is a reference to this conjecture in Question 5 on page 36 of [9]. Progress with this problem has been slow and sporadic and it was in 1998 that Cossey and Stonehewer [107] have produced examples to show that this conjecture is false in general: there are dinilpotent groups $G = AB$, with A of class m and B of class n and G of derived length $d > m + n$ for the following triples $(m, n, d) = (1, 2, 4)$, $(2, 2, 5)$, $(2, 2, 6)$. We include one of them here. One can see that these examples work only for small derived lengths and become increasingly difficult to find as the derived length increases; they are also consistent with a linear bound in the derived length of the factors. A bound for the derived length remains elusive. The above remark is telling us the problem of bounding $\mathrm{d}(G)$ is reduced to p-groups, p a prime.

Example 3.3.5 (Cossey and Stonehewer [107]). Let P be the non-abelian group of order 27 and exponent 3. The natural action of $\mathrm{SL}_2(3) = \langle \left(\begin{smallmatrix} 1 & 1 \\ 0 & 1 \end{smallmatrix}\right), \left(\begin{smallmatrix} 0 & 1 \\ 2 & 0 \end{smallmatrix}\right) \rangle$ on P/P' can be extended to give an action on P such that P' is centralised by $\mathrm{SL}_2(3)$ (note that if $P = \langle x, y \rangle$, $x^\alpha = xy$, $y^\alpha = y$, $x^\beta = y$ and $y^\beta = x^2$, then $[x, y]^\alpha = [x, y]^\beta = [x, y]$). Now fix a Sylow 3-subgroup of $\mathrm{SL}_2(3)$, R say (which is cyclic of order 3), and let Q denote the Sylow 2-subgroup of $\mathrm{SL}_2(3)$ (which is isomorphic to the quaternion group of order 8). Then let P_0 be isomorphic to P, and let $\mathrm{SL}_2(3)$ act on $P \times P_0$ by acting as $\mathrm{SL}_2(3)$ on the first factor and as R on the second. Since the action of R on the two copies of P is the same, we may choose an R-invariant diagonal subgroup D of $P \times P_0$. Now set $H = (P \times P_0)\mathrm{SL}_2(3)$. Again we have that D' is a normal subgroup of H, and we set $G = H/D'$, $A = DR/D'$, and $B = (Q \times P_0)D'/D'$. Note that $|A| = 3^3$, and since A is non-abelian, it is of class 2. Clearly, B is of class 2. Further, $DR \cap (Q \cap P_0)D' = D'$, and so $A \cap B = 1$. Again by order considerations we have $G = AB$. It is also easy to check that G has derived length precisely 5.

It is possible to establish a bound on the derived length of a dinilpotent group in terms of the nilpotency classes of the factors for some classes of dinilpotent groups as Cossey and Stonehewer showed in [108].

Lemma 3.3.6. *Let the group G be the product of two normal subgroups M and N. Suppose that $M = AB$ with A a p-group for some prime p and B is a nilpotent p'-group. Suppose further that N and $\mathrm{F}(M)$ are p-groups. Let c be the nilpotency class of A and d the maximum of the nilpotency classes of B and $N/(N \cap M)$. Then the derived length of G is at most $c + d$.*

Proof. We will prove the result by induction on c. First we note that M has derived length at most $c + d$ by Theorem 3.3.1 and that $M \cap N$ is a normal p-subgroup of M

and hence $M \cap N \leq \mathrm{F}(M)$. Since M has derived length at most $c + d$, we have that $M^{(c+d-1)}$ is abelian and hence contained in $\mathrm{F}(M)$. Moreover, $N/(N \cap M)$ has derived length at most d and so

$$N^{(c+d-1)} \leq N \cap M \leq \mathrm{F}(M).$$

Since $G/\mathrm{F}(M)$ is a central product of $M/\mathrm{F}(M)$ and $N/\mathrm{F}(M)$, it follows that

$$G^{(c+d-1)} \leq \mathrm{F}(M).$$

Note that if $\mathrm{F}(M)$ is abelian, then $G^{(c+d)} = 1$.

Assume first that $c = 1$. From the previous paragraph, it follows that $G^{(d)} \leq \mathrm{F}(M)$ and then, since $\mathrm{F}(M)$ is abelian, we have $G^{(d+1)} = 1$. Thus the result is true for $c = 1$.

Now suppose that $c > 1$ and that the result is true for groups in which A has smaller nilpotency class. Note that since $\mathrm{F}(M)$ is a p-group, we have $\mathrm{Z}(A) \leq \mathrm{Z}(\mathrm{F}(M))$, where $\mathrm{Z}(A)$ is the centre of A. We set $Z = \mathrm{Z}(\mathrm{F}(M))$ and note that Z is a normal subgroup of G. We chose T/Z to be a normal subgroup of G/Z which is maximal in M/Z satisfying $(T/Z) \cap (\mathrm{F}(M)/Z) = 1$. This implies that $\mathrm{F}(M/T)$ is a p-group and G/Z is isomorphic to a subdirect product of G/T and $G/\mathrm{F}(M)$. We know that $G/\mathrm{F}(M)$ has derived length at most $c + d - 1$. Now consider the derived length of G/T. We have that $G/T = (M/T)(NT/T)$, with $M/T = (AT/T)(BT/T)$, AT/T a p-group of class at most $c - 1$, BT/T a nilpotent p'-group, and NT/T and $\mathrm{F}(M/T)$ both p-groups. Moreover $(NT/T) \cap (M/T) = (NT \cap M)T = (N \cap M)T/T$, and hence

$$\begin{aligned}(NT/T)/((NT/T) \cap (M/T)) \cong NT/(N \cap M)T &\cong N/(N \cap (N \cap M)T) \\ &\cong N/(N \cap M)(N \cap T).\end{aligned}$$

Hence the nilpotency classes of BT/T and $(NT/T)/((NT/T) \cap (M/T))$ are at most d. By our inductive hypothesis, we now have that the derived length of G/T is at most $c + d - 1$. But then the derived length of G/Z is at most $c + d - 1$ and so $G^{(c+d-1)} \leq Z$. Since Z is abelian, we see that G has derived length at most $c + d$, as required. □

Theorem 3.3.7. *Let $G = AB$ be a dinilpotent with A and B of nilpotency classes c and d, respectively. Then if G satisfies that $\mathrm{F}(G)$ is a p-group for some prime p, and $\mathrm{F}(G)/\Phi(\mathrm{F}(G))$ contains no central chief factors of G, then the derived length of G is at most $c + d + 1$.*

Proof. If one of A and B is a p-group, by Lemma 3.2.34 the other is a p'-group and the theorem follows from Theorem 3.3.1. Hence we may suppose that G satisfies that $G = AB$ is a dinilpotent group, $\mathrm{F}(G)$ is a p-group for some prime p, and $\mathrm{F}(G)/\Phi(\mathrm{F}(G))$ contains no central chief factors of G, and neither A nor B is a p-group.

By Theorem 3.2.32, we see that $O^p(G)$ is the central product of $X = \langle (A_{p'})^G \rangle \neq 1$ and $Y = \langle (B_{p'})^G \rangle \neq 1$ and, by Theorem 3.2.36, $G/\Phi(G) = (U/\Phi(G)) \times (V/\Phi(G))$, with $U = A_{p'} B_p \Phi(G)$ and $V = A_p B_{p'} \Phi(G)$. Since U is normal in G and $A_{p'} \leq U$, we have $X = O^p(U)$ and similarly $Y = O^p(V)$.

We wish to apply Lemma 3.3.6 to G/Y, where $M = UY/Y$ and $N = VY/Y$, and so we check that the hypotheses are satisfied. We have that UY/Y is normal in G/Y and V/Y is a normal p-subgroup of G/Y. Also, since UY contains $A_{p'}$ and B, it follows that $V/(V \cap UY)$, which is isomorphic to G/UY, is isomorphic to a quotient of A_p. Next observe that we have $UY/Y = (A_{p'}Y/Y)(B_p\Phi(G)Y/Y)$, a product of a p'-group and a p-group. Now we prove that $F(UY/Y) \cong F(U/(U \cap Y))$ is a p-group. If $F(U/(U \cap Y))$ is not a p-group, then $U/(U \cap Y)$ has a non-trivial p'-subgroup $T/(U \cap Y)$ with T normal in G. But then $G/\Phi(G)$ would have a non-trivial normal p'-subgroup $T\Phi(G)/\Phi(G)$. This contradicts our hypotheses. Thus $F(UY/Y)$ is a p-group. Finally we show that $B_p\Phi(G)Y/Y$ is isomorphic to a quotient of B_p. Recall that

$$F(G) = (A \cap F(G))(B \cap F(G)) = (A_p \cap F(G))(B_p \cap F(G)).$$

Since $F(G)/\Phi(F(G))$ contains no central chief factors of G, we have $F(G) \leq O^p(G)$ and $F(G) = F(X)F(Y)$, $\Phi(F(G)) = \Phi(F(X))\Phi(F(Y))$. In particular, $\Phi(F(G)) \leq \Phi(F(X))Y \cap X\Phi(F(Y))$ and so $F(G)/\Phi(F(X))Y$ contains no central chief factors of G. Since $(A_p \cap F(G))\Phi(F(X))Y/\Phi(F(X))Y$ is centralised by $A_{p'}$, by Lemma 3.2.5 we must have $A_p \cap F(G) \leq \Phi(F(X))Y$. Then we have $\Phi(G)Y \leq (B_p \cap F(G))Y$ and hence $B_p\Phi(G)Y = B_pY$ and $B_p\Phi(G)Y/Y \cong B_p/(B_p \cap Y)$, as required.

We can now apply Lemma 3.3.6 with $M = UY/Y$ and $N = V/Y$ to deduce that G/Y has derived length at most $c + d$. Similarly, G/X has derived length at most $c + d$. Hence $G/(X \cap Y)$ has derived length at most $c + d$, and $X \cap Y$ is central in XY and so abelian. Consequently we have that G has derived length at most $c+d+1$, as desired. □

John Cossey informed us that the counterexamples to Kegel's conjecture came from analysing the structure of a group of derived length $c + d + 1$ in the above theorem.

Our next object is to give some bounds for the derived length of a nilpotent group in terms of the order of the derived subgroups of the factors. A couple of preparatory lemmas are needed. They were proved by Morigi [201].

Lemma 3.3.8. *Let the group $G = AB$ be the product of two subgroups A and B with A abelian and $Z(G) \neq 1$. Then there exists a non-trivial normal subgroup N of G such that $N \leq A$ or $N \leq B$.*

Proof. Let ab be a non-trivial element of $Z(G)$ with $a \in A$ and $b \in B$. We may assume that $a \neq 1 \neq b$. Then for each $x \in A$, we have $1 = [ab, x] = [a, x]^b[b, x] = [b, x]$. Hence $[b, A] = 1$ and so $\langle \langle b \rangle^G \rangle = \langle \langle b \rangle^{AB} \rangle = \langle \langle b \rangle^B \rangle \leq B$ is a normal subgroup of G contained in B, as desired. □

Remark 3.3.9. Lemma 3.3.8 is no longer true if A is not abelian, even for p-groups, as an example of Gillam [138] shows:

Let p be a prime and let V be an elementary abelian p-group with basis set $\{a_1, a_2, b_1, b_2\}$. Let a_3 and b_3 be the automorphisms of V whose matrices with respect to (a_1, a_2, b_1, b_2) are

$$\begin{pmatrix} 1 & 0 & 0 & 0 \\ 1 & 1 & 0 & 0 \\ -1 & 0 & 1 & 0 \\ 0 & 1 & 1 & 1 \end{pmatrix} \quad \text{and} \quad \begin{pmatrix} 1 & -1 & -1 & 0 \\ 0 & 1 & 0 & -1 \\ 0 & 0 & 1 & 1 \\ 0 & 0 & 0 & 1 \end{pmatrix},$$

respectively. Let $H = \langle a_3, b_3 \rangle$. Then a direct computation shows that H is an elementary abelian group of order p^2. Let P be the semidirect product $P = [V]H$. Then the order of P is p^6. Let $A = \langle a_1, a_2, a_3 \rangle$ and $B = \langle b_1, b_2, b_3 \rangle$. Then $A \cap B = 1$ and $|A| = |B| = p^3$. Hence $P = AB$. Since $\mathrm{C}_P(V) = V$, $\mathrm{Z}(P) \leq V$. Hence $\mathrm{Z}(P) \cap A = \mathrm{Z}(P) \cap A \cap V = \mathrm{Z}(P) \cap \langle a_1, a_2 \rangle = 1$. Therefore $A_P = 1$. An analogous argument shows that $B_P = 1$.

Next comes a very useful result about factorised p-groups with one of the factors abelian.

Lemma 3.3.10. *Let the p-group $G = AB$ be the product of an abelian subgroup A and a non-abelian subgroup B. Then there exists a subgroup $W \leq \mathrm{Z}(B)$ such that $AW = WA$ and $AW \cap B' \neq 1$.*

Proof. Let $W \leq \mathrm{Z}(B)$ be chosen maximal such that AW is a subgroup and assume by contradiction that $AW \cap B' = 1$. Let $K = (AW)_G$, the core of AW in G. Then $G/K = (AK/K)(BK/K)$ and AK/K is core-free. By Lemma 3.3.8, there exists a minimal normal subgroup N/K of G/K such that $N \leq BK$ and so $N = ZK$, where $Z = B \cap N$ is not contained in K. Of course, $N/K \leq \mathrm{Z}(G/K)$ since G is a p-group. Moreover $[Z, B] \leq K \cap B' \leq AW \cap B' = 1$ and so $Z \leq \mathrm{Z}(B)$. It follows that $AWZ = AWKZ = AWN$ is a subgroup because N is a normal subgroup of G. Moreover Z is not contained in W, otherwise we would have $N = ZK \leq (AW)_G = K$. As $WZ \leq \mathrm{Z}(B)$, this contradicts the maximality of W. The proof of the lemma is now complete. □

We are now in a position to undertake the proof of an interesting particular case due to Morigi [201].

Theorem 3.3.11. *Let $G = AB$ be a p-group, p a prime, where A is abelian and $|B'| = p^n$. Then the derived length of G is at most $n + 2$.*

Proof. We will prove the result by induction on n. For $n = 0$, the result is true by Theorem 3.1.7.

We now assume that the result holds for each natural number smaller than $n-1$. If $A \cap B$ is not contained in $Z(B)$, then $N = \langle (A \cap B)^B \rangle = \langle (A \cap B)^{AB} \rangle = \langle (A \cap B)^B \rangle = (A \cap B)[A \cap B, B]$ is a normal subgroup of G contained in B, and since $[A \cap B, B] \neq 1$, $N \cap B' \neq 1$. If N is abelian, then $d(G) \leq n+2$ because $|(B/N)'| \leq p^{n-1}$ and so by induction $d(G/N) \leq n+1$. If N is non-abelian, then N' is a normal subgroup of G of order p^t, $t \geq 1$. Moreover $|(B/N')'| = p^{n-t}$, so by induction $d(G/N') \leq n-t+2$. Since $d(N') \leq t$, the result follows. Hence we may assume that $A \cap B \leq Z(B)$.

By Lemma 3.3.10, there exists $W \leq Z(B)$ such that $AW = WA$ and $1 \neq T = AW \cap B'$. We have that $T \leq AW \cap B = (A \cap B)W \leq Z(B)$, and moreover $[A, T] \leq [A, AW] = [A, W] \leq Z([A, B])$ by Lemma 3.1.6. Consider $\langle A^G \rangle = A(\langle A^G \rangle \cap B)$. Suppose that $T \leq (\langle A^G \rangle \cap B)'$. Then $T \leq [A, B]$ because $A[A, B]/[A, B] \cong A/(A \cap [A, B])$ is abelian and so $\langle A^G \rangle' = (A[A, B])' \leq [A, B]$. We conclude that $\langle T^G \rangle = T[A, T]$ is abelian. By induction, $d(G/\langle T^G \rangle) \leq n+1$. It follows that $d(G) \leq n+2$. Hence we may assume that T is not contained in $(\langle A^G \rangle \cap B)'$ and then we consider $\langle A^G \rangle T$, which is a normal subgroup in G as $T \leq Z(B)$. We have $\langle A^G \rangle T = A(\langle A^G \rangle T \cap B) = A(\langle A^G \rangle \cap B)T$, and $((\langle A^G \rangle \cap B)T)' = (\langle A^G \rangle \cap B)'$ as $T \leq Z(B)$. Let $p^m = |((\langle A^G \rangle \cap B)T)'|$. Note that $(\langle A^G \rangle \cap B)'$ is a proper subgroup of B' because it does not contain T. By induction, $d(\langle A^G \rangle T) \leq m+2$. Let now $|\langle A^G \rangle T \cap B'| = p^{m+t}$, where $t \geq 1$ because $T \leq \langle A^G \rangle T \cap B'$ and $T \not\leq ((\langle A^G \rangle \cap B)T)'$. The derived subgroup of $G/\langle A^G \rangle T \cong B/(B \cap \langle A^G \rangle T)$ has order $p^{n-(m+t)}$, so $d(G/\langle A^G \rangle T) \leq n-(m+t)+1$ by Lemma 1.7.5. It follows that $d(G) \leq n+2-t+1 \leq n+2$ and the proof of the theorem is complete. □

The case when B' has order p is of interest.

Theorem 3.3.12 (Morigi [201]). *Let $G = AB$ be a p-group, p a prime, where A is abelian and B' has order p. Then one of the following must occur:*

1. *$\langle A^G \rangle B'$ is the product of two abelian groups.*

2. *$\langle (B')^G \rangle$ is abelian.*

In both cases, the derived length of G is at most 3.

Proof. The last part of the thesis of this theorem follows from Theorem 3.3.11 with $n = 1$.

We adhere closely to the pattern of the proof of Theorem 3.3.11, taking into account that $n = 1$ and that the subgroup T considered there is actually B'. If $A \cap B$ is not contained in $Z(B)$ and N is abelian, we obtain that $B' \leq N$ and if N is not abelian, then $B' = N'$; in both cases we conclude that B' is contained in an abelian normal subgroup of G and so $\langle (B')^G \rangle$ is abelian. Suppose that $A \cap B \leq Z(B)$ and $B' \leq (\langle A^G \rangle \cap B)'$, then $\langle (B')^G \rangle$ is abelian. In the last part of the proof, we note that $|\langle A^G \rangle B' \cap B'| = |B'| = p^{m+t} = p$, which implies that $m = 0$ and $t = 1$.

Consequently $|((\langle A^G\rangle \cap B)B')'| = 1$ and $\langle A^G\rangle B'$ is the product of the abelian groups A and $(\langle A^G\rangle \cap B)B'$. □

The nilpotency class of the commutator $[A, B]$ of A and B in a p-group $G = AB$ which is the product of an abelian group A and a nilpotent group B can be bounded in terms of the order of B' (Mann [197]).

Theorem 3.3.13. *Assume that the p-group $G = AB$ is the product of an abelian group A and a group B satisfying $|B'| = p^n$. Then* $\mathrm{c}([A, B]) \le 2n + 1$.

Proof. We prove the result by induction on n. The case $n = 0$ is just Theorem 3.1.7. Hence we can assume that B is not abelian.

Suppose first that $T = A \cap B' \neq 1$. Then $\langle T^G\rangle = \langle T^{AB}\rangle = \langle T^B\rangle \le B'$. Let N be a minimal normal subgroup of G contained in $\langle T^G\rangle$. Then $N \le \mathrm{Z}(G)$ and $|(B/N)'| = p^{n-1}$. By induction, $\mathrm{c}([A, B]N/N) \le 2(n-1) + 1$ and so $\mathrm{c}([A, B]) \le 2n$.

Assume now that $A \cap B' = 1$. Let $Z = \mathrm{Z}([A, B])$ and consider

$$G/Z = (AWZ/Z)(BZ/Z),$$

where W is the subgroup given by Lemma 3.3.10. As in the proof of Theorem 3.3.11, we note that taking $V = A$ in Lemma 3.1.6 we have that $[A, W] \le Z$. Therefore AWZ/Z is abelian. We also note that $AWZ \cap (BZ)' \ge AW \cap B'$. Suppose that $AW \cap B'$ is contained in Z. Then $|(BZ/Z)'| < |B'|$, so by induction $\mathrm{c}([A, B]/Z) \le 2(n-1) + 1$ and $\mathrm{c}([A, B]) \le 2n$. Assume that $AW \cap B'$ is not contained in Z. Then the factorisation $G/Z = (AWZ/Z)(BZ/Z)$ satisfies the assumption of the case $A \cap B' \neq 1$. Hence

$$\mathrm{c}([A, B]/Z) \le 2n$$

and so $\mathrm{c}([A, B]) \le 2n + 1$. □

Mann [197] proved that if G is a group as in Theorem 3.3.13, then

$$\mathrm{d}(G) < 2\log_2(n+2) + 3.$$

The following result provides a slight improvement of this bound.

Corollary 3.3.14. *Let G be a group as in Theorem* 3.3.13. *Then*

$$\mathrm{d}(G) < 2\log_2(n+1) + 3.$$

Proof. By Theorem 3.1.7, we can assume that $n \ge 1$. We know that $[A, B]$ is a normal subgroup of G by [119, A, 7.4]. Since $G/[A, B]$ is a central product of the subgroups $A[A, B]/[A, B]$ and $B[A, B]/[A, B]$ and A is abelian, the derived length of $G/[A, B]$ is bounded by the derived length of B. On the other hand, B has nilpotency class at most $n + 1$ by Lemma 1.7.5. This implies that $\mathrm{d}(G/[A, B]) \le \mathrm{d}(B) \le$

$\log_2(n+1)+1$ by Theorem 1.7.4. Moreover, $[A, B]$ has nilpotency class at most $2n+1$ by Theorem 3.3.13. Hence $\mathrm{d}([A, B]) \leq \log_2(2n+1)+1$. Consequently

$$\begin{aligned}\mathrm{d}(G) &\leq \mathrm{d}(G/[A,B]) + \mathrm{d}([A,B]) \leq \log_2(n+1) + \log_2(2n+1) + 2 \\ &= \log_2((n+1)(2n+1)) + 2 < \log_2((n+1)^2) + 3 = 2\log_2(n+1) + 3,\end{aligned}$$

as desired. □

For a factorised p-group, p a prime, Mann [197] gave the following asymptotic bound for the derived length.

Corollary 3.3.15. *Given $\varepsilon > 0$ there exists $C = C(\varepsilon)$ such that if $G = AB$ is a p-group and $|A'| = p^m$ and $|B'| = p^n$, then* $\mathrm{d}(G) \leq \varepsilon n + (m+1)C$.

Proof. We choose C such that $2\log_2(k+1) + 3 \leq \varepsilon k + C$ holds for all k. If A is abelian, the result follows from Corollary 3.3.14. Assume that A is non-abelian. Then we repeat the construction of Kazarin [170]. Let $H = \langle A', B\rangle$. Then $H = A_1 B$, where $A_1 = H \cap A$. Here $A' \leq A_1$, so A_1 is a normal subgroup of A, and $N = \langle A_1^G \rangle = \langle A_1^B \rangle \leq H$, so $N = A_1 B_1$, where $B_1 = N \cap B$. Let $|(B/B_1)'| = p^k$. Suppose first that $A_1' \neq A'$. Then by induction $\mathrm{d}(N) \leq mC + \varepsilon(n-k)$, while $\mathrm{d}(G/N) \leq \varepsilon k + C$ by Corollary 3.3.14. Hence the result follows. Finally, if $A_1' = A'$, then $H = \langle A', B\rangle = \langle H', B\rangle = B$, so $A' \leq B$ and $R = \langle (A')^G \rangle = \langle (A')^B \rangle \leq B$. Let $p^t = |(B/R)'|$. Then R' is contained in $B' \cap R$ and so $|R'| \leq p^{n-t}$. Applying induction and Corollary 3.3.14, we have

$$\mathrm{d}(G) \leq \mathrm{d}(G/R) + \mathrm{d}(R) \leq \varepsilon t + C + \varepsilon(n-t) + C \leq \varepsilon n + (m+1)C,$$

by the choice of C and the assumption $m \geq 1$. □

Now we are in a position to give bounds for the derived length of a factorised nilpotent group in terms of the orders of the derived subgroups of the factors. We will use the following notation.

Definition 3.3.16 (Kazarin [170]). Given a natural number $n = p_1^{\alpha_1} p_2^{\alpha_2} \cdots p_r^{\alpha_r}$, we denote by $\gamma(n) = \max_{1 \leq i \leq k} \alpha_i$, and for a nilpotent group X, $\gamma(X) = \gamma(|X'|)$.

Kazarin proved in [170] that the derived length of G is bounded by $2\gamma(A) + 2\gamma(B) + 2$. This bound was sharpened by Morigi in [201] for p-groups by using the same techniques and Theorem 3.3.11.

Theorem 3.3.17. *Let the nilpotent group $G = AB$ be the product of two subgroups A and B such that $\gamma(A) = m$ and $\gamma(B) = n$ with $m \geq n$. Then the derived length of G is at most $m + 2n + 2$.*

Proof. Since the derived length of a nilpotent group is the maximum of the derived lengths of its Sylow subgroups, we can assume without loss of generality that G is a p-group. Note that if $m \geq n$, then $m+2n+2 \leq 2m+n+2$. We argue by induction on $n+m$. If $n+m=0$, then $A' = 1 = B'$ and the result is true by Theorem 3.1.7. If B is abelian the result follows from Theorem 3.3.11. We may assume that $B' \neq 1$ and $A \neq G$. Then $B' \leq G' \leq \Phi(G)$. Hence $H = \langle A, B' \rangle \neq G$. Moreover, $H = A(H \cap B) \neq G$ and $BH = BA = G$ with $B' \leq B \cap H$. Since $B \cap H$ is a normal subgroup of B, then $N = H_G = H_{AB} = H_B$ contains $H \cap B$. Hence $N = N \cap H = N \cap A(H \cap B) = (N \cap A)(H \cap B)$ is a normal subgroup of G. Let $k = \gamma(N \cap A)$. Since $G/N = (AN/N)(BN/N)$ and $BN/N \cong B/(B \cap N)$ is abelian, we have by Theorem 3.3.11 that $\mathrm{d}(G/N) \leq \gamma(AN/N) + 2$. Now $(AN/N)' \cong A'/(A' \cap N)$ and $A' \cap N \geq (A \cap N)'$, which implies that $\gamma(AN/N) + \gamma(N \cap A) \leq \gamma(A)$ and $\mathrm{d}(G/N) \leq \gamma(A) - k + 2$. Assume first that $r = \gamma(B \cap H) \leq \gamma(B) - 1$. By the induction hypothesis and the remark made at the beginning of the proof, $\mathrm{d}(N) \leq k + 2r + 2$. We conclude that

$$\begin{aligned}\mathrm{d}(G) &\leq \mathrm{d}(G/N) + \mathrm{d}(N) \leq \gamma(A) - k + 2 + k + 2r + 2 \\ &= \gamma(A) + 2r + 4 \leq \gamma(A) + 2(\gamma(B) - 1) + 4 = \gamma(A) + 2\gamma(B) + 2,\end{aligned}$$

as desired. Thus we can assume that $\gamma(B \cap H) = \gamma(B)$. This means that $(B \cap H)' = B'$. Since then $B' \leq H' \leq \Phi(H)$, we obtain that $H = \langle A, B' \rangle = A$. In particular, $B' \leq A$. In this case $N = A_B$. Hence $\mathrm{d}(N) \leq k + 1$ by Lemma 1.7.5. Therefore

$$\mathrm{d}(G) \leq \mathrm{d}(G/N) + \mathrm{d}(N) \leq \gamma(A) - k + 2 + k + 1 = \gamma(A) + 3 \leq \gamma(A) + 2\gamma(B) + 2.$$

The proof of the theorem is now complete. □

In the case of a product $G = AB$ of an extraspecial p-group and a group B of class at most 2 and $|B'| = p^n$, Theorem 3.3.17 gives that the derived length of G is bounded by $n+4$. The following result of McCann [198] improves this bound in one unit.

Theorem 3.3.18. *Let the group $G = AB$ be the product of an extraspecial p-group A and a group B such that $B' \leq \mathrm{Z}(B)$ (so that B has class two or is abelian) and $|B'| = p^n$. Then $G^{(n+3)} = 1$.*

Proof. We argue by induction on n. For $n = 0$, the group B is abelian and the result follows from Theorem 3.3.11. If $B' \cap B_G \neq 1$, then either B_G is an abelian normal subgroup of G or we have $1 \neq (B_G)' \leq \mathrm{Z}(B_G)$, which is a normal subgroup of G. In both cases there exists an element $1 \neq g \in B'$ such that $\langle \langle g \rangle^G \rangle$ is an abelian subgroup of B_G. Now $G/\langle \langle g \rangle^G \rangle$ is the product of $A\langle \langle g \rangle^G \rangle/\langle \langle g \rangle^G \rangle$ and $B/\langle \langle g \rangle^G \rangle$ and

$$|(B/\langle \langle g \rangle^G \rangle)'| = |B'\langle \langle g \rangle^G \rangle/\langle \langle g \rangle^G \rangle| = |B'/(B' \cap \langle \langle g \rangle^G \rangle)| \leq p^{n-1}.$$

Furthermore $A\langle\langle g\rangle^G\rangle/\langle\langle g\rangle^G\rangle$ is extraspecial or abelian. By the induction hypothesis or Theorem 3.3.11, we obtain that $G^{(n-1+3)} \leq \langle\langle g\rangle^G\rangle$, which implies that $G^{(n+3)} = 1$.

We now assume that $B' \cap B_G = 1$. Since $[B, B_G] \leq B' \cap B_G = 1$, we have that $B \leq \mathrm{C}_G(B_G)$, which is a normal subgroup of G, and so $\langle B^G\rangle \leq \mathrm{C}_G(B_G)$. Assume that B is normal in G. We see that $G/B \cong A/(A \cap B)$ and so $G^{(2)} \leq B$. Since B has class 2, $G^{(4)} = 1$. Since $n \geq 1$, we have $G^{(n+3)} \leq G^{(4)} = 1$. Hence we may also assume that B is a proper subgroup of $\langle B^G\rangle$.

Now $\langle B^G\rangle = B(\langle B^G\rangle \cap A)$ and $\langle B^G\rangle \cap A$ is a non-trivial normal subgroup of A. Since A is extraspecial, we conclude that $A' \leq \langle B^G\rangle \cap A$. If $\langle B^G\rangle \cap A$ is abelian then, by Theorem 3.3.11, $(\langle B^G\rangle)^{(n+2)} = 1$. Since $G/\langle B^G\rangle \cong A/(A \cap \langle B^G\rangle)$, which is abelian, we have $G' \leq \langle B^G\rangle$ and so $G^{(n+3)} = 1$, as desired. Hence we must only consider the case that $\langle B^G\rangle \cap A$ is non-abelian. If $\langle (A')^G\rangle$ were abelian, then the derived length of $G/\langle (A')^G\rangle$ would be at most $n+2$ by Theorem 3.3.11, and it would follow that $G^{(n+3)} = 1$. Consequently, we may assume that $\langle (A')^G\rangle$ is not abelian. This implies that $A \cap B_G = 1$ (otherwise, $A' \leq B_G$ and so $\langle (A')^G\rangle$ would be abelian). Thus we have $AB_G/B_G \cong A$.

Suppose that the core of AB_G/B_G in G/B_G is non-trivial. Then $A'B_G/B_G$ is a normal subgroup of G/B_G. Thus $A'B_G$ is a normal subgroup of G. But $A' \leq \langle B^G\rangle$, and so A' centralises B_G. Since B_G is abelian and A' is cyclic, $A'B_G$ is an abelian normal subgroup of G. Hence $\langle (A')^G\rangle$ is abelian which is not the case.

Hence we can assume that both AB_G/B_G and B/B_G are core-free subgroups of G/B_G. In order to make the notation easier, we now work modulo B_G and assume $B_G = 1$. There exists an element $x = ab$, where $1 \neq a \in A$ and $1 \neq b \in B$, such that $x \in \mathrm{Z}(G)$ and $\langle x\rangle$ of order p. If $a \in \mathrm{Z}(A)$ then, for all $a_1 \in A$,

$$[x, a_1] = [ab, a_1] = [a, a_1]^b [b, a_1] = [b, a_1] = 1$$

because $[a, a_1] = 1$. Thus $1 \neq b \in \mathrm{C}_B(A)$. Now

$$\langle \mathrm{C}_B(A)^G\rangle = \langle \mathrm{C}_B(A)^B\rangle,$$

which is a normal subgroup of contained in B. This contradiction shows that $a \notin \mathrm{Z}(A)$. A similar argument shows that $b \notin \mathrm{Z}(B)$.

To prove the theorem it is enough to show that $G^{(n+2)} = 1$. Let $C = \mathrm{C}_G(b)$. Since $x = ab \in \mathrm{Z}(G)$ and $a = abb^{-1} = xb^{-1}$, we conclude that

$$C = \mathrm{C}_G(b) = \mathrm{C}_G(a) = \mathrm{C}_G(\langle a, b\rangle).$$

Since B has class two, $B' \leq \mathrm{Z}(B) \leq C$, and so $B \cap C$ is a normal subgroup of B. In addition, $A \cap C$ is a normal subgroup of A. Now let $N = \mathrm{N}_G(C)$. Assume that $G = NB$. Then

$$C_G = \bigcap_{\substack{n \in N \\ b \in B}} C^{nb} = \bigcap_{b \in B} C^b.$$

Since $B \cap C$ is a normal subgroup of B, we then have that $b \in B \cap C \leq C_G$, and, moreover, $x \in C_G \cap \mathrm{Z}(G)$. Hence $\langle a, b\rangle \leq C_G$. Moreover $\langle a, b\rangle$ is a subgroup of $\mathrm{Z}(C_G)$, which is normal in G. Hence $\langle \langle a, b\rangle^G\rangle$ is abelian. Since $a \notin \mathrm{Z}(A)$, $A' = [A, a] \leq \langle \langle a, b\rangle^G\rangle$. In addition, $1 \neq [B, b] \leq \langle \langle a, b\rangle^G\rangle$. Let $H = \langle \langle a, b\rangle^G\rangle$. Furthermore, G/H is the product of AH/H, which is abelian, and BH/H. Since $(BH/H)' = B'H/H \cong B'/(B' \cap H)$ and $1 \neq [B, b] \leq B' \cap H$, we see that $|(BH/H)'| \leq p^{n-1}$. By Theorem 3.3.11, $G^{(n-1+2)} = G^{(n+1)} \leq H$ and, since H is abelian, we see that $G^{(n+2)} = 1$.

We may now assume that $G \neq NB$. Suppose that $B \cap N = B \cap C$. Then

$$|BN| = \frac{|B|\,|N|}{|B \cap N|} = \frac{|B|\,|N|}{|B \cap C|} = |N|\frac{|B|}{|B \cap C|}.$$

If $C = N$, then $C = G = NB$ since G is a p-group. Therefore C is a proper subgroup of N and so $|N| \geq p|C|$. Hence

$$|BN| \geq p|C|\frac{|B|}{|B \cap C|} = p\frac{|B|\,|C|}{|B \cap C|} = p|BC|.$$

Now $|A : A \cap C| = p$. Hence for the subsets BC and $B(A \cap C)$ we have

$$\begin{aligned} |BC| \geq |B(A \cap C)| &= \frac{|B|\,|A \cap C|}{|B \cap A \cap C|} = \frac{1}{p}\frac{|B|\,|A|}{|B \cap A \cap C|} \\ &\geq \frac{1}{p}\frac{|B|\,|A|}{|B \cap A|} = \frac{1}{p}|BA| = \frac{1}{p}|G|. \end{aligned}$$

Thus

$$|BN| \geq p|BC| \geq p\Big(\frac{1}{p}|G|\Big) = |G|,$$

that is, $BN = NB = G$, case which has been discarded. Hence we have that $B \cap C$ is a proper subgroup of $B \cap N$.

Let $b_1 \in (B \cap N) \setminus (B \cap C)$. Thus b_1 normalises C but does not commute with b. Hence, since $b \in \mathrm{Z}(C) \trianglelefteq N$,

$$1 \neq [b_1, b] \in \mathrm{Z}(C).$$

Let $C_1 = \mathrm{C}_G([b_1, b])$. Then $C \leq C_1$ and, since B has class two, $B \leq C_1$. If $C_1 = G$, then we have $1 \neq \langle [b_1, b]\rangle$ is s normal subgroup of G contained in B. This has been ruled out. Since from the above argument

$$|B(A \cap C)| \geq \frac{1}{p}|G|,$$

we conclude that $C_1 = B(A \cap C)$. In particular, $BC = B(A \cap C) = C_1$ is a normal subgroup of G.

If $\langle b_1\rangle C = BC$, then, since C is not normal in G (because otherwise $G = N = BN$), $N = \langle b_1\rangle C = BC$ and so $|G : N| = p$. Thus C has exactly p conjugates in G. Let $a_1 \in A \setminus (A \cap C)$. If a_1 normalises C then, by order considerations, $A = (A \cap C)\langle a_1\rangle$ normalises C. But now B (which is contained in $BC = N$) and A normalise C, which implies that C is a normal subgroup of G, which has been ruled out. Hence a_1 does not normalise C and the conjugates of C in G are $C, C^{a_1}, \dots, C^{a_1^{p-1}}$. Since $A \cap C$ is a normal subgroup of A, we then have

$$A \cap C \leq C_G = \bigcap_{i=0}^{p-1} C^{a_1^i}.$$

We obtain again that $\langle a, b\rangle \leq C_G$ and, as above, $G^{(n+2)} = 1$.

If $\langle b_1\rangle C$ is a proper subgroup of BC, then there exists an element c_2b_2, with $c_2 \in C$ and $b_2 \in B$, such that $c_2b_2 \in BC \setminus \langle b_1\rangle C$ and c_2b_2 normalises $\langle b_1\rangle C$. We have

$$(\langle b_1\rangle C)^{c_2b_2} = (\langle b_1\rangle C)^{b_2},$$

and so b_2 normalises $\langle b_1\rangle C$ but does not centralise b. Moreover, $b_2 \notin \langle b_1\rangle C$.

Suppose that $[b_2, b] \in \langle [b_1, b]\rangle$, say $[b_2, b] = [b_1, b]^\alpha$ with $\alpha \in \mathbb{Z}$. Then, since B has class two,

$$[b_2^{-1}b_1^\alpha, b] = [b_2, b]^{-1}[b_1, b]^\alpha = [b_2, b]^{-1}[b_2, b] = 1,$$

and so $b_2^{-1}b_1^\alpha \in C$. Then

$$b_2 = b_1^\alpha b_1^{-\alpha} b_2 = b_1^\alpha (b_2^{-1}b_1^\alpha)^{-1} \in \langle b_1\rangle C,$$

case which has been ruled out. We conclude that $[b_2, b] \notin \langle [b_1, b]\rangle$.

Now $[b_1, b] \in \mathrm{Z}(C_1) = \mathrm{Z}(BC)$, and so $\langle [b_1, b]\rangle$ is a normal subgroup of BC. We have $b\langle [b_1, b]\rangle \in \mathrm{Z}(\langle b_1\rangle C/\langle [b_1, b]\rangle)$. Hence, from above,

$$1 \neq [b_2, b]\langle [b_1, b]\rangle \in \mathrm{Z}(BC/\langle [b_1, b]\rangle).$$

Thus $\langle [b_2, b], [b_1, b]\rangle$ is an abelian normal subgroup of B of order at least p^2. Let W be a subgroup of $\langle [b_2, b], [b_1, b]\rangle$ such that W is a normal subgroup of BC and $|W| = p^2$.

We know that $\langle B^G\rangle \cap A \leq B(A \cap C) \cap A = (B \cap A)(A \cap C)$. Assume that $A \cap C$ is abelian. Since $\langle B^G\rangle \cap A$ is non-abelian, it follows that $A \cap B$ is not contained in $A \cap C$ and so $A = (B \cap A)(A \cap C)$ because $|A : A \cap C| = p$. Thus $G = BA = B(A \cap C) = BC$, contrary to our supposition. Thus $A \cap C$ is non-abelian. Now $A' \leq C$ and so $A \cap C$ is a normal subgroup of A. Let $A_2 = \mathrm{C}_{A\cap C}(W)$ and suppose, for a contradiction, that A_2 is not a normal subgroup of A. Then $A_2 \cap A' = 1$ and so A_2 is abelian. Since p is the largest power of p dividing $|\mathrm{Aut}(W)|$ and $(A \cap C)/A_2$ is a p-subgroup of $\mathrm{Aut}(W)$, it follows that $|A \cap C : A_2| = p$. Since A'

is not contained in A_2, we conclude that $A \cap C = A_2 A'$. This implies that $A \cap C$ is abelian, a contradiction which yields $A' \leq A_2$ and so A_2 is a normal subgroup of A. Now B centralises B', which contains W. Since ab also centralises W, we have $a \in A_2$. In addition,

$$\mathrm{C}_{BC}(W) = B(\mathrm{C}_{BC}(W) \cap (A \cap C)) = BA_2.$$

Thus BA_2 is a subgroup and

$$\langle A_2^G \rangle = \langle A_2^{AB} \rangle = \langle A_2^B \rangle \leq BA_2.$$

Now $a \in \langle A_2^B \rangle$ and so $b \in \langle ab \rangle \langle A_2^B \rangle$, which is a normal subgroup of G. Therefore,

$$\langle [b_2, b], [b_1, b] \rangle \leq \langle ab \rangle \langle A_2^B \rangle.$$

Thus $W \leq \langle ab \rangle \langle A_2^B \rangle$. From above, $\langle ab \rangle \langle A_2^B \rangle \leq \mathrm{C}_{BC}(W)$, and so

$$W \leq \mathrm{Z}(\langle ab \rangle \langle A_2^B \rangle) \trianglelefteq G.$$

We conclude that $\langle W^G \rangle$ is abelian. Since $|\langle W^G \rangle| \geq |W| = p^2$, we see that $G/\langle W^G \rangle$ is the product of $A\langle W^G \rangle / \langle W^G \rangle$, which is either extraspecial or abelian, and $B\langle W^G \rangle / \langle W^G \rangle$, where

$$|(B\langle W^G \rangle / \langle W^G \rangle)'| = |B'/(B' \cap \langle W^G \rangle)| \leq p^{n-2}.$$

Thus, either by our induction hypothesis or by Theorem 3.3.11,

$$G^{(n-2+3)} \leq \langle W^G \rangle,$$

and we conclude that $G^{(n-2)} \leq \langle W^G \rangle' = 1$, as desired. □

Corollary 3.3.19. *Let the group $G = AB$ be the product of two extraspecial subgroups A and B. Then $G^{(4)} = 1$.*

Proof. If A and B are extraspecial p-groups, then $G = AB$ has derived length at most 4 by Theorem 3.3.18. Assume that A is an extraspecial p-group and B is an extraspecial q-group for primes $p \neq q$. By Theorem 3.3.1, then $G = AB$ has derived length at most 4. This completes the proof of the corollary. □

Another more recent result about the derived length of a dinilpotent group that covers a special case is due to Jabara [166]. In order to state and prove it we need some preliminary observations.

Definition 3.3.20. Let the class $\mathfrak{A}_1$ be the class of all abelian p-groups for a prime p. We say that a group p-group B of order p^r belongs to the class $\mathfrak{A}_n$, $n > 1$, if for every chief series

$$1 = K_0 < K_1 < \cdots < K_r = B$$

there exists an abelian term K_i with $B/K_i \in \mathfrak{A}_{n-1}$.

Notation 3.3.21. If $1 = K_0 < K_1 < \cdots < K_r = B$ is a chief series for a p-group B, we will denote by K_* the largest abelian term of the series and K_{*+1} the next term of the series.

It is clear then that, in order to prove that $B \in \mathfrak{A}_n$, it is enough to check that for every chief series $1 = K_0 < K_1 < \cdots < K_n = B$ of B, $B/K_* \in \mathfrak{A}_{n-1}$.

Lemma 3.3.22. *The class $\mathfrak{A}_n$ is closed under taking homomorphic images.*

Lemma 3.3.23. *Let B be a p-group of order p^r. If B contains a subgroup E of index p^k such that $E \in \mathfrak{A}_n$, then $B \in \mathfrak{A}_{n+k}$.*

Proof. We argue by induction on k.

Assume first that $k = 1$ and $1 = K_0 < K_1 < \cdots < K_r = B$ is a chief series of B; it is enough to show that $B/K_* \in \mathfrak{A}_n$.

We can distinguish three cases:

1. Suppose that $K_{*+1} \leq E$. We analyse this case arguing by induction on r. If $r = 1$, then $|B| = p$ and the initial step is trivial. Since K_{*+1} is not abelian and $E \in \mathfrak{A}_n$, it is clear that $n > 1$. Since $E \in \mathfrak{A}_n$, it follows that $E/K_* \in \mathfrak{A}_{n-1}$. In $\bar{B} = B/K_*$, the subgroup $\bar{E} = E/K_*$ has index p. Thus, by induction on r, $\bar{B} \in \mathfrak{A}_n$.

2. Assume that $K_* \not\leq E$. Since E is a maximal subgroup of B, we have $B = EK_*$ and so $B/K_* = EK_*/K_* \cong E/(E \cap K_*) \in \mathfrak{A}_n$.

3. Assume now that $K_* \leq E$ and $K_{*+1} \not\leq E$. Then $K_{*+1} = K_*\langle t\rangle$, where $t \notin E$ and $t^p \in K_*$. If H is a subgroup of G, we write $\bar{H} = HK_*/K_*$. Then $\bar{t} \in \mathrm{Z}(\bar{B})$ and $\bar{t}^p = \bar{1}$ so that $\bar{B} = \bar{E}\langle\bar{t}\rangle = \bar{E} \times \langle\bar{t}\rangle$. Since $\bar{E} \in \mathfrak{A}_n$, it is clear that $\bar{B} \in \mathfrak{A}_n$.

Suppose now that $k > 1$ and $x \in \mathrm{N}_B(E)$, $x \notin E$, and $x^p \in E$. Let $E_1 = E\langle x\rangle$. By induction, we get $E_1 \in \mathfrak{A}_{n+1}$. Since $|B : E_1| = p^{k-1}$, the induction hypothesis gives $B \in \mathfrak{A}_{(n+1)+(k-1)} = \mathfrak{A}_{n+k}$. □

It follows from Lemma 3.3.23 that if B_0 is an abelian group and $\langle b\rangle$ is a cyclic group of prime order p, then the standard wreath product $B = B_0 \wr \langle b\rangle$ belongs to the class $\mathfrak{A}_2$. Hence neither the nilpotency class nor the order of the derived subgroup of the groups in $\mathfrak{A}_2$ is bounded. Moreover, there are groups in $\mathfrak{A}_2$ with no abelian maximal subgroups, for example

$$B = \langle x, y : x^{p^4} = 1 = y^{p^2}, x^y = x^{1+p^2}\rangle.$$

Theorem 3.3.24. *Let $G = AB$ be a p-group, where A is abelian and $B \in \mathfrak{A}_n$, then G has derived length at most $2n$.*

Proof. We argue by induction on n. The first step follows from Theorem 3.1.7. We distinguish two cases.

Assume first that $X = A \cap B = 1$. Let

$$1 = G_0 < G_1 < \cdots < G_t = G$$

be a chief series of G constructed as follows: if in $\bar{G} = G/G_i$ there exists an element $\bar{a} \in \mathrm{Z}(\bar{G}) \cap \bar{A}_{\bar{G}}$ of order p, then we define $G_{i+1} = \langle a, G_i \rangle$. Otherwise, $\bar{B}_{\bar{G}} \neq 1$ by Theorem 3.1.12 and we can define $G_{i+1} = \langle b, G_i \rangle$, where $\bar{b}$ is an element of order p of $\mathrm{Z}(\bar{G}) \cap \bar{B}_{\bar{G}}$.

Since for every $i \in \{1, 2, \ldots, t\}$, we have $\bar{A} \cap \bar{B} = \bar{1}$ in $\bar{G} = G/G_i$, we can show by induction that each G_i is factorised, namely $G_i = (A \cap G_i)(B \cap G_i)$.

Let G_* be the maximal element of the above series such that $B \cap G_*$ is abelian. Then $G_* = (A \cap G_*)(B \cap G_*)$ is metabelian by Theorem 3.1.7. Since in the chief series $K_i = B \cap G_i$ of B we have $K_* = B \cap G_*$, then in $\bar{G} = G/G_*$ we have $\bar{B} = BG_*/G_* \cong B/(B \cap G_*) = B/K_* \in \mathfrak{A}_{n-1}$ (note that $\bar{A}$ is abelian). By the induction hypothesis, $\bar{G}$ has derived length at most $2(n-1)$. Therefore G has derived length at most $2n$.

Assume now that $X = A \cap B \neq 1$. Then $\langle X^G \rangle = \langle X^{AB} \rangle = \langle X^B \rangle \leq B$. Therefore $\langle X^G \rangle$ is factorised and in $\bar{G} = G/\langle X^G \rangle$ we have $\bar{A} \cap \bar{B} = \bar{1}$. Let

$$1 = G_0 < G_1 < \cdots < G_k = \langle X^G \rangle$$

be a part of a chief series of G. Such a series can be extended to a chief series of G by constructing a chief series of $G/\langle X^G \rangle$ as in the case $A \cap B = 1$. With the same notation as before, if G_* contains $\langle X^G \rangle$, then G_* is factorised and the conclusion follows. Otherwise, if $G_* < \langle X^G \rangle \leq B$, then G_* is an abelian subgroup of B and, since the term $G_{*+1} \leq \langle X^G \rangle \leq B$ is non-abelian, we must have $B/G_* \in \mathfrak{A}_{n-1}$. Therefore G/G_* has derived length at most $2(n-1)$ and G has derived length at most $1 + 2(n-1) < 2n$. □

Corollary 3.3.25. *Let $G = AB$ be a p-group. If A is abelian and B contains an abelian subgroup of index p^{n-1}, then G has derived length at most $2n$.*

Proof. This result follows as a consequence of Theorem 3.3.24 and Lemma 3.3.23. □

When the factors of a dinilpotent group are of coprime orders, the bound of Hall and Higman is best possible for small values of their nilpotency classes. However it seemed likely that for larger values of the nilpotency classes this bound is too large and should be replaced by a function of the derived lengths of the factors. Note that, by Theorem 1.7.4, a bound involving the derived lengths of the factors implies a bound involving their nilpotency classes.

The following results of Kazarin [176] exhibit bounds for the Fitting length and derived length of soluble products of two subgroups of coprime orders.

Theorem 3.3.26. *Let the soluble group $G = AB$ be the product of its subgroups A and B of coprime orders. Then the Fitting length* $\mathrm{h}(G)$ *of G does not exceed* $\mathrm{d}(A) + \mathrm{d}(B) + \mathrm{l}_2(G)$. *In particular, if $2 \in \pi(A)$, then* $\mathrm{h}(G) \le 2\mathrm{d}(A) + \mathrm{d}(B)$.

Proof. We shall derive a contradiction by supposing that the result is false. Let G be a factorised soluble group of minimal order subject to being a product of the subgroups A and B of coprime orders and $\mathrm{h}(G) > \mathrm{d}(A) + \mathrm{d}(B) + \mathrm{l}_2(G)$. By familiar reasoning, G possesses a unique minimal normal subgroup, N say, and there exists a prime p such that $N \le \mathrm{O}_p(G) = \mathrm{F}(G) = \mathrm{O}_{p',p}(G)$. Let $\bar{G} = G/\mathrm{F}(G)$, $\bar{A} = A\mathrm{F}(G)/\mathrm{F}(G)$ and $\bar{B} = B\mathrm{F}(G)/\mathrm{F}(G)$. Obviously $\mathrm{h}(\bar{G}) = \mathrm{h}(G) - 1$ and $\mathrm{h}(\bar{G}) \le \mathrm{d}(\bar{A}) + \mathrm{d}(\bar{B}) + \mathrm{l}_2(\bar{G})$ by minimality of G. We reach a contradiction by proving that at least one of the parameters on the right of the above inequality is strictly lower for $\bar{G}$ than for G.

If $p = 2$, then $\bar{G} = G/\mathrm{O}_{2',2}(G)$ and $\mathrm{l}_2(\bar{G}) < \mathrm{l}_2(G)$. Therefore we can suppose that $p > 2$. Without loss of generality, we may assume that p divides $|A|$. Let $\pi = \pi(A)$, $t = \mathrm{d}(A)$, and $K = A^{(t-1)}$. Obviously $[\mathrm{O}_{p'}(K), \mathrm{F}(G)] = 1$. Since $\mathrm{C}_G(\mathrm{F}(G)) \le \mathrm{F}(G)$ by [119, A, 10.6(a)], we have that K is a p-group. If $2 \in \pi$, then from Lemma 1.7.16 it follows that $K \le \mathrm{O}_{p',p}(G) \le \mathrm{F}(G)$ and if $2 \notin \pi$, $K \le \mathrm{F}(G)$ by Corollary 1.7.23. In both cases, $\mathrm{d}(\bar{A}) \le \mathrm{d}(A) - 1$. This completes the proof of the first part of the theorem. Now, by Theorem 1.7.2, $\mathrm{l}_2(G) \le \mathrm{d}_2(A)$ and obviously $\mathrm{d}_2(A) \le \mathrm{d}(A)$. This gives the last bound of the theorem. □

We are now in a position to undertake the proof of the bounded derived length theorem.

Theorem 3.3.27. *Let the soluble group $G = AB$ be the product of its subgroups A and B with coprime orders. Then* $\mathrm{d}(G) \le 2\mathrm{d}(A)\mathrm{d}(B) + \mathrm{d}(A) + \mathrm{d}(B)$. *Moreover:*

1. *If A is abelian, then* $\mathrm{d}(G) \le 2\mathrm{d}(B) + 1$.

2. *If A is abelian of even order, then* $\mathrm{d}(G) \le 2 + \mathrm{d}(B)$.

3. *If G is of odd order, then* $\mathrm{d}(G) \le \mathrm{d}(A)\mathrm{d}(B) + \max\{\mathrm{d}(A), \mathrm{d}(B)\}$.

Proof. Suppose that G has even order. There is no loss of generality in assuming that $2 \in \pi = \pi(A)$ and $\pi' = \pi(G) \setminus \pi(A) = \pi(B)$. Note that $d_\pi = \mathrm{d}(A) = \mathrm{d}_\pi(G)$ and $d_{\pi'} = \mathrm{d}(B) = \mathrm{d}_{\pi'}(G)$. We consider the following canonical normal series of G:

$$1 = N_0 \le P_1 \le \cdots \le P_{l-1} \le N_{l-1} \le P_l \le N_l \le P_{l+1} = G,$$

where $N_i/P_i = \mathrm{O}_{\pi'}(G/P_i)$ for $1 \le i \le l$ and $P_i/N_{i-1} = \mathrm{O}_\pi(G/N_{i-1})$ for $1 \le i \le l+1$. Obviously $l \le \mathrm{l}_{\pi'}(G)$ and $l \le \mathrm{l}_\pi(G)$. Note that $\mathrm{d}(P_i/N_{i-1}) \le \mathrm{d}(A)$. Let d_i be the derived length of BP_i/P_i, where $1 \le i \le l$. By Corollary 1.7.23, $B^{(d_i-1)}P_i/P_i \le \mathrm{F}(G/P_i) \le \mathrm{O}_{\pi'}(G/P_i) = N_i/P_i$. It follows that $\mathrm{d}(BP_{i+1}/P_{i+1}) \le \mathrm{d}(BN_i/N_i) < \mathrm{d}(BP_i/P_i)$. Consequently

$$0 = \mathrm{d}(BP_{l+1}/P_{l+1}) < \mathrm{d}(BP_l/P_l) < \cdots < \mathrm{d}(BP_1/P_1) = \mathrm{d}(B)$$

and so $\mathrm{d}(BP_i/P_i) + i - 1 \leq \mathrm{d}(B)$ for $1 \leq i \leq l$. It follows that $\mathrm{d}(N_i/P_i) \leq \mathrm{d}(BP_i/P_i) \leq \mathrm{d}(B) - i + 1$ for $1 \leq i \leq l$ and $\mathrm{d}(P_i/N_{i-1}) \leq d_\pi$. Therefore we have

$$\begin{aligned} \mathrm{d}(G) &\leq (l+1)d_\pi + \sum_{i=0}^{l-1}(d_{\pi'} - i) \\ &= (l+1)d_\pi + \frac{1}{2}(2d_{\pi'} - l + 1)l \\ &= d_\pi + \frac{1}{2}(2d_\pi + 2d_{\pi'} - l + 1)l = \varphi(l). \end{aligned}$$

If we consider this as a function of l, we see that its derived function is $\varphi'(l) = \frac{1}{2}(2d_\pi + 2d_{\pi'} - 2l + 1)$. Since $l \leq l_{\pi'} \leq d_{\pi'}$ by Theorem 1.7.20, we have that $\varphi'(l) \geq 0$ for each value of l and the maximum of $\varphi(l)$ is reached when l has maximal value. Assume first that $d_{\pi'} \leq 2d_\pi$. In this case,

$$\mathrm{d}(G) \leq d_\pi + \frac{1}{2}(2d_\pi + d_{\pi'} + 1)d_{\pi'} \leq 2d_\pi d_{\pi'} + d_\pi + d_{\pi'}$$

and the result is proved. Now assume that $d_{\pi'} > 2d_\pi$. By Theorem 1.7.19, we have that $l \leq \mathrm{l}_\pi(G) \leq 2d_\pi$ and so

$$\mathrm{d}(G) \leq d_\pi + \frac{1}{2}(2d_{\pi'} + 1)2d_\pi = 2d_\pi d_{\pi'} + 2d_\pi \leq 2d_\pi d_{\pi'} + d_\pi + d_{\pi'}.$$

This completes the proof of the first statement.

Now assume that one of the factors is abelian. Then $l = 1$. Therefore the maximum of the expression $d_\pi + \frac{1}{2}(2d_\pi + 2d_{\pi'} - l + 1)l$ is achieved for $l = 1$. Hence $\mathrm{d}(G) \leq 2d_\pi + d_{\pi'}$. If A is abelian, then $d_\pi = 1$ and $\mathrm{d}(G) \leq 2 + d_{\pi'} \leq 2d_{\pi'} + 1$. If B is abelian, then $d_{\pi'} = 1$ and $\mathrm{d}(G) \leq 2d_\pi + 1$. Therefore Assertions 1 and 2 are proved.

Now suppose that G has odd order and is a counterexample of minimal order to the Assertion 3. Then G has a unique minimal normal subgroup N and $N \leq \mathrm{O}_p(G) = \mathrm{F}(G)$ for some prime p. Suppose that $p \in \pi = \pi(A)$. If $t = d_\pi$, then by Corollary 1.7.23 it follows that $A^{(t-1)} \leq \mathrm{F}(G)$. Thus $\mathrm{d}_\pi(G/\mathrm{O}_\pi(G)) \leq d_\pi - 1$. A similar argument shows that $\mathrm{d}_{\pi'}(G/\mathrm{O}_{\pi,\pi'}(G)) \leq d_{\pi'} - 1$. By minimality of G, we have that $\mathrm{d}(G/\mathrm{O}_{\pi,\pi'}(G)) \leq (d_\pi - 1)(d_{\pi'} - 1) + \max\{d_\pi - 1, d_{\pi'} - 1\}$. Note that $\mathrm{d}(\mathrm{O}_{\pi,\pi'}(G)) \leq d_\pi + d_{\pi'}$ and so

$$\begin{aligned} \mathrm{d}(G) &\leq \mathrm{d}(G/\mathrm{O}_{\pi,\pi'}(G)) \\ &= d_\pi d'_\pi - d_\pi - d_{\pi'} + 1 + \max\{d_\pi, d_{\pi'}\} - 1 + d_\pi + d_{\pi'} \\ &= d_\pi d_{\pi'} + \max\{d_\pi, d_{\pi'}\}. \end{aligned}$$

The symmetry between A and B gives the proof in the general case. □

If the first factor is abelian and the second one is nilpotent, Kazarin [176] gives a bound based on the nilpotency class of the second factor.

Corollary 3.3.28. *Let the group $G = AB$ be the product of an abelian group A and a nilpotent group B of nilpotency class c. If $\pi(A) \cap \pi(B) = \emptyset$, then G is soluble and the derived length of G does not exceed $3 + 2\log_2(c)$. If G is of odd order, then $\mathrm{d}(G) \leq 2 + 2\log_2(c)$.*

Proof. We know that the group is soluble by Theorem 3.2.1. By Theorem 3.3.27, $\mathrm{d}(G) \leq 2\mathrm{d}(B) + 1$ and $\mathrm{d}(G) \leq 2\mathrm{d}(B)$ if G is of odd order. The result follows by Theorem 1.7.4. □

Our next result concerns the derived length of a dinilpotent group $G = AB$ when $A \cap B = 1$ and both subgroups A and B are permutable in G. It was obtained by Dixon and Stonehewer in [118].

Theorem 3.3.29. *Let $G = AB$ be a dinilpotent group with $A \cap B = 1$ and A and B permutable in G. Then $\mathrm{d}(G) \leq \max\{2, \mathrm{d}(A), \mathrm{d}(B)\}$.*

It is clear that in this case A and B are subnormal by Theorem 1.2.14 (3) and hence G is a nilpotent group. Therefore the proof of Theorem 3.3.29 reduces to the study of the p-group case, p a prime. We begin with a preliminary lemma.

Lemma 3.3.30. *Let p be a prime and let the p-group $G = AB$ be the product of the subgroups A and B such that A and B are permutable in G, $A \cap B = 1$, and B is cyclic. Then A' is a normal subgroup of G.*

Proof. Suppose that the result is false. Assume that G is a counterexample of least order to the lemma. Let x be an element of order p of A. Since B is permutable in G, it turns out that $B\langle x\rangle$ is a subgroup of G and $|B\langle x\rangle : B| = p$. Therefore x normalises B. It follows that the subgroup $\Omega_1(A)$ generated by all elements of order p of A normalises B. Since A' is not normal in G, A cannot be cyclic. Moreover, $\Omega_1(A)/\mathrm{C}_{\Omega_1(A)}(B)$ is isomorphic to a subgroup of $\mathrm{Aut}(B)$. Assume first that p is odd. By [157, I, 4.6 and 13.19], $\mathrm{Aut}(B)$ is cyclic. By [157, III, 8.2], since A is not cyclic, $\Omega_1(A)$ is not cyclic. Hence A contains a non-trivial element centralising B. Assume now that $p = 2$. We will show that all elements of order 4 of A normalise B. Let z be the involution in $B = \langle b\rangle$ and let a be an element of order 4 in A. By Theorem 1.2.14 (4), $\langle z\rangle$ is permutable in $\langle a, z\rangle$. Since $\langle a, z\rangle$ is a group of order 8, $\langle a, z\rangle$ is abelian, or isomorphic to a dihedral group of order 8 or to a quaternion group of order 8. However, neither of these two non-abelian groups has a permutable subgroup of order 2 complementing a subgroup of order 4. It follows that $\langle a, z\rangle$ is abelian. Now assume that $\langle b^{2^r}\rangle$ is normalised by an element $a \in A$ of order 4. The same argument used before shows that a centralises $\langle b^{2^{r-1}}\rangle/\langle b^{2^r}\rangle$. Hence $(b^{2^{r-1}})^a = b^{2^{r-1}}(b^{2^r})^t$ for some integer t and so a normalises $\langle b^{2^{r-1}}\rangle$. By induction, we conclude that a normalises B. Consequently all elements of order 4 in A normalise B. Now B must have order at least 4. Assume that x is an element of A inverting B. Then $\langle x\rangle$ must be permutable in $H = \langle x, b\rangle$ and since $H_0 = \langle x^2, b^4\rangle$ is a normal subgroup of H, we can argue with H/H_0 as in the previous paragraph to reach the contradiction that H/H_0 must be abelian. Hence A cannot contain an element x inverting B. By [157, I, 4.6 and 13.19], it follows that either A contains a non-trivial element centralising B, or A contains a unique subgroup of order 4. In the latter case, by [157, III, 8.4] A is cyclic, against the hypothesis.

Thus in all cases A contains a non-trivial element centralising B. Let $C = C_A(B)$, then $1 \neq C = \langle C^B \rangle \leq \langle C^{BA} \rangle = \langle C^G \rangle \leq A_G$. Therefore $N = A_G$ is non-trivial. By the choice of G, $(A/N)'$ is normal in G/N. It follows that $A'N$ is normal in G. Hence $[A', B] \leq [A'N, B] \leq A'N \leq A$. But since B is permutable in G, $[A', B] \leq A'B$. Therefore $[A', B] \leq A \cap A'B = A'(A \cap B) = A'$ and A' is a normal subgroup of G, as required. □

We now remove the hypothesis that B is cyclic.

Proposition 3.3.31. *Let the p-group $G = AB$ be the product of the subgroups A and B permutable in G and suppose that $A \cap B = 1$. Then A' and B' are normal in G.*

Proof. By symmetry, it is enough to prove that A' is a normal subgroup of G. Let b be an element of B and let $H = A\langle b \rangle$. Then $B \cap H = \langle b \rangle$ is permutable in H by Theorem 1.2.14 (4). By Lemma 3.3.30, b normalises A'. Since this is true for all $b \in B$, B normalises A'. The result follows from the normality of A' in A. □

Lemma 3.3.32. *Let the group G have a permutable subgroup A and an abelian normal subgroup N such that $A \cap N = 1$. Then $[A, G]$ centralises N.*

Proof. Let $a \in A$, $g \in G$, and $x \in N$. Since A is permutable in G, by the remark at the beginning of Section 1.3 we obtain that A induces a group of power automorphisms on N. Since N is abelian, there exists an integer m such that a^{-1} conjugates each element of N to its mth power by Theorem 1.3.2. Then

$$x^{a^{-1}g^{-1}ag} = (x^m)^{g^{-1}ag} = ((x^{g^{-1}})^m)^{ag} = (x^{g^{-1}})^g = x,$$

and so N is centralised by $[A, G]$. □

Now we are in a position to prove Theorem 3.3.29.

Proof of Theorem 3.3.29 (Cossey and Ezquerro [106]). Suppose that G is a counterexample of least order to the theorem. Assume that N_1 and N_2 are two different minimal normal subgroups of G. Then the minimality of G implies that for $i \in \{1, 2\}$,

$$G/N_i \leq \max\{2, \mathrm{d}(AN_i/N_i), \mathrm{d}(BN_i/N_i)\} \leq \max\{2, \mathrm{d}(A), \mathrm{d}(B)\}.$$

It follows that $\mathrm{d}(G) \leq \max\{2, \mathrm{d}(A), \mathrm{d}(B)\}$ and G is not a counterexample. Therefore G has a unique minimal normal subgroup. From Theorem 1.2.14 (3) we obtain the subnormality of A and B and this implies that G itself is nilpotent. Since G has a unique minimal normal subgroup, G is a p-group for some prime p. By Proposition 3.3.31, A' and B' are normal subgroups of G. It follows that A' or B' must be trivial, assume that $A' = 1$. Hence A is abelian. Then B cannot be abelian by Theorem 3.1.7. Note that $[A, B]$ and B' are normal subgroups of G by [119, A, 7.4] and Proposition 3.3.31, respectively. Since $G/[A, B]B'$ is a central product of the abelian groups $A[A, B]B'/[A, B]B'$ and $B[A, B]/[A, B]B'$, it follows that $G' = [A, B]B'$.

Next we prove that $[A, B]$ is abelian. We know that $B' \leq B_G$ by Proposition 3.3.31. Let $N = B_G$, $\bar{G} = G/N$, $\bar{A} = AN/N$, and $\bar{B} = B/BN$. Then $\bar{G} = \bar{A}\bar{B}$. We also have that $\bar{A}$ and $\bar{B}$ are abelian. As in the proof of Lemma 3.3.30, $\bar{B}_1 = \Omega(\bar{B})$ normalises all cyclic subgroups of $\bar{A}$ and, if $p = 2$, then all elements of order 4 of $\bar{B}$ normalise all cyclic subgroups of $\bar{A}$. Therefore such elements act as universal power automorphisms on $\bar{A}$ by Theorem 1.3.2. The same arguments of the proof of Lemma 3.3.30 show that if $\bar{B}$ is not cyclic, then there is a non-trivial element of $\bar{B}$ centralising $\bar{A}$ and so $\bar{B}$ has non-trivial core in $\bar{G}$. Since this contradicts the construction of $\bar{B}$, $\bar{B}$ is cyclic. Hence there exists an element $b \in B$ such that $B = N\langle b\rangle$. Since B is not abelian, $N \neq 1$ by Proposition 3.3.31. Now by the remark at the beginning of Section 1.3, A induces a group of power automorphisms on N. In particular, A induces a group of power automorphisms on $\mathrm{Z}(N)$. By Theorem 1.3.3, $[A, N] \leq \mathrm{Z}(N)$. Consequently

$$[A, B] = [A, N\langle b\rangle] = [A, N][A, \langle b\rangle] \leq \mathrm{Z}(N)[A, \langle b\rangle].$$

By Lemma 3.3.32, $\mathrm{Z}(N)$ is centralised by $[A, G]$. Hence $\mathrm{Z}(N)$ centralises $[A, \langle b\rangle]$, which is abelian by Theorem 3.1.7. Therefore $[A, B]$ is abelian.

Assume that B is metabelian. Then B' is abelian and by Lemma 3.3.30 it is a normal subgroup of G. Therefore B' is centralised by $[A, G]$ by Lemma 3.3.32. Consequently $G' = [A, B]B'$ is an abelian group and G is metabelian. This contradiction shows that B is not metabelian.

Since B' is a normal subgroup of G contained in B, A induces a group of power automorphisms on B' and so, by Theorem 1.3.3, $[A, B'] \leq \mathrm{Z}(B')$. Now $G'' = ([A, B]B')' = [[A, B], B']B''$. But the group $\mathrm{PAut}(B')$ of power automorphisms of B is a normal subgroup of the group of automorphisms of B'. It follows that $[\mathrm{PAut}(B'), \mathrm{Aut}(B')] \leq \mathrm{PAut}(B')$. Since B' is normal in B, elements of B induce a group of automorphisms of B'. Thus the group of automorphisms induced by $[A, B]$ by conjugation on B' is a group of power automorphisms. Hence $[[A, B], B'] \leq \mathrm{Z}(B')$ by Theorem 1.3.3 again. Consequently $G'' \leq \mathrm{Z}(B')B''$. It follows that $G''' \leq B'''$ and so $G''' = B'''$. By induction, $G^{(n)} = B^{(n)}$ for all $n \geq 3$. This contradicts the fact that G is a counterexample and the proof of the theorem is complete. □

Now assume that $G = AB$ is a product of an abelian group A and a metabelian nilpotent group B. According to Theorem 3.3.27, $\mathrm{d}(G) \leq 5$ and if G is of odd order, then $\mathrm{d}(G) \leq 4$. In the following we will present a result of Cossey and Wang [112] in which these bounds are improved to 4 and 3, respectively, and that these bounds are best possible. Moreover, we will show that the bound will be three in most situations and that the groups with derived length 4 have a well defined structure.

Example 3.3.33. If A is abelian and B is metabelian, then the wreath product $G = A \wr B$ has derived length 3. Therefore the bound of 3 cannot be improved.

Example 3.3.34. Let $G = \mathrm{GL}_2(3)$. Then $G = AB$, where A is a Sylow 3-subgroup and B is a Sylow 2-subgroup. We have that G has derived length 4, A is abelian, and B is metabelian. Hence the bound 4 for products of abelian and metabelian nilpotent groups of even order cannot be improved.

Definition 3.3.35. We say that the group G is an *E-group* when it has the following structure: $\mathrm{F}(G)$ is an extraspecial 2-group, $\mathrm{F}(G)/\Phi(\mathrm{F}(G))$ is either a minimal normal subgroup of $G/\Phi(\mathrm{F}(G))$ or the product of two minimal normal subgroups of $G/\Phi(\mathrm{F}(G))$ and $G/\mathrm{F}(G)$ is a dihedral group of order $2n$ for some odd positive integer n. Moreover, n is prime provided all chief factors of G between $\Phi(\mathrm{F}(G))$ and $\mathrm{F}(G)$ are G-isomorphic.

Example 3.3.36 (Cossey and Wang [112]). Let q be an odd prime. Let D denote the dihedral group of order $2q$ and let U be a faithful irreducible module for D over the field of 2 elements. Then $|U| = 2^r$, where r is the order of 2 modulo q if this order is even and twice the order of 2 modulo q if this order is odd. It is not difficult to see that U is isomorphic to its dual V. It then follows that the trivial module is a quotient of $U \otimes V$ by [158, VII, 8.6]. The construction of [119, B, 12.17] gives an extraspecial group F of order 2^{2r+1} on which acts so that F' is trivial and $F/F' \cong U \oplus V$ as D-modules. The semidirect product $G = FD$ is an E-group.

There are four main ingredients in the proof of Cossey and Wang's result. The first ingredient is a beautiful theorem of Berger and Gross. Since the machinery needed for its proof goes beyond the limits we have set for this section, we will simply state the result without proof and direct the reader to [69] for full details.

Theorem 3.3.37 (Berger and Gross [69]). *Assume that G is a soluble group of even order such that $\mathrm{O}_2(G) = 1$. Suppose that F is a field of characteristic 2 and that V is a faithful FG-module. Let P be a Sylow 2-subgroup of G and assume that $\mathrm{d}(VP) = \mathrm{d}(P) = n$. Then $\mathrm{O}_{2'}(G)$ contains a non-abelian 3-subgroup, and the quaternion group of order 8 is an epimorphic image of $P^{(n-2)}$.*

The second ingredient is an application of Corollary 1.7.23 and Theorem 3.3.37.

Lemma 3.3.38. *Let p be a prime and let K be a field of characteristic p. Let G be a p-nilpotent group, P a Sylow p-subgroup of G, and $Q = \mathrm{O}_{p'}(G)$ a Hall p'-subgroup of G. Suppose that U is a faithful and irreducible KG-module. Then if Q is abelian and P is non-abelian, the semidirect product of U and P has derived length at least 3.*

Proof. Let $X = [U]G$ be the semidirect product of U and G and let $A = [U]P$. Clearly A is a Sylow p-subgroup of G. Since U is faithful, $O_{p'}(X) = 1$. Let $t = d(A)$. Suppose that $p > 2$. We can apply Corollary 1.7.23 for $\pi = \{p\}$ and conclude that $A^{(t-1)} \leq F(X) = U$. Note that $t \geq 2$. If $t = 2$, then $A' \leq U$ and so P is abelian, which contradicts the hypothesis. We can therefore deduce that $t \geq 3$. Now suppose that $p = 2$. Since P is non-abelian, $d(P) \geq 2$. If $t < 3$, then $d(A) = d(P) = 2$, but then, by Theorem 3.3.37, $O_{2'}(G) = Q$ contains a non-abelian 3-subgroup, which contradicts the fact that Q is abelian. Therefore $t \geq 3$. □

Note that Cossey and Wang have offered a direct proof of the previous lemma in [112].

The third of the preliminary results needed is a generalisation of a result from modular representation theory.

Lemma 3.3.39. *Let p be a prime, G a group with U an abelian normal p-subgroup and G/U a p-nilpotent group. Then $U = U_1 \times \cdots \times U_t$, where each U_i is normal in G, all chief factors of G contained in U_i are isomorphic as G-modules and if $i \neq j$ no chief factor of U_i is isomorphic to a chief factor of U_j.*

Proof. When U is elementary abelian, we can regard U as G/U-module and the result is then essentially a restatement of a theorem of Srinivasan ([158, VII, 16.10]). We proceed by induction on the length of a G-chief series from U to 1. The result is clearly true for 1. By the previous observation we can assume that U is not elementary abelian. If U has exponent p^a, then $U^{p^{a-1}}$ is elementary abelian and isomorphic (as a G-module) to a quotient of $U/\Phi(U)$. Since $U^{p^{a-1}} \leq \Phi(U)$, we have that for some minimal normal subgroup V of G contained in U, U/V contains a G-chief factor G-isomorphic to V. Now by our inductive hypothesis, U/V can be written as a direct product $U/V = (U_1^*/V) \times (U_2^*/V) \times \cdots \times (U_t^*/V)$, where the U_j^*/V satisfy the requirements of the lemma and U_1^*/V has been chosen so that each chief factor of G contained in U_1^*/V is isomorphic to V. For $i > 1$ we have that the length of a G-series of U_i^* is smaller than the length of a G-series of U and so $U_i^* = V \times U_i$, because no chief factor of U_i^*/V is isomorphic to V. Set $U_1 = U_1^*$. Then it is easy to see that $U = U_1 \times \cdots \times U_t$ and that the U_i satisfy the requirements of the lemma. □

The final ingredient needed in the proof of the theorem of Cossey and Wang is the following technical lemma, which may be interpreted in the following way. Suppose that a soluble group G is the semidirect product of a minimal normal p-subgroup U and a group $H = [V]W$, where V is a minimal normal q-subgroup of H and W is cyclic of p-power order, p and q distinct primes. Suppose also that $C_H(U) = C_W(V) = 1$. Then $U = [U, W][U, W]^v$ for some $v \in V$.

Lemma 3.3.40. *Let p be a prime. Suppose that $G = AB$ is a non-nilpotent factorised group, where A is an abelian normal Hall subgroup of G and B is cyclic of order p^i.*

Let U be a faithful irreducible FG-module, where F is a finite field of characteristic p. Let V be the radical of U_B, U regarded as an FB-module by restriction. Then U_B is a free FB-module (of rank t say). Then there exists an element $a \in A$ such that $V + Va = U$, and $U/(V \cap Va)$ has dimension at most $2t$.

Proof. Recall that the radical of a module is the smallest submodule with completely reducible quotient (see [119, B, 3.7] and the remarks following this definition). If F is a splitting field for G, then U is induced from a 1-dimensional irreducible submodule for A by Clifford's theorem [119, B, 7.3] and [119, B, 9.2] and so by the Mackey subgroup theorem [119, B, 6.20] U_B is a free FB-module. It then follows easily that U_B is free for any field F of characteristic p by [119, B, 5.25]. If the rank of U_B is t, then $\dim_F(U) = p^i t$. Note that if $a \in A$, then Va is the radical of U_{B^a}. There are now two cases to consider.

Suppose first that U_A is reducible, so that $U_A = U_0 \oplus \cdots \oplus U_{p^s-1}$, where $s \leq i$ and each U_j is a distinct irreducible FA-submodule of dimension $p^{i-s}t$. If $B = \langle b \rangle$, then B permutes the U_j, say $U_0^{b^j} = U_j$. We then have that $Y_j = \{u - ub^j : u \in U_0\}$ is a subspace of U contained in V. Moreover $V \leq Y_1 \oplus \cdots \oplus Y_{p^s-1}$. Hence $V = Y_1 \oplus \cdots \oplus Y_{p^s-1}$.

Applying [119, B, 9.8 (a)], we have that $A/\mathrm{Ker}(A \text{ on } U_j)$ is cyclic. Since G is not nilpotent, we can find an element $1 \neq a \in A$ of maximal order such that $[a, b^{-1}] \neq 1$. Then $[a, b^{-1}] \notin \mathrm{Ker}(A \text{ on } U_j)$ for some j. Since $A/\mathrm{Ker}(A \text{ on } U_j)$ is cyclic, $[a, b^{-1}]$ acts fixed point freely on U_j. We may suppose that $j = 0$ and so $[a, b^{-1}]$ acts fixed point freely on U_0. Let $W = \{u - ub^a : u \in U_0\}$. Note that both ub and ub^a are in U_1 and so Y_1 and W are subspaces of $U_0 \oplus U_1$. Suppose now that $v - vb \in Y_1 \cap W$. Then $v - vb = u - ub^a$ for some $u \in U_0$ and so $v = u$ and $ub = ub^a$. But then $u[b^{-1}, a] = u$, which implies $u = 0$. Thus we have $Y_1 \cap W = 0$ and then $V \cap W = 0$. Since $V + U_0 = U$, $V \cap U_0 = 0$ and $\dim W = \dim U_0$, it follows from dimension considerations that $V + W = U$. Since $W \leq Va$, we then have $V + Va = U$.

Next suppose that U_A is irreducible. Then A is cyclic. Let E be a field of order $|F|^{p^i t}$. Then we can regard U as the additive group of E and A as a subgroup of the multiplicative group of E, and B can be seen as a subgroup of the Galois group of E over F (see [119, B, 9.7, 9.8]). Let D denote the subfield of E fixed element-wise by B. Then B has dimension p^i as a vector space over D (see for example [153, V, 2.5]). Since U_B is free, it is free of rank one as a DB-module. Since a has maximal order, it follows that $A = \langle a \rangle$. We cannot have $Va \leq V$ because then V would be G-invariant, which contradicts the irreducibility of U as an FG-module. Since V has co-dimension 1 as a DB-module and Va is a D-subspace, it follows that $V < V + Va = U$.

The final statement of this lemma follows from the fact that the free FB-module of rank t modulo its radical has dimension t. □

We are now in a position to undertake the proof of Cossey and Wang's theorem ([112]).

Theorem 3.3.41. *Let the group $G = AB$ be the product of the abelian subgroup A and the metabelian and nilpotent subgroup B. Suppose further that the order of A and the order of B are coprime. Then G is soluble of derived length at most 4. Further, the derived length is at most 3 unless G has a section which is an E-group. In this last case, the derived length of G is 4.*

Proof. Suppose that G satisfies the following hypothesis:

$$G = AB \text{ with } A \text{ abelian, } B \text{ metabelian and nilpotent, } A \text{ and } B \text{ have coprime orders, and } G \text{ has no section isomorphic to an } E\text{-group.} \tag{3.3}$$

We want to show that if G satisfies (3.3), then G has derived length at most 3. Hence we suppose that G is a group of order as small as possible with derived length greater than 3 and satisfying (3.3).

Since any quotient of a group satisfying (3.3) also satisfies (3.3), we obtain that G has a unique minimal normal subgroup N whose quotient G/N has derived length 3. Moreover $F = \mathrm{F}(G)$ is a p-group for some prime p. Moreover G has $\pi(A)$-length 1. Assume that $p \in \pi(A)$. Then A centralises F, which implies that $A \leq F$ by [157, III, 4.2]. Thus $A = F$, $G/F \cong B$ and so G has derived length 3. This contradiction shows that $p \in \pi(B)$. Let H be the Hall p'-subgroup of B. Then H centralises F and so $H \leq F$ by [157, III, 4.2]. This gives $H = 1$. Thus B is a p-group. If $B = F$, then $G/B \cong A$ and again G has derived length at most 3. This contradiction shows that $B \neq F$.

Let $M = FA$. Since G has $\pi(A)$-length 1, we obtain that $M = \mathrm{O}_{p,p'}(G)$ is a normal subgroup of G (we can reach the same conclusion by applying Theorem 3.2.28). Suppose first that G/M is non-abelian. Then G' is not contained in M and, in particular, G' is not nilpotent. Let C/J be a self-centralising p-chief factor of G such that G' is not contained in C. Then, since G/J has p-length at most 1 and $\mathrm{O}_p(G/J) = 1$ by [119, A, 13.6], we deduce that G/J is p-nilpotent and so the Sylow p-subgroup of G/J cannot be abelian. In particular, G/J satisfies the hypothesis of Lemma 3.3.38. Hence it has a Sylow p-subgroup of derived length at least 3. This contradicts the fact that B is metabelian. Therefore G/M is abelian. Note that this gives a bound of 4 for the derived length, because G/M and M/F is abelian and F is metabelian. Our aim is to show that F is abelian unless G has a section isomorphic to an E-group.

From the minimality of G and the fact it has derived length greater than 3, we have that $N = G''' < G'' \leq F$. Let L denote the $\mathfrak{N}\mathfrak{A}$-residual of G, the residual of G for the class of nilpotent-by-abelian groups. Note that $L \neq 1$ and so N is contained in L. Moreover, every chief factor X/Y of G such that $L \leq X < Y \leq G$ satisfies that $G/\mathrm{C}_G(X/Y)$ is abelian (see [119, IV, 3.4 (b)]). Note that $L \leq G''$. Since G/M is abelian, then $G' \leq M$. Hence $B' \leq F$. This implies that B/F is abelian. Let $T/F = (G/F)'$. Then $T/F \leq M/F$, which is a p'-group. Let T_0/L be a Hall p'-subgroup of K/L. It follows that $T/L = (T_0L/L)(F/L)$.

Consider a chief factor of G between L and F. Note that G induces an abelian group of automorphisms on such a chief factor. Therefore $(G/F)' = T/F$ centralises such factor, and so does T_0/L. Since T_0/L and F/L have coprime orders, by [119, A, 12.3] we conclude that T_0/L centralises F/L. In particular, $T/L = (F/L) \times (T_0/L)$.

Assume that L is abelian. Let D a maximal abelian normal subgroup of G containing L. Then $D \leq F$. Suppose further that F/D is abelian. Then $T/D = (F/D) \times (T_0D/D)$ is abelian. Consequently G/D is a metabelian group and so G has derived length 3. This contradiction shows that F/D is non-abelian. Let $E/D = (F/D)'$. Suppose that L is not contained in $\mathrm{Z}(E)$. Then $L \cap \mathrm{Z}(E)$ is a proper subgroup of L and we can find a chief factor L/R of G such that $\mathrm{Z}(E) \cap L \leq R$. The group G cannot induce an abelian automorphism group on L/R. Then $G/\mathrm{C}_G(L/R)$ cannot be a p'-group and so its order is divisible by p. Thus there exists an element $x \in B$ of order p^α not centralising L/R. Since L/R is non-trivial as a module for $\langle x \rangle$, there exists an element $y \in L$ such that $[x, y] \notin R$. On the other hand, $[x, y] \in L$. Let $c \in F'$. Then c and $[x, y]$ both belong to B'. Thus $[c, [x, y]] = 1$. Since $E = F'D$ and D is abelian, we obtain that $[x, y]$ centralises E. But this implies that $[x, y] \in \mathrm{Z}(E) \cap L \leq R$. This contradiction shows that $L \leq \mathrm{Z}(E)$. Since F/D is non-abelian and $E = F'D$, it follows that D is a proper subgroup of E. Since L is abelian, by [119, IV, 5.18] there exists a subgroup S of G such that $G = LS$ and $L \cap S = 1$. Thus $E = L(S \cap E)$ is the direct product of L and $S \cap E$ because $L \leq \mathrm{Z}(E)$ and $L \cap S = 1$. Consequently $E' = L' \times (S \cap E)' = (S \cap E)'$ is a non-trivial normal subgroup of G, which implies that $N \leq (S \cap E)'$. But then $N \leq (S \cap E)' \cap L = 1$. This contradiction shows that L is not abelian.

Suppose, arguing by contradiction, that $L \neq F$. Let F/K be a chief factor of G with $L \leq K$ and let H/K be a complement for F/K in G/K. Then H satisfies (3.3). Let

$$\begin{aligned} G_0 = 1 \trianglelefteq G_1 = N \trianglelefteq G_2 \trianglelefteq \cdots \trianglelefteq G_r = L \trianglelefteq \cdots \trianglelefteq G_{s-1} = K \\ \trianglelefteq G_s = F \trianglelefteq G_{s+1} \trianglelefteq \cdots \trianglelefteq G_t = G \end{aligned} \tag{3.4}$$

be a chief series of G passing through L, K, and F. Note that G induces an abelian group of automorphisms on the chief factors G_i/G_{i-1} for $i \in \{r+1, \dots, t\}$. Moreover, since $N = G'''$, L/N is abelian. By [47, 4.2.19], G cannot induce an abelian group of automorphisms by conjugation on any of the chief factors of G between L and N. For every $i \in \{0, 1, 2, \dots, t\}$, set $H_i = H \cap G_i$. Then H_i is a normal subgroup of H for every $i \in \{0, 1, 2, \dots, t\}$. Since F centralises all chief factors of the series (3.4), and H_j/H_{j-1} is H-isomorphic to G_j/G_{j-1} for $1 \leq i \leq t, i \neq s$, it turns out that

$$\begin{aligned} H_0 = 1 \trianglelefteq H_1 = N \trianglelefteq H_2 \trianglelefteq \cdots \trianglelefteq H_r = L \\ \trianglelefteq \cdots \trianglelefteq H_{s-1} = K \trianglelefteq H_{s+1} \trianglelefteq H_t = H \end{aligned}$$

is a chief series of H. Note that since F centralises all chief factors of G and so $G/\mathrm{C}_G(G_i/G_{i-1}) \cong H/\mathrm{C}_H(H_i/H_{i-1})$ for $i \in \{1, 2, \dots, t\} \setminus \{s\}$. Hence H induces an abelian group of automorphisms on H_i/H_{i-1} for $i \in \{r+1, \dots, t\} \setminus \{s\}$ and a non-abelian group of automorphisms on H_i/H_{i-1} for $i \in \{2, \dots, r\}$. It follows that L/N is the nilpotent-by-abelian residual of H/N, that is, $H^{\mathfrak{N}\mathfrak{A}}N/N = L/N$ and then $H^{\mathfrak{N}\mathfrak{A}}N = L$. By minimality of H, $H''' = 1$ and so $H^{\mathfrak{N}\mathfrak{A}} \leq H''$ is abelian. Since N is contained in the centre of G'', it is contained in the centre of L. This shows that L is abelian, which contradicts the assumption. Hence $L = F$. Since $L \leq G'' \leq F$, we have $G'' = F$. Then $F' = N = G''' \leq \mathrm{Z}(F)$ and so F is of nilpotency class 2. Moreover, since F' is abelian, we have that for all $x, y \in F$, $[x, y]^p = [x^p, y] = 1$ and so p-powers of elements of F are central in F. This gives that $\Phi(F)$ is central in F.

We know that G/F is p-nilpotent, and so by Lemma 3.3.39 $F/N = (U_1/N) \times \cdots \times (U_n/N)$, where all chief factors of G between U_i and N are G-isomorphic and if $i \neq j$, no chief factor between U_i and N is isomorphic to a chief factor of G between U_j and N. Note that no chief factor F/K of G can have $G/\mathrm{C}_G(F/K)$ nilpotent: since this group has no non-trivial normal p-subgroups, it would a p'-subgroup and so abelian. It follows that F/N is the metanilpotent residual of G/N, and so it is complemented by H/N, say, by [47, 4.2.17]. Let F/K be a chief factor of G and let $E = KH$. If K is non-abelian, then E satisfies (3.3) and has derived length 4. This contradiction shows that K must be abelian. Suppose that $F/\Phi(G) = (V_1/\Phi(G)) \times \cdots \times (V_m/\Phi(G))$ with $V_i/\Phi(G)$ a chief factor of G. Suppose that $m > 2$. Then the product of $m - 1$ of the V_i is abelian. In particular, $[V_i, V_j] = 1$ and then, since F is generated by the V_i, we have that F is abelian. This contradicts the fact that G has derived length 4. Thus $F/\Phi(G)$ is either a minimal normal subgroup of G or the product of two minimal normal subgroups of G.

We consider now the structure of G/F. We have that B is a Sylow p-subgroup of G. Let K be a maximal subgroup of B containing F. Then KA is a normal subgroup of G of index p and also satisfies (3.3). It follows that KA has derived length at most 3 and hence $(KA)''$ is properly contained in F. As a $\mathrm{GF}(p)(KA)$-module, $F/\Phi(G)$ is completely reducible by Clifford's theorem ([119, B, 7.3]). Thus if F/V is a chief factor of G with $(KA)'' \leq V$, then KA acts on each composition factor of F/V as an abelian group. Therefore K centralises F/V. If $F/\Phi(G)$ is irreducible or the direct sum of two isomorphic irreducible modules, then $K \leq F$ and hence $K = F$. Now suppose that $F/\Phi(G) = (U/\Phi(G)) \times (V/\Phi(G))$ with $U/\Phi(G)$ and $V/\Phi(G)$ irreducible and not isomorphic. If B has two distinct maximal subgroups containing F, it has at least $p + 1$ maximal subgroups containing F. Let K_1, K_2, and K_3 be three of these maximal subgroups. One of the factors $U/\Phi(G)$ or $V/\Phi(G)$ must be centralised by two of these maximal subgroups. Suppose, without loss of generality, that K_1 and K_2 centralise $U/\Phi(G)$. Thus B centralises $U/\Phi(G)$ and so B centralises F/V. This is impossible, because then G would induce an abelian automorphism group on this chief factor. Hence B contains a unique maximal subgroup containing F. It now follows that B/F is cyclic.

By the Frattini argument [119, A, 6.3 (b)] applied to the normal subgroup FA, we obtain that $G = F\mathrm{N}_G(A)$. Since B/F is cyclic, there exists an element $b \in B$

such that b normalises A and $G = FA\langle b\rangle$. Since A is an abelian p'-group and $\langle b\rangle$ is a p-group normalising A, we can apply [119, A, 12.5] to decompose $A = A_0 \times A_1$ where $A_0 = \mathrm{C}_A(\langle b\rangle)$ and $A_1 = [A, \langle b\rangle]$. Suppose that $A_1 = 1$. Then $B = F\langle b\rangle$ is normalised by A and so G has derived length 3, which contradicts the choice of G. Therefore $A_1 \neq 1$. Assume now that $A_0 \neq 1$. Under the action of A_1, we can decompose $F/F' = [F/F', A_1] \times \mathrm{C}_{F/F'}(A_1)$. Arguing as above, we have $\mathrm{C}_{F/F'}(A_1) = 1$, $[F/F', A_1] = F/F'$ and so $[F, A_1]F' = F$, which implies that $[F, A_1] = F$ and $FA_1\langle b\rangle$ has derived length 4, against the minimal choice of G. Hence $A_0 = 1$. In particular, $[A_q, \langle b\rangle] = A_q$ for every Sylow subgroup A_q of A.

Recall that F is the $\mathfrak{N}\mathfrak{A}$-residual of G. By [47, 4.2.19], we have that $G/\mathrm{C}_G(S/Z)$ is not abelian for all chief factor S/Z of G between F' and F. Let K be a normal subgroup of G such that F/K is a chief factor of G. Clearly, F/K is complemented in G. Assume that K is not abelian. Then $F' = K'$ because F' is the only minimal normal subgroup of G. Assume that $\mathrm{C}_{K/K'}(A_1) \neq 1$. Let S/K' be a minimal normal subgroup of G/K' contained in $\mathrm{C}_{K/K'}(A_1)$. Then A_1 centralises S/K' and so $G/\mathrm{C}_G(S/K')$ is abelian. This contradiction yields $\mathrm{C}_{K/K'}(A_1) = 1$ and, by [119, A, 12.5], it follows that $K = [K, A_1]K' = [K, A_1]$. Let H be a complement for F/K in G. By replacing H by a suitable conjugate if needed, we can assume that A is contained in H. The same argument used to find the element b can be used to find an element $h \in H$ such that h normalises A and $H = KA\langle h\rangle$, and then there is no loss of generality in assuming that $h = b$, that is, $H = KA\langle b\rangle$. Hence $H' = (KA\langle b\rangle)' = [K, A\langle b\rangle]K'(A\langle b\rangle)' = [K, A][K, \langle b\rangle]K'[A, \langle b\rangle] = KA_1$ and $H'' = (KA_1)' = K'[K, A_1] = K$. Since K is non-abelian, we get that H has derived length 4. This contradiction shows that K must be abelian.

Suppose first that $F/\Phi(G)$ is a direct product of two minimal normal subgroups of $G/\Phi(G)$. Then

$$F/\Phi(G) = (U/\Phi(G)) \times (V/\Phi(G)),$$

where $(U/\Phi(G))$ and $(V/\Phi(G))$ are two distinct chief factors of G. Suppose that $|U/\Phi(G)| = p^{p^i s}$ and $|V/\Phi(G)| = p^{p^i t}$, where $p^i = |BF/F|$. Then U and V are abelian and so $\Phi(G) = U \cap V \leq \mathrm{Z}(F)$. Now consider $U/\Phi(G)$ and $V/\Phi(G)$ as modules for $\langle b\rangle$. Set $U_0 = [U, \langle b\rangle]\Phi(G)$ and $V_0 = [V, \langle b\rangle]\Phi(G)$. Note that $U_0/\Phi(G)$ and $V_0/\Phi(G)$ are, respectively, the radicals of $(U/\Phi(G))_B$ and $(V/\Phi(G))_B$. By Lemma 3.3.40, there exists an element a of A such that $U_0^a U_0 = U$ and an element a_1 of A such that $V_0^{a_1} V_0 = V$. Now $|F : U_0 \cap U_0^a \cap V_0 \cap V_0^{a_1}| \leq p^{2(s+t)}$ and $|F : \Phi(G)| = p^{p^i(s+t)}$. However, $\Phi(G) \leq U_0 \cap U_0^a \cap V_0 \cap V_0^{a_1} \leq U \cap V = \Phi(G)$. By index considerations, we get a contradiction unless $p^i = 2$. Consequently, $p = 2$ and $\langle b\rangle$ is a cyclic group of order 2.

Assume that $F/\Phi(G)$ is a chief factor of G. Applying Lemma 3.3.40, $F/\Phi(G)$ is faithful and free B/F-module. By [158, VII, 7.14], $F/\Phi(G)$ is projective as G/F-module and hence $\Phi(G)/\Phi(F)$ is a direct summand of $F/\Phi(F)$ by [119, B, 2.5]. Since every chief factor F/K of G is complemented in G, it follows that

$\Phi(G) = \Phi(F) \leq Z(F)$. By Lemma 3.3.40, there exists an element a of A such that $U_0^a U_0 = F$, where $U_0/\Phi(G)$ is the radical of $F/\Phi(G)$ as B/F-module. Since $U_0 = [U, \langle b\rangle]\Phi(G)$, $[U, \langle b\rangle] \leq B'$ is abelian, and $\Phi(G)$ is central in F, it follows that U_0 and U_0^a are abelian and $|F : U_0 \cap U_0^a| \leq p^{2s}$ and $|F : \Phi(G)| = p^{p^i s}$. If p were odd, then $\Phi(G) < U_0 \cap U_0^a \leq Z(F)$. This would imply that $Z(F) = F$, contrary to supposition. Therefore $p = 2$, and $\langle b\rangle$ is also a cyclic group of order 2. On the other hand, we may choose generators $u_1, u_2, \dots, u_k \in U_0^a$, $v_1, v_2, \dots, v_k \in U_0$ for F with $[u_i, u_j] = [v_i, v_j] = 1$ for $i, j \in \{1, 2, \dots, k\}$. Thus F' is generated by the commutators $[u_i, v_j]$, $i, j \in \{1, 2, \dots, k\}$. Consider a fixed u_i. Since $\Phi(G) \leq Z(F)$, the map $x\Phi(G) \mapsto [u_i, x]$ defines a group homomorphism $\tau\colon F/\Phi(G) \to F'$. We prove that it is a $\langle b\rangle$-module homomorphism. It suffices to check that if $x \in F$, then $[u_i, x^b] = [u_i, x]^b$, that is, $[u_i, x^b] = [u_i^b, x^b]$. Since V is generated by the elements $[z, b]$ with $z \in U_0$, it suffices to check that $[z, b]^{-1}[z, b]^b \in \Phi(G)$. Let $w = [z, b]$ for $z \in U_0$. Now let $z \in U_0$, then $z^2 \in \Phi(G)$ and so $w\Phi(G) = z^{-1}z^b\Phi(G) = zz^b\Phi(G)$. Then $(zz^b\Phi(G))^b\Phi(G) = z^b z^{b^2}\Phi(G)$. But $b^2 = 1$ and so $z^{b^2}\Phi(G) = z\Phi(G)$. Hence $w^b\Phi(G) = z^b z\Phi(G) = zz^b[z^b, z]\Phi(G) = zz^b\Phi(G) = w\Phi(G)$ since $[z^b, z] \in F' \leq \Phi(G)$. Consequently τ is a $\langle b\rangle$-module homomorphism and $V/\Phi(G)$ is contained in the kernel of τ. Thus the image is a completely reducible $\langle b\rangle$-submodule of F'. It follows that F' is a completely reducible $\langle b\rangle$-module by [119, A, 4.4]. In particular, B centralises F'. Since F' is irreducible as a G/F-module, $C_A(F')$ is a normal subgroup of G, and F' cannot be faithful by Lemma 3.3.40, we conclude that $A = C_A(F')$. Thus we have F' central in G. Now suppose that $\Phi(G) \neq F'$. Then there is a chief factor of G between F' and F which is central in G. This contradicts [47, 4.2.19]. Thus $F' = Z(G) = \Phi(G) = \Phi(F)$ and so F is extraspecial. A similar argument applies if $F/\Phi(G)$ is a direct product of two minimal normal subgroups of $G/\Phi(G)$, again leading to F being extraspecial.

Assume that F/F' is either a chief factor of G or a direct product of two G-isomorphic chief factors of G. Let q be a prime dividing $|A|$ and a Sylow q-subgroup A_q of A. Suppose that A_q is a proper subgroup of A. Note that A_q is normal in $A\langle b\rangle$ and so $C_{F/F'}(A_q)$ is a normal subgroup of G/F'. Since F/F' is faithful module for $A\langle b\rangle$, it follows that $C_{F/F'}(A) = C_{F/F'}(A_q) = 1$. Therefore $F/F' = [F/F', A_q]$ and $F = [F, A_q]F'$ by [119, A, 12.5]. But now $(FA_q\langle b\rangle)' = F'[F, A_q\langle b\rangle](A_q\langle b\rangle)' = F'[F, A_q][F, \langle b\rangle][A_q, \langle b\rangle] = FA_q$ and $(FA_q\langle b\rangle)'' = (FA_q)' = F'[F, A_q] = F$. We conclude that $FA_q\langle b\rangle$ is a proper subgroup of G of derived length 4. This contradicts the choice of G. Consequently, $A = A_q$ and A is cyclic by [119, B, 9.4]. In this case, every subgroup of A is normal in $A\langle b\rangle$ and so, arguing as above, we deduce that A is of prime order and $A\langle b\rangle$ is dihedral of order $2q$ for some odd prime q. It follows at once that G is an E-group, which yields the final contradiction.

We are now left with proving that if a group $G = AB$, with A abelian, B metabelian and nilpotent and A and B of coprime order and G has a section of type E, then G has derived length 4. It is enough to show that every group of type E satisfies these conditions and is of derived length 4. That a group of type E has derived length 4 is clear. If G is of type E, then we can write $G = AB$, where A is an abelian Hall $2'$-subgroup and B is a Sylow 2-subgroup. We show that B is metabelian to complete the proof. The proof is similar to the argument above. If b is chosen so that $G = \mathrm{F}(G)A\langle b\rangle$, we then let $V/\Phi(G)$ be the radical of $\mathrm{F}(G)/\Phi(G)$ as a $\langle b\rangle$-module. Let v be a fixed element of V. The map $x\Phi(G) \mapsto [v,x]$, $x \in \mathrm{F}(G)$, is a $\langle b\rangle$-module homomorphism from $\mathrm{F}(G)/\Phi(G)$ to $\Phi(G)$. Since the image is completely reducible, we have $V/\Phi(G)$ in the kernel, which gives $[v,x] = 1$ for all $x \in V$. Since this is true for all $v \in V$, we have that V is abelian. That B/V is abelian comes immediately from the definition of V because B acts trivially on $\mathrm{F}(G)/V$. Hence B is metabelian as required. □

It remains as an unanswered question *whether it is possible to give examples of E-groups G for which $\mathrm{F}(G)/\Phi(\mathrm{F}(G))$ is a product of two non-G-isomorphic minimal normal subgroups of $G/\Phi(\mathrm{F}(G))$.*

We round this section off with some results of Heineken [150] concerning the Fitting length of dinilpotent groups.

The following definition is motivated by the relation between the nilpotency class of Sylow p-subgroups and the p-length of p-soluble groups shown in [158, IX, 5.5] and Theorem 1.7.14.

Definition 3.3.42. The nilpotent group G belongs to the class $\mathfrak{L}_m$ if and only if the following statements hold:

1. the nilpotency class of the Sylow 2-subgroup of G is smaller than or equal to m,
2. the nilpotency class of he Sylow p-subgroups of G for primes p that are not Fermat primes is smaller than p^m,
3. the nilpotency class of the Sylow 3-subgroups of G is smaller than 2^m, and
4. for the remaining primes p, the nilpotency class of the Sylow p-subgroups of G is smaller than
$$\frac{(p-2)^{m+1}-1}{p-3}.$$

It is clear that the class $\mathfrak{L}_m$ is a formation for all m.

Theorem 3.3.43. *Let $G = AB$ be a dinilpotent group with $A \in \mathfrak{L}_m$ and $B \in \mathfrak{L}_n$, where $m \le n$. Then the Fitting length of G is at most $2m$ if $m = n$ and at most $2m+1$ otherwise.*

Proof. Since the class of all groups of Fitting length at most r is a saturated formation, we may assume that G is primitive and so $\mathrm{F}(G) = N$ be a self-centralising minimal normal subgroup of G with complement M. We distinguish two cases, according to Theorem 3.2.8, namely $N \le A$ and $N \le B$.

Suppose first that $N \leq A$. Without loss of generality we may assume that $B \leq M$ by Theorem 3.2.8. If N is a p-group, then $\mathrm{F}(M)$ is a p'-group contained in B, also $\mathrm{F}_2(M)/\mathrm{F}(M)$ is a p-group contained in $(A \cap M)\mathrm{F}(M)/\mathrm{F}(M)$. By induction, $\mathrm{F}_{2i}(M)/\mathrm{F}_{2i-1}(M)$ is a p-group contained in $(M \cap A)\mathrm{F}_{2i-1}(M)/\mathrm{F}_{2i-1}(M)$ and $\mathrm{F}_{2i+1}(M)/\mathrm{F}_{2i}(M)$ is a p'-group contained in $B\mathrm{F}_{2i}(M)/\mathrm{F}_{2i}(M)$, because A and B are nilpotent. The Fitting length of G is obviously bounded by $2t$, where t is the p-length of G. Now we apply Theorem 1.7.14 for $p = 2$ and [158, IX, 5.5] for the other primes to obtain that t is bounded by m. Hence the Fitting length of G is bounded by $2m$ in this case.

Now assume that $N \leq B$. Proceeding as before, we find that the Fitting length of G is bounded by $2n$. If $m = n$, we obtain the desired result. If $n > m$, we remember that the quotient $\mathrm{F}_{2i+1}(M)/\mathrm{F}_{2i}(M)$ is contained in $A\mathrm{F}_{2i}(M)/\mathrm{F}_{2i}(M)$, and that the set of primes dividing the order of this quotient always contains the set of primes dividing the order of the next following quotient $\mathrm{F}_{2i+3}(M)/\mathrm{F}_{2i+2}(M)$ because A is nilpotent and the centraliser of the Fitting subgroup of a soluble group is contained in the Fitting subgroup by [119, A, 10.6 (a)]. Since $A \in \mathfrak{L}_m$, there can only be m such quotients by Theorem 1.7.14 and [158, IX, 5.5]. Hence the Fitting length of G is bounded by $2m + 1$, as stated. □

The following result extends Theorem 3.2.28.

Theorem 3.3.44. *Let $G = AB$ be a dinilpotent group. The subgroup $A\mathrm{F}(G)$ is normal in G if one of the following two conditions is satisfied:*

1. *B' is of odd order and the Sylow p-subgroups of A have nilpotency class at most $p - 1$ for every prime p,*

2. *A belongs to the class $\mathfrak{L}_1$.*

Proof. Let M be a maximal subgroup of G and let $D = M_G$. Arguing like in the proof of Theorem 3.3.43 (and by Corollary 1.7.11 for the first case) we see that AD/D is contained in $\mathrm{F}_2(G/D)$. Therefore A is contained in $\mathrm{F}_2(G)$, and the result follows from Theorem 3.2.30. □

Corollary 3.3.45. *Let $G = AB$ be a dinilpotent group all of whose Sylow p-subgroups have nilpotency class at most $p-1$ for every prime p. Then G is metanilpotent.*

Proof. Note that in this case A and B belong to $\mathfrak{L}_1$. □

Postscript

N. S. Chernikov presented in [96] some interesting results on dinilpotent groups. The attention is focused there on the subgroup $K/\Phi(G) = \mathrm{Z}(G/\Phi(G))$ and the natural epimorphism $\varphi\colon G \to G/K$, where G is a dinilpotent group. He studies the inclusion of Hall subgroups H of one of the factors, A say, in members of the Fitting series of G in terms of the exponents of the Sylow subgroups and the derived length of H^{φ}. This can be regarded as an extension of Theorem 3.2.28. Information about the inclusion of normal p-subgroups of A as a function of the Fitting series of G as a function of the structure of U^{φ} is also obtained. Moreover, the exponents of $\langle A_p^G\rangle \mathrm{F}_4(G)/\mathrm{F}_4(G)$, $A\mathrm{F}_4(G)/\mathrm{F}_4(G)$, $\langle A_p^G\rangle \mathrm{F}_2(G)/\mathrm{F}_2(G)$, and $A\mathrm{F}_2(G)/\mathrm{F}_2(G)$ are studied as a function of the p-exponent of A^{φ} for all primes p, as well as the p-exponents of $G/\mathrm{F}_4(G)$ and $G/\mathrm{F}_2(G)$. The statements of some of these results are long since they take into account all possibilities given in Corollary 1.7.11. The reader can see the details in [96].

The same author has proved in [99] the following result:

Theorem 3.3.46. *Let $G = AB$ be a dinilpotent group, $H = [A, B]\mathrm{F}(G)$, and let $\varphi\colon G \to G/\mathrm{F}(G)$ be the natural epimorphism. Given two primes p and q, let $D_{p,q} = H^{\varphi} \cap \langle A_p^G\rangle^{\varphi} \cap \langle B_q^G\rangle^{\varphi}$. The following assertions hold:*

1. $D_{p,q} = (D_{p,q} \cap A_p^{\varphi})(D_{p,q} \cap B_q^{\varphi})$, *and* $D_{p,p} = 1$. *In particular,* $D_{p,q}$ *is a* $\{p,q\}$*-group.*
2. $\langle A^G\rangle^{\varphi} = \mathsf{X}_{p\in\pi(A^{\varphi})} \langle A_p^G\rangle^{\varphi}$.
3. $\langle A_p^G\rangle^{\varphi} = A_p^{\varphi}(\langle A_p^G\rangle^{\varphi} \cap B_{p'}^{\varphi})$.
4. $H^{\varphi} = \mathsf{X}_{p,q\in\pi(G^{\varphi})} D_{p,q}$.
5. *The nilpotent residual of* G^{φ} *coincides with* $H^{\varphi} = [A^{\varphi}, B^{\varphi}]$.
6. $A \cap B \le H = (H \cap A)(H \cap B)$.

The interest of N. S. Chernikov in [98] is the study of quotients of factorised soluble groups $G = AB$ by the the Frattini subgroup $\Phi(G)$, the Fitting subgroup $\mathrm{F}(G)$, and the $\pi(A) \cap \pi(B)$-Hall subgroup of $\mathrm{F}(G)$ when these quotients are dinilpotent. He obtains information about the derived length and the exponent of the Sylow subgroups of the quotients in terms of the derived lengths and the nilpotency classes of the images of A and B under the corresponding natural epimorphisms. We refer the reader to [98] for the details.

Chapter 4

Totally and mutually permutable products of groups – structural results

Let the group $G = G_1G_2\cdots G_r$ be the product of its pairwise permutable subgroups $G_1, G_2, \dots, G_r$. It is rather clear that not many properties carry over from the factors G_i to the group itself. Indeed if one thinks about properties such as solubility, supersolubility, or nilpotency, one soon realises the difficulty of using factorisation to obtain information about the structure of the whole group: there exist non-abelian (respectively non-nilpotent) groups which are products of two abelian subgroups and every soluble group is the product of pairwise permutable nilpotent subgroups. In the much more special case when G_i, $1 \leq i \leq r$, are normal nilpotent subgroups of G, the product $G_1G_2\cdots G_r$ is nilpotent. This is a well-known result of Fitting. However, if $G_1, G_2, \dots, G_r$ are normal supersoluble subgroups of G, the product $G_1G_2\cdots G_r$ is not supersoluble in general even in the finite case (see Example 4.1.32). Consequently it seems reasonable to look into these problems under additional assumptions. In this context, assumptions on permutability connections between the factors turn out to be very useful.

Some interesting considerations in that direction were made by Asaad and Shaalan in [24]: they studied subgroups A and B of a group G such that A permutes with every subgroup of B and B permutes with every subgroup of A and subgroups A and B such that every subgroup of A permutes with every subgroup of B and vice-versa. Carocca [84] called A and B mutually permutable in the first case, and totally permutable in the second one. If $G = AB$ and A and B are mutually (respectively totally) permutable, then G is called a mutually (respectively totally) permutable product of A and B. More generally, a group $G = G_1G_2\cdots G_r$ is said to be the product of the pairwise mutually (respectively totally) permutable subgroups $G_1, G_2, \dots, G_r$ if G_i and G_j are mutually (respectively totally) permutable subgroups of G for all $i, j \in \{1, 2, \dots, r\}$.

It was Shunkov [228] (see also the book of N. S. Chernikov [95, Theorem 3.18]), following the suggestion of his advisor S. N. Chernikov, who first studied totally permutable products in the p-group case (he called them uniform products). Asaad and Shaalan [24] proved that if G is a product of the subgroups A and B and A and B are supersoluble, then G is supersoluble if A and B are totally permutable. When A and B are mutually permutable, then G is supersoluble provided that either G', the derived subgroup of G, is nilpotent or A or B is nilpotent.

Since Asaad and Shaalan's 1989 paper the theory of mutually and totally permutable products has been thoroughly explored and richly generalised, and we shall give details of many of these developments in the last sections of the chapter. The first attempt to broaden the study of these products was made by Carocca [83]. He localised Asaad and Shaalan's results to the p-supersoluble case. The next significant step in this development was prompted by Maier [194]. He shows that, for totally permutable products, infinitely many analogues of Asaad and Shaalan's result are true. In fact, he proved that saturated formations containing the class of all supersoluble groups are closed under the operation of forming totally permutable products. Ballester-Bolinches and Pérez-Ramos [55], answering a question posed by Maier in the above paper, were able to remove the hypothesis of saturation from his result. The study of the behaviour of totally permutable products with respect to Fitting classes containing the class of all supersoluble groups was initiated by Hauck, Martínez-Pastor, and Pérez-Ramos [144] and was subsequently elaborated and refined by them and others. Beidleman and Heineken [59] were able to present a number of interesting results about totally permutable products of torsion groups which generalised some of the results first established for finite groups. By the way of contrast, Stonehewer [230] constructed a non-soluble infinite group G which is the product of two metabelian subgroups that are permutable in G. This shows that soluble groups are not closed under the operation of forming totally permutable products in the general case.

On the other hand, a mutually permutable product of two subgroups intersecting trivially is a totally permutable one. Thus, in such case, if both are supersoluble, the group is supersoluble. Bearing in mind this observation, Alejandre, Ballester-Bolinches, and Cossey [3] proved that mutually permutable products of two supersoluble subgroups with core-free intersection are again supersoluble. This result was localised by Beidleman and Heineken in [65], where important theorems about the behaviour of some relevant subgroups of mutually permutable products are proved, and generalised to SC-groups by Ballester-Bolinches, Cossey, and Pedraza-Aguilera in [41]. In this paper extensions of Asaad and Shaalan's results to SC-groups are also proved. These three authors, together with Beidleman and Heineken, undertook the study of mutually permutable products of nilpotent groups in [30]. That work was continued in [29] and, together with [39], gives a rather complete picture of mutually and totally permutable products of nilpotent groups. A nice characterisation of soluble PST-groups due to Shunkov in terms of total and mutual permutability will follow from our study. Mutually and totally permutable products of SC-groups, PST-groups, PT-groups, and T-groups were also presented in [37], [66], and further remarkable developments in the theory of mutually permutable products are dealt with in Bochtler's doctoral thesis [74], written under supervision of Hauck.

The extension of many of the relevant theorems on mutually and totally permutable products of two subgroups to general pairwise mutually and totally permutable products turns out to be difficult in many cases, and some of them are indeed open problems.

The classical results motivated Carocca ([84], [86]) and Carocca and Maier [89] to study more general situations. It is largely their approach that we base the first section of this chapter, where we establish some preliminary machinery whose motivation become evident when it is applied in Section 4.2. There we proceed to develop the subgroup structure of totally and mutually permutable products that play a central part in Chapter 5, which deals with the behaviour of these products with respect to classes of groups.

4.1 Elementary properties

We begin with the concepts of mutually and totally f-permutable subgroups, where f is a subgroup embedding functor (see Definition 1.2.4). They have been introduced by Carocca in his doctoral thesis ([84]).

Definition 4.1.1. Let f be a subgroup embedding functor, and let A and B be two subgroups of a group G.

1. We say that A *f-permutes* with B (or A is *f-permutable* with B) if A permutes with every element of $f(B) \cup \{B\}$.

2. A and B are called *mutually f-permutable* if A f-permutes with B and B f-permutes with A.

3. A and B are said to be *totally f-permutable* if every element of $f(A) \cup \{A\}$ permutes with every element of $f(B) \cup \{B\}$. When f is s, it is usually omitted and we say that A and B are *mutually permutable* or *totally permutable* instead of mutually s-permutable or totally s-permutable.

4. If $G = G_1 G_2 \cdots G_r$ is the product of the pairwise mutually (totally) f-permutable subgroups $G_1, G_2, \dots, G_r$, we say that G is the *mutually (totally) f-permutable product* of $G_1, G_2, \dots, G_r$.

Illustrations 4.1.2.

1. If f_1 and f_2 are two subgroup embedding functors and $f_1 \subseteq f_2$, then f_2-permutability implies f_1-permutability but the converse is not true in general.

2. If A is permutable subgroup of G, then A is f-permutable with all subgroups of G, for every subgroup embedding functor f.

3. If A and B are totally f-permutable, then they are mutually f-permutable. If $G = \mathrm{Dih}(8)$ is the dihedral group of order 8 and A and B are its non-cyclic subgroups of order 4, we have and A and B are mutually permutable, but not totally permutable.

4. If A and B are two cyclic subgroups of a group G and A is normal in G, then A and B are totally permutable. In particular, the symmetric group $G = \mathrm{Sym}(3)$ of degree 3 is a totally permutable of a cyclic group of order 3 and a cyclic group of order 2, which is not a direct product.

5. If A and B are two permutable cyclic subgroups of a group G of prime power order, then A and B are totally permutable. This is a restatement of Corollary 3.1.9.

Lemma 4.1.3. *Let f be a subgroup embedding functor and let the group $G = AB$ be the product of the subgroups A and B such that A f-permutes with B. Then, for any $x, y \in G$, we have that $G = A^x B^y$, and A^x f-permutes with B^y. In particular, A permutes with every conjugate of X in B for any $X f B$.*

Proof. Let $x \in G$. Write $x = ab$ with $a \in A$ and $b \in B$. Then

$$A^x B = A^{ab} B = A^b B = (AB^{b^{-1}})^b = (AB)^b = G.$$

For each $X f B$ we have that $X^{b^{-1}} f B$. Hence $AX^{b^{-1}} = X^{b^{-1}} A$. It follows that $A^x X = A^b X = XA^b = XA^x$. Consequently A^x f-permutes with B. Let $y \in G$. Then, by the above argument, $G = A^x B^y$ and A^x f-permutes with B^y. □

In a certain sense some f-permutable products are close to being non-simple. This is mainly due to the following lemma proved by Carocca [82].

Lemma 4.1.4. *Let the group $G = AB$ be the product of the subgroups A and B. Let $1 \neq X \leq A$. If B permutes with every conjugate of X in A, then there exists a non-trivial subnormal subgroup S of G such that $S \leq A$ or $S \leq XB$.*

Proof. Let $g \in G$. Write $g = ab$ with $a \in A$ and $b \in B$. Then

$$X^g B = X^{ab} B = (X^a B^{b^{-1}})^b = (X^a B)^b = (BX^a)^b = BX^g.$$

Therefore B permutes with every conjugate of X in G. By Theorem 1.1.9 (2), we have that $R = \langle X^B \rangle \cap \langle B^X \rangle$ is a subnormal subgroup of G. If $R = 1$, then $\langle X^B \rangle = X$ and $1 \neq S = \langle X^G \rangle = X^{BA} = X^A \leq A$. If $R \neq 1$, then R is a non-trivial subnormal subgroup of G contained in XB. □

This result finds immediate application in the following theorem which was also proved by Carocca [82].

Theorem 4.1.5. *Let f be a subgroup embedding functor such that* $\mathrm{Syl} \subseteq f$ *or* $\mathrm{Max} \subseteq f$ *and let $G = G_1 G_2 \cdots G_r$ be a group such that $G_1, G_2, \dots, G_r$ are pairwise mutually f-permutable subgroups of G. Suppose that G_i is non-trivial for all $i \in \{1, 2, \dots, r\}$. If G is simple, then $G = G_1 = G_2 = \cdots = G_r$.*

Proof. If the theorem is false, let G be a counterexample in which r is least possible. Then G is a non-abelian simple group and there exists $i \in \{1, 2, \dots, r\}$ such that $G_i \neq G$. There is no loss of generality in assuming that $i = 1$. Assume that $G_j = G$ for some $j \in \{2, \dots, r\}$. Then G_1 is S-permutable in G if $\mathrm{Syl} \subseteq f$ and G_1 is M-permutable in G if $\mathrm{Max} \subseteq f$. In the first case, G_1 is subnormal in G by Theorem 1.2.14 (3) and, in the second case, G_1 has non-trivial core in G by Lemma 1.2.9. This contradicts the simplicity of G. Consequently, G_j is a proper subgroup of G for all $j \in \{1, 2 \dots, r\}$. Moreover, by Corollary 3.2.2, there exists $k \in \{1, 2, \dots, r\}$ such that G_k does not have a prime power order. We can assume without loss of generality that $|G_1|$ is not a prime power.

Let $H_1 = G_2 \cdots G_r$ and $1 \neq X f G_1$. Since H_1 f-permutes with G_1, we can apply Lemma 4.1.4 and Lemma 4.1.3 and the simplicity of G to conclude that $G = XH_1$. Denote by $D_1 = G_1 \cap H_1$. Then $G_1 = G_1 \cap XH_1 = XD_1$. By minimality of r, $D_1 \neq G_1$. Suppose first that $\mathrm{Max} \subseteq f$. If X is a maximal subgroup of G_1 such that $D_1 \leq X$, we have $G_1 = XD_1 = X$. Suppose now that $\mathrm{Syl} \subseteq f$. For every prime divisor p of $|G_1|$ and $X \in \mathrm{Syl}_p(G_1)$, we have $G_1 = XD_1$. Hence $|G_1 : D_1|$ is a power of p. Since there are at least two possibilities for the choice of p, we have $G_1 = D_1$. In any case we have a contradiction, and therefore no counterexample exists. □

Remark 4.1.6. The alternating group $G = \mathrm{Alt}(5)$ of degree 5 can be written as the product $G = G_1G_2$, where G_1 is an alternating group of degree 4 and G_2 is a Sylow 5-subgroup of G. We have that G_1 permutes with all subgroups of G_2, but G_2 does not permute with all Sylow subgroups of G_1.

Definition 4.1.7. Let f be a subgroup embedding functor.

1. We say that f is *compatible with quotients* when for each group G and each normal subgroup N of G, if $X/N f G/N$, then there exists $Y f G$ such that $YN = X$.

2. We say that f is *compatible with subgroups* when for every group G and every subgroup H of G, if $X f H$, then there exists $Y f G$ such that $X = Y \cap H$.

Lemma 4.1.8. 1. *The subgroup embedding functors* Syl, Max, sn, sfn, *and* s *are compatible with quotients.*

2. *The subgroup embedding functors* Syl, sfn, *and* s *are compatible with subgroups, but* Max *and* sn *are not.*

Proof. 1. Let $X/N f G/N$. Suppose that $f = \mathrm{Syl}$. If X/N is a Sylow p-subgroup of G/N, then $|G/N : X/N| = |G : X|$ is not divisible by p. Let Y be a Sylow p-subgroup of X, then $|G : Y| = |G : X||X : Y|$ is not divisible by p and so Y is a Sylow p-subgroup of G; furthermore, $YN = X$. Assume that $f = \mathrm{Max}, f = \mathrm{sn}$, or

$f = \mathrm{s}$. It is clear that $Y = X$ satisfies the condition. Suppose now that $f = \mathrm{sfn}$. If $\mathrm{N}_{G/N}(X/N) = X/N$ and $g \in \mathrm{N}_G(X)$, then $X^g = X$ and so $(X/N)^{gN} = X/N$. This implies that $gN \in \mathrm{N}_{G/N}(X/N) = X/N$. Therefore $g \in X$. Consequently $Y = X$ also satisfies the condition.

2. The result is clear for Syl and s. Let X sfn $H \leq G$. We argue as in Lemma 1.2.8. Let $U_0 = X$, $U_{i+1} = \mathrm{N}_G(U_i)$ for $i = 0, 1, 2, \ldots$. There exists an n such that $U_{n+1} = U_n$. Let $Y = U_n$. Then Y sfn G, X is a subnormal subgroup of Y, and X is a subnormal subgroup of $Y \cap H$. Since X is subnormal in H, we conclude that $X = Y \cap H$.

Max and sn are not compatible with subgroups. For Max, witness the subgroups $H \in \mathrm{Syl}_2(G)$ and $X \leq H$ with $|X| = 2$ of the alternating group $G = \mathrm{Alt}(4)$ on 4 letters. For sn we consider the symmetric group $G = \mathrm{Sym}(4)$ on 4 letters, $H \in \mathrm{Syl}_2(G)$ and $X \leq H$ with X cyclic of order 4. Then X is normal in H and so it is subnormal in H, but there is no subnormal subgroup Y of G such that $X = Y \cap H$. □

The next results are particularly useful when an inductive argument involving subgroups or quotient groups is applied.

Lemma 4.1.9. *Let $G = G_1 G_2 \cdots G_r$ be the product of the pairwise f-permutable subgroups $G_1, G_2, \ldots, G_r$. Suppose that $f \in \{\mathrm{Syl}, \mathrm{sfn}, \mathrm{s}\}$. For each $i \in \{1, 2, \ldots, r\}$, write $H_i = G_1 G_2 \cdots G_{i-1} G_{i+1} \cdots G_r$. If $H_i \leq L \leq G$, then L is the product of the pairwise mutually f-permutable subgroups $G_1, G_2, \ldots, G_{i-1}, L \cap G_i, G_{i+1}, \ldots, G_r$ (with the obvious adaptations of the notation in the cases $i = 1$ and $i = r$).*

Proof. Let $H_i \leq L \leq G$. We have that $L = H_i(L \cap G_i)$. Clearly, for each $j \in \{1, 2, \ldots, i-1, i+1, \ldots, r\}$ and $X f G_j$, $X(L \cap G_i) = L \cap XG_i = L \cap G_i X = (L \cap G_i)X$. It follows that $L \cap G_i$ f-permutes with each G_j. To finish the proof of the lemma it is enough to check that G_j f-permutes with $L \cap G_i$.

Let $Y f (L \cap G_i)$. By Lemma 4.1.8, there exists $W f G_i$ such that

$$Y = W \cap (L \cap G_i) = W \cap L.$$

Thus

$$G_j Y = G_j(W \cap L) = G_j W \cap L = WG_j \cap L = (W \cap L)G_j = YG_j.$$

This completes the proof. □

Lemma 4.1.10. *Let f be a subgroup embedding functor which is compatible with quotients. Assume that the subgroups A and B of the group G are mutually (respectively, totally) f-permutable and that N is a normal subgroup of G. Then the subgroups AN/N and BN/N are mutually (respectively, totally) f-permutable in G/N.*

Proof. Let N be a normal subgroup of G. Let $X/N f BN/N$. Then $X = X \cap BN = (X \cap B)N$. The standard isomorphism from BN/N to $B/(B \cap N)$ transforms X/N into $(X \cap B)/(B \cap N)$. Therefore $(X \cap B)/(B \cap N) f B/(B \cap N)$. Since f is compatible with quotients, there exists $Y f B$ such that $Y(B \cap N) = X \cap B$. Since A f-permutes with B, AY is a subgroup of G. Therefore $ANY = AN(B \cap N)Y = AN(X \cap B) = AX$ and so AN/N f-permutes with X/N. We can conclude that AN/N and BN/N are f-permutable in G/N.

A similar argument is valid for totally f-permutable products. □

Corollary 4.1.11. *Let f be a subgroup embedding functor which is compatible with quotients and let the group $G = G_1G_2 \cdots G_r$ be the product of the pairwise mutually (respectively, totally) f-permutable subgroups $G_1, G_2, \dots, G_r$. Then G/N is the product of the mutually (respectively, totally) f-permutable subgroups G_1N/N, $G_2N/N, \dots, G_nN/N$.*

The next result can be seen as a partial converse of Lemma 4.1.10.

Lemma 4.1.12. *Assume that $f \in \{\mathrm{Syl}, \mathrm{Max}, \mathrm{s}\}$. Let A and B be two subgroups of a group G. Let N be a normal subgroup of G such that $N \leq A \cap B$. If A/N and B/N are mutually f-permutable, then A and B are mutually f-permutable.*

Proof. Let $Y f B$. Then $YN/N f G/N$ and so A/N and YN/N permute because A/N and B/N are mutually f-permutable. Consequently $A(YN) = A(NY) = (AN)Y = AY$ is a subgroup of G and so A and Y f-permute. We conclude that A and B are mutually f-permutable. □

Corollary 4.1.13. *Assume that $f \in \{\mathrm{Syl}, \mathrm{Max}, \mathrm{s}\}$. Let the group $G = G_1G_2 \cdots G_r$ be the product of the pairwise permutable subgroups $G_1, G_2, \dots, G_r$. Suppose that N is a normal subgroup of G contained in $\bigcap_{i=1}^{r} G_i$ such that G/N is the product of the mutually f-permutable subgroups $G_1/N, G_2/N, \dots, G_r/N$. Then G is the mutually f-permutable product of $G_1, G_2, \dots, G_r$.*

We may now apply the preceding results to show that for some subgroup embedding functors f, the class of all p-soluble groups, p a prime, is closed under taking mutually f-permutable products. This was proved by Carocca in [83].

Theorem 4.1.14. *Let f be a subgroup embedding functor such that $\mathrm{Syl} \subseteq f$ and let the group $G = G_1G_2 \cdots G_r$ be the mutually f-permutable product of $G_1, G_2, \dots, G_r$. If for all $i \in \{1, 2, \dots, r\}$, G_i is p-soluble for some prime number p, then G is p-soluble. In particular, if all G_i are soluble, then G is soluble.*

Proof. Suppose that the theorem is false. Let G be a counterexample of the smallest possible order with r least possible. Clearly we can assume that $f = \mathrm{Syl}$.

For each normal subgroup N of G, we have that

$$G/N = (G_1N/N)(G_2N/N)\cdots(G_rN/N)$$

is the product of pairwise mutually Syl-permutable p-soluble subgroups G_1N/N, $G_2N/N, \ldots, G_rN/N$ by Lemma 4.1.10. Thus, if $N \neq 1$, then G/N is p-soluble. We will reach a contradiction by proving that there is a non-trivial normal subgroup N of G such that N is p-soluble.

By Corollary 3.2.2, there exists $k \in \{1, 2, \ldots, r\}$ such that G_k does not have a prime power order. We can assume without loss of generality that $|G_1|$ is not a prime power. Let $H_1 = G_2G_3\cdots G_r$. By Lemma 4.1.9 and the minimal choice of G, we have that L is p-soluble for any subgroup L such that $H_1 \leq L \leq G$. Applying Lemmas 4.1.4 and 4.1.3, we conclude that $G = H_1Q$ for some Sylow q-subgroup Q of G_1. Since G_1 does not have a prime power order, there exists a non-trivial Sylow t-subgroup T of G_1 for a prime $t \neq q$ such that $T \leq H_1$. Now H_1 permutes with every conjugate of T in G_1. Therefore, according to Lemma 4.1.4, there is a non-trivial subnormal subgroup S of G such that $S \leq G_1$ or $S \leq H_1$. In both cases, S is p-soluble. Therefore $N = \langle S^G \rangle$ is a non-trivial normal p-soluble subgroup of G, as desired. □

An immediate consequence of this result is:

Theorem 4.1.15. *Let the group $G = G_1G_2\cdots G_r$ be the product of the pairwise mutually permutable subgroups $G_1, G_2, \ldots, G_r$. If all the factors G_i are p-soluble for a prime p, then G is p-soluble. In particular, if all G_i are soluble, then G is soluble.*

It is impossible to discuss mutually and totally permutable products at any depth without analysing early the embedding of the intersection of the factors. An indication of the important role of this subgroup is the following result.

Proposition 4.1.16. *Let f be a subgroup embedding functor and $G = AB$ a group which is the mutually f-permutable product of A and B. Denote by $D = A \cap B$.*

1. *For each $X f A$ and $Y f B$ we have that $DX = XD$ and $DY = YD$. In particular, D is f-permutable in both A and B.*
2. *If $D \leq X f A$ and $Y f B$, then $XY = YX$. If $D \leq Y f B$ and $X f A$, then $XY = YX$.*
3. *If $D = 1$, then A and B are totally f-permutable.*
4. *If $\mathrm{Syl} \subseteq f$, then D is S-permutable in A and B and D is subnormal in G.*
5. *If $\mathrm{sfn} \subseteq f$, then D is a subnormal subgroup of G.*
6. *If A and B are totally permutable, then $D \leq \mathrm{F}(G)$, D is a nilpotent Iwasawa group and its Sylow subgroups have the structure described in Theorem 1.4.3.*

Proof. 1. Let $X f A$. Then

$$XD = X(A \cap B) = A \cap XB = A \cap BX = (A \cap B)X = DX.$$

2. Let $D \leq X f A$ and $Y f B$. We have:

$$XY = XDY = X(A \cap B)Y = AY \cap XB = YA \cap BX = YDX = YX.$$

3. By Statement 2, every element of $f(A) \cup \{A\}$ permutes with every element of $f(B) \cup \{B\}$.

4. By Statement 1, D is S-permutable in A and D is S-permutable in B. Applying Theorem 1.2.14 (3), D is subnormal in both A and B. This implies that D is a subnormal subgroup of G by Theorem 1.1.7.

5. By Statement 1, D permutes with every self-normalising subgroup of A and B. Hence D is a subnormal subgroup of A and D is a subnormal subgroup of B by Lemma 1.2.10. It follows that D is subnormal in G by Theorem 1.1.7.

6. Let X be any subgroup of D. Then X is permutable in D. In particular, X is subnormal in D by Theorem 1.2.14 (2). Hence D is a nilpotent Iwasawa group. Therefore D is contained in $\mathrm{F}(G)$ by Statement 4. □

Corollary 4.1.17. *Let f be a subgroup embedding functor and let the group $G = G_1G_2 \cdots G_r$ be the product of the pairwise mutually f-permutable subgroups G_1, $G_2, \ldots, G_r$. If either* Syl $\subseteq f$ *or* sfn $\subseteq f$, *then $\bigcap_{i=1}^{r} G_i$ is a subnormal subgroup of G.*

Proof. Let T denote the intersection of the subgroups $G_1, G_2, \ldots, G_r$. If we fix an index $i \in \{1, 2, \ldots, r\}$, it follows that $T = \bigcap_{j \neq i}(G_j \cap G_i)$. By Proposition 4.1.16, we have that $G_j \cap G_i$ is subnormal in G_i for every $j \in \{1, 2, \ldots, r\}$. Consequently T is subnormal in G_i by [119, A, 14.2]. By Corollary 1.1.8, we conclude that T is subnormal in G. □

In particular, it follows that the intersection of the factors in mutually permutable products is subnormal. This fact was already observed by Ballester-Bolinches, Beidleman, Heineken, and Pedraza-Aguilera [35]. Somewhat more can be said of the structure of the intersection of two mutually permutable subgroups.

Lemma 4.1.18. *Let the group $G = AB$ be the product of the mutually permutable subgroups A and B. Then $\langle (A \cap B)^G \rangle / (A \cap B)_G$ is nilpotent.*

Proof. Let $D = A \cap B$. By Proposition 4.1.16 (1), D is permutable in A. By Corollary 1.5.6, we have that D/D_A is nilpotent. Therefore the nilpotent residual $D^{\mathfrak{N}}$ of D is contained in D_A. Hence $\langle (D^{\mathfrak{N}})^A \rangle \leq D_A \leq D$. Moreover $\langle (D^{\mathfrak{N}})^G \rangle = \langle \langle (D^{\mathfrak{N}})^A \rangle^B \rangle \leq \langle D^B \rangle \leq B$ and arguing in the same way $\langle (D^{\mathfrak{N}})^G \rangle \leq A$. It follows that $\langle (D^{\mathfrak{N}})^G \rangle \leq D$. This yields $\langle (D^{\mathfrak{N}})^G \rangle \leq D_G$ and so D/D_G is nilpotent. Since D/D_G is subnormal in G/D_G, it follows that $\langle D^G \rangle / D_G$ is nilpotent. □

The intersection of two mutually permutable subgroups is not nilpotent in general.

Example 4.1.19. Let $S = \mathrm{Sym}(3) = \langle a, b : a^3 = b^2 = (ab)^2 = 1\rangle$ denote the symmetric group of degree 3, $C_3 = \langle c\rangle$ be a cyclic group of order 3, and $C_2 = \langle d\rangle$ be a cyclic group of order 2. Then $G = S \times C_3 \times C_2$ is the product of the normal subgroups $A = S \times C_3 = \langle a, b, c\rangle$ and $B = S \times C_2 = \langle a, b, d\rangle$. The subgroups A and B are not totally permutable, because $\langle bc\rangle \leq A$ does not permute with $\langle a\rangle \leq B$. Note that $A \cap B = S$ is not nilpotent.

In the rest of the section, we focus our attention on mutually and totally permutable products. The next lemma, due to Carocca [85, Lemma 1] (see also Maier [194, Lemma 2]), shows that it is not possible to construct totally permutable products with core-free factors. It will be used several times in subsequent proofs.

Lemma 4.1.20. *Let the group $G = G_1G_2\cdots G_r$ be the product of the pairwise totally permutable subgroups $G_1, G_2, \dots, G_r$. If G is non-trivial, then there exists $i \in \{1, 2, \dots, r\}$ such that G_i contains a non-trivial normal subgroup of G.*

Proof. Let p denote the largest prime divisor of $|G|$. It is clear that p divides at least one of $|G_1|, |G_2|, \dots, |G_r|$. Let x be a p-element of the union set $G_1 \cup G_2 \cup \cdots \cup G_r$ of maximal order and suppose $x \in G_1$. Let R be the subgroup of order p in $\langle x\rangle$. Let $i \in \{2, \dots, r\}$ and let y be a q-element of G_i, q a prime number. Since G_1 and G_i are totally permutable, we have that $\langle x\rangle\langle y\rangle = \langle y\rangle\langle x\rangle$ is supersoluble by Lemma 3.1.1.

If $q \neq p$, then $\langle x\rangle$ is a normal subgroup of $\langle x\rangle\langle y\rangle$ because $p > q$ (see [119, VII, 2.1]). If $q = p$, then $|\langle x\rangle| \geq |\langle y\rangle|$ and there is a non-identity normal subgroup of $\langle x\rangle\langle y\rangle$ contained in $\langle x\rangle$ by Lemma 3.1.1. In any case, the unique subgroup R of order p in $\langle x\rangle$ is normalised by y. Since this can be done for all elements y of G_i whose order is a power of a prime, we conclude that G_i normalises R. Consequently, the normal closure

$$\langle R^G\rangle = \langle R^{G_1G_2\cdots G_r}\rangle = \langle R^{G_2G_3\cdots G_rG_1}\rangle = \langle R^{G_1}\rangle \leq G_1$$

is a non-trivial normal subgroup of G contained in G_1. □

Our next goal is to get information about subgroups of mutually permutable products. In this context, the following result proved by Beidleman and Heineken [65] is very helpful.

Lemma 4.1.21. *Let the group $G = AB$ be the product of the mutually permutable subgroups A and B. Then:*

1. *If U is a subgroup of G, then $(U \cap A)(U \cap B)$ is a subgroup and $U \cap A$ and $U \cap B$ are mutually permutable.*
2. *If U is a normal subgroup of G, then $(U \cap A)(U \cap B)$ is a normal subgroup of G.*

3. *If $X \leq A$, $Y \leq B$, and $X \cap Y = A \cap B$, then X and Y are mutually permutable. Moreover $XY = (XY \cap A)(XY \cap B)$.*

4. *If N is a normal subgroup of G and $A \cap B \leq N$, then $G/(N \cap A)(N \cap B)$ is a totally permutable product of $A(N \cap B)/(N \cap A)(N \cap B) \cong A/(N \cap A)$ and $B(N \cap A)/(N \cap A)(N \cap B) \cong B/(N \cap B)$.*

Proof. 1. Let X be a subgroup of $U \cap A$. Then $X(U \cap B) = U \cap XB = U \cap BX = (U \cap B)X$. Therefore X permutes with $U \cap B$. An analogous argument with A and B interchanged proves the statement.

2. Note that $U \cap A(U \cap B) = (U \cap A)(U \cap B) = U \cap (U \cap A)B$. Therefore $(U \cap A)(U \cap B)$ is normalised by both A and B and so by G.

3. If $X \cap Y = A \cap B$, then $AY \cap XB = (A \cap XB)Y = X(A \cap B)Y = XY$ and likewise $YA \cap BX = YX$. Since A and B are mutually permutable, we have that $XY = YX$. Moreover $XY = X(Y \cap X)(Y \cap X)Y \leq X(Y \cap A)(B \cap X)Y = (XY \cap A)(XY \cap B) \leq XY$, from which it follows that $XY = (XY \cap A)(XY \cap B)$. Let U be a subgroup of X. Then $U(XY \cap B)$ is a subgroup of G and $U(XY \cap B) = XY \cap UB = (X \cap UB)Y = U(X \cap B)Y = UY$ since $X \cap B \leq A \cap B = X \cap Y$.

4. By Statement 2, $(N \cap A)(N \cap B)$ is a normal subgroup of G. Moreover,

$$\begin{aligned} A(N \cap B) \cap (N \cap A)B &= (A \cap (N \cap A)B)(N \cap B) \\ &= (N \cap A)(A \cap B)(N \cap B) = (N \cap A)(N \cap B) \end{aligned}$$

since $A \cap B \leq N \cap A$. Applying Proposition 4.1.16 (3), $G/(N \cap A)(N \cap B)$ is the totally permutable product of $A(N \cap B)/(N \cap A)(N \cap B)$ and $B(N \cap A)/(N \cap A)(N \cap B)$. □

Statements 1 and 2 of Lemma 4.1.21 can be easily extended to mutually permutable products of arbitrary factors as Ballester-Bolinches, Beidleman, Heineken, and Pedraza-Aguilera showed in [35].

Corollary 4.1.22. *Let the group $G = G_1G_2 \cdots G_r$ be the product of the pairwise mutually permutable subgroups $G_1, G_2, \dots, G_r$. Then:*

1. *If U is a subgroup of G, then $(U \cap G_1)(U \cap G_2) \cdots (U \cap G_r)$ is the pairwise mutually permutable product of the subgroups $U \cap G_1, U \cap G_2, \dots, U \cap G_r$.*

2. *If U is a normal subgroup of G, then $(U \cap G_1)(U \cap G_2) \cdots (U \cap G_r)$ is a normal subgroup of G.*

Proof. 1. Let U be a subgroup of G and $i, j \in \{1, 2, \dots, r\}, i \neq j$. Then $U \cap G_iG_j$ is a subgroup of the mutually permutable product G_iG_j of the subgroups G_i and G_j. By Lemma 4.1.21 (1) we have that $(U \cap G_i)(U \cap G_j)$ is the mutually permutable product of the subgroups $U \cap G_i$ and $U \cap G_j$. This implies that $(U \cap G_i)(U \cap G_2) \cdots (U \cap G_r)$

is a subgroup of G which is the pairwise mutually permutable product of the subgroups $U \cap G_1, U \cap G_2, \dots, U \cap G_r$. This proves Statement 1.

2. Assume now that U is a normal subgroup of G. Then (with the obvious adaptations of the notation in the cases $j = 1$ and $j = r$) G_j normalises

$$\begin{aligned} &U \cap ((U \cap G_1)(U \cap G_2) \cdots (U \cap G_{j-1})G_j(U \cap G_{j+1}) \cdots (U \cap G_r)) \\ &\qquad = (U \cap G_1)(U \cap G_2) \cdots (U \cap G_{j-1})(U \cap G_j)(U \cap G_{j+1}) \cdots (U \cap G_r) \end{aligned}$$

for all $j \in \{1, 2, \dots, n\}$. Consequently $G = G_1 G_2 \cdots G_r$ normalises $(U \cap G_1)(U \cap G_2) \cdots (U \cap G_r)$ and Statement 2 holds. □

The preceding results are useful in finding interesting subnormal subgroups in mutually permutable products. We need a preliminary definition.

Definition 4.1.23. If n is a natural number, we define

$$\mathrm{m}(n) = \mathrm{lcm}(\{p - 1 : p \text{ is a prime dividing } n\}).$$

Theorem 4.1.24 (Beidleman and Heineken [65, Theorem 1]). *Let the group $G = AB$ be the product of the mutually permutable subgroups A and B. Then $B'B^m$ is a subnormal subgroup of G, where m denotes* $\mathrm{m}(|A : A \cap B|)$.

Proof. We can assume that B is a proper subgroup of G. By Proposition 4.1.16 (1), $A \cap B$ is a permutable subgroup of both A and B and is subnormal in G. Set $B^+ = B'B^m$, we will show that $(A \cap B)B^+$ is subnormal in G.

Since $A \cap B$ is a proper subnormal subgroup of A, there exists a composition series of A containing $A \cap B$. Let

$$\cdots \trianglelefteq A \cap B = X_n \trianglelefteq \cdots \trianglelefteq X_2 \trianglelefteq X_1 \trianglelefteq X_0 = A$$

be such a composition series. For every i, X_i/X_{i+1} is a simple group. Let $0 \le i < n$. Then, by Proposition 4.1.16 (2), X_{i+1} and X_i permute with B and $X_{i+1}B < X_iB$ (otherwise we would have $X_i \le X_{i+1}B \cap A = X_{i+1}(B \cap A) = X_{i+1}$). Let $K = (X_{i+1}B)_{X_iB} = (X_{i+1}B)_{X_i}$. Then $X_{i+1} \le K = X_{i+1}(K \cap B)$. According to Lemma 4.1.21 (3), X_iB is the mutually permutable product of the subgroups X_i and B. Moreover, $BK \cap X_iK = (BK \cap X_i)K = (BX_{i+1}(K \cap B) \cap X_i)K = (BX_{i+1} \cap X_i)K = (B \cap X_i)X_{i+1}K \le (B \cap A)X_{i+1}K = K$. Hence, by Proposition 4.1.16 (3), X_iB/K is the totally permutable product of the subgroups X_iK/K and BK/K. Moreover $X_iK/K \cong X_i/(K \cap X_i)$ and this is simple, because $K \cap X_i$ is a proper normal subgroup of X_i containing X_{i+1}. Therefore the first factor is simple. Let L be a normal subgroup of X_iB such that $K \le L \le KB$. Then $L = K(L \cap B) = X_{i+1}(L \cap B)$. Hence L is contained in $X_{i+1}B$ and so $L = K$. Therefore BK/K does not contain non-trivial normal subgroups of X_iB/K. Applying Lemma 4.1.20, X_iK/K must contain a normal subgroup of X_iB/K. It follows that

X_iK/K is normal in X_iB/K and BX_i/K is in fact a semidirect product of X_iK/K by BK/K. Note that $\mathrm{C}_{BK/K}(X_iK/K)$ is centralised by X_iK/K and normalised by BK/K. Consequently BK/K acts faithfully on X_iK/K.

Assume first that X_i/X_{i+1} is non-abelian. Let H/K be a subgroup of X_iK/K. Then $(H/K)(BK/K) \cap (X_iK/K) = H/K$ is normal in $(H/K)(BK/K)$, this implies that BK/K normalises H/K. In particular, BK/K acts on X_iK/K as a group of power automorphisms. Since X_iK/K is non-abelian, this implies that BK/K centralises X_iK/K. Hence BK/K is trivial because it acts faithfully on X_iK/K. It follows that $B \leq K$.

Assume now that X_i/X_{i+1} is abelian. Then it has prime order p and BK/K must be cyclic of exponent dividing $p-1$. It follows that $B^+ = B'B^m \leq B'B^{p-1} \leq K$.

In both cases $B^+ \leq K$. Since K is normal in X_iB, we have that $K \cap X_iB^+ = (K \cap X_i)B^+ = (X_{i+1}(K \cap B) \cap X_i)B^+ = X_{i+1}(K \cap B \cap X_i)B^+ = X_{i+1}B^+$ is a normal subgroup of X_iB^+.

We obtain that

$$(A \cap B)B^+ = X_nB^+ \trianglelefteq X_{n-1}B^+ \trianglelefteq \cdots \trianglelefteq X_2B^+ \trianglelefteq X_1B^+ \trianglelefteq X_0B^+ = AB^+$$

is a part of a composition series of $(A \cap B)B^+$ and in particular $(A \cap B)B^+$ is subnormal in AB^+. Since B^+ is normal in B, it is also normal in $(A \cap B)B^+$ and subnormal in AB^+. By Theorem 1.1.7, we have that B^+ is also subnormal in the product $(AB^+)B = AB = G$. This proves the theorem. □

Corollary 4.1.25. *Let the group $G = AB$ be the product of the mutually permutable subgroups A and B. Then A' and B' are subnormal in G. In particular, they belong to the Fitting class generated by G.*

Proof. By Theorem 4.1.24, $A'A^m$ is subnormal in G, where $m = \mathrm{m}(|A : A \cap B|)$. Since $A'A^m$ contains A', it follows that A' is normal in $A'A^m$. This implies that A' is subnormal in G. □

Corollary 4.1.26. *Let the group $G = G_1G_2 \cdots G_r$ be the product of the pairwise mutually permutable subgroups $G_1, G_2, \ldots, G_r$. Then G_i' is a subnormal subgroup of G for each $i \in \{1, 2, \ldots, r\}$.*

Proof. By Corollary 4.1.25, it follows that G_i' is a subnormal subgroup of the mutually permutable product G_iG_j of the subgroups G_i and G_j for all $i, j \in \{1, 2, \ldots, r\}$. This implies that G_i' is a subnormal subgroup of G by Corollary 1.1.8 and so the lemma is proved. □

Our next main objective will be to prove that some relevant classes of groups defined by sets of primes which are extensions of the class of all supersoluble groups are closed under the operation of forming pairwise mutually permutable products.

Definition 4.1.27 ([65]). Let p be a prime. The set π of primes is said to be *p-special* if $q \notin \pi$ whenever p divides $q(q-1)$.

Theorem 4.1.28 ([65, Theorem 2]). *Let p be a prime and let π be a p-special set of primes. Let the group G be the product of the mutually permutable subgroups A and B which are normal extensions of p-groups by π-groups. Then the same is true for G.*

Proof. By definition of π, we observe that $\mathrm{m}(|A : A \cap B|)$ and $\mathrm{m}(|B : A \cap B|)$ are relatively prime to p. Let us define $A^+ = A' A^{\mathrm{m}(|A:A\cap B|)}$ and $B^+ = B' B^{\mathrm{m}(|B:A\cap B|)}$ as in the proof of Theorem 4.1.24. Then the Sylow p-subgroup A_p of A is contained in A^+ and, since A_p is normal in A, we obtain that A_p is normal in A^+. By symmetry, B_p is normal in B^+. On the other hand, there exists a Sylow p-subgroup P of G such that $P = A_p B_p$ by Theorem 1.1.19. By Theorem 4.1.24, A^+ and B^+ are subnormal in G and hence so are A_p and B_p. Consequently G_p is subnormal in G by a result of Wielandt (see [119, A, 14.4]). It follows that P is normal in G by Theorem 1.5.2. Therefore G is an extension of the normal Sylow p-subgroup P by a π-group, as desired. □

The extension of the above result to finitely many pairwise mutually permutable subgroups was obtained by Ballester-Bolinches, Beidleman, Heineken, and Pedraza-Aguilera in [35]

Corollary 4.1.29. *Let the group $G = G_1 G_2 \cdots G_r$ be the product of the pairwise mutually permutable subgroups $G_1, G_2, \dots, G_r$. If $G_1, G_2, \dots, G_r$ are normal extensions of p-groups by π-groups for a p-special set of primes π, then the same is true for G.*

Proof. Let P_i be the Sylow p-subgroup of G_i for every $i \in \{1, 2, \dots, r\}$. Then P_i is a normal subgroup of G_i. It follows from Theorem 4.1.28 that $G_i G_j$ has a normal Sylow p-subgroup. In particular P_i is a subnormal subgroup of G by Corollary 1.1.8. Consequently either G is a π-group or $\mathrm{O}_p(G) \neq 1$. Assume that $M = \mathrm{O}_p(G) \neq 1$. Then

$$G/M = (G_1 M/M)(G_2 M/M) \cdots (G_r M/M)$$

is the product of pairwise mutually permutable subgroups $G_1 M/M$, $G_2 M/M, \dots,$ $G_r M/M$ by Lemma 4.1.10. In addition, $G_i M/M$ is a normal extension of a p-group by a π-group for all i. Consequently, G/M satisfies the hypotheses of the theorem. Arguing by induction on the group order, it follows that G/M is a normal extension of a p-group by a π-group. This implies that G is a normal extension of a p-group by a π-group. □

Corollary 4.1.30. *Let the group $G = G_1 G_2 \cdots G_r$ be the product of the pairwise mutually permutable subgroups $G_1, G_2, \dots, G_r$. If G_i, $1 \leq i \leq r$, is a Sylow tower group with respect to a Sylow tower where all primes q dividing $p - 1$ appear on top of p, then G is a Sylow tower group of the same form.*

Note that supersoluble groups belong to all of these classes of Sylow tower groups.

The next result shows that the class of all p-supersoluble groups, p a prime, is closed under the formation of totally permutable products. This is an extension of the Asaad and Shaalan classical result due to Carocca [84].

Theorem 4.1.31. *Let the group $G = G_1G_2 \cdots G_r$ be the product of the pairwise totally permutable subgroups $G_1, G_2, \ldots, G_r$. If all G_i are p-supersoluble for a prime p, then G is p-supersoluble.*

The proof of Theorem 4.1.31 will be presented after some previous lemmas.

A mutually permutable product of p-supersoluble groups is not p-supersoluble in general.

Example 4.1.32. If $G = AB$ is the product of two mutually permutable supersoluble subgroups, G need not be supersoluble. Let Q be the quaternion group of order 8 and let V be a faithful and irreducible Q-module over GF(5). Let $G = [V]Q$ be the corresponding semidirect product. Let A and B be two maximal subgroups of G containing V. Then $A = A \cap VQ = V(A \cap Q)$ and $B = V(B \cap Q)$ are two maximal subgroups of Q and so A and B are normal in G. In particular, A and B are mutually permutable. Moreover, A and B are 5-supersoluble, because $A \cap Q$ and $B \cap Q$ are cyclic groups of order 4. However, G is not 5-supersoluble.

As we have said in the introduction of the chapter, Asaad and Shaalan [24] proved that a group G factorised by two supersoluble mutually permutable subgroups is supersoluble provided G' is nilpotent or at least one of the factors is nilpotent. One can wonder whether these results remain true for mutually permutable products of more than two factors. Carocca [84] proved the following:

Theorem 4.1.33. *Assume the group $G = G_1G_2 \cdots G_r$ is the pairwise mutually permutable product of the p-supersoluble subgroups $G_1, G_2, \ldots, G_r$. If the commutator subgroup G' of G is p-nilpotent, then G is p-supersoluble.*

However the answer to the second question is negative as the following example shows:

Example 4.1.34. Let the group $T = AB$ be a mutually permutable product of two supersoluble subgroups A and B. Suppose that T is not supersoluble. Consider $G = T \times C$, the direct product of T with a cyclic group C of order p, p a prime. Then $G = ABC$ is the product of the pairwise mutually permutable supersoluble subgroups A, B and C, G is not supersoluble, and C is nilpotent.

Again, if we demand p-nilpotency of at least $r - 1$ factors of a pairwise mutually permutable product $G = G_1G_2 \cdots G_r$ of the p-supersoluble subgroups $G_1, G_2, \ldots, G_r$, then we obtain the p-supersolubility of G. This was proved in [35].

Theorem 4.1.35. *Let the group $G = G_1 G_2 \cdots G_r$ be the product of the pairwise mutually permutable subgroups $G_1, G_2, \ldots, G_r$. If G_1 is p-supersoluble and G_i is p-nilpotent for every $i \in \{2, \ldots, r\}$, then G is p-supersoluble.*

The information given in the following lemma comes in useful when analysing pairwise mutually permutable factorised groups which are not p-supersoluble, but every proper epimorphic image is p-supersoluble.

Lemma 4.1.36. *Let the group $G = G_1 G_2 \cdots G_r$ be the product of the pairwise mutually permutable p-supersoluble subgroups $G_1, G_2, \ldots, G_r$. Assume that G is not p-supersoluble, but every proper epimorphic image of G is p-supersoluble. Then:*

1. *G is p-soluble, $N = \mathrm{Soc}(G) = \mathrm{F}(G)$ is a minimal normal p-subgroup of G of rank greater than 1, p is the largest prime dividing $|G|$, and N is the unique Sylow p-subgroup of G.*
2. *If p divides the order of G_i for some $i \in \{1, 2, \ldots, r\}$, then N is contained in G_i. Otherwise, G_i is an abelian p'-group. In this case, G_i normalises each subgroup of N.*
3. *If $G_{i_1} G_{i_2} \cdots G_{i_k}$ is a p-supersoluble group for every proper subset $\{i_1, i_2, \ldots, i_k\}$ of $\{1, 2, \ldots, r\}$, then N is contained in $\bigcap_{i=1}^{r} G_i$.*

Proof. Since p-supersoluble groups form a saturated formation, it follows that G has a unique minimal normal subgroup, N say. Applying Theorem 4.1.15, G is p-soluble and so N is an elementary abelian p-group of rank greater than 1. Moreover $N = \mathrm{C}_G(N) = \mathrm{F}(G) = \mathrm{O}_{p',p}(G)$.

Suppose that $G_i' \neq 1$ for some $i \in \{1, 2, \ldots, r\}$. Then G_i' is a p-nilpotent subnormal subgroup of G by Corollary 4.1.26. In particular, G_i' is contained in N. This implies that G_i is supersoluble. If G_i is abelian for some $i \in \{1, 2, \ldots, r\}$, then G_i is supersoluble. It follows then that G_i is supersoluble for all $i \in \{1, 2, \ldots, r\}$. Consequently G_j is a Sylow tower group with respect to the reverse natural ordering of the prime numbers for all $j \in \{1, 2, \ldots, r\}$. By Corollary 4.1.30, the same is true for G. In particular, p is the largest prime dividing $|G|$ and N is the Sylow p-subgroup of G. Hence Statement 1 holds.

Assume that p divides the order of G_i for some $i \in \{1, 2, \ldots, r\}$. Then $1 \neq N \cap G_i$ is the Sylow p-subgroup of G_i. For every $k \in \{1, \ldots, r\}$, let H_j be a Hall p'-subgroup of G_i. Since $|G_i : N \cap G_i|$ and $|G_i H_j : N \cap G_i|$ are not divisible by p, we have $N \cap G_i H_j = N \cap G_i$ and $N \cap G_i$ is normalised by $N\langle H_1, \ldots, H_r\rangle = G_1 G_2 \cdots G_r = G$. The minimality of N yields that $N \leq G_i$. Suppose now that p does not divide the order of G_i. Then G_i' is a subnormal p'-subgroup of G and so $G_i' \leq \mathrm{O}_{p'}(G) = 1$. Hence G_i is abelian. Since G is not a p'-group, there exists $j \in \{1, 2, \ldots, r\}$ such that p divides the order of G_j. Let X be a subgroup of N. Then since the product $G_i G_j$ is mutually permutable, it follows that XG_i is a subgroup of G. Therefore $X = N \cap XG_i$ is normal in XG_i. This confirms Statement 2.

Suppose now that $G_{i_1} G_{i_2} \cdots G_{i_k}$ is a p-supersoluble group for every proper subset $\{i_1, i_2, \dots, i_k\}$ of $\{1, 2, \dots, r\}$. If N is not contained in $\bigcap_{i=1}^{r} G_i$, there exists $j \in \{1, 2, \dots, r\}$ such that G_j normalises every subgroup of N. Without loss of generality we may assume that $j = 1$. Then N is contained in $G_2 \cdots G_r$ and there is a cyclic subgroup Y of N which is normal in $G_2 \cdots G_r$ because $G_2 \cdots G_r$ is p-supersoluble. This means that Y is normal in G, contrary to Statement 1. Consequently, N is contained in $\bigcap_{i=1}^{r} G_i$. □

Next we make one elementary and extremely useful observation about mutually permutable products.

Lemma 4.1.37. *Let the group $G = NB$ be the product of the mutually permutable subgroups N and B. Assume that N is a normal subgroup of G and $N \cap B = 1$. Then B acts as a group of power automorphisms on N.*

Proof. Let T be a subgroup of N. Then TB is a subgroup of G and so $TB \cap N = T(B \cap N) = T$ is a normal subgroup of TB. It follows that B normalises T. □

Proof of Theorem 4.1.31. The theorem is true is $G = 1$. Therefore suppose that $G \neq 1$, $r > 1$, and by induction that the theorem is true for either totally permutable products of less than r factors or totally permutable products of smaller order. Then, for each normal subgroup N of G, we have that

$$G/N = (G_1N/N)(G_2N/N) \cdots (G_rN/N)$$

is the product of the pairwise totally permutable p-supersoluble subgroups G_1N/N, $G_2N/N, \dots, G_rN/N$ by Lemma 4.1.10. Thus, if $N \neq 1$, then G/N is p-supersoluble. Moreover, the choice of G implies that $G_{i_1} G_{i_2} \cdots G_{i_k}$ is p-supersoluble for every proper subset $\{i_1, i_2, \dots, i_k\}$ of $\{1, 2, \dots, r\}$. Applying Lemma 4.1.36, G has a unique Sylow p-subgroup $N = \mathrm{F}(G)$, N is the unique minimal normal subgroup of G, G/N is p-supersoluble, and N is contained in $\bigcap_{i=1}^{r} G_i$. Let $j \in \{1, \dots, r\}$, and consider a complement U_j of N in G_j. Let X be a subgroup of N. Since $n > 1$ and the product is totally permutable, we have that $XU_j = U_jX$ and so $X = X(U_j \cap N) = N \cap XU_j$ is a normal subgroup of XU_j. Since N is abelian, we have that X is a normal subgroup of G_j. Hence X is a normal subgroup of G. This implies that $|N| = p$, and G is p-supersoluble. □

Proof of Theorem 4.1.33. We suppose that the theorem is false and derive a contradiction. Let G be a counterexample of minimal order with n least possible. Then, arguing as in Theorem 4.1.31, every epimorphic image of G is p-supersoluble and so, by Lemma 4.1.36, G is p-soluble and it has a unique minimal normal subgroup, N say; N is an elementary abelian p-group of rank greater than 1, p is the largest prime dividing $|G|$ and N is the Sylow p-subgroup of G contained in $\bigcap_{i=1}^{r} G_i$. Moreover,

$1 \neq G'$ is p-nilpotent and normal in G. Hence G' is a p-group and so it is contained in $N = \mathrm{F}(G)$. Let H be a Hall p'-subgroup of G. Then H is abelian and so it is cyclic because N is an irreducible and faithful H-module over $\mathrm{GF}(p)$ ([119, B, 9.4 (b)]). Since G_1 and $G_2 \cdots G_r$ are both p-supersoluble and N is self-centralising in G, it follows that the Hall p'-subgroups of G_1 and $G_2 \cdots G_r$ are cyclic of exponent dividing $p - 1$. Applying Theorem 1.1.19, there exist Hall p'-subgroups C and D of G_1 and $G_2 \cdots G_r$ respectively such that CD is a Hall p'-subgroup of G. Then CD has exponent $p - 1$. By [119, B, 9.8 (d)], N is cyclic. This contradiction proves the theorem. □

Proof of Theorem 4.1.35. We adhere to the pattern of the proof of Theorem 4.1.33. Assume that the theorem is false, and let G be a counterexample of the smallest possible order with n least possible. Then every epimorphic image of G is p-supersoluble and so by Lemma 4.1.36, G is p-soluble and $N = \mathrm{F}(G)$ is the unique minimal normal subgroup. N is an elementary abelian p-group of rank greater than 1, p is the largest prime dividing $|G|$ and N is the Sylow p-subgroup of G contained in $\bigcap_{i=1}^{r} G_i$. Since G_i is p-nilpotent, $i \in \{2, \dots, r\}$, and N is self-centralising in G, it follows that G_i is a p-group. Consequently $G_i = N$ and so $G = G_1$ is p-supersoluble, contrary to assumption. □

Corollary 4.1.38. *Let the group $G = G_1 G_2 \cdots G_r$ be the product of the pairwise totally permutable subgroups $G_1, G_2, \dots, G_r$. If all G_i are supersoluble, then G is supersoluble.*

Corollary 4.1.39. *Let the group $G = G_1 G_2 \cdots G_r$ be the product of the pairwise mutually permutable subgroups $G_1, G_2, \dots, G_r$. If all G_i are supersoluble, and either G' is nilpotent or G_i are nilpotent for all $i \neq 1$, then G is supersoluble.*

Similar arguments to those used in the above proofs allow us to establish the following.

Theorem 4.1.40. *Let p be a prime. Assume the group $G = G_1 G_2 \cdots G_r$ is the pairwise mutually permutable product of the p-supersoluble subgroups $G_1, G_2, \dots, G_r$. Then the p-length of G is less or equal than* 1.

Corollary 4.1.41. *Let the group $G = G_1 G_2 \cdots G_r$ be the product of the pairwise mutually permutable subgroups $G_1, G_2, \dots, G_r$. If all G_i are supersoluble, then the p-length of G is less or equal than* 1 *for all primes p.*

In the rest of the section we shall discuss prefactorised and factorised subgroups in the context of mutual permutability concentrating on results which depend specifically on factorisations of some distinguished subgroups (see Definitions 1.1.12 and 1.1.13). It turns out that prefactorised and factorised subgroups of products of pairwise permutable subgroups of more than two factors are most comfortably handled in the mutually permutable setting. This is mainly due to the fact that they are again products of pairwise permutable subgroups, which are mutually permutable by Corollary 4.1.22 (1). It is not a comprehensive treatment of this theme and further results of this nature will be presented at the appropriated places.

Some facts particularly relevant to these concepts are worth noting. The first is that if $G = G_1G_2 \cdots G_r$ is a product of the pairwise mutually permutable subgroups G_1, $G_2, \dots, G_r$, then, by Lemma 1.1.14, a subgroup S of G is factorised with respect to this factorisation if and only if S is prefactorised and $G_i \cap \prod_{j \neq i} G_i \leq S$ for all $i \in \{1, 2, \dots, r\}$. The second is that the set of all factorised subgroups is a sublattice of the subgroup lattice of G. This is a consequence of Lemma 1.1.17 (3) and the following result proved by Ballester-Bolinches, Beidleman, Heineken, and Pedraza-Aguilera [34].

Lemma 4.1.42. *Let the group $G = G_1G_2 \cdots G_r$ be the product of the pairwise mutually permutable subgroups G_1, $G_2, \dots, G_r$. Then:*

1. *If N is a normal prefactorised subgroup of $G = G_1G_2 \cdots G_r$ and N is contained in S, then S/N is a prefactorised subgroup of the quotient group $G/N = (G_1N/N)(G_2N/N) \cdots (G_rN/N)$ if and only if S is a prefactorised subgroup of $G = G_1G_2 \cdots G_r$.*

2. *If U is a factorised (prefactorised) subgroup of G and V is a factorised (respectively, prefactorised) subgroup of U (with respect to the factorisation $U = (U \cap G_1)(U \cap G_2) \cdots (U \cap G_r)$), then V is factorised (respectively, prefactorised) subgroup of G (with respect to the factorisation $G = G_1G_2 \cdots G_r$).*

3. *If U and V are factorised (respectively, prefactorised) with respect to G, then also $\langle U, V\rangle$ is factorised (respectively, prefactorised).*

Proof. 1. If S is prefactorised, it follows by Lemma 1.1.17 (1) that S/N is prefactorised. Conversely, assume that S/N is prefactorised. Then using the fact that N is prefactorised and $N \leq S$, we have

$$\begin{aligned} S &= (S \cap G_1N)(S \cap G_2N) \cdots (S \cap G_rN) \\ &= (S \cap G_1)(S \cap G_2) \dots (S \cap G_r)N \\ &= (S \cap G_1)(S \cap G_2) \dots (S \cap G_r)(N \cap G_1)(N \cap G_2) \dots (N \cap G_r) \\ &= (S \cap G_1)(S \cap G_2) \dots (S \cap G_r) \leq S, \end{aligned}$$

which shows that S is prefactorised.

2. We have that $U = (U \cap G_1)(U \cap G_2) \cdots (U \cap G_r)$ is the product of the pairwise mutually permutable subgroups $(U \cap G_1), (U \cap G_2), \dots, (U \cap G_r)$ by Corol-

lary 4.1.22 (1) and V is a factorised subgroup of U. Then

$$\begin{aligned} V &= (V \cap (U \cap G_1))(V \cap (U \cap G_2)) \cdots (V \cap (U \cap G_r)) \\ &= (V \cap G_1)(V \cap G_2) \cdots (V \cap G_r), \end{aligned}$$

that is, V is a prefactorised subgroup of G. Assume now that U is factorised in G and V is factorised in U. Then V is prefactorised in G. Moreover $G \cap \prod_{j \neq 1} G_j \leq U$ and $(U \cap G_i) \cap \prod_{j \neq i}(U \cap G_j) \leq V$. Note that the fact that U is factorised in G yields $U \cap \prod_{j \neq i} G_j \leq \prod_{j \neq i}(U \cap G_j)$. Consequently $G_i \cap \prod_{j \neq i} G_j \leq V$ and V is a factorised subgroup of G.

3. If $U = (U \cap G_1)(U \cap G_2) \cdots (U \cap G_r)$ and $V = (V \cap G_1)(V \cap G_2) \cdots (V \cap G_r)$, then

$$\begin{aligned} \langle U, V \rangle &= \langle (U \cap G_1)(U \cap G_2) \cdots (U \cap G_r), (V \cap G_1)(V \cap G_2) \cdots (V \cap G_r) \rangle \\ &\leq \langle (\langle U, V \rangle \cap G_1), (\langle U, V \rangle \cap G_2), \ldots, (\langle U, V \rangle \cap G_r) \rangle \\ &= (\langle U, V \rangle \cap G_1)(\langle U, V \rangle \cap G_2) \cdots (\langle U, V \rangle \cap G_r) \\ &\leq \langle U, V \rangle. \end{aligned}$$

Thus $\langle U, V \rangle$ is a prefactorised subgroup of G. Moreover, in the case that U and V are factorised subgroups of G, we have that $G_i \cap \prod_{j \neq i} G_j \leq U$ and $G_i \cap \prod_{j \neq i} G_j \leq V$ for all $i \in \{1, 2, \ldots, r\}$. Consequently $G_i \cap \prod_{j \neq i} G_j \leq \langle U, V \rangle$ for all $i \in \{1, 2, \ldots, r\}$ and $\langle U, V \rangle$ is a factorised subgroup of G. □

Normal subgroups of pairwise mutually permutable products are not prefactorised in general as the following examples appeared in [34] show.

The first one confirms that $\mathrm{Z}(G)$ is not prefactorised in general.

Example 4.1.43. Let $G = A \times B$, where $A = \langle u, v \rangle$ is a non-abelian group of order p^3 and $B = \langle x \rangle$ is a group of order p with p an odd prime. Then A and $\langle u, vx \rangle$ are mutually permutable subgroups of G and $G = A\langle u, vx \rangle$. We have $\mathrm{Z}(G) = \langle [u, v], x \rangle$ and $\mathrm{Z}(G) \cap A = \mathrm{Z}(G) \cap \langle u, vx \rangle = \langle [u, v] \rangle \neq \mathrm{Z}(G)$. It follows that $\mathrm{Z}(G)$ is not prefactorised.

The second one is an example of a p-group, p a prime, in which the derived subgroup is not prefactorised.

Example 4.1.44. Consider the group $M = (\langle a \rangle \times \langle b \rangle \times \langle x, y \rangle)/\langle a^2 b^2 x^2 \rangle$ with $N = \langle a^2 b^2 x^2 \rangle$, a and b elements of order 4, and $\langle x, y \rangle \cong Q_8$, the quaternion group of order 8. Denote by u^* the class uN of M. We show that $M = \langle a^* \rangle \langle a^* x^* \rangle \langle b^* \rangle \langle b^* y^* \rangle$ is a pairwise mutually permutable product: Since two of the factors are normal subgroups, we have to check only one case, and we obtain $(a^* x^*)(b^* y^*) = a^* b^* x^* y^* = a^* b^* y^* x^* (a^* b^*)^2 = b^* y^* a^* x^* (b^* y^*)^2 (a^* x^*)^2 = (b^* y^*)^{-1} (a^* x^*)^{-1}$ because

$(a^*x^*)^2 = (b^*)^2$ and $(b^*y^*)^2 = (a^*)^2$. Note that all squares belong to the centre of M, so the given relation leads to all other further relations to be checked, so that if $\{e, f\} \subseteq \{-1, 1\}$, we find $(a^*x^*)^e(b^*y^*)^f = (b^*y^*)^{-f}(a^*x^*)^{-e}$. Hence the subgroup $M' = \langle [x, y]N \rangle$ is not prefactorised because it intersects trivially with all factors.

In a positive direction we have the following results, all of them proved in [34]. We begin with some prominent conjugacy classes of subgroups.

Proposition 4.1.45. 1. *Let the group $G = G_1G_2 \cdots G_r$ be the product of the pairwise mutually permutable subgroups $G_1, G_2, \dots, G_r$. For each prime dividing $|G|$, there exists a Sylow p-subgroup of G, P say, such that P is prefactorised, that is, $P = (P \cap G_1)(P \cap G_2) \cdots (P \cap G_r)$. Moreover $P \cap G_i$ is a Sylow p-subgroup of G_i for all $i \in \{1, 2, \dots, r\}$.*

2. *Let the group $G = G_1G_2 \cdots G_r$ be the product of the pairwise mutually permutable soluble subgroups $G_1, G_2, \dots, G_r$. For each set of primes π dividing the order of G, there exists a Hall π-subgroup H of G such that H is prefactorised, that is, $H = (H \cap G_1)(H \cap G_2) \cdots (H \cap G_r)$. Moreover $H \cap G_i$ is a Hall π-subgroup of G_i for all $i \in \{1, 2, \dots, r\}$.*

Proof. We prove Statement 2. Since all G_i are soluble, by Theorem 4.1.15, G is soluble. We proceed by induction on r. The result is clear if $r = 1$. Suppose that $r > 1$ and the result is clear for all products of less than r factors. Consider the product $G_1G_2 \cdots G_{r-1}$. Then there exists a Hall π-subgroup T of $G_1G_2 \cdots G_{r-1}$ such that $T = (T \cap G_1)(T \cap G_2) \cdots (T \cap G_{r-1})$. Moreover $T \cap G_i$ is a Hall π-subgroup of G_i for all $i \in \{1, 2, \dots, r-1\}$. Since the product is pairwise mutually permutable, it follows that TG_k is a subgroup of G. Let H be a Hall π-subgroup of TG_r containing T. Then $H = H \cap TG_r = T(H \cap G_r)$. Also $T \cap G_i \leq H \cap G_i$ for all $i \in \{1, 2, \dots, r-1\}$. But $T \cap G_i$ is a Hall π-subgroup of G for all $i \in \{1, 2, \dots, r-1\}$. Therefore $T \cap G_i = H \cap G_i$ for all $i \in \{1, 2, \dots, r-1\}$ and $H = (H \cap G_1)(H \cap G_2) \cdots (H \cap G_r)$. Note that $|H : TG_r| = |G_r : H \cap G_r|$ is a π'-number. Consequently $H \cap G_r$ is a Hall π-subgroup of G_r. Now an argument on the orders shows that H is a Hall π-subgroup of G.

Using the same arguments just given above we obtain Statement 1 for non-soluble groups. □

The following example shows that in a mutually permutable product (even in a totally permutable one) there is not necessarily a factorised Sylow subgroup for each prime p dividing the order of G.

Example 4.1.46. Let $X = \langle x, y : x^3 = y^2 = 1, x^y = x^{-1} \rangle$ be a group isomorphic to the symmetric group of degree 3 and $Y = \langle a, b : a^5 = b^2 = [a, b] = 1 \rangle \cong C_5 \times C_2$. Denote by $G = X \times Y$, $A = \langle y \rangle \times \langle a \rangle \times \langle b \rangle$ and $B = \langle x \rangle \times \langle b \rangle$. Then $G = AB$ is the totally permutable product of A and B. Moreover $A \cap B = \langle b \rangle$ and $\langle x \rangle$ is a Sylow 3-subgroup of G which is clearly prefactorised, but not factorised.

The next remark describes a minimal configuration often encountered in the study of normal prefactorised subgroups.

Remark 4.1.47. Assume that a certain normal subgroup N of a pairwise mutually permutable product $G = G_1G_2 \cdots G_r$ satisfies that N/M is prefactorised in $G/M = (G_1M/M)(G_2M/M) \cdots (G_rM/M)$ for each minimal normal subgroup M of G contained in N. Then either N is prefactorised in $G = G_1G_2 \cdots G_r$ or N is a minimal normal subgroup of G.

Assume that N is not prefactorised in G. Then $N \neq 1$. If N is not a minimal normal subgroup of G, then there exists a minimal normal subgroup M of G with $M \leq N$. By hypothesis, N/M is prefactorised in G/M, that is, $N/M = ((G_1M/M) \cap (N/M))((G_2M/M) \cap (N/M)) \cdots ((G_rM/M) \cap (N/M))$ and from here $N = (N \cap G_1)(N \cap G_2) \cdots (N \cap G_r)M$. On the other hand, by Corollary 4.1.22 (2), $(N \cap G_1)(N \cap G_2) \cdots (N \cap G_r)$ is a normal subgroup of G. If $(N \cap G_1)(N \cap G_2) \cdots (N \cap G_r) \neq 1$, then it contains a minimal normal subgroup of G, R say, and this implies $N = (N \cap G_1)(N \cap G_2) \cdots (N \cap G_r)$, that is, N is prefactorised as desired. Therefore we may assume that $(N \cap G_1)(N \cap G_2) \cdots (N \cap G_r) = 1$. Thus $N = M$, that is, N is a minimal normal subgroup of G.

We shall now investigate some important prefactorised normal subgroups.

Theorem 4.1.48. *Let the group $G = G_1G_2 \cdots G_r$ be the product of the pairwise mutually permutable subgroups $G_1, G_2, \ldots, G_r$. Then:*

1. *The subgroup of G generated by all π-elements of G (where π denotes a set of primes dividing the order of G) is prefactorised in G.*
2. *$G^p = \langle x^p : x \in G \rangle$ is prefactorised in G for each prime p.*
3. *If M and N are prefactorised normal subgroups of G and N has exponent p, p a prime, then also $[M, N]$ is prefactorised in G.*

Proof. 1. Let N denote the subgroup generated by all π-elements of G, where π is a set of primes dividing the order of G. Assume that N is not prefactorised and choose for G a counterexample of least order. Then N/M is prefactorised in $G/M = (G_1M/M)(G_2M/M) \cdots (G_rM/M)$ for each minimal normal subgroup M of G contained in N. By Remark 4.1.47, N is a minimal normal subgroup of G and so $G_i \cap N = 1$ for all $i \in \{1, 2, \ldots, r\}$. Since G/N is a π'-group, it follows that G_iN/N is also a π'-group for all i. Therefore G is a π'-group. This implies that $N = 1$, contrary to assumption. Hence N is prefactorised.

2. Let $N = G^p$. Assume that $N \neq (N \cap G_1)(N \cap G_2) \cdots (N \cap G_r)$ and let G be of least order. Arguing like in Statement 1, N is a minimal normal subgroup of G and

so $N \cap G_i = 1$ for all $i \in \{1, 2, \ldots r\}$. In particular, G_i has exponent p for every i. Hence G is a p-group and $|N| = p$. Choose $x \in G_i$. Then $\langle x\rangle G_i = G_i\langle x\rangle$ and G_i is a subgroup of index at most p in the product $G_i\langle x\rangle$. Hence G_i is normalised by all elements x of G_j and so by G.

Let t be an element of G. Then $t = u_1 u_2 \cdots u_r$ with $u_i \in G_i$, $i \in \{1, 2, \ldots, r\}$. Assume now that $t^p \neq 1$ and choose t with least number of non-trivial factors. Let u_w be the first non-trivial factor. Then $t = u_w s$ and $s^p = 1$ by our choice of t. Since G_w is normal in G, it follows that every conjugate of u_w belongs to G_w. We have $t^p = (u_w s)^p s^{-p} = u_w(s u_w s^{-1})(s^2 u_w s^{-2}) \cdots (s^{p-1} u_w s^{1-p}) \in G_w$. This means that $t^p \in N \cap G_w = 1$ against the choice of t. Consequently G has exponent p and $N = G^p = 1$, the final contradiction.

3. Assume that the result is not true and let G be a minimal counterexample. By Lemma 4.1.42 (3), MN is a prefactorised subgroup of G. If we suppose that MN is a proper subgroup of G, then it is clear that MN satisfies the same hypotheses as G and, by the minimal choice of G, $[M, N]$ is prefactorised in MN. By Lemma 4.1.42 (2), $[M, N]$ is prefactorised in G. Therefore we may assume that $G = MN$. Then $[M, N]$ is a normal subgroup of G and by Remark 4.1.47, we may assume that $[M, N]$ is a minimal normal subgroup of G such that $[M, N] \cap G_i = 1$ for every $i \in \{1, 2, \ldots, r\}$. Let $1 \neq x \in N \cap G_j$. Then x has order p since $N^p = 1$. In particular, it normalises G_i since $|G_i\langle x\rangle : G_i| \in \{1, p\}$. This implies that $N \cap G_j$ normalises $N \cap G_i$ for all $i \in \{1, 2, \ldots, r\}$. On the other hand, by hypothesis we have $M = (M \cap G_1)(M \cap G_2) \cdots (M \cap G_r)$ and $N = (N \cap G_1)(N \cap G_2) \cdots (N \cap G_r)$. Hence $[M, N] = [(N \cap G_1)(N \cap G_2) \cdots (N \cap G_r), M]$. Applying [119, A, 7.4 (f)], $[M, N] = [N \cap G_1, M][N \cap G_2, M] \cdots [N \cap G_r, M]$. Let $z \in [N \cap G_j, M]$. By [119, A, 7.2], $z = x_1 x_2 \cdots x_r$ with $x_i \in [N \cap G_j, M \cap G_i]^g$, $g \in G$. Therefore $x_i \in G_i^g \cap [N, M]$. Now $x_i^{g^{-1}} \in G_i \cap [N, M] = 1$ and $x_i = 1$. Consequently $z = 1$ and from here $[M, N] = 1$. This final contradiction proves the claim. □

Corollary 4.1.49. 1. *If the group $G = G_1 G_2 \cdots G_r$ of prime exponent is the product of the pairwise permutable subgroups $G_1, G_2, \ldots, G_r$, then every term of the lower central series and every term of the derived series of G is prefactorised in G. Furthermore, for every prefactorised normal subgroup M of G, we have that $[M, G]$ and M' are prefactorised subgroups of G.*

2. *If p is a prime number, then $G^p G'$ is prefactorised in G.*

3. *If m is a square-free integer, then $G^m G'$ is prefactorised in G.*

Proof. 1. This is a direct consequence of Statement 3 of Theorem 4.1.48.

2. To prove Statement 2, note that by Statement 1, $(G/G^p)' = G^p G'/G^p$ is prefactorised in G/G^p and G^p is prefactorised in G by Theorem 4.1.48 (2). The result now follows from Statement 1 of Lemma 4.1.42.

3. We proceed by induction on the number k of prime factors of m. The result is true if m is a prime by Statement 2. Assume that the result is true for all square-free integers which are products of $k-1$ primes and choose a prime dividing m. Then $m = pn$ and $H_n = G^n G'$ and $H_p = G^p G'$ are prefactorised by the induction hypothesis. Hence $(H_n)^p(H_n)'$ and $(H_p)^n(H_p)'$ are prefactorised, too. Now we have the following inclusions:

$$G^m(G')^p \leq (H_n)^p(H_n)' \leq G^m G'$$
$$G^m(G')^n \leq (H_p)^n(H_p)' \leq G^m G'$$

and therefore $G^m G' = (H_n)^p(H_n)'(H_p)^n(H_p)'$ is prefactorised in G by virtue of Lemma 4.1.42 (3). □

Note that the hypothesis about m in the above corollary is essential (see Example 4.1.44).

The following statement gives an indication that there is quite a range of prefactorised normal subgroups.

Corollary 4.1.50. *Let the soluble group $G = G_1G_2 \cdots G_r$ be the product of the pairwise mutually permutable subgroups $G_1, G_2, \ldots, G_r$. Then there exists a descending sequence of characteristic subgroups $G = A_0 > A_1 > A_2 > \cdots > A_m = 1$ such that every quotient A_s/A_{s+1} is an elementary abelian p-group for all $s \in \{1, \ldots, m-1\}$ and some prime p.*

Proof. Since $G \neq 1$ is soluble, we have $G' \neq G$. Choose a prime dividing $|G/G'|$. Then also $G^pG' \neq G$ and we may take $A_1 = G^pG'$ by Corollary 4.1.49 (2). The corollary now follows by induction on $|G|$. □

4.2 Nilpotent and supersoluble residuals of totally permutable products

Our final objective in this section is to investigate the nilpotent and supersoluble residuals of totally permutable products. One of the main results is that in totally permutable products every factor centralises the nilpotent residual of the others. This important fact was first proved by Beidleman and Heineken in [59] for totally permutable torsion subgroups and plays an essential part in the subsequent development of the theory. Our analysis leading up to the proof of this theorem in the finite case calls for a study of the behaviour of the supersoluble residual in totally permutable products. This was made by Ballester-Bolinches and Pérez-Ramos in [55]. In the rest of the section, we derive some structural information about totally permutable products which is used to prove that pairwise totally permutable products of soluble groups have prefactorised Carter subgroups and system normalisers. These results support the viewpoint that, concerning formation subgroups, totally permutable products are quite close to central products. Further justification will be provided by some other results in Section 5.2, where residuals, projectors, and normalisers of totally permutable products associated to saturated formations will be studied in detail.

We begin with some preparatory lemmas before coming to the main result of the section. They were proved by Ballester-Bolinches and Pérez-Ramos in [55] and give some interesting structural information about totally permutable products.

Lemma 4.2.1. *Let the group $G = NB$ be the product of the totally permutable subgroups N and B. Assume that N is a minimal normal subgroup of G.*

1. *If N is abelian, then N is a cyclic group of prime order.*
2. *If N is not abelian, then B centralises N.*

Proof. First of all, note that every subgroup L of B such that $L \cap N = 1$ normalises every subgroup of N by Lemma 4.1.37.

1. Suppose that N is abelian. Then N is a p-group for some prime p. Let P be a Sylow p-subgroup of B. Since N is a normal subgroup of the p-group NP, we have that $N \cap \mathrm{Z}(NP) \neq 1$. Let g be an element of $N \cap \mathrm{Z}(NP)$ of order p. Then $\langle g \rangle$ is a normal subgroup of G, because P and every Sylow q-subgroup of H for every prime $q \neq p$ normalise $\langle g \rangle$. Consequently $N = \langle g \rangle$ is a cyclic group of prime order.

2. Assume now that N is non-abelian. By Proposition 4.1.16 (6), we have that $N \cap B$ is a nilpotent subnormal subgroup of G. In particular, $N \cap B \leq \mathrm{F}(N) = 1$. Therefore B normalises all subgroups of N. By Theorem 1.3.3, $[N, B] \leq \mathrm{Z}(N) = 1$. It follows that B centralises N. □

Lemma 4.2.2. *Let the group $G = AB$ be the product of the totally permutable subgroups A and B. Assume that $A \cap B = 1$. Then $[A, B] \leq \mathrm{F}(G)$.*

Proof. We argue by induction on $|G|$. Clearly we may assume that G is not trivial. By Lemma 4.1.20, we may also assume that A contains a minimal normal subgroup N of G. By Lemma 4.1.10, G/N satisfies the hypothesis of the theorem. Hence $[A, B]N/N$ is contained in $\mathrm{F}(G/N)$ and so $[A, B]$ centralises all chief factors H/K of G such that $N \leq K$ by [119, A, 13.8]. Assume that N is non-abelian. Then N is a direct product of minimal normal subgroups of the totally permutable product NB, and all of them are centralised by B by Lemma 4.2.1. Therefore B centralises N. Hence if $n \in N$, $a \in A$, and $b \in B$, we have that $n^{[a,b]} = n^{a^{-1}b^{-1}ab} = n$ and so $[A, B] \leq \mathrm{C}_G(N)$. Suppose now that N is abelian. Then N is a p-group for a prime p and N can be regarded as a $G/\mathrm{C}_G(N)$-module over the field of p elements. Since $N \cap B = 1$, B acts as a group of power automorphisms on B by Lemma 4.1.37. In particular, $G/\mathrm{C}_G(N) = (A\mathrm{C}_G(N)/\mathrm{C}_G(N))(B\mathrm{C}_G(N)/\mathrm{C}_G(N))$ is abelian. Consequently $[A, B] \leq \mathrm{C}_G(N)$. This implies that $[A, B]$ centralises all chief factors of a given chief series of G. Therefore $[A, B] \leq \mathrm{F}(G)$ by [119, A, 13.8]. □

Lemma 4.2.3. *Let the group G be the product of the totally permutable subgroups A and B. Assume that $A \cap B = 1$ and that $\mathrm{F}(G)$ is a p-group for a prime p. Then:*

1. $[\mathrm{O}^p(A), \mathrm{O}^p(B)] = 1$, *and*
2. $\mathrm{O}^p(A)$ *normalises each p-subgroup of B.*

Proof. 1. Let q and r two primes different from p. Let Q be a Sylow q-subgroup of A and let R be a Sylow r-subgroup of B. By Lemma 4.2.2, we have that $[Q, R] \leq \mathrm{F}(G)$ and so $[Q, R]$ is a p-group. On the other hand, since QR is a subgroup of G, we have that $[Q, R] \leq QR$ and so $[Q, R]$ is a $\{q, r\}$-group. We conclude that $[Q, R] = 1$. Since $\mathrm{O}^p(A)$ is generated by all Sylow q-subgroups of A for all primes $q \neq p$ and $\mathrm{O}^p(B)$ is generated by all Sylow r-subgroups of B for all primes $r \neq p$, we conclude that $[\mathrm{O}^p(A), \mathrm{O}^p(B)] = 1$.

2. Let P be a p-subgroup of B and let $1 \neq Q$ be a Sylow q-subgroup of A for a prime $q \neq p$. Then QP is a subgroup of G. Since $[Q, P]$ is contained in $\mathrm{F}(G)$ by Lemma 4.2.2, we conclude that $[Q, P]$ is a normal p-subgroup of PQ by applying [157, III, 1.6]. But P is a Sylow p-subgroup of QP. Hence $[Q, P] \leq P$ and so Q normalises P. It follows that $\mathrm{O}^p(A)$ normalises P. □

The next result [55, Lemma 5] is valid for the formation of all supersoluble groups.

Lemma 4.2.4. *Let the group $G = AB$ be the product of the totally permutable subgroups A and B. If B is supersoluble, then $G^{\mathfrak{U}} = A^{\mathfrak{U}}$.*

Proof. Since $G/G^{\mathfrak{U}} \in \mathfrak{U}$, it follows that $AG^{\mathfrak{U}}/G^{\mathfrak{U}}$ is supersoluble and so $A^{\mathfrak{U}} \leq G^{\mathfrak{U}}$. We argue by induction on $|G|$. By Lemma 4.1.10, we can assume that $G^{\mathfrak{U}}N = A^{\mathfrak{U}}N$ for every minimal normal subgroup N of G. Consequently $G^{\mathfrak{U}} = A^{\mathfrak{U}}(G^{\mathfrak{U}} \cap N)$. This allows us to assume that $\mathrm{Soc}(G) \leq G^{\mathfrak{U}}$ and that $G^{\mathfrak{U}} = A^{\mathfrak{U}}N$ for any minimal normal subgroup N of G. If one such N were contained in $A^{\mathfrak{U}}$, then $G^{\mathfrak{U}} = A^{\mathfrak{U}}$. On the other hand, applying Lemma 4.1.20, there exists a minimal normal subgroup N of G such that $N \leq A$ or $N \leq B$. We analyse both possibilities.

Assume first that $N \leq A$. Then N normalises $A^{\mathfrak{U}}$ and so $A^{\mathfrak{U}}$ is a normal subgroup of $G^{\mathfrak{U}}$. If N is non-abelian, then N is a direct product of non-abelian simple groups. Since $N/(N \cap A^{\mathfrak{U}}) \cong A^{\mathfrak{U}}N/A^{\mathfrak{U}} = G^{\mathfrak{U}}/A^{\mathfrak{U}}$ is a subgroup of the supersoluble group $A/A^{\mathfrak{U}}$, it follows that $N = N \cap A^{\mathfrak{U}}$, that is, that $N \leq A^{\mathfrak{U}}$ and so $G^{\mathfrak{U}} = A^{\mathfrak{U}}$. Hence we can assume that N is a p-group for a prime p. Suppose that $G^{\mathfrak{U}}$ is not a p-group. Then we can consider a non-trivial Sylow q-subgroup Q of $G^{\mathfrak{U}}$ for a prime $q \neq p$. In this case, Q is contained in $A^{\mathfrak{U}}$ and, by the Frattini argument [119, A, 6.3 (b)], $G = G^{\mathfrak{U}}\mathrm{N}_G(Q)$. Consequently, $1 \neq \langle Q^G \rangle = \langle Q^{G^{\mathfrak{U}}} \rangle \leq A^{\mathfrak{U}}$. This implies that there exists a minimal normal subgroup of G contained in $A^{\mathfrak{U}}$ and so $G^{\mathfrak{U}} = A^{\mathfrak{U}}$. Therefore we can assume that $G^{\mathfrak{U}}$ is a p-group. If the Frattini subgroup $\Phi(G^{\mathfrak{U}})$ is non-trivial, then $G^{\mathfrak{U}} = A^{\mathfrak{U}}\Phi(G^{\mathfrak{U}}) = A^{\mathfrak{U}}$ bearing in mind that $\Phi(G^{\mathfrak{U}})$ is a normal subgroup of G. If $\Phi(G^{\mathfrak{U}}) = 1$, then $G^{\mathfrak{U}}$ is elementary abelian. In this case,

G is soluble. Let F be a supersoluble projector of A, then $A = A^{\mathfrak{U}}F$. Therefore $G = A^{\mathfrak{U}}(FB) = G^{\mathfrak{U}}(FB)$. Since F and B are supersoluble, FB is supersoluble by Corollary 4.1.38. But since $G^{\mathfrak{U}}$ is abelian, there exists a supersoluble projector E of G such that $FB \leq E$ by [119, III, 3.14]. Applying [119, IV, 5.18], we have that $G^{\mathfrak{U}} \cap E = 1$. Hence $G^{\mathfrak{U}} = A^{\mathfrak{U}}(FB \cap G^{\mathfrak{U}}) = A^{\mathfrak{U}}$, as desired.

Assume now that $N \leq B$. Then N is abelian. If AN is a proper subgroup of G, by induction we have that $(AN)^{\mathfrak{U}} = A^{\mathfrak{U}}$. This implies that $N \leq \mathrm{N}_G(A^{\mathfrak{U}})$ and so $A^{\mathfrak{U}}$ is a normal subgroup of $G^{\mathfrak{U}}$. Arguing as above, we obtain the result. It follows that $AN = G$, A is a maximal subgroup of G and N is cyclic by Lemma 4.2.1. If $A_G \neq 1$, then we can take a minimal normal subgroup of G contained in A and the result holds. Consequently $A_G = 1$, G is a primitive group and $N = \mathrm{Soc}(G)$ is self-centralising in G. Since $A/\mathrm{C}_A(N)$ is cyclic, it follows that $A^{\mathfrak{U}} \leq \mathrm{C}_A(N) = 1$ and so A is supersoluble. By Corollary 4.1.38, G is supersoluble and the result holds. The induction argument is therefore complete. □

The following lemma is a special case of [55, Lemma 8].

Lemma 4.2.5. *Let the group $G = AB$ be the product of the totally permutable subgroups A and B. Assume that B is cyclic of prime power order. Then B centralises $G^{\mathfrak{U}} = A^{\mathfrak{U}}$.*

Proof. By Lemma 4.2.4, we have that $G^{\mathfrak{U}} = A^{\mathfrak{U}}$. Assume that the lemma is false and that (G, B) is a counterexample with least $|G| + |B|$. We reach a contradiction by following the next steps:

1. *G has a unique minimal normal subgroup N and $N \leq A^{\mathfrak{U}} = G^{\mathfrak{U}}$.*

Let N be a minimal normal subgroup of G. Then G/N is the totally permutable product of AN/N and BN/N by Lemma 4.1.10, and BN/N is cyclic of prime power order. The choice of G implies that $BN/N \leq \mathrm{C}_{G/N}(G^{\mathfrak{U}}N/N)$. Hence $[B, G^{\mathfrak{U}}] \leq N$. If G had two different minimal normal subgroups of G, then it would follow $[B, G^{\mathfrak{U}}] \leq N_1 \cap N_2 = 1$, against the choice of G.

2. *B is a cyclic group of prime order.*

Let M be the unique maximal subgroup of B and assume that $M \neq 1$. Then M centralises $A^{\mathfrak{U}}$ by minimality of G. Let X be a supersoluble projector of A. Then XB is supersoluble by Corollary 4.1.38. Moreover $G = A^{\mathfrak{U}}(XB)$. Let U be a $\mathfrak{U}$-maximal subgroup of G containing XB. Then $G = A^{\mathfrak{U}}U$. By [119, IV, 6.14], $\mathrm{Z}_{\mathfrak{U}}(G) = \mathrm{C}_U(G^{\mathfrak{U}})$. It follows that M is a non-trivial subgroup of $\mathrm{Z}_{\mathfrak{U}}(G)$, which is a normal subgroup of G. Consequently, $N = \mathrm{Soc}(G)$ is contained in $\mathrm{Z}_{\mathfrak{U}}(G)$. This implies that N is cyclic and $\mathrm{Z}_{\mathfrak{U}}(G/N) = \mathrm{Z}_{\mathfrak{U}}(G)/N$. On the other hand, $G/N = (A^{\mathfrak{U}}/N)(UN/N)$. Suppose that $G = UN$. Then G/N is supersoluble and so is G. This contradiction implies that UN/N is a proper subgroup of G/N. Let U_1/N be a $\mathfrak{U}$-maximal subgroup of G/N containing UN/N. Then $G/N = (A^{\mathfrak{U}}/N)(U_1/N)$ and $\mathrm{C}_{U_1/N}(A^{\mathfrak{U}}/N) =$

$Z_{\mathfrak{U}}(G)/N$ by [119, IV, 6.14]. The minimality of G implies that $BN/N \leq Z_{\mathfrak{U}}(G)/N$ and so $B \leq Z_{\mathfrak{U}}(G) \leq C_G(G^{\mathfrak{U}})$, a contradiction which shows that M is trivial and so B is cyclic of prime order.

3. *The final contradiction.*

Let $q = |B|$. Assume now that B is not contained in A. Then $A \cap B = 1$. Therefore $[A, B] \leq F(G)$ by Lemma 4.2.2. Now $[A, B] \neq 1$ and so $F(G) \neq 1$. Since G has a unique minimal normal subgroup, $F(G) = O_p(G)$ for a prime p. Assume first that $p = q$. Then $O^p(A) \leq N_G(B)$ by Lemma 4.2.3. Since $A^{\mathfrak{U}} \leq O^p(A)$, it follows that $[B, A^{\mathfrak{U}}] \leq A \cap B = 1$. This yields the contradiction $B \leq C_G(A^{\mathfrak{U}})$. Therefore $q \neq p$. By Lemma 4.2.3, $[O^p(A), O^p(B)] = 1$. Since $O^p(B) = B$, we have that $A^{\mathfrak{U}} \leq O^p(A) \leq C_G(B)$. Therefore $B \leq C_G(A^{\mathfrak{U}})$, which contradicts the choice of G.

Suppose now that $B \leq A = G$. Then B is a permutable subgroup of G. If $B_G = 1$, by Corollary 1.5.6 and [119, IV, 6.14], $B \leq Z_\infty(G) \leq Z_{\mathfrak{U}}(G) \leq C_G(G^{\mathfrak{U}})$. It follows that B centralises $A^{\mathfrak{U}}$, a contradiction which shows that $B_G \neq 1$. In this case, B is a normal subgroup of G. Since $|B| = q$, we obtain that $B \leq Z_{\mathfrak{U}}(G)$. This final contradiction proves the theorem. □

From now on we prepare the way for the main result of the section. A major step in its proof is to show that the nilpotent residuals of the factors of a totally permutable product are normal.

Lemma 4.2.6. *Let the group $G = AB$ be the product of the totally permutable subgroups A and B. Then the nilpotent residual $A^{\mathfrak{N}}$ of A is a normal subgroup of G.*

Proof. Assume that the result is false and that G is a counterexample of the smallest possible order with least $|A|+|B|$. If B_0 is a proper subgroup of B, then B_0 normalises $A^{\mathfrak{N}}$ by minimality of G and so B has a unique maximal subgroup, that is, B is a cyclic group of order a power of a prime p. Suppose that $A^{\mathfrak{N}}$ contains a minimal normal subgroup of G, N say. Since G/N is a totally permutable product of A/N and BN/N by Lemma 4.1.10, the choice of G implies that BN/N is contained in $N_{G/N}(A^{\mathfrak{N}}/N)$, and so $BN \leq N_G(A^{\mathfrak{N}})$. It follows that $A^{\mathfrak{N}}$ is a normal subgroup of G, contrary to supposition. Therefore $A^{\mathfrak{N}}$ is core-free in G. On the other hand, $A^{\mathfrak{N}}$ contains the supersoluble residual $A^{\mathfrak{U}}$ and $B \leq C_G(A^{\mathfrak{U}})$ by Lemma 4.2.5. Consequently $A^{\mathfrak{U}}$ is a normal subgroup of G. Therefore $A^{\mathfrak{U}} = 1$ and A is supersoluble. By Corollary 4.1.38, G is supersoluble.

Assume that N_1 and N_2 are two minimal normal subgroups of G of respective orders p_1 and p_2, $p_1 \neq p_2$. Then $A^{\mathfrak{N}}N_1 \cap A^{\mathfrak{N}}N_2$ is a normal subgroup of G. But note that $A^{\mathfrak{N}}N_1 \cap A^{\mathfrak{N}}N_2 = A^{\mathfrak{N}}$, a contradiction which shows that $F(G)$ is a q-group for some prime q. Since G is supersoluble and $C_G(F(G)) \leq F(G)$, it follows that q is the largest prime dividing $|G|$.

Suppose that $q \neq p$. Since $A \cap B$ is nilpotent and subnormal by Proposition 4.1.16 (6), we have that $A \cap B = 1$. By Lemma 4.2.3, $[O^q(A), O^q(B)] = 1$. Since

$O^q(B) = B$, we can deduce that $A^{\mathfrak{N}} \leq O^q(A) \leq C_G(B)$. Therefore $B \leq C_G(A^{\mathfrak{N}})$. This contradiction shows that $q = p$. Let $r \neq p$ be a prime and let R be a Sylow r-subgroup of A. If B_1 is a subgroup of B, then RB_1 is supersoluble by Corollary 4.1.38 and B_1 is normal in RB_1 because B_1 is a Sylow p-subgroup of RB_1. Therefore $O^p(A)$, which is generated by all Sylow r-subgroups of A for $r \neq p$, acts as a group of power automorphisms on B. If $O^p(A) = A$, then every subgroup of B is normal in G, which implies that $A/C_A(B)$ is abelian. Thus $A^{\mathfrak{N}} \leq C_G(B)$, which contradicts the choice of G. Hence $O^p(A)$ is a proper subgroup of A. By minimality of $|A| + |B|$, $(O^p(A))^{\mathfrak{N}}$ is normalised by B. Consequently, $(O^p(A))^{\mathfrak{N}}$ is a normal subgroup of G contained in $A^{\mathfrak{N}}$. In particular, $O^p(A)$ is nilpotent. Clearly, A has a normal Sylow p-subgroup. Now, if r is a prime different from p, every Sylow r-subgroup of A is also a Sylow r-subgroup of $O^p(A)$ and so it is normal in A. Therefore A is nilpotent and $A^{\mathfrak{N}} = 1$. This final contradiction proves the lemma. □

This completes our preparations, and now we can prove:

Theorem 4.2.7. *Let the group $G = AB$ be the product of the totally permutable subgroups A and B. Then $B \leq C_G(A^{\mathfrak{N}})$. In particular, if X is a normal subgroup of A contained in $A^{\mathfrak{N}}$, then X is normal in G.*

Proof. Assume that the result is false and that G is a counterexample of least order with $|A| + |B|$ minimal. Then $B_0 \leq C_G(A^{\mathfrak{N}})$ for every proper subgroup B_0 of B. This implies that B has a unique maximal subgroup. Hence B is cyclic of prime power order, that is, $B = \langle x \rangle$ and $o(x) = p^n$ for some prime p.

Let N be a minimal normal subgroup of G. Since G/N is a totally permutable product of A/N and BN/N by Lemma 4.1.10, the choice of G implies that $[A^{\mathfrak{N}}, B] \leq N$. Consequently, G has a unique minimal normal subgroup, N say. Moreover, $N \leq A^{\mathfrak{N}}$ because $A^{\mathfrak{N}}$ is a normal subgroup of G by Lemma 4.2.6.

Suppose that there exists a prime q dividing $|A|$, $q > p$, such that B does not centralise a Sylow q-subgroup Q of A. Since QB is a subgroup of G and QB is a totally permutable product, it follows that QB is supersoluble by Corollary 4.1.38. Hence Q is a normal subgroup of QB. By Lemma 4.1.37, B acts as a group of power automorphisms on Q. Since B does not centralise Q, we can deduce from Lemma 1.3.4 that Q is abelian. Moreover, by [119, A, 12.5], $Q = C_Q(x) \times [x, Q]$. Since B acts as a group of universal power automorphisms on Q by Theorem 1.3.2, we conclude that $C_Q(x) = 1$ and $[x, Q] = Q$. Let S be a Sylow q-subgroup of A and let u be an element of Q. Since the subgroup $S^{u^{-1}}B$ is again a totally permutable subgroup, it follows that it is supersoluble by Corollary 4.1.38, and $S^{u^{-1}}$ is normalised by B. Hence $x \in N_G(S^{u^{-1}})$ and so $x^u \in N_G(S)$. Since $x \in N_G(S)$, it follows that $x^{-1}x^u = [x, u] \in N_G(S)$ and $[x, u]$ is an element of Q. Therefore $[x, u] \in Q \cap N_G(S)$. Consequently $Q = \langle [x, u] : u \in Q \rangle$ is contained in $Q \cap N_G(S) \leq S$. It follows that $S = Q$ and Q is a normal subgroup of A. Furthermore, since Q is a normal subgroup of QB, we have that Q is normal in G.

Consider now a prime r dividing $|A|$ such that $r < p$. If R is a Sylow r-subgroup of A, then RB is a supersoluble group and B is a normal subgroup of RB. By Lemma 4.1.37, R acts as a group of power automorphisms on B. Assume that R does not centralise B. Then there exists an element $y \in R$ such that $[y, B] \neq 1$. Since $B = [y, B] \times \mathrm{C}_B(y)$ by [119, A, 12.5], and y acts as a universal power automorphism on B by Theorem 1.3.2, it follows that $\mathrm{C}_B(y) = 1$. Consequently $B = [y, B]$. Let $y^{-1}xy = x^j$. We have that

$$[x^i, y] = x^{-i}y^{-1}x^i y = x^{-i}(y^{-1}xy)^i = x^{-i}(x^j)^i = (x^{-1}x^j)^i = [x, y]^i.$$

Thus $B = \langle [x, y] \rangle$. Let z be an element of Q, and assume that $z^x = z^k$. Then $z = (z^x)^{x^{-1}} = (z^k)^{x^{-1}} = (z^{x^{-1}})^k$. On the other hand, $z^{x^{-1}y^{-1}}$ is an element of Q and, since Q is abelian, x induces a universal power automorphism on Q by Theorem 1.3.2. This implies that $(z^{x^{-1}y^{-1}})^x = (z^{x^{-1}y^{-1}})^k$. Consequently

$$z^{[x,y]} = z^{x^{-1}y^{-1}xy} = ((z^{x^{-1}y^{-1}})^k)^y = ((z^{x^{-1}})^k)^{y^{-1}y} = z.$$

That is, $[x, y]$ centralises Q. Hence $B \leq \mathrm{C}_G(Q)$, but this contradicts our choice of Q.

Therefore, if $r < p$, then the Sylow r-subgroups of A centralise B. Hence if r is a prime and $r < p$, A_r is a Sylow r-subgroup of A, and $a \in A$, we have that $A_r^a \leq \mathrm{C}_G(B)$. If $r > p$ and $A_r \leq \mathrm{C}_G(B)$, then $A_r^a \leq \mathrm{C}_G(B)$ (otherwise, A_r^a would be a normal Sylow r-subgroup of A and so $A_r^a = A_r$, against our choice of A_r). Let $A_0 = \langle A_s : s > p, A_s \not\leq \mathrm{C}_G(B) \rangle$. We note that A_0 is an abelian normal subgroup of G because its Sylow subgroups are abelian normal subgroups of G. Since $A_0 = \mathrm{C}_{A_0}(x) \times [A_0, x]$ and x acts as a group of power automorphisms on A_0, we conclude that $\mathrm{C}_{A_0}(x) = 1$. Let $T = \langle A_r : r \neq p, A_r \leq \mathrm{C}_G(B) \rangle$. We have that $\mathrm{O}^p(A) = A_0 T$, and T is a normal subgroup of A such that $T \leq \mathrm{C}_G(B)$. In particular, T is a normal subgroup of G. Since $\mathrm{C}_{A_0}(x) = 1$, we have that $A_0 \cap T = 1$.

On the other hand, $A^{\mathfrak{N}} \leq \mathrm{O}^p(A)$. Since $A^{\mathfrak{N}} \cap A_0$ and $A^{\mathfrak{N}} \cap T$ are normal subgroups of G with trivial intersection, and G has a unique minimal normal subgroup, we have that either $A^{\mathfrak{N}} \cap A_0 = 1$ or $A^{\mathfrak{N}} \cap T = 1$. Moreover, $A^{\mathfrak{N}} = (A^{\mathfrak{N}} \cap A_0)(A^{\mathfrak{N}} \cap T)$ since both factors are Hall subgroups of $A^{\mathfrak{N}}$. If $A^{\mathfrak{N}} \cap A_0 = 1$, then $A^{\mathfrak{N}} \leq T \leq \mathrm{C}_G(B)$. Since this contradicts the choice of G, we conclude that $A^{\mathfrak{N}} \cap T = 1$. Moreover, $N \leq A^{\mathfrak{N}}$. It follows that $T = 1$. Hence $\mathrm{O}^p(A) = A_0$ and $A^{\mathfrak{N}} = A^{\mathfrak{N}} \cap A_0$ is abelian. Now $A^{\mathfrak{N}} = \mathrm{C}_{A^{\mathfrak{N}}}(x) \times [A^{\mathfrak{N}}, x]$ by [119, A, 12.5]. Since x acts as a universal power automorphism on $A^{\mathfrak{N}}$, we conclude that $\mathrm{C}_{A^{\mathfrak{N}}}(x) = 1$. Hence $A^{\mathfrak{N}} = [A^{\mathfrak{N}}, B] = [A^{\mathfrak{N}}, x]$ is contained in N, and $A^{\mathfrak{N}} = N$. Applying Lemma 4.2.5, $A^{\mathfrak{U}} \leq \mathrm{C}_{A^{\mathfrak{N}}}(x) = 1$, and so A is supersoluble. By Corollary 4.1.38, G is supersoluble. This implies that G/C is a cyclic group, where $C = \mathrm{C}_G(N)$. Since $\mathrm{O}^p(A)$ and every maximal subgroup of B centralise N and $G = (\mathrm{O}^p(A))(A_p B)$, we deduce that $G/C = (A_p C/C)(BC/C)$ is a cyclic p-group and BC/C is cyclic of order p. It follows then $BC/C \leq A_p C/C$. Therefore $G = A_p \mathrm{C}_G(N)$. Consider the totally permutable product $X = NA_p$. If X is a proper subgroup of G, then A_p centralises N

by the choice of G. This implies that N is central in G, a contradiction which shows that $G = NA_p = A$. Therefore B is a permutable subgroup of G. Assume that B_G is not trivial. Then $N \leq B$, against the fact that B does not centralise N. Therefore $B_G = 1$. It follows that $B \leq \mathrm{Z}_\infty(G)$ by Corollary 1.5.6. Hence G has non-trivial centre and so $N \leq \mathrm{Z}(G)$. This is also a contradiction.

Consequently, there exists a prime $r < p$ such that a Sylow r-subgroup A_r of A does not centralise B. The previous arguments show that the Sylow s-subgroups of A centralise B for all primes $s > p$.

Let t be a prime dividing $|A|$ such that $t < p$. If A_t is a Sylow t-subgroup of A, then the totally permutable product $A_t B$ is supersoluble by virtue of Corollary 4.1.38. Hence B is a normal subgroup of $A_t B$ and so A_t acts as a group of power automorphisms on B. It follows that $\mathrm{O}^p(A)$ acts as a group of power automorphisms on B. In particular, $A^{\mathfrak{N}}$ acts as a group of power automorphisms on B. Let $z \in B$ of order p. If $B \neq \langle z \rangle$, it follows by the choice of B that $\langle z \rangle \leq \mathrm{C}_G(A^{\mathfrak{N}})$ and so $A^{\mathfrak{N}} \leq \mathrm{C}_{A^{\mathfrak{N}}}(z) = \mathrm{C}_{A^{\mathfrak{N}}}(B)$. This contradiction shows that B has order p. Consequently $A^{\mathfrak{N}}/\mathrm{C}_{A^{\mathfrak{N}}}(B)$ is a cyclic group of order dividing $p-1$. Let t be a prime dividing $|A^{\mathfrak{N}}|$ and assume that a Sylow t-subgroup $(A^{\mathfrak{N}})_t$ of $A^{\mathfrak{N}}$ is not contained in $\mathrm{C}_G(B)$. Let a be an element of $(A^{\mathfrak{N}})_t$ such that $a \notin \mathrm{C}_G(B)$. By [119, A, 12.15], $B = \mathrm{C}_B(a) \times [a, B]$. Since $\mathrm{C}_B(a) = 1$ because a acts on B as a power automorphism, we conclude that $B = [a, B] \leq [A^{\mathfrak{N}}, B] \leq N$. Consequently $B \leq N \leq A^{\mathfrak{N}}$ and so $G = A$. It follows that B is a permutable subgroup of G. If $B_G = 1$, then $B \leq \mathrm{Z}_\infty(G)$ by Corollary 1.5.6 and so $\mathrm{Z}(G)$ contains the unique minimal subgroup N of G. This means that $B \leq \mathrm{Z}(G) \leq \mathrm{C}_G(A^{\mathfrak{N}})$. Therefore $B_G \neq 1$. This implies that $N = B$ and so $G/\mathrm{C}_G(N)$ is abelian of exponent dividing $p-1$. In particular, $A^{\mathfrak{N}} \leq \mathrm{C}_G(N)$. It follows that $N = B \leq \mathrm{C}_G(A^{\mathfrak{N}})$. This final contradiction proves the result. □

Corollary 4.2.8. *Let the group $G = G_1 G_2 \cdots G_r$ be the product of the pairwise totally permutable subgroups $G_1, G_2, \dots, G_r$. Then $\prod_{j \in \{1,\dots,r\} \setminus \{i\}} G_j$ centralises $G_i^{\mathfrak{N}}$. In particular, if X_i is a normal subgroup of G_i contained in $G_i^{\mathfrak{N}}$, then X_i is a normal subgroup of G.*

Proof. The result follows from Theorem 4.2.7 taking into account that if $i \neq j$, G_j centralises $G_i^{\mathfrak{N}}$ because G_i and G_j are totally permutable. □

Let $\mathfrak{F}$ be a formation containing the class of all nilpotent groups. Then $G^{\mathfrak{F}}$ is contained in $G^{\mathfrak{N}}$ for all groups G. Therefore we have:

Corollary 4.2.9. *Let the group $G = G_1 G_2 \cdots G_r$ be the product of the pairwise totally permutable subgroups $G_1, G_2, \dots, G_r$ and let $\mathfrak{F}$ be a formation containing the class of all nilpotent groups. Then $\prod_{j \in \{1,\dots,r\} \setminus \{i\}} G_j$ centralises $G_i^{\mathfrak{F}}$.*

We use the above result to show that saturated formations containing the class of all supersoluble groups are closed under the formation of totally permutable products. This nice extension of Corollary 4.1.38 was proved by Maier [194] ($r = 2$) and Carocca [85]. The hypothesis $\mathfrak{U} \subseteq \mathfrak{F}$ comes from the fact that the product of two cyclic groups is supersoluble, but not nilpotent in general (Lemma 3.1.1).

Theorem 4.2.10. *Let the group $G = G_1G_2 \cdots G_r$ be the product of the pairwise totally permutable subgroups $G_1, G_2, \dots, G_r$. Let $\mathfrak{F}$ be a saturated formation containing the class $\mathfrak{U}$ of all supersoluble groups. If G_i belongs to $\mathfrak{F}$, $i \in \{1, 2, \dots, r\}$, then $G \in \mathfrak{F}$.*

Proof. Assume that the result is false, and let G be a counterexample of least order. Then, by Lemma 4.1.10, every proper epimorphic image of G belongs to $\mathfrak{F}$. Therefore, since $\mathfrak{F}$ is a saturated formation, G has a unique minimal normal subgroup N, $C_G(N)$ is contained in N, and $G/N \in \mathfrak{F}$. By Corollary 4.1.38, if all G_i are nilpotent, then $G \in \mathfrak{U} \subseteq \mathfrak{F}$. Consequently one of the factors G_i is not nilpotent. There is no loss of generality in assuming that $i = 1$. Hence $G_1^{\mathfrak{N}}$ is a non-trivial normal subgroup of G by Lemma 4.2.6. Thus N is contained in $G_1^{\mathfrak{N}}$. Let $j \in \{2, \dots, r\}$. Then G_j centralises $G_1^{\mathfrak{N}}$ by Corollary 4.2.8 and so $G_j \leq C_G(N) \leq N$. It follows that $G = G_1G_2 \cdots G_r \leq G_1N \cdots N = G_1$ and so $G = G_1 \in \mathfrak{F}$. This final contradiction proves the theorem. □

Next we derive two corollaries from Theorem 4.2.7, the first is an improvement of Lemma 4.2.2 due to Hauck, Martínez-Pastor, and Pérez-Ramos [147] and the second a deduction made from it and proved by the same authors in [145].

Corollary 4.2.11. *Let the group $G = G_1G_2 \cdots G_r$ be the product of the pairwise totally permutable subgroups $G_1, G_2, \dots, G_r$. Let $\{I, J\}$ be a partition of $\{1, 2, \dots, r\}$. Then $[\prod_{i \in I} G_i, \prod_{j \in J} G_j]$ is a nilpotent normal subgroup of G.*

Proof. The normality of $[\prod_{i \in I} G_i, \prod_{j \in J} G_j]$ follows from [157, III, 1.6 (b)]. Denote by C_i a nilpotent projector of G_i for each $i \in \{1, 2, \dots, r\}$. Then $G_i = G_i^{\mathfrak{N}} C_i$ for each $i \in \{1, 2, \dots, r\}$. Since the group G_j centralises $G_i^{\mathfrak{N}}$ for each i, $j \in \{1, 2, \dots, r\}$, $i \neq j$, by Theorem 4.2.7, we have that

$$\Bigl[\prod_{i \in I} G_i, \prod_{j \in J} G_j\Bigr] = \Bigl[\prod_{i \in I} G_i^{\mathfrak{N}} C_i, \prod_{j \in J} G_j^{\mathfrak{N}} C_j\Bigr] = \Bigl[\prod_{i \in I} C_i, \prod_{j \in J} C_j\Bigr],$$

which is contained in $(\prod_{i \in I} C_i \prod_{j \in J} C_j)' = (\prod_{i=1}^{r} C_i)'$. Since $\prod_{i=1}^{r} C_i$ is a product of pairwise totally permutable nilpotent subgroups, it is supersoluble by Corollary 4.1.38 and so its derived subgroup is nilpotent. □

Lemma 4.2.12. *Let the group $G = G_1 G_2 \cdots G_r$ be the product of the pairwise totally permutable subgroups $G_1, G_2, \dots, G_r$. Then*

$$\Big[\prod_{i\in I} G_i, \prod_{j\in J} G_j\Big] \le \mathrm{Z}_{\mathfrak{U}}(G)$$

for any $I, J \subseteq \{1, 2, \dots, r\}$ such that $I \cap J = \emptyset$. In particular,

$$\Big(\prod_{i\in I} G_i\Big) \cap \Big(\prod_{j\in J} G_j\Big) \le \mathrm{Z}_{\mathfrak{U}}(G).$$

Proof. Assume that the lemma is false and let G be a counterexample with $|G| + |G_1| + |G_2| + \cdots + |G_r|$ as small as possible. Let $I, J \subseteq \{1, 2, \dots, r\}$ such that $I \cap J = \emptyset$ and $C = [\prod_{i\in I} G_i, \prod_{j\in J} G_j] \not\le \mathrm{Z}_{\mathfrak{U}}(G)$. Obviously, $G_i \ne 1$ for $1 \le i \le r$. By Corollary 4.2.11, C is a normal subgroup of G. By hypothesis, $C \ne 1$ and so there exists a minimal normal subgroup N of G contained in C. By Lemma 4.1.10, G/N is the product of the pairwise totally permutable subgroups $G_1N/N, G_2N/N, \dots, G_rN/N$ and so $CN/N \le \mathrm{Z}_{\mathfrak{U}}(G/N)$ by the choice of G. Therefore it is enough to show that N is cyclic in order to reach a final contradiction. Assume that N is not cyclic. Suppose that N_1 is a minimal normal subgroup of G different from N. Since G/N_1 is the product of the pairwise totally permutable subgroups $G_1N_1/N_1, G_2N_1/N_1, \dots, G_rN_1/N_1$ by Lemma 4.1.10, and NN_1/N_1 is a minimal normal subgroup of C/N_1 contained in

$$\Big[\prod_{i\in I}(G_iN_1/N_1), \prod_{j\in J}(G_jN_1/N_1)\Big] = \Big[\prod_{i\in I} G_i, \prod_{j\in J} G_j\Big]N_1/N_1,$$

it follows that NN_1/N_1 is cyclic and so N is cyclic. This contradiction shows that N is the unique minimal normal subgroup of G.

If all G_i are supersoluble for $1 \le i \le r$, then G is supersoluble by Corollary 4.1.38. In this case, $\mathrm{Z}_{\mathfrak{U}}(G) = G$, which contradicts the choice of G. Therefore we can assume without loss of generality that G_1 is not supersoluble. Hence $G_1^{\mathfrak{U}} \ne 1$. Let X_1 be a supersoluble projector of G_1. Then $G_1 = G_1^{\mathfrak{U}} X_1$. For $2 \le j \le r$, denote $X_j = G_j$. By Corollary 4.2.9 we have that $[G_1^{\mathfrak{U}}, G_j] = 1$ for $2 \le j \le r$. Consequently

$$N \le \Big[\prod_{i\in I} G_i, \prod_{j\in J} G_j\Big] = \Big[\prod_{i\in I} X_i, \prod_{j\in J} X_j\Big] \le \prod_{i=1}^{r} X_i.$$

If $i \ne 1$, then $1 \ne G_i \le \mathrm{C}_G(G_1^{\mathfrak{U}})$. Since $\mathrm{C}_G(G_1^{\mathfrak{U}})$ is a normal subgroup of G, it follows that $N \le \mathrm{C}_G(G_1^{\mathfrak{U}})$. Besides, we have that $G = G_1^{\mathfrak{U}} \prod_{i=1}^{r} X_i$. Since N is contained in $\prod_{i=1}^{r} X_i$ and is centralised by $G_1^{\mathfrak{U}}$, N is a minimal normal subgroup of $\prod_{i=1}^{r} X_i$ contained in $[\prod_{i\in I} X_i, \prod_{j\in J} X_j]$. By minimality of G, we have that $N \le \mathrm{Z}_{\mathfrak{U}}(\prod_{i=1}^{r} X_i)$ and so N is cyclic. This final contradiction proves the first part of the lemma.

For the second part, we denote by $A = \prod_{i \in I} G_i$ and $B = \prod_{j \in J} G_j$. Since $[A, B] \leq Z_{\mathfrak{U}}(G)$, it follows that $[A, A \cap B] \leq Z_{\mathfrak{U}}(G)$. Therefore

$$AZ_{\mathfrak{U}}(G)/Z_{\mathfrak{U}}(G) \leq C_{G/Z_{\mathfrak{U}}(G)}((A \cap B)Z_{\mathfrak{U}}(G)/Z_{\mathfrak{U}}(G)).$$

By symmetry,

$$B\,Z_{\mathfrak{U}}(G)/Z_{\mathfrak{U}}(G) \leq C_{G/Z_{\mathfrak{U}}(G)}((A \cap B)\,Z_{\mathfrak{U}}(G)/Z_{\mathfrak{U}}(G))$$

and so $G/Z_{\mathfrak{U}}(G) \leq C_{G/Z_{\mathfrak{U}}(G)}((A \cap B)\,Z_{\mathfrak{U}}(G)/Z_{\mathfrak{U}}(G))$. Therefore

$$(A \cap B)\,Z_{\mathfrak{U}}(G)/Z_{\mathfrak{U}}(G) \leq Z(G/Z_{\mathfrak{U}}(G)) = 1.$$

We conclude that $A \cap B \leq Z_{\mathfrak{U}}(G)$. □

An immediate consequence is:

Corollary 4.2.13. *Let the group $G = G_1G_2 \cdots G_r$ be the product of the pairwise totally permutable subgroups $G_1, G_2, \ldots, G_r$. Then*

$$G/Z_{\mathfrak{U}}(G) = (G_1Z_{\mathfrak{U}}(G)/Z_{\mathfrak{U}}(G)) \times (G_2\,Z_{\mathfrak{U}}(G)/Z_{\mathfrak{U}}(G)) \times \cdots \times (G_rZ_{\mathfrak{U}}(G)/Z_{\mathfrak{U}}(G)).$$

Another consequence is:

Corollary 4.2.14 (Hauck et al. [145]). *Let the group $G = G_1G_2 \cdots G_r$ be the product of the pairwise totally permutable subgroups $G_1, G_2, \ldots, G_r$. Then $Z_{\mathfrak{U}}(G_i) = Z_{\mathfrak{U}}(G) \cap G_i$ for $1 \leq i \leq r$ and $Z_{\mathfrak{U}}(G) = Z_{\mathfrak{U}}(G_1)Z_{\mathfrak{U}}(G_2) \cdots Z_{\mathfrak{U}}(G_r)$.*

Proof. By Corollary 4.2.13, we have that

$$G/Z_{\mathfrak{U}}(G) = (G_1Z_{\mathfrak{U}}(G)/Z_{\mathfrak{U}}(G)) \times (G_2\,Z_{\mathfrak{U}}(G)/Z_{\mathfrak{U}}(G)) \times \cdots \times (G_rZ_{\mathfrak{U}}(G)/Z_{\mathfrak{U}}(G)).$$

Consider $i \in \{1, 2, \ldots, r\}$. Since $Z_{\mathfrak{U}}(G) \cap G_i$ is a supersolubly embedded normal subgroup of G_i, we conclude that $Z_{\mathfrak{U}}(G) \cap G_i \leq Z_{\mathfrak{U}}(G_i)$. Since $Z_{\mathfrak{U}}(G/Z_{\mathfrak{U}}(G)) = 1$, it follows that $Z(G_iZ_{\mathfrak{U}}(G)/Z_{\mathfrak{U}}(G)) = 1$. Consequently $G_i/(Z_{\mathfrak{U}}(G) \cap G_i)$ has trivial supersoluble hypercentre. Thus $Z_{\mathfrak{U}}(G) \cap G_i = Z_{\mathfrak{U}}(G_i)$.

By Corollary 4.1.22, we have that $T = Z_{\mathfrak{U}}(G_1)Z_{\mathfrak{U}}(G_2) \cdots Z_{\mathfrak{U}}(G_r)$ is a normal subgroup of G. We also know that $T \leq Z_{\mathfrak{U}}(G)$. Hence $T \cap G_i$ is contained in $Z_{\mathfrak{U}}(G) \cap G_i$ and, by definition of T, this is in turn contained in $T \cap G_i$. Hence $T \cap G_i = Z_{\mathfrak{U}}(G) \cap G_i$. On the other hand, G/T is the product of the totally permutable subgroups $G_1T/T, G_2T/T, \ldots, G_rT/T$. Hence $|G/T| \leq \prod_{i=1}^r |G_iT/T| = \prod_{i=1}^r |G_i/(T \cap G_i)| = \prod_{i=1}^r |G_i/(Z_{\mathfrak{U}}(G) \cap G_i)| = \prod_{i=1}^r |G_iZ_{\mathfrak{U}}(G)/Z_{\mathfrak{U}}(G)|$. But this coincides with the order of $G/Z_{\mathfrak{U}}(G)$. Since $T \leq Z_{\mathfrak{U}}(G)$ and $|G/T| \leq |G/Z_{\mathfrak{U}}(G)|$, we conclude that both relations are in fact equalities and so $T = Z_{\mathfrak{U}}(G)$. This completes the proof of the result. □

We bring the section to a close with a couple of results about the structure of a primitive group which is a totally permutable product. They will be needed in the next chapter.

Lemma 4.2.15. *Let the soluble group $G = G_1G_2 \cdots G_r$ be the product of the pairwise totally permutable subgroups $G_1, G_2, \dots, G_r$. If G is a primitive group, then G is supersoluble or $\prod_{j \neq i} G_j = 1$ for some $i \in \{1, 2, \dots, r\}$.*

Proof. Let N be the unique minimal normal subgroup of G. By [47, 1.1.7] N is abelian and $C_G(N) = N$. Applying Corollary 4.2.13, we have that

$$G/Z_{\mathfrak{U}}(G) = (G_1Z_{\mathfrak{U}}(G)/Z_{\mathfrak{U}}(G)) \times (G_2 Z_{\mathfrak{U}}(G)/Z_{\mathfrak{U}}(G)) \times \cdots \times (G_rZ_{\mathfrak{U}}(G)/Z_{\mathfrak{U}}(G)).$$

Assume that $Z_{\mathfrak{U}}(G)$ is non-trivial. Then N is contained in $Z_{\mathfrak{U}}(G)$ and so N is cyclic. Hence G/N is also cyclic and G is supersoluble.

Suppose that $Z_{\mathfrak{U}}(G) = 1$. Then G is the direct product of the subgroups $G_1, G_2, \dots, G_r$. Since N is the unique minimal normal subgroup of G, it follows that $\prod_{j \neq i} G_j = 1$ for some $i \in \{1, 2, \dots, r\}$, and the lemma is proved. □

Lemma 4.2.16. *Let the group $G = G_1G_2 \cdots G_k$ be the pairwise totally permutable product of the subgroups $G_1, G_2, \dots, G_k$. If G is supersoluble and it is a primitive group with unique minimal normal subgroup N, then one of the following cases holds:*

1. *If $N \leq G_i$ for all $i \in \{1, 2, \dots, k\}$ then there exists $r \in \{1, 2, \dots, k\}$ such that $G = G_r \neq N$ and either $G_i = N$ for all $i \neq r$ or $G = G_1 = G_2 = \cdots = G_k = N$.*

2. *If $N \leq G_i$ for $i \in \{1, 2, \dots, r\}$, $r < k$, and $G_j \neq 1$, for some $j > r$, then $G_1 = G_2 = \cdots = G_r = N$ and $G_{r+1}G_{r+2} \cdots G_k \neq 1$ is a complement of N in G.*

3. *If $N \leq G_i$ for $i \in \{1, 2, \dots, r\}$, $k > r \geq 2$ and $G_j = 1$ for all $j > r$, then there exists $1 \leq s \leq r$ such that $G = G_s \neq N$ and either $G_i = N$ for all $s \neq i \leq r$, or $G = G_1 = G_2 = \cdots = G_r = N$.*

4. *There exists $i \in \{1, 2, \dots, k\}$ such that $G = G_i$ and $G_j = 1$ for all $j \neq i$.*

Proof. Let p be the prime dividing $|N|$. Since G is supersoluble, we have that N is of order p and $C_G(N) = N = \langle x \rangle$ is the Sylow p-subgroup of G. Let M denote a core-free complement of N in G. Then $M = \langle y \rangle$ is an abelian maximal subgroup of G with exponent dividing $p - 1$. In particular, M is a Hall p'-subgroup of G. Therefore N has to be contained in at least one of the factors G_i, $i \in \{1, 2, \dots, k\}$ by Lemma 4.1.20.

1. Assume first that $N \leq G_i$ for all $i \in \{1, 2, \dots, k\}$. Consider the factor G_i and take G_j with $G_j \neq G_i$. We see that $G_i = N$ or $G_j = N$. It is clear that

$G_i = \langle x\rangle\langle y^{\alpha_i}\rangle$ and in the same way $G_j = \langle x\rangle\langle y^{\alpha_j}\rangle$ with $\langle y^{\alpha_i}\rangle \neq \langle y^{\alpha_j}\rangle$. Assume $y^{\alpha_i} \neq 1 \neq y^{\alpha_j}$. Write $H = \langle xy^{\alpha_i}\rangle$. If p divides the order of H, then $N \leq H$. Since H is abelian and N is self-centralising in G, $H = N$ and so $y^{\alpha_i} \in N$ and $G_i = N$. This contradiction implies that $H = \langle xy^{\alpha_i}\rangle$ is a p'-group. Now the fact that G_i and G_j are totally permutable yields that $\langle xy^{\alpha_i}, y^{\alpha_j}\rangle = \langle xy^{\alpha_i}\rangle\langle y^{\alpha_j}\rangle$ is a p'-group and therefore it is abelian. Consequently $[xy^{\alpha_i}, y^{\alpha_j}] = 1$. Moreover $[y^{\alpha_i}, y^{\alpha_j}] = 1$. Let $s = y^{\alpha_i}$. By [119, A, 7.2 (c)], $1 = [xy^{\alpha_i}, y^{\alpha_j}] = [x, y^{\alpha_j}]^s[y^{\alpha_i}, y^{\alpha_j}] = [x, y^{\alpha_j}]^s$. Hence $y^{\alpha_j} = 1$, which contradicts our assumption. Therefore if $G_i \neq G_j$ we have $G_i = N$ or $G_j = N$. Assume that $M \neq 1$. Then at least one of the factors has a non-trivial Hall p'-subgroup. Without loss of generality, we can suppose that $M \cap G_1 \neq 1$. The above argument implies that either $G_i = G_1$ or $G_i = N$ for every $i \neq 1$. Assume that $G_j = G_1$ for some $j \neq 1$. Applying the above argument, $1 \neq y^{\alpha_j}$ centralises x. This contradiction shows that $G_2 = \cdots = G_k = N$ and then $G = G_1$.

2. Suppose that $N \leq G_i$ for $i \in \{1, 2, \ldots, r\}$, $r < k$. Then $N \cap G_j = 1$ for all $j > r$. Assume that $G_j \neq 1$, for some $j > r$. Then $G_j = \langle z\rangle$ is a non-trivial p'-subgroup of G. Suppose there exists $i \leq r$ such that $G_i \neq N$. Then $G_i = N\langle y^{\alpha_i}\rangle$ and $y^{\alpha_i} \neq 1$. Consider $H = \langle xy^{\alpha_i}\rangle$. Arguing as in case 1, H is a p'-group. Moreover, as G_i and G_j are totally permutable, H permutes with G_j and HG_j is a p'-subgroup of G and hence is abelian. Therefore $[xy^{\alpha_i}, z] = 1$. It is also clear that $\langle y^{\alpha_i}\rangle$ permutes with $\langle z\rangle$. Therefore $\langle y^{\alpha_i}\rangle\langle z\rangle$ is an abelian p'-group. Hence $[y^{\alpha_i}, z] = 1$. Therefore z centralises x. This contradiction yields $G_i = N$ for all $i \in \{1, 2, \ldots, r\}$. Therefore $G_{r+1}G_{r+2}\cdots G_k$ is a complement of N in G.

3. If $N \leq G_i$ for $i \in \{1, 2, \ldots, r\}$, $k > r \geq 2$, and $G_j = 1$ for all $j > r$, then we are in case 1. Hence there exists $1 \leq s \leq r$ such that $G = G_s \neq N$ and $G_i = N$ for all $s \neq i \leq r$, or $G = G_1 = G_2 = \cdots = G_r = N$.

4. The remaining possibility is case 4. □

4.3 Minimal normal subgroups and nilpotent and soluble residuals of mutually permutable products

The topic of this section may be said to be the study of the normal structure of mutually permutable products. The results of the preceding sections show that these products have a great deal of normality built into their structure, so it is not surprising that minimal normal subgroups, maximal normal ones, and residuals associated to the formations of nilpotent and soluble groups exhibit special features. Some of the topics explored in the section confirm the viewpoint that mutually permutable products are quite close to normal products.

We begin with a result about totally permutable products.

Lemma 4.3.1. *Let the group $G = AB$ be the product of the totally permutable cyclic subgroups A and B, $|A| = q$, $|B| = r$, where q is a prime and r is a natural number. Then A is a normal subgroup of G or B is a normal subgroup of G.*

Proof. We will obtain a contradiction by supposing that the result is not true and choosing a counterexample G of least order. By Corollary 4.1.38, G is supersoluble. Moreover, neither A nor B is a normal subgroup of G. In particular, $A_G = 1$. Clearly, G is not nilpotent because otherwise B would be a normal subgroup of G. By Lemma 4.1.20, G has a minimal normal subgroup X contained in B. The minimality of G and Lemma 4.1.10 force that AX is a normal subgroup of G. Assume that $|X| = s$. Since G/AX is nilpotent, we have that $G^{\mathfrak{N}} \leq AX$. Since G is supersoluble, $G^{\mathfrak{N}}$ is nilpotent.

Assume that $s \neq q$. If A were contained in $G^{\mathfrak{N}}$, then A would be a normal subgroup of G because $G^{\mathfrak{N}}$ is nilpotent and A is a Sylow subgroup of $G^{\mathfrak{N}}$. Hence q does not divide the order of $G^{\mathfrak{N}}$ and so $G^{\mathfrak{N}} = X$. Since $G^{\mathfrak{N}} = X$ is abelian, it is complemented by a Carter subgroup C of G by [47, 4.1.18 and 4.2.17]. Therefore $B \cap C = \mathrm{C}_C(X) = \mathrm{Z}_\infty(G)$ by [47, 4.1.22]. Hence $B = XC \cap B = X(B \cap C)$. Obviously A cannot centralise X because this would imply the nilpotency of AX and so the normality of A in G. Hence B is the product of two normal subgroups of G and so B is normal in G, against the choice of G.

Therefore all minimal normal subgroups of G contained in B have order q. Let X be one of them. Then AX is an elementary abelian group of order q^2 and so $G^{\mathfrak{N}} \leq AX$ is abelian. Moreover $X \leq \mathrm{Z}(G)$. Since $1 \neq T = G^{\mathfrak{N}}$ contains no central chief factors of G by [47, 4.2.19], we conclude that $|T| = q$ and $TX = AX = AT$. In particular, $A \leq \mathrm{C}_G(T)$ and $1 \neq B/\mathrm{C}_B(T)$ is a cyclic group of order dividing $q-1$. Then $\mathrm{C}_B(T)$ is a Sylow q-subgroup of B because it cannot contain minimal normal subgroups of G of order different from q. Let $B_{q'}$ be the Hall q'-subgroup of B. Then $B_{q'}$ is cyclic of order dividing $q - 1$. Since the product of A and B is totally permutable, $AB_{q'}$ is a supersoluble subgroup of G. This implies that $B_{q'}$ normalises A. Furthermore, $A \leq TX \leq TB_q$, which is abelian, and so $B_q \leq \mathrm{C}_G(A)$. It follows that A is a normal subgroup of G, against the choice of G. This proves the result. □

Remark 4.3.2. The totally permutable character of the product is necessary in the proof of Lemma 4.3.1. Let E be an elementary abelian group $E = \langle b, c\rangle$ of order 7^2 and let a be an automorphism of E of order 3 such that $b^a = b$ and $c^a = c^2$. The corresponding semidirect product $G = [E]\langle a\rangle$ can be factorised as the product of the cyclic subgroups $A = \langle bc\rangle$ and $B = \langle ab\rangle$, of respective orders 7 and 21. None of the factors is a normal subgroup of G. Note that the product is not totally permutable (even mutually permutable) because A does not permute with $\langle a\rangle = \langle (ab)^7\rangle$.

It is clearly of importance in the study of a mutually permutable product of the subgroups A and B to know the relationship between the minimal normal subgroups of AB and the factors A and B. The first result we prove in this direction is due to Beidleman and Heineken [65].

Lemma 4.3.3. *Let the group $G = AB$ be the product of the mutually permutable subgroups A and B. Then:*

1. *If N is a maximal normal subgroup of G, then $\{AN, BN, (A \cap B)N\} \subseteq \{N, G\}$.*

2. *If N is a non-abelian minimal normal subgroup of G, then $\{A \cap N, B \cap N\} \subseteq \{N, 1\}$ and $N = (N \cap A)(N \cap B)$ (that is, N is prefactorised with respect to $G = AB$).*

3. *If N is a minimal normal subgroup of G, then $N \leq A \cap B$ or $[N, A \cap B] = 1$.*

4. *If N is a minimal normal subgroup of G, then $\{A \cap N, B \cap N\} \subseteq \{N, 1\}$.*

5. *If N is a minimal normal subgroup of G contained in A and $B \cap N = 1$, then $N \leq \mathrm{C}_G(A)$ or $N \leq \mathrm{C}_G(B)$. If furthermore N is not cyclic, then $N \leq \mathrm{C}_G(B)$.*

Proof. 1. The simple group G/N is a mutually permutable product of AN/N and BN/N by Lemma 4.1.10. By Theorem 4.1.5, $AN = G$ and $BN = N$, $AN = N$ and $BN = G$, or $AN = BN = G$. Since $(A \cap B)N$ is a subnormal subgroup of G containing N by Corollary 4.1.17, $(A \cap B)N = N$ or $(A \cap B)N = G$.

2. Assume that N is non-abelian and that N is not contained in $A \cap B$. Suppose that $A \cap B \cap N \neq 1$. Then the normal closure of $A \cap B \cap N$ in G coincides with N and is contained in $\langle (A \cap B)^G \rangle$. Let $C = (A \cap B)_G$. Then NC/C is nilpotent by Lemma 4.1.18 and so $N/(C \cap N)$ is nilpotent. Since N is not contained in C, we obtain that N has a non-trivial nilpotent quotient. This contradicts the fact that N is a non-abelian minimal normal subgroup of G. Therefore $A \cap B \cap N = 1$. Assume that $1 \neq A \cap N \neq N$. By Lemma 4.1.21 (2), $(N \cap A)(N \cap B)$ is a normal subgroup of G. The minimality of N implies that $N = (N \cap A)(N \cap B)$. By Lemma 4.1.21 (1), this is a mutually permutable product. Since the intersection is trivial, in fact this is a totally permutable product by Lemma 4.1.21 (3). Since N has trivial supersoluble hypercentre, we can apply Corollary 4.2.13 to conclude that $N \cap A$ and $N \cap B$ are direct factors of N. Let x be an element of $B \cap N$ and let $X = \langle x \rangle$. Then x centralises $A \cap N$. It follows that $AX \cap N = (A \cap N) \times X$ and so $X = \mathrm{Z}(AX \cap N)$. Since $AX \cap N$ is normal in AX, it follows that X is a normal subgroup of AX and so the elements of A induce power automorphisms in $B \cap N$ by conjugation. By Theorem 1.3.3 shows that $g^{-1}g^a \in \mathrm{Z}(B \cap N) = 1$ for every $g \in B \cap N$. Therefore $B \cap N$ is centralised by A. Consequently $B \cap N$ is normalised by $G = AB$. This contradiction shows that if $N \cap A \neq 1$, then $N \leq A$. Assume now that $A \cap N = B \cap N = 1$. In this case, since, by Corollary 4.1.25 A' and B' are subnormal in G, it follows, by [119, I, 14.3], that N normalises A' and B' and so A' and B' are contained in $\mathrm{C}_G(N)$. Consequently $G/\mathrm{C}_G(N)$ is a product of the abelian groups $A\mathrm{C}_G(N)/\mathrm{C}_G(N)$ and $B\mathrm{C}_G(N)/\mathrm{C}_G(N)$, which is soluble by Theorem 3.1.7. In particular, G induces by conjugation a soluble group of automorphisms in N. This is impossible because N is a non-abelian minimal normal subgroup.

3. Denote by $D = A \cap B$ and assume that N is not contained in D. Hence N is not contained in D_G. Note that D is subnormal in G by Corollary 4.1.17. Assume that $N \cap D = 1$. Then N normalises D by [119, A, 14.3]. Therefore $[N, D] \leq N \cap D = 1$

as desired. Assume now that $N \cap D \neq 1$. Then $\langle (N \cap D)^G \rangle = N$ is contained in $\langle D^G \rangle$. By Lemma 4.1.18, we have that $\langle D^G \rangle / D_G$ is nilpotent. Hence ND_G/D_G, which is a minimal normal subgroup of G/D_G contained in $\langle D^G \rangle / D_G$, is contained in $\mathrm{Z}(\langle D^G \rangle / D_G)$. This implies that $[N, D] \leq D_G$, but since $[N, D] \leq N$, we also obtain that $[N, D] \leq D_G \cap N = 1$. Therefore $N \leq A \cap B$ or $[N, A \cap B] = 1$.

4. Assume that N is a counterexample to the statement. By Statement 2, we have that N is abelian p-group for some prime p. By Statement 3, $N \leq A \cap B$ or $[N, A \cap B] = 1$. Since N is a counterexample, $[N, A \cap B] = 1$ and we can assume without loss of generality that $1 \neq A \cap N \neq A$. By Lemma 4.1.21 (2) and the minimality of N, we have that $N = (N \cap A)(N \cap B)$ and so $N \cap B \neq 1$.

Suppose now that $N \leq B$. Then $A \cap B$ is permutable in B by Corollary 4.1.17. It follows that $A \cap B$ is hypercentrally embedded in B by Corollary 1.5.6, that is, $\langle (A \cap B)^B \rangle / (A \cap B)_B$ is contained in $\mathrm{Z}_\infty(B/(A \cap B)_B)$. By [222, 5.2.1], we have that

$$(\langle (A \cap B)^B \rangle \cap N)/((A \cap B)_B \cap N) \leq \mathrm{Z}_\infty(B/((A \cap B)_B \cap N)).$$

Since $\langle (A \cap B)^B \rangle \cap N$ is a normal subgroup of B containing $A \cap B \cap N$ and $(A \cap B)_B \cap N$ is a normal subgroup of B contained in $A \cap B \cap N$, it follows that

$$\langle (A \cap B \cap N)^B \rangle / (A \cap B \cap N)_B \leq \mathrm{Z}_\infty(B/(A \cap B \cap N)_B)$$

and so $A \cap N = A \cap B \cap N$ is hypercentrally embedded in B. This implies that the p'-elements of $B/(A \cap N)_G$ centralise all chief factors of B between $(A \cap N)_B = (A \cap N)_G = 1$ and $\langle (A \cap N)^B \rangle$. By [119, A, 12.4], $B/\mathrm{C}_B(\langle (A \cap N)^B \rangle)$ is a p-group. Since

$$\langle (A \cap N)^B \rangle = \langle (A \cap N)^{AB} \rangle = \langle (A \cap N)^G \rangle = N,$$

we have that $B/\mathrm{C}_B(N)$ is a p-group. Let N_0 be a minimal normal subgroup of B contained in N. Then $B/\mathrm{C}_B(N_0)$ is a p-group. By [119, B, 9.3], N_0 is centralised by B. Let x be a non-trivial element of N_0. Then $\langle \langle x \rangle^G \rangle = \langle \langle x \rangle^{BA} \rangle = \langle \langle x \rangle^A \rangle = N$. It follows that N is contained in $\langle x, A \rangle$, which coincides with $\langle x \rangle A$ because the product is mutually permutable. Hence $N = \langle x \rangle A \cap N = \langle x \rangle (A \cap N)$. Now set $C = \mathrm{C}_G(N)$. By Lemma 4.1.21 (2), $E = (C \cap A)(C \cap B)$ is a normal subgroup of G. Since

$$\begin{aligned} AE \cap BE = A(C \cap B) \cap (C \cap A)B &= (A \cap (C \cap A)B)(C \cap B) \\ &= (C \cap A)(A \cap B)(C \cap B) = E \end{aligned}$$

because $A \cap B \leq C$, we obtain that G/E is a totally permutable product of AE/E and BE/E by Proposition 4.1.16 (3). By Theorem 4.2.7, the nilpotent residual $(AE/E)^{\mathfrak{N}} = A^{\mathfrak{N}}E/E$ is a normal subgroup of G/E. Note that $E \cap A = (C \cap A)(C \cap B) \cap A = (C \cap A)(C \cap B \cap A) = C \cap A$, and, by symmetry, $E \cap B = C \cap B$. It follows that $AC/C \cong A/(A \cap C) = A/(A \cap E) \cong AE/E$ and $BC/C \cong BE/E$. Since $N/(A \cap N)$ is cyclic, we have that $A/\mathrm{C}_A(N/(A \cap N))$

is isomorphic to a subgroup of $\mathrm{Aut}(C_p)$ and, in particular, the nilpotent residual $A^{\mathfrak{N}}$ of A is contained in $\mathrm{C}_A(N/(A \cap N))$. It follows that $[A^{\mathfrak{N}}, N] \leq N \cap A$. Moreover, $[A^{\mathfrak{N}}, N] = [A^{\mathfrak{N}}E, N]$ is a normal subgroup of G. By minimality of N, $[A^{\mathfrak{N}}, N] = 1$ and so $A^{\mathfrak{N}} \leq C$. This implies that $AC/C \cong AE/E$ is nilpotent. Since BC/C is a p-group, so is BE/E. Therefore G/E is a supersoluble group by Corollary 4.1.38. On the other hand, since $E \leq C$, we have that G/C is also a totally permutable subgroup of AC/C and BC/C by Lemma 4.1.10. Let q be the largest prime divisor of $|G/C|$. Suppose that $q > p$. Let S be a Sylow q-subgroup of A. Then q does not divide $|BC/C|$ and so SC/C is a Sylow q-subgroup of G/C. Then SC is a normal subgroup of G. Therefore $[SC, N]$ is a normal subgroup of G. Moreover, $[SC, N] = [S, N]$ and the q-group S acts trivially on the p-group $N/(A \cap N)$. This implies that $[S, N] \leq A \cap N$. By minimality of N, $[S, N] = 1$. Hence $S \leq C$. This contradicts the choice of q. It follows that $q \leq p$. Moreover, by [119, B, 9.3], G/C cannot possess non-trivial normal p-subgroups because N is a p-group. It follows that $|G/C|$ can only be divisible by primes smaller than p and so $B \leq C$. But then $N \cap A$ is a normal subgroup of G, which contradicts the minimality of N. We have deduced that N is not contained in B.

We repeat part of the above process, using the subgroup $E = (A \cap C)(B \cap C)$ and the nilpotent residuals of A and B; G/E is a totally permutable product of AE/E and BE/E and G/C is a totally permutable product of AC/C and BC/C. Since $(AC/C)^{\mathfrak{N}} = A^{\mathfrak{N}}C/C$ is a normal subgroup of G/C by Theorem 4.2.7, we have that $A^{\mathfrak{N}}C$ is a normal subgroup of G. Let x be an element of $N \cap B$. Then $\langle x \rangle$ permutes with A and so $\langle \langle x \rangle^A \rangle \leq \langle x \rangle A \cap N = \langle x \rangle (A \cap N)$. This implies that A induces power automorphisms on the cyclic quotient $N/(A \cap N)$. Hence $[N, A^{\mathfrak{N}}C] = [N, A^{\mathfrak{N}}] \leq A \cap N$ because A induces an abelian group of automorphisms in $N/(A \cap N)$. By minimality of N, we obtain that $[N, A^{\mathfrak{N}}] = 1$. By symmetry, $[N, B^{\mathfrak{N}}] = 1$, too. It follows that AC/C and BC/C are nilpotent and therefore G/C is supersoluble by Corollary 4.1.38. Let q denote the largest prime dividing $|G/C|$ and let $K/C = \mathrm{O}_{q'}(G/C)$. Then G/K is a totally permutable product of AK/K and BK/K by Lemma 4.1.10. Hence one of the factors contains a minimal normal subgroup L/K of G/K by Lemma 4.1.20. We assume without loss of generality that $L \leq AK$. We obtain that L/K is abelian by the supersolubility of G/K and, by the definition of K, L/K is a q-group. Since G/C is supersoluble, the Sylow q-subgroup of L/C is normal in G/C and so $L/C = (U/C) \times (K/C)$, where U is contained in AC. This argument shows that G/C has a non-trivial normal q-subgroup U/C which is contained in AC/C.

Now G/C does not have non-trivial normal p-subgroups by [119, B, 9.3]. We have then that $q \neq p$. Assume now that $q > p$. In this case, $(U/C)^{p-1} = U/C$ and A^{p-1} centralises $N/(A \cap N)$ because A acts as a group of power automorphisms on $N/(A \cap N)$. It follows that $[N, U] = [N, U^{p-1}C] = [N, U^{p-1}] \leq [N, A^{p-1}] \leq N \cap A$. The minimality of N shows again that $[N, U] = 1$ and so $U \leq C$. This is contrary to the definition of U. We have proved then that all prime divisors of $|G/C|$ are smaller than p.

Since $A \cap N$ is not normal in G, there is an element x such that $(A \cap N)^x \neq A \cap N$. This element x can be chosen from B and of order prime to p, because all prime divisors of $|G/C|$ are smaller than p. Now $(A \cap N)(A \cap N)^x \leq \langle A, A^x \rangle \cap N \leq A\langle x \rangle \cap N$. Let Y be a Sylow p-subgroup of A. Then Y is also a Sylow p-subgroup of $A\langle x \rangle$ and $A \cap N = Y \cap N = \langle x \rangle A \cap N$. It follows that $A \cap N$ is a normal subgroup of $A\langle x \rangle$ and so x normalises $A \cap N$. This contradiction shows that there is no counterexample to the claim.

5. By Lemma 4.1.37, the elements of B induce power automorphisms on N. If N is non-abelian, then it is a direct product of non-abelian simple groups and Theorem 1.3.3 shows that the only possible power automorphism of N is the identity. Hence $N \leq C_G(B)$ and the conclusion holds. Therefore we can assume that N is a p-group for a prime p. Assume now that N is not contained in $C_G(B)$. Let $C = C_G(N)$. Then $B/(B \cap C)$ is a non-trivial cyclic group because B induces power automorphisms on N. Consequently $B = \langle b \rangle (B \cap C)$ for some $b \in B$. We will show that $N \leq C_G(A)$. Assume the contrary. We have that G/C does not possess non-trivial normal p-subgroups by [119, A, 13.6]. Now N is abelian and, by Theorem 1.3.2, b acts as a universal power automorphism on N. Therefore b centralises all automorphisms of N, in particular, $BC/C \leq Z(G/C)$. Hence there is an element $y(A \cap C) \in A/(A \cap C)$ of order $q \neq p$; y can be assumed to have as order a power of q. Since $|\langle y \rangle B : B| = |\langle y \rangle : \langle y \rangle \cap B|$ and $|NB : B| = |N|$, we have that $NB \cap \langle y \rangle B = B$ and $N \cap \langle y \rangle B \leq N \cap B = 1$. Let u be an element of N such that $[y, u] \neq 1$. Since $A \cap B \leq C$ by Statement 3, we obtain that $y \notin B$. By Lemma 4.1.21 (2), $T = (C \cap A)(C \cap B)$ is a normal subgroup of G. Moreover, $\langle y \rangle B \cap (C \cap A)(C \cap B) = (\langle y \rangle B \cap C \cap A)(C \cap B) = (\langle y \rangle (B \cap A) \cap C)(C \cap B) = (\langle y \rangle \cap C)(B \cap A)(C \cap B) = (\langle y \rangle \cap C)(C \cap B) = \langle y^q \rangle (C \cap B)$ is a normal subgroup of $\langle y \rangle B$. On the other hand,

$$\langle y \rangle B / \langle y^q \rangle (B \cap C) = (\langle y \rangle (B \cap C) / \langle y^q \rangle (B \cap C))(\langle y^q \rangle B / \langle y^q \rangle (B \cap C)) \quad (4.1)$$

is a product of the cyclic subgroups $\langle y \rangle (B \cap C)/\langle y^q \rangle (B \cap C)$, which is isomorphic to $\langle y \rangle / (\langle y \rangle \cap \langle y^q \rangle (B \cap C)) = \langle y \rangle / \langle y^q \rangle (\langle y \rangle \cap B \cap C) = \langle y \rangle / \langle y^q \rangle$ because $y \notin B$, and $\langle y^q \rangle B / \langle y^q \rangle (B \cap C)$, which is isomorphic to $B/(\langle y^q \rangle (B \cap C) \cap B) = B/((\langle y^q \rangle \cap B)(B \cap C)) = B/(B \cap C)$. Note that this product is totally permutable because it is mutually permutable and the intersection of the factors is trivial (Lemma 4.1.21 (3) and Proposition 4.1.16 (3)). We are now in a position to apply Lemma 4.3.1 to conclude that one of the factors of (4.1) is normal in the product and the corresponding quotient is abelian. On the other hand, $\langle y \rangle B / (\langle y \rangle B \cap C)$, which is isomorphic to $\langle y \rangle BC/C$, is a product of the cyclic subgroup $\langle y \rangle C/C$ and the central subgroup BC/C, turns out to be abelian. Assume that the first factor of (4.1) is normal. Then

$$\langle y \rangle (B \cap C) \cap \langle y \rangle B \cap C = \langle y \rangle (B \cap C) \cap C = (\langle y \rangle \cap C)(B \cap C) = \langle y^q \rangle (B \cap C).$$

If the second factor of (4.1) is normal, then

$$\langle y^q \rangle B \cap \langle y \rangle B \cap C = \langle y^q \rangle B \cap C = \langle y^q \rangle (B \cap C).$$

In both cases, $\langle y^q \rangle (B \cap C)$ is the intersection of two normal subgroups of $\langle y \rangle B$ which give abelian quotient. It follows that $\langle y \rangle B / \langle y^q \rangle (B \cap C)$ is abelian. The same is true for $\langle y^u \rangle B / \langle y^q \rangle (B \cap C)$. Now $[b, y^u] = [b, y[y, u]] = [b, [y, u]][b, y]^{[y,u]} = [y, u]^t [b, y]$, where $[y, u]^t \neq 1$. But $[b, y^u]$ and $[b, y]$ belong to $\langle y^q \rangle (B \cap C)$. Hence $[y, u]^t$ also does. It follows that $[y, u]^t \in N \cap \langle y^q \rangle (B \cap C) \leq N \cap \langle y \rangle B = 1$. This contradiction shows that A is a subgroup of C if B is not.

If A centralises N, then N must be cyclic because B induces power automorphisms on N. The proof of Statement 5 is now complete. □

Recall that a subgroup U of a group G *covers* the chief factor H/K of G if $HU = KU$ and that U *avoids* H/K if $U \cap H = U \cap K$. If U covers or avoids every chief factor of G, then we say that U has the *cover-avoidance property* and call U a CAP-subgroup of G.

An interesting application of the above lemma is the following remarkable property.

Corollary 4.3.4. *Let the group $G = AB$ be the mutually permutable product of the subgroups A and B. Then A and B are* CAP-*subgroups of G.*

Proof. If H/K is a chief factor of G, then H/K is a minimal normal subgroup of G/K which is a mutually permutable product of the subgroups AK/K and BK/K by Lemma 4.1.10. By Lemma 4.3.3 (4), AK/K covers or avoids H/K. Hence A covers or avoids H/K. The same holds for B. □

We break off the discussion on minimal normal subgroups to get information about the soluble residuals of the factors of a mutually permutable product. The following theorem was proved by Ballester-Bolinches, Cossey, and Pedraza-Aguilera [41].

Recall that, by [47, 6.5.5], if $\mathfrak{F}$ is a Fitting formation, the $\mathfrak{F}$-residual respects the operation of forming subnormal products. We shall used henceforth this property without further comment.

Theorem 4.3.5. *Let the group $G = AB$ be the product of the mutually permutable subgroups A and B. Then $[A, B^{\mathfrak{S}}]$ and $[A^{\mathfrak{S}}, B]$ are contained in $(A \cap B)_G$.*

Proof. Assume that the result is not true and that G is a minimal counterexample to the theorem. Then, for instance, $[A, B^{\mathfrak{S}}]$ is not contained in $X = (A \cap B)_G$. Suppose that $X \neq 1$. By Lemma 4.1.10, we have that $G/X = (A/X)(B/X)$ is the mutually permutable product of the subgroups AX/X and BX/X. Moreover, $((A/X) \cap (B/X))_{G/X} = 1$. By minimality of G, we have that $[A/X, B^{\mathfrak{S}}X/X] = 1$, that is, $[A, B^{\mathfrak{S}}] \leq X$, a contradiction to our assumption. Hence $X = (A \cap B)_G = 1$.

Denote by $T = A \cap B$. We distinguish three cases:

Case 1. Suppose that $G = AB^{\mathfrak{S}}$. In particular, $B = B^{\mathfrak{S}}(A \cap B)$. We prove that $B^{\mathfrak{S}}$ is a normal subgroup of G. First of all, T is a nilpotent subnormal subgroup of G by Corollary 4.1.17 and Lemma 4.1.18. Consider the normal closure $\langle T^G \rangle$ of T in G. If

$T = 1$, G is a totally permutable product of A and B by Proposition 4.1.16 (3). By Corollary 4.2.9, we have that $[A, B^{\mathfrak{S}}] = 1$, contrary to supposition. Therefore $T \neq 1$. Since $G/\langle T^G\rangle$ is a totally permutable product of $A\langle T^G\rangle/\langle T^G\rangle$ and $B\langle T^G\rangle/\langle T^G\rangle$, we have that $[A, B^{\mathfrak{S}}\langle T^G\rangle] \leq \langle T^G\rangle$ by Corollary 4.2.9. This means that A normalises $B^{\mathfrak{S}}\langle T^G\rangle$. Since $B^{\mathfrak{S}}\langle T^G\rangle = B^{\mathfrak{S}}(A\cap B)\langle T^G\rangle = B\langle T^G\rangle$, it follows that A normalises $B\langle T^G\rangle$ and $B\langle T^G\rangle$ is a normal subgroup of G. Now we consider the subgroup $\langle T^A\rangle B$ of G. By Lemma 4.1.21 (3), it is a mutually permutable product of $\langle T^A\rangle$ and B. Denote $S = T_{\langle T^A\rangle B}$. Note that T is a proper subnormal subgroup of A. Hence $\langle T^A\rangle B$ is a proper subgroup of G. The minimal choice of G implies that $[B^{\mathfrak{S}}, \langle T^A\rangle] \leq S$. Therefore $SB^{\mathfrak{S}}$ is a normal subgroup of $\langle T^A\rangle B$. On the other hand, $B^{\mathfrak{S}}$ is the soluble residual of $SB^{\mathfrak{S}}$. Hence $B^{\mathfrak{S}}$ is normal in $\langle T^A\rangle B = \langle T^G\rangle B$. The quotient $B\langle T^G\rangle/B^{\mathfrak{S}} = (\langle T^G\rangle B^{\mathfrak{S}}/B^{\mathfrak{S}})(B/B^{\mathfrak{S}})$ is a product of a normal soluble group and a soluble group. Hence it is soluble. This means that the soluble residual of $B\langle T^G\rangle$ is contained in $B^{\mathfrak{S}}$ and then we actually have $B^{\mathfrak{S}} = (B\langle T^G\rangle)^{\mathfrak{S}}$. Since $B\langle T^G\rangle$ is normal in G, so is $B^{\mathfrak{S}}$. Moreover, since $[B^{\mathfrak{S}}, \langle T^A\rangle]$ is contained in S, we have that $B^{\mathfrak{S}}$ normalises T. Therefore $\langle T^B\rangle = \langle T^{B^{\mathfrak{S}}(A\cap B)}\rangle = \langle T^{A\cap B}\rangle = T$, which implies $\langle T^G\rangle = \langle T^A\rangle$. Then $[A, B^{\mathfrak{S}}]$ is a normal subgroup of G contained in $\langle T^G\rangle\cap B^{\mathfrak{S}} = \langle T^A\rangle\cap B^{\mathfrak{S}} \leq A\cap B$. Since $(A\cap B)_G = 1$, we have that $[A, B^{\mathfrak{S}}] = 1$, which contradicts the choice of G.

Case 2. Suppose that $G = BA^{\mathfrak{S}}$. Then $A = A^{\mathfrak{S}}(A\cap B) = A^{\mathfrak{S}}T$. With the same arguments to those used in Case 1, we get $[B, A^{\mathfrak{S}}] = 1$. Moreover, repeating the argument of the Case 1 with the subgroup $\langle T^A\rangle B$, we obtain that $[T, B^{\mathfrak{S}}] \leq S$, where $S = T_{\langle T^A\rangle B}$. Then, by [157, III, 1.10 (a)] $[A, B^{\mathfrak{S}}] = [A^{\mathfrak{S}}T, B^{\mathfrak{S}}] = [A^{\mathfrak{S}}, B^{\mathfrak{S}}][T, B^{\mathfrak{S}}] = [T, B^{\mathfrak{S}}]$. Hence $\langle [A, B^{\mathfrak{S}}]^G\rangle = \langle [T, B^{\mathfrak{S}}]^B\rangle$ is contained in $S \leq T$ and so in $X = 1$, a contradiction to the choice of G.

Case 3. Assume that $AB^{\mathfrak{S}}$ and $BA^{\mathfrak{S}}$ are both proper subgroups of G. Consider the subgroup $W = A(TB^{\mathfrak{S}})$ of G and denote $M = T_W$. Then, by the minimal choice of G, $[A, B^{\mathfrak{S}}] \leq M$. In particular, $B^{\mathfrak{S}} = (B^{\mathfrak{S}}M)^{\mathfrak{S}}$ is normal in W because $B^{\mathfrak{S}}M$ is. Therefore $B^{\mathfrak{S}}$ is a normal subgroup of G since A normalises it. Arguing as above with the subgroup $Y = BA^{\mathfrak{S}}$, we obtain that $A^{\mathfrak{S}}$ is a normal subgroup of G.

Let N be a minimal normal subgroup of G contained in $B^{\mathfrak{S}}$. Set $D = (NT)_G$. Then $D = N(D\cap A)$. Note that $D\cap A$ is a nilpotent subnormal subgroup of G and so N normalises $D\cap A$. Suppose first that N is non-abelian. Then $D\cap A$ is the soluble radical of D. Therefore $D\cap A$ is normal in G. Since $X = 1$, we have that $D\cap A = 1$ and $D = N$. By minimality of G, we have that $[A, B^{\mathfrak{S}}] \leq N$. Since $[A, B^{\mathfrak{S}}]$ is also contained in $A\cap B$, it follows that $[A, B^{\mathfrak{S}}]$ is a normal nilpotent subgroup of N. This implies that $[A, B^{\mathfrak{S}}] = 1$, a contradiction. Therefore N is elementary abelian and so it is a p-group for some prime p. On the other hand, D', the derived subgroup of D, is contained in $D\cap A$ and so D is abelian. With the same argument, we get that D is a p-group.

Since, by Corollary 4.3.4, A covers or avoids N, it follows that $N \cap A = 1$ because $X = 1$. Moreover, applying Lemma 4.3.3 (5), N is a minimal normal subgroup of B. Denote by $C = T_B$. Then $D \cap C$ is a normal subgroup of B and so we have that $\langle (D \cap C)^G \rangle = \langle (D \cap C)^{BA} \rangle = \langle (D \cap C)^A \rangle \le D \cap A \le A \cap B$, because $D \cap A$ is normal in A. Thus $D \cap C = 1$. Denote by Z the normal closure of $D \cap A$ in B. Since $D \cap A$ is normalised by A, we have that Z is actually the normal closure of $D \cap A$ in G and $Z = (D \cap A)(Z \cap N)$ is a normal subgroup of G. Now $Z \cap N$ is a normal subgroup of B which is contained in N. Hence we have $Z \cap N = 1$ and $Z = D \cap A$, or $Z \cap N = N$ and $Z = D$. If $Z = D \cap A$, then $D \cap A$ is a normal subgroup of G contained in $A \cap B$. Then $D \cap A = 1$, and so $D = N$. Now $G/N = (AN/N)(B/N)$ satisfies the hypothesis in the theorem by Lemma 4.1.10. The minimality of G implies that $[A, B^{\mathfrak{S}}] \le N$. Since $[A, B^{\mathfrak{S}}] \le A$, we conclude that $[A, B^{\mathfrak{S}}] \le N \cap A = 1$, a contradiction to our assumption. Therefore we have that $Z = \langle (D \cap A)^G \rangle = D$. Now, by Corollary 1.5.6, T is hypercentrally embedded in B. Therefore, ZC/C is contained in the hypercentre of B/C. But $Z \cap C = D \cap C = 1$. Consequently D is contained in the hypercentre of B and then $B/C_B(D)$ is a p-group. It follows that $[D, B^{\mathfrak{S}}] = 1$. We also have $[A, B^{\mathfrak{S}}] \le D$ by the minimal choice of G. Therefore $B^{\mathfrak{S}}$ stabilises the series $1 < D < AD$. Applying [119, IV, 6.9], $B^{\mathfrak{S}}/C_{B^{\mathfrak{S}}}(AD)$ is nilpotent. Since $B^{\mathfrak{S}}$ is perfect, we have that $B^{\mathfrak{S}}$ centralises AD. In particular, $[A, B^{\mathfrak{S}}] = 1$. This is the final contradiction. □

Corollary 4.3.6. *Let the group $G = AB$ be product of the mutually permutable subgroups A and B. Then $A^{\mathfrak{S}}$ and $B^{\mathfrak{S}}$ are normal subgroups of G.*

Proof. Denote $N = (A \cap B)_G$. Applying Theorem 4.3.5, $B^{\mathfrak{S}}N$ is a normal subgroup of G. Moreover $B^{\mathfrak{S}} = (B^{\mathfrak{S}})^{\mathfrak{S}} \le (B^{\mathfrak{S}}N)^{\mathfrak{S}} = B^{\mathfrak{S}}N^{\mathfrak{S}} \le B^{\mathfrak{S}}$. Therefore $B^{\mathfrak{S}} = (B^{\mathfrak{S}}N)^{\mathfrak{S}}$. Consequently $B^{\mathfrak{S}}$ is a characteristic subgroup of $B^{\mathfrak{S}}N$, which is a normal subgroup of G. Hence $B^{\mathfrak{S}}$ is normal in G. Analogously $A^{\mathfrak{S}}$ is a normal subgroup of G. □

The extension of Corollary 4.3.6 to products of finitely many pairwise mutually permutable subgroups is due to Ballester-Bolinches, Beidleman, Heineken, and Pedraza-Aguilera [35].

Corollary 4.3.7. *Let the group $G = G_1G_2 \cdots G_r$ be the product of the pairwise mutually permutable subgroups $G_1, G_2, \dots, G_r$. Then the soluble residuals of all the factors G_i are normal subgroups of G and their product is the soluble residual of G.*

Proof. Let E be the soluble residual of G and D_i the soluble residual of G_i, $1 \le i \le r$. It follows from Corollary 4.3.6 that D_i is a normal subgroup of G_iG_j for all $i, j \in \{1, 2, \dots, r\}$. Hence D_i is a normal subgroup of G for every $i \in \{1, 2, \dots r\}$. Write $D = D_1D_2 \cdots D_r$. We must show that $D = E$. It is clear that D is contained in E. On the other hand, the factor G/D is the product of the pairwise mutually permutable soluble subgroups $G_i D/D$, $1 \leq i \leq r$. By Theorem 4.1.15, we see that G/D is soluble. In particular, E is contained in D and the equality holds. □

Non-abelian minimal normal subgroups have a nice behaviour in products of finitely many pairwise mutually permutable products as the following lemma proved by Ballester-Bolinches, Beidleman, Heineken, and Pedraza-Aguilera [35] shows.

Lemma 4.3.8. *Let the group $G = G_1 G_2 \cdots G_r$ be the product of the pairwise mutually permutable subgroups $G_1, G_2, \dots, G_r$. If M is a non-abelian minimal normal subgroup of G, then there exists $j \in \{1, 2, \dots, r\}$ such that M is contained in G_j. Moreover, if M is not contained in G_i for $i \in \{1, 2, \dots, r\}$, then $M \cap G_i = 1$.*

Proof. Let D be the soluble residual of G. By Corollary 4.3.7, $D = D_1 D_2 \cdots D_n$, where D_i is the soluble residual of G_i for $i \in \{1, 2, \dots, r\}$. Clearly M is contained in D. Since D_i is a normal subgroup of G, it follows that $M \cap D_i = M$ or $M \cap D_i = 1$ for every $i \in \{1, 2, \dots, r\}$.

Assume that $M \cap D_i = 1$ for all i. Then D centralises M, which contradicts the fact that M is non-abelian. Consequently there exists $j \in \{1, 2, \dots, r\}$ such that M is contained in G_j. Consider $i \neq j$. Then M is a direct product of non-abelian minimal normal subgroups of $G_i G_j$, $M_1, M_2, \dots, M_t$ say. It follows from Lemma 4.3.3 (4) that either $M_s \cap G_i = 1$ or $M_s \leq G_i$ for every $s \in \{1, 2, \dots, t\}$. Assume that $M_s \cap G_i = 1$ for all s. Then by Lemma 4.3.3 (5), $M = M_1 M_2 \cdots M_t \leq C_G(G_i)$ and so $M \cap G_i = 1$ since $Z(M) = 1$. Now assume that there exists an s such that $M_s \leq G_i$. Then $M_s \leq D_i$. Therefore $M \cap D_i \neq 1$ so that $M \leq D_i$. Thus $M \leq G_i$. This completes the proof. □

Now we turn again our attention to the minimal normal subgroups of mutually permutable products of two factors with the following result due to Beidleman and Heineken [65].

Lemma 4.3.9. *Let the group $G = AB$ be the product of the mutually permutable subgroups A and B. Let N be a minimal normal subgroup of G. Assume that $N \cap A = N \cap B = 1$. Then $|N| = p$, where p is a prime, and $N \leq C_G(A)$ or $N \leq C_G(B)$.*

Proof. From Lemma 4.3.3 (2), we have that N is abelian. Hence there exists a prime p such that $|N| = p^k$ for some positive integer k. We have that $AN = AB \cap AN = A(AN \cap B)$ and $|AN : A| = |A(AN \cap B) : A| = |AN \cap B : A \cap (AN \cap B)| = |AN \cap B : A \cap B| = p^k$. Since $A \cap B$ is subnormal in B by Corollary 4.1.17, there exists a series

$$A \cap B = W_0 \trianglelefteq W_1 \trianglelefteq \cdots \trianglelefteq W_k = AN \cap B$$

such that $|W_{i+1} : W_i| = p$ for all i. We obtain a new sequence

$$A = AW_0 \leq AW_1 \leq \cdots \leq AW_k = AN.$$

Like in the proof of Theorem 4.1.24, if $K_i = (AW_i)_{AW_{i+1}}$, we have that AW_{i+1}/K_i is isomorphic to a subgroup of the holomorph of C_p. It follows that AK_i/K_i is cyclic of exponent dividing $p-1$ like in the proof of Corollary 4.1.17. In particular, $A'A^{p-1} \leq K$. On the other hand, $[K_i, AW_{i+1}] \leq K_i \leq AW_i$ because K_i is normal in AW_{i+1}. Thus $[A'A^{p-1}, AW_{i+1}] \leq AW_i$. Since $[A'A^{p-1}, N] \leq N$, we obtain that $[A'A^{p-1}, AW_{i+1} \cap N] \leq AW_i \cap N$. In other words, $A'A^{p-1}$ induces the identity in all quotients $(AW_{i+1} \cap N)/(AW_i \cap N)$. By [119, A, 12.4], $A'A^{p-1}$ acts as a p-group on N. Set $C = \mathrm{C}_G(N)$. In particular, $A'A^{p-1}/(A'A^{p-1} \cap C)$ is a p-group. It follows that $A/(A \cap C)$ is an extension of a p-group by an abelian group of exponent dividing $p-1$ and therefore supersoluble. By symmetry, the same holds for $B/(B \cap C)$. By Lemma 4.3.3 (3), $A \cap B \leq C$. Put $R = (A \cap C)(B \cap C)$. Then G/R is a totally permutable product of the two supersoluble groups AR/R and BR/R. Moreover $AR/R \cong AC/C \cong A/(A \cap C)$ and $BR/R \cong BC/C \cong B/(B \cap C)$ by Lemma 4.1.21 (4). Thus G/R is supersoluble by Corollary 4.1.38. Since N is a minimal normal subgroup, G/C does not have any non-trivial normal p-subgroup by [119, A, 13.6]. Therefore AC/C and BC/C are abelian of exponent dividing $p-1$.

For the rest of the proof, we will assume that $G = AB$ is a minimal counterexample with a minimal normal subgroup N such that $N \cap A = 1 = N \cap B$ and N not contained in $\mathrm{C}_G(A)$ nor $\mathrm{C}_G(B)$. Let d be the smallest prime dividing G/R. There is a normal subgroup K/R of index d in G. If A and B are not contained in K, then both $A(K \cap B)$ and $B(K \cap A)$ are normal subgroups of G of index d. Hence we may assume without loss of generality that $A \leq K$ and so $K = A(K \cap B)$. Since $d \neq p$, the normal subgroup N can be expressed as a direct product of minimal normal subgroups M_i of K. We may apply the induction hypothesis for each M_i in the mutually permutable product $A(K \cap B)$ to conclude that $M_i \leq \mathrm{C}_G(A)$ or $M_i \leq \mathrm{C}_G(B)$ for each i. Let L denote the product of all M_i which are centralised by $K \cap B$. By taking into account the action of B on the M_i, it is clear that L is a normal subgroup of K which is normalised by B. We deduce that L is a normal subgroup of G. The minimality of N implies that $L = 1$ or $L = N$. Assume first that $L = 1$. Then all minimal normal subgroups M_i are centralised by A and so $A \leq \mathrm{C}_G(N)$. Now $R = A(C \cap B)$ and $G/C \cong B/(B \cap C)$ is abelian of exponent dividing $p-1$. The conclusion holds by [119, B, 9.8 (d)]. Suppose now that $L = N$. In this case, N centralises $K \cap B$. Since B is not contained in C, it follows that B is not contained in K and BK/K is a group of order d which is isomorphic to BR/R. Now, by Proposition 4.1.16 (3), G/R is a totally permutable product of the subgroup BR/R of order d and the normal abelian subgroup AR/R of exponent dividing $p-1$. Let $BR = \langle b \rangle R$, where b is an element whose order is a power of d. Then the element bR of G/R induces a power automorphism on AR/R and all subgroups of AR/R are invariant by the action of bR. Hence $A \cap C = A \cap \mathrm{C}_G(M_i)$ for all i because all M_i are permuted by b, and $A/(A \cap C)$ must be cyclic. Let $A = \langle a \rangle (A \cap C)$ for some element $a \in A$. Put $H = \langle b \rangle A = A\langle b \rangle$. We have that $|H : A|$ and $|AN : A| = |N|$ are coprime. This implies that $AN \cap H = A$ and so $A(N \cap H) = A$. Hence $N \cap H = 1$. We have that $H \cap K = A(\langle b^d \rangle \cap K) = A\langle b^d \rangle$ because $BK/K \cong BR/R = \langle b \rangle R/R$

has order d and $x = [a,b] \in H \cap K$. On the other hand, let y be a non-trivial element of N. Then $y = vu$ with $v \in B$ and $u \in A$. Hence $y^{-1}By = u^{-1}Bu$ and G is also a mutually permutable product of $y^{-1}By$ and A. Moreover $\langle y^{-1}by\rangle A = A\langle y^{-1}by\rangle$ and $\langle y^{-1}by\rangle A \cap K = \langle y^{-1}b^d y\rangle A = H \cap K$. Let us choose y such that $byb^{-1} = y^{1+k} \neq y$. Now $[a, y^{-1}by] = [a, y^k b] = [a,b][a,y^k]^b$. Now $[a,b]$ and $[a,y^k][a,y^k]^b$ belong to $H \cap K$. This implies that $[a,y^k]^b \in N \cap H = 1$. But then $y^k \neq 1$ centralises a, and so y centralises a. It follows that $N = \mathrm{C}_N(A) \cup \mathrm{C}_N(B)$. This implies that $N = \mathrm{C}_N(A)$ or $N = \mathrm{C}_N(B)$, because a group cannot be written as a union of two proper subgroups. This final contradiction proves the result. □

According to Lemma 4.1.20, there do not exist pairwise totally permutable products of core-free factors. Our next result, due to Beidleman and Heineken [66], confirms that the same is true for mutually permutable products of two factors. We need an elementary lemma.

Lemma 4.3.10. *The group of all automorphisms α of an elementary abelian group $E = \langle a,b\rangle$ of order p^2, p a prime, such that α leaves the subgroup $\langle a\rangle$ invariant is isomorphic to $C_{p-1} \times \mathrm{Hol}(C_p)$.*

Proof. These automorphisms have the form $a^\alpha = a^{m_{1,1}}$, $b^\alpha = a^{m_{1,2}}b^{m_{2,2}}$, where p does not divide $m_{1,1}$ nor $m_{2,2}$. The isomorphism between the automorphisms of E and the automorphisms of the vector space of dimension 2 over the field of p elements allows to see these automorphisms as a subgroup of $\mathrm{GL}_2(p)$ composed by the matrices of the form

$$\begin{pmatrix} m_{1,1} & m_{1,2} \\ 0 & m_{2,2} \end{pmatrix}$$

with $m_{1,1}, m_{2,2} \in \mathrm{GF}(p) \setminus \{0\}$ and $m_{1,2} \in \mathrm{GF}(p)$. This group is the product of the subgroups

$$J = \left\{\begin{pmatrix} a & 0 \\ 0 & a \end{pmatrix} : a \in \mathrm{GF}(p) \setminus \{0\}\right\} \cong C_{p-1},$$

$$K = \left\{\begin{pmatrix} 1 & 0 \\ 0 & b \end{pmatrix} : b \in \mathrm{GF}(p) \setminus \{0\}\right\} \cong C_{p-1},$$

$$L = \left\{\begin{pmatrix} 1 & c \\ 0 & 1 \end{pmatrix} : c \in \mathrm{GF}(p)\right\} \cong C_p,$$

where J centralises K and L, and since

$$\begin{pmatrix} 1 & 0 \\ 0 & b^{-1} \end{pmatrix}\begin{pmatrix} 1 & c \\ 0 & 1 \end{pmatrix}\begin{pmatrix} 1 & 0 \\ 0 & b \end{pmatrix} = \begin{pmatrix} 1 & bc \\ 0 & 1 \end{pmatrix}$$

we obtain that $[L]K$ is isomorphic to $\mathrm{Hol}(C_p)$ and the result follows. □

Theorem 4.3.11. *Let the non-trivial group $G = AB$ be the product of the mutually permutable subgroups A and B. Then $A_G B_G$ is not trivial.*

Proof. We proceed by induction on $|G|$. Assume that $A_G = B_G = 1$ and derive a contradiction. In this case, $(A \cap B)_G = 1$. By Corollary 4.3.6, the soluble residual $A^{\mathfrak{S}}$ of A is normal in G. Therefore $A^{\mathfrak{S}} \leq A_G = 1$ and A is soluble. A similar argument shows that B is soluble. Therefore G is soluble by Theorem 4.1.15.

Let N be a minimal normal subgroup of G. Then $(N \cap A)(N \cap B)$ is a normal subgroup of G by Lemma 4.1.21 (2) and $N \cap A = N$ or $N \cap A = 1$ by Lemma 4.3.3 (4). Since $A_G = 1$, $N \cap A = 1$. By symmetry, $N \cap B = 1$. By Lemma 4.3.9, there exists a prime p such that $|N| = p$ and $A \leq \mathrm{C}_G(N)$ or $B \leq \mathrm{C}_G(N)$. By Lemma 4.1.10, G/N is the mutually permutable product of AN/N and BN/N. Clearly we can assume that G/N is non-trivial. Then the induction hypothesis applies to G/N and so there exists a minimal normal subgroup M/N of G/N that is contained in one of the factors AN/N or BN/N. Assume that $M/N \leq AN/N$. Then $M = M \cap AN = (M \cap A)N$. Moreover, $M \cap A \neq M$ because $A_G = 1$. Therefore $|M : M \cap A| = |N : N \cap M \cap A| = p$. Now $(M \cap A)(M \cap B)$ is a normal subgroup of G by Lemma 4.1.21 (2). Since $M \cap A$ is not a normal subgroup of G, it follows that $M \cap B \neq 1$. Hence $(M/N) \cap (BN/N) \neq 1$. Moreover, $M \cap B \neq M$. By Corollary 4.3.4, we obtain that $M/N \leq BN/N$. Thus $M = (M \cap B)N$ and $|M : M \cap B| = p$ with a similar argument to the one used with A. In particular, $M \cap B$ is a maximal subgroup of M. Hence $M/N = (M \cap A)N/N$ is isomorphic to $(M \cap A)/(M \cap A \cap N) \cong M \cap A$ and to $M \cap B$. Since $N \leq \mathrm{C}_G(A)$ or $N \leq \mathrm{C}_G(B)$, it follows that $[N, AN] = 1$ or $[N, BN] = 1$. Consequently $[N, M] = 1$ and so $N \leq \mathrm{Z}(M)$. This implies that $M \cap A$ and $M \cap B$ are normal subgroups of M. Consequently $M' \leq N \cap (M \cap A) = 1$. It follows that M is abelian. Moreover, $M \cap A$ is not contained in $M \cap B$ nor $M \cap B$ is contained in $M \cap A$ because none of $M \cap A$, $M \cap B$ is a normal subgroup of G. It follows that $M = (M \cap A)(M \cap B)$ and so $|M|$ divides $|M \cap A|^2$. It follows that $|N|$ divides $|M \cap A| = |M/N|$ and so M is a p-group. Since $N \leq \mathrm{Z}(M)$ and $M = (M \cap A)N = (M \cap A)\mathrm{Z}(M)$, we conclude that M is elementary abelian.

Assume first that M/N is not cyclic. Denote by $D = M \cap A \cap B = (M \cap A) \cap (M \cap B)$. If $D = 1$, then $|M| = |M \cap A|^2 = |N||M \cap A|$ and so $|M \cap A| = |N|$. This implies that M/N is cyclic of order p, against the assumption. Consequently $D \neq 1$. Clearly $B_G = B_A = A \cap B_A = (A \cap B)_A = 1$ and so $A \cap B \leq \mathrm{Z}_\infty(A)$ because $A \cap B$ is a permutable subgroup of A (see Corollary 1.5.6). Hence D is a p-group which is contained in $\mathrm{Z}_\infty(A)$. Therefore $\mathrm{O}^p(A) \leq \mathrm{C}_G(D)$ by Lemma 1.2.22 and, analogously, $\mathrm{O}^p(B) \leq \mathrm{C}_G(D)$. Now

$$|M| = \frac{|M \cap A| \cdot |M \cap B|}{|D|} = |M \cap B| \cdot |N|,$$

from which we have that

$$|M \cap A| \cdot |M \cap B| = |D| \cdot |M \cap B| \cdot |N|$$

and, since $|M \cap A : D| = p$, D is a maximal subgroup of $M \cap A$. Since $D_A \leq (A \cap B)_A = 1$, D is not a normal subgroup of A. In particular, there exists $a \in A$ such that $D^a \neq D$. Since $M \cap A$ is abelian, $M \cap A = DD^a$. It follows that $\mathrm{O}^p(A) \leq \mathrm{C}_G(M \cap A)$. Analogously, $\mathrm{O}^p(B) \leq \mathrm{C}_G(M \cap B)$. Since M/N is A-isomorphic to $M \cap A$, we have that $\mathrm{O}^p(A) \leq \mathrm{C}_G(M/N)$ and, symmetrically, $\mathrm{O}^p(B) \leq \mathrm{C}_G(M/N)$. Therefore $(G/N)/\mathrm{C}_{G/N}(M/N)$ is a product of p-groups, and so it is a p-group, too. Hence $G/N = \mathrm{C}_{G/N}(M/N)$ and M/N is cyclic. This contradicts the hypothesis about M/N. Therefore, M/N is cyclic and $|M| = p^2$.

Let $C = \mathrm{C}_G(M)$. Then $G/C = (AC/C)(BC/C)$ is isomorphic to a subgroup of $C_{p-1} \times \mathrm{Hol}(C_p)$ by Lemma 4.3.10. Suppose that p divides $|G/C|$. We can assume then that there exists $a \in A \setminus (A \cap C)$ such that $a^p \in A \cap C$. If $a\mathrm{C}_G(N) \neq \mathrm{C}_G(N)$, then $\mathrm{o}(a\mathrm{C}_G(N)) \neq p$ because $|N| = p$ and so $a \in \mathrm{C}_G(N)$. Then $N = [M, a] = [N(M \cap A), a]$. Therefore $[N(M \cap A), a] = [M \cap A, a] = 1$ because $M \cap A$ has order p and so $M \cap A$ cannot have automorphisms of order p. This contradiction shows that p does not divide $|G/C|$. Consequently G/C is abelian of exponent dividing $p - 1$.

Note that not all elements of G induce power automorphisms on M, because otherwise $M \cap A$ and $M \cap B$ would be normal subgroups of B. Let $b \in B$ such that b does not induce a power automorphism on M. Then b induces different power automorphisms on N and on $M \cap B$. Now since G/C is abelian, if $g \in G$, then b and $g^{-1}bg$ induce the same automorphism on M. Hence $g^{-1}bg \in \mathrm{N}_G(M \cap B)$ and so $b \in \mathrm{N}_G(g(M \cap B)g^{-1})$. It follows that b induces the same power automorphism on $M \cap B$ and on $g(M \cap B)g^{-1}$. This implies that $M \neq (M \cap B)(g(M \cap B)g^{-1})$. Consequently $M \cap B = g(M \cap B)g^{-1}$ for all $g \in G$ and so $M \cap B$ is a normal subgroup of G. This contradiction proves the theorem. □

Remark 4.3.12. We note that we have not used the full force of the mutual permutability in the proof of Theorem 4.3.11. In fact, the arguments just given above allow us to prove that if π is a set of primes dividing $|G|$ and $(A \cap \mathrm{O}_\pi(G))(B \cap \mathrm{O}_\pi(G)) \neq 1$, then $\mathrm{O}_\pi(A)_G \mathrm{O}_\pi(B)_G \neq 1$. It is enough to check that if N is a minimal normal π-subgroup of G such that $A \cap N = B \cap N = 1$, then $(AN/N \cap \mathrm{O}_\pi(G/N))(BN/N \cap \mathrm{O}_\pi(G/N)) \neq 1$, in order to apply the inductive hypothesis to G/N. This holds because $AN/N \cap \mathrm{O}_\pi(G/N) = (AN \cap \mathrm{O}_\pi(G))/N = (A \cap \mathrm{O}_\pi(G))N/N \cong (A \cap \mathrm{O}_\pi(G))/(N \cap A \cap \mathrm{O}_\pi(G)) \cong A \cap \mathrm{O}_\pi(G)$ and, symmetrically, $BN/N \cap \mathrm{O}_\pi(G/N) \cong B \cap \mathrm{O}_\pi(G)$.

It is not known whether Theorem 4.3.11 holds for general mutually permutable products of more than two factors, that is, whether some factor of a product of pairwise mutually permutable subgroups contains a minimal normal subgroup. We know, by Lemma 4.3.8, that the result is true if the group contains non-abelian minimal normal subgroups.

We round the section off with some results that confirm the good behaviour of the factors of a mutually permutable product with respect to normal subgroups. We begin with a theorem which will be used in Section 5.4.

Theorem 4.3.13. *Let the* SC*-group* $G = G_1 G_2 \cdots G_r$ *be the product of the pairwise mutually permutable subgroups* $G_1, G_2, \ldots, G_r$. *Then* $\mathrm{S}(G_i) = G_i \cap \mathrm{S}(G)$ *for every* $i \in \{1, 2, \ldots, r\}$, *where* $\mathrm{S}(X)$ *is the soluble radical of a group* X.

Proof. We argue by induction on $|G|$. Let $i \in \{1, 2, \ldots, r\}$. Let N be a minimal normal subgroup of G. Assume first that N is abelian. By induction, $\mathrm{S}(G_i N/N) = \mathrm{S}(G/N) \cap G_i N/N$. Since $\mathrm{S}(G_i N/N) = \mathrm{S}(G_i)N/N$ and $\mathrm{S}(G/N) = \mathrm{S}(G)/N$, we have that $\mathrm{S}(G_i)N = \mathrm{S}(G) \cap G_i N$. Hence $\mathrm{S}(G) \cap G_i = \mathrm{S}(G_i)N \cap G_i = \mathrm{S}(G_i)(N \cap G_i) = \mathrm{S}(G_i)$ because $N \cap G_i \leq \mathrm{S}(G_i)$.

Assume now that N is not abelian. By Lemma 4.3.8, either $N \leq G_i$ or $N \cap G_i = 1$, and there exists $j \in \{1, 2, \ldots, r\}$ such that $N \leq G_j$. Let $C = C_G(N)$. Since $\mathrm{S}(G) \leq C$, we have that $\mathrm{S}(G) = \mathrm{S}(C)$. Suppose that $N \leq G_i$. Then $\mathrm{S}(G_i) = \mathrm{S}(C \cap G_i)$. Suppose that $N \cap G_i = 1$. Since G is an SC-group, N is simple and so N is a minimal normal subgroup of $G_i G_j$. By Lemma 4.3.3 (5), $G_i \leq C$. Hence $\mathrm{S}(G_i) = \mathrm{S}(C \cap G_i)$. Consider now the product $Z = (C \cap G_1)(C \cap G_2) \cdots (C \cap G_r)$. By Corollary 4.1.22, Z is a normal subgroup of G, Z is a pairwise mutually permutable product of the subgroups $C \cap G_1, C \cap G_2, \ldots, C \cap G_r$. Clearly $Z \leq C$ and Z is a proper subgroup of G. Moreover, $\mathrm{S}(Z) = \mathrm{S}(G) \cap Z = \mathrm{S}(C) \cap Z$. By induction, $\mathrm{S}(Z) \cap C \cap G_i = \mathrm{S}(C \cap G_i) = \mathrm{S}(G_i)$. Furthermore $\mathrm{S}(Z) \cap C \cap G_i = \mathrm{S}(C) \cap C \cap G_i = \mathrm{S}(C) \cap G_i = \mathrm{S}(G) \cap G_i$. This shows that $\mathrm{S}(G_i) = \mathrm{S}(G) \cap G_i$ for every $i \in \{1, 2, \ldots, r\}$, and the proof of the theorem is complete. □

When $r = 2$, then obviously each minimal normal subgroup of G contained in $G_1 G_2 = G$ is a minimal normal subgroup of G. Hence the argument just given above yields the following.

Theorem 4.3.14. *Let the group* $G = AB$ *be the product of the mutually permutable subgroups* A *and* B. *Then* $\mathrm{S}(A) = \mathrm{S}(G) \cap A$ *and* $\mathrm{S}(B) = \mathrm{S}(G) \cap B$.

Whether or not the above result is true for general mutually permutable products of more than two factors remains open.

Example 4.3.15. We might ask whether $\mathrm{S}(G) = \mathrm{S}(A)\,\mathrm{S}(B)$ in a mutually permutable product. The answer is negative:

Let $G = \mathrm{Sym}(5) \times C_2$, where $\mathrm{Sym}(5)$ is the symmetric group on five letters and $C_2 = \langle x \rangle$. Let $A = \mathrm{Sym}(5) = \mathrm{Alt}(5)\langle (1,2) \rangle$, which is a normal subgroup of G, and $B = \langle \mathrm{Alt}(5), (1,2)x \rangle$, which is also a normal subgroup of G. Then $G = AB$ is a mutually permutable product, $\mathrm{S}(A) = \mathrm{S}(B) = 1$, but $\mathrm{S}(G) = C_2 \neq \mathrm{S}(A)\,\mathrm{S}(B)$.

Theorem 4.3.16. *Let the group* $G = AB$ *be the product of the mutually permutable subgroups* A *and* B. *Let* $\mathcal{K}_X$ *denote the set of all composition factors of a group* X. *Then* $\mathcal{K}_G = \mathcal{K}_A \cup \mathcal{K}_B$.

Proof. The theorem is true if $G = 1$. We may assume that $G \neq 1$ and by induction that the theorem is true for mutually permutable products of smaller order. Applying Theorem 4.3.11, there exists a minimal normal subgroup N of G which is contained in A_G or B_G. There is no loss of generality in assuming that $N \leq A_G$. Then, by Lemma 4.1.10, G/N is a mutually permutable product of A/N and BN/N. By induction, we have that $\mathcal{K}_{G/N} = \mathcal{K}_{A/N} \cup \mathcal{K}_{BN/N}$. Moreover, $\mathcal{K}_G = \mathcal{K}_{G/N} \cup \mathcal{K}_N$ and $\mathcal{K}_A = \mathcal{K}_{A/N} \cup \mathcal{K}_N$. Since $\mathcal{K}_{BN/N} \subseteq \mathcal{K}_B \subseteq \mathcal{K}_{BN/N} \cup \mathcal{K}_N$, it follows that $\mathcal{K}_G = \mathcal{K}_A \cup \mathcal{K}_B$. □

An unsolved problem is *whether the above theorem remains valid if* 2 *is replaced by an arbitrary natural number* r. These open questions are closely related to the behaviour of the abelian minimal normal subgroups in products of finitely many pairwise mutually permutable products. A possible counterexample might be found by looking for a mutually permutable product with an abelian minimal normal subgroup which intersects trivially with all the factors.

4.4 Mutually permutable products: residuals and radicals

Our first object here is to give an account of extensions of the main result of Section 4.2 to mutually permutable products. The first principal aim is to generalise Theorem 4.2.7 to the following: *in a mutually permutable product of the subgroups A and B, the nilpotent residual of B normalises A*. This result was found by Bochtler in his PhD Thesis [74], and its proof will be our first main preoccupation in this section.

Our second object will be to study the behaviour of residuals associated to saturated and Fitting formations and radicals associated to Fitting classes in mutually permutable products.

There are several ingredients in the proof of the generalisation of Theorem 4.2.7. All of them are statements about the normality of some residuals in products. We begin with an extremely useful lemma.

Lemma 4.4.1. *Let A and B be two mutually permutable subgroups of a group G and π a set of prime numbers. Assume that $X \leq A$ and that $A \cap B \leq Y \leq B$. If Y is subnormal in XY ($= YX$), then $\mathrm{O}^{\pi}(Y)$ is normalised by $\mathrm{O}^{\pi'}(X)$.*

In particular, $\mathrm{O}^{\pi'}(A) \leq \mathrm{N}_G(\mathrm{O}^{\pi}(A \cap B))$.

Proof. Let x be a π-element of X. By Lemma 4.1.21 (3), X and Y are mutually permutable subgroups. Hence $\langle x \rangle Y$ is a subgroup of G and by hypothesis Y is a subnormal subgroup of $\langle x \rangle Y$ whose index is a π-number. From Lemma 1.1.11, we obtain that $\mathrm{O}^{\pi}(Y) = \mathrm{O}^{\pi}(\langle x \rangle Y)$ is a normal subgroup of $\langle x \rangle Y$. Consequently x normalises $\mathrm{O}^{\pi}(Y)$ and so $\mathrm{O}^{\pi'}(X) = \langle x \in X : x \text{ is a } \pi\text{-element} \rangle$ normalises $\mathrm{O}^{\pi}(Y)$. □

The following result was proved by Bochtler [74, 3.2.3]. The particular result $\pi = \rho$ was presented in [73].

Theorem 4.4.2. *Let A and B be mutually permutable subgroups of a group G and let π and ρ be two sets of prime numbers such that for all $p \in \pi$ and $q \in \pi \cup \rho$, p does not divide $q - 1$. Then $O^{\pi'}(A)$ is a subnormal subgroup of $AO^{\rho'}(B)$. If, moreover, $\pi \cap \rho = \emptyset$, then $O^{\pi'}(A)$ is normal in $AO^{\rho'}(B)$.*

Proof. We denote by $\tilde{A} = O^{\pi'}(A)(A \cap B)$ and by $\tilde{B} = O^{\rho'}(B)(A \cap B)$. By Lemma 4.1.21 (3), $\tilde{A}$ and $\tilde{B}$ are mutually permutable subgroups of G. Our eventual goal is to show that $\tilde{A}$ is a subnormal subgroup of $\tilde{A}\tilde{B}$. We can then deduce that $O^{\pi'}(A)$, which is a normal subgroup of $\tilde{A}$, is subnormal in $\tilde{A}\tilde{B}$ and finally appeal to Theorem 1.1.7 to conclude that $O^{\pi'}(A)$ is also subnormal in the product $A(\tilde{A}\tilde{B}) = AO^{\rho'}(B)$, for the desired result.

By Corollary 4.1.17, we know that $A \cap B$ is subnormal in A. Therefore $\tilde{A}$ is subnormal in A. Since $|A : \tilde{A}|$ is a π'-number, we can apply Lemma 1.1.11 to deduce that $O^{\pi'}(\tilde{A}) = O^{\pi'}(A)$. Analogously, we obtain that $O^{\rho'}(\tilde{B}) = O^{\rho'}(B)$. Consequently $\tilde{A}$ and $\tilde{B}$ are two mutually permutable subgroups of G satisfying $\tilde{A} = O^{\pi'}(\tilde{A})(\tilde{A} \cap \tilde{B})$ and $\tilde{B} = O^{\rho'}(\tilde{B})(\tilde{A} \cap \tilde{B})$. In order to see that $\tilde{A}$ is subnormal in $\tilde{A}\tilde{B}$, we may assume without loss of generality that

$$\begin{aligned} A = \tilde{A} = O^{\pi'}(A)(A \cap B) \\ B = \tilde{B} = O^{\rho'}(B)(A \cap B) \end{aligned} \tag{4.2}$$

and, of course, that $G = AB$.

Let $N = \langle (A \cap B)^G \rangle$ and $R = A^{\mathfrak{N}} N B^{\mathfrak{N}}$. Since the normal subgroup N contains the intersection $A \cap B$ one can easily deduce from Proposition 4.1.16 (3) that G/N is the totally permutable product of the subgroups AN/N and BN/N. Since $A^{\mathfrak{N}}N/N$ and $B^{\mathfrak{N}}N/N$ are the nilpotent residuals of AN/N and BN/N respectively, we appeal to Theorem 4.2.7 to deduce R is a normal subgroup of G. The subnormality of A will follow from the next three steps:

1. *A is a subnormal subgroup of AN.*

From Lemma 4.4.1, we obtain that $O^{\pi'}(A) \leq O^{(\pi\cup\rho)'}(A) \leq N_G(O^{(\pi\cup\rho)}(A \cap B))$ and $O^{\rho'}(B) \leq N_G(O^{(\pi\cup\rho)}(A \cap B))$. We deduce, bearing in mind (4.2), that

$$O^{(\pi\cup\rho)}(A \cap B) \trianglelefteq O^{\pi'}(A)(A \cap B)O^{\rho'}(B) = AB = G,$$

as stated above. Since $N/O^{(\pi\cup\rho)}(A \cap B)$ is generated by subnormal $(\pi \cup \rho)$-subgroups of $G/O^{(\pi\cup\rho)}(A \cap B)$, we deduce that it is a $(\pi \cup \rho)$-group.

Furthermore, by Corollary 4.3.4, A and B have the cover and avoidance property. Here A centralises every $(\pi \cup \rho)$-chief factor of G that it avoids: Let H/K be such a chief factor and suppose that it is not centralised by A. It follows from Lemma 4.3.3 (5) and Lemma 4.3.9 that H/K is cyclic and centralised by B. Let $q = |H/K| \in \pi \cup \rho$. Since the index $|G : C_G(H/K)|$ divides $q - 1$, $C_G(H/K)$ contains also every π-element of G due to the choice of π and ρ. This yields

$$G = AB = O^{\pi'}(A)B = C_G(H/K),$$

which contradicts the hypothesis $A \nleq C_G(H/K)$.

Now we consider a series

$$O^{(\pi\cup\rho)}(A\cap B) = K_r \trianglelefteq K_{r-1} \trianglelefteq \cdots \trianglelefteq K_1 = N,$$

where all K_i/K_{i+1} are $(\pi \cup \rho)$-chief factors of G. If K_i/K_{i+1} is avoided by A, then it centralises AK_{i+1}/K_{i+1} as seen before. It follows that

$$AK_{i+1}/K_{i+1} \trianglelefteq (AK_{i+1}/K_{i+1})(K_i/K_{i+1}) = AK_i/K_{i+1}$$

and so AK_{i+1} is a normal subgroup of AK_i. Otherwise, A covers K_i/K_{i+1}, that is, $AK_{i+1} = AK_i$. In every case, we have that AK_{i+1} is a normal subgroup of AK_i for all i, which leads to the series

$$A = AK_r \trianglelefteq AK_{r-1} \trianglelefteq \cdots \trianglelefteq AK_1 = AN.$$

It follows that A is subnormal in AN.

2. *AN is a normal subgroup of AR.*

Theorem 4.2.7 yields

$$AN/N \trianglelefteq (AN/N)(BN/N)^{\mathfrak{N}} = AR/N,$$

that is, AN is a normal subgroup of AR.

3. *AR is a subnormal subgroup of G.*

It is clear that $A/(R \cap A)$ possesses a normal Hall π-subgroup and $B/(R \cap B)$ possesses a normal Hall ρ-subgroup because $A^{\mathfrak{N}}$ and $B^{\mathfrak{N}}$ are contained in R. But due to the equalities (4.2), there cannot exist normal subgroups of A and B containing $A \cap B$ with non-trivial π'- and ρ'-factor groups, respectively. Thus AR/R is a π-group and BR/R is a ρ-group, and so their product G/R is a $(\pi \cup \rho)$-group.

Arguing like in Step 4.4, we obtain that A covers or centralises every $(\pi \cup \rho)$-chief factor K_i/K_{i+1} of G in a series $R = K_r \trianglelefteq K_{r-1} \trianglelefteq \cdots \trianglelefteq K_1 = G$. It follows that AK_{i+1} is a normal subgroup of AK_i for all i. We deduce

$$AR = AK_r \trianglelefteq AK_{r-1} \trianglelefteq \cdots \trianglelefteq AK_1 = G$$

and so AR is a subnormal subgroup of G.

Combining our three steps, we conclude that A is subnormal in G.

Suppose now that the sets of primes π and ρ satisfy $\pi \cap \rho = \emptyset$. We have shown that $\tilde{A}$ is a subnormal subgroup of $\tilde{A}O^{\rho'}(B)$. Since π and ρ are disjoint, π is contained in ρ' and we obtain that $O^{\pi}(O^{\rho'}(B)) = O^{\rho'}(B)$. By Lemma 4.4.1, $O^{\rho'}(B) \leq N_G(O^{\pi'}(\tilde{A}))$. By Lemma 1.1.11, it follows that $O^{\pi'}(\tilde{A}) = O^{\pi'}(A)$. This completes the proof of the theorem. □

Remarks 4.4.3. 1. Let p and q be two primes with $p \geq q$. The hypothesis about the sets π and ρ of Theorem 4.4.2 is satisfied with $\pi = \{p\}$ and $\rho = \{p, q\}$. Hence given two mutually permutable subgroups A and B of a group G, $\mathrm{O}^{p'}(A)$ is subnormal in $A\mathrm{O}^{\{p,q\}'}(B)$ and, in particular, $\mathrm{O}^{p'}(A)$ is a subnormal subgroup of $A\mathrm{O}^{q'}(B)$. If, moreover, $p > q$, then $\pi = \{p\}$ and $\rho = \{q\}$ satisfy the condition $\pi \cap \rho = \emptyset$ and so Theorem 4.4.2 gives that $\mathrm{O}^{p'}(A)$ is a normal subgroup of $A\mathrm{O}^{q'}(B)$.

2. Let p be a prime. The sets $\pi = \rho = \{p\}$ fulfil the condition of Theorem 4.4.2. Thus we have that $\mathrm{O}^{p'}(A)$ is a subnormal subgroup of $A\mathrm{O}^{p'}(B)$.

3. There exist mutually permutable products $G = AB$ where A is not subnormal in G. If we take the set π of all prime divisors of $|G|$ for such a group G, we see that in general the result "$\mathrm{O}^{\pi'}(A)$ is subnormal in $A\mathrm{O}^{\pi'}(B)$" is not true for an arbitrary set of primes π.

4. Let π and ρ be two sets of prime numbers satisfying the hypothesis of Theorem 4.4.2 but with $\pi \cap \rho \neq \emptyset$. The result "$\mathrm{O}^{\pi'}(A)$ is a normal subgroup of $A\mathrm{O}^{\rho'}(B)$" is false in general, as the following example shows. In the dihedral group

$$G = \mathrm{Dih}(8) = \langle a, b : a^2 = b^4 = 1, b^a = b^{-1} \rangle$$

of order 8, the subgroups $A = \langle a \rangle$ and $B = \langle b \rangle$ are mutually permutable, but $A = \mathrm{O}^{2'}(A)$ is not normal in $A\mathrm{O}^{2'}(B) = G$.

The following lemma will be needed in the proof of Theorem 4.4.5. It has appeared in Bochtler [74, 3.3.3].

Lemma 4.4.4. *Let G be a p-group and p a prime number. Then*

$$G^{\mathfrak{N}} = (\mathrm{O}^p(G))^{\mathfrak{N}} \mathrm{O}^p(G^{\mathfrak{N}}).$$

Proof. There is no loss of generality in assuming that $\mathrm{O}^p(G^{\mathfrak{N}}) = 1$. Therefore we must only prove that $G^{\mathfrak{N}} = (\mathrm{O}^p(G))^{\mathfrak{N}}$ when $G^{\mathfrak{N}}$ is a p-group. The inclusion $(\mathrm{O}^p(G))^{\mathfrak{N}} \leq G^{\mathfrak{N}}$ is clear. Assume that $(\mathrm{O}^p(G))^{\mathfrak{N}}$ is a proper subgroup of $G^{\mathfrak{N}}$. In this case, there exists a p-chief factor $G^{\mathfrak{N}}/R$ of G with $(\mathrm{O}^p(G))^{\mathfrak{N}} \leq R < G^{\mathfrak{N}}$. Under the action of the normal subgroup $\mathrm{O}^p(G)/R$ of G/R, this factor is decomposed as a direct product $G^{\mathfrak{N}}/R = (T_1/R) \times (T_2/R) \times \cdots \times (T_n/R)$ of minimal normal subgroups of $\mathrm{O}^p(G)/R$ by Clifford's theorem [119, B, 7.3]. All factors T_i/R are centralised by $\mathrm{O}^p(G)/R$ because $(\mathrm{O}^p(G))^{\mathfrak{N}} \leq R$. Hence $\mathrm{O}^p(G)/R$ centralises $G^{\mathfrak{N}}/R$. In particular, $G^{\mathfrak{N}}/R$ is a central chief factor of G. Hence G/R is nilpotent and $G^{\mathfrak{N}} \leq R < G^{\mathfrak{N}}$. This contradiction shows that $G^{\mathfrak{N}} = (\mathrm{O}^p(G))^{\mathfrak{N}}$, as desired. □

We now come to the first significant theorem of this section.

Theorem 4.4.5. *Let the group $G = AB$ be the product of the mutually permutable subgroups A and B. Then the nilpotent residual $B^{\mathfrak{N}}$ of B normalises A.*

Proof. Assume that the result is false and let $G = AB$ be a counterexample with least $|G| + |A| + |B|$. We reach the contradiction after the following steps:

1. *Let N be a minimal normal subgroup of G and $Z = AB^{\mathfrak{N}}$. Then $A(N \cap Z)$ is a normal subgroup of Z.*

We know, by Lemma 4.1.10, that G/N is the mutually permutable product of AN/N and BN/B. The minimal choice of G then implies that AN/N is a normal subgroup of ZN/N, so that AN is a normal subgroup of ZN. Hence $AN \cap Z = A(N \cap Z)$ is a normal subgroup of Z.

2. $A_G = (A \cap B)_G = 1$.

Otherwise, if N is a minimal normal subgroup of G contained in A, then $A(N \cap Z) = A$ is a normal subgroup of Z, contrary to assumption.

3. *A is soluble.*

By Corollary 4.3.6, we have that $A^{\mathfrak{S}}$ is a normal subgroup of G and so $A^{\mathfrak{S}} \leq A_G = 1$.

4. *$A \cap B$ is a subnormal nilpotent subgroup of G. In particular, $A \cap B \leq \mathrm{F}(G)$.*

Since $A \cap B$ is subnormal in G by Corollary 4.1.17 and $(A \cap B)_G = 1$, we conclude that $A \cap B$ is nilpotent by virtue of Lemma 4.1.18.

5. *$A \cap B \neq 1$ and $\mathrm{F}(G)$ is a non-trivial p-group for a prime p.*

Suppose that $A \cap B = 1$. Then, by Proposition 4.1.16 (3), A and B are totally permutable subgroups of G. Applying Theorem 4.2.7, $B^{\mathfrak{N}}$ centralises A, which is not the case. Hence $A \cap B \neq 1$. By Statement 4, $\mathrm{F}(G) \neq 1$. Suppose that p and q are two different primes dividing $|\mathrm{F}(G)|$. Let N and L be minimal normal subgroups of G such that $|N| = p^{\alpha}$ and $|L| = q^{\beta}$ for some natural numbers α and β. Since $A(N \cap Z)$ is a normal subgroup of Z, the normal closure $\langle A^Z \rangle$ of A in Z is contained in $A(N \cap Z)$. Consequently $\langle A^Z \rangle = A(N \cap \langle A^Z \rangle)$. A similar argument with L shows that $\langle A^Z \rangle = A(N \cap \langle A^Z \rangle) = A(L \cap \langle A^Z \rangle)$. It follows that $|\langle A^Z \rangle : A|$ divides p^{α} and q^{β}. Therefore $A = \langle A^Z \rangle$ and so A is a normal subgroup of Z, contrary to assumption. This allows us to conclude that $\mathrm{F}(G)$ is a p-group for a prime p.

6. *There exists a prime q dividing $|A|$ such that $A = A_q(A \cap B)$, where A_q denotes a Sylow q-subgroup of A.*

Let q be a prime dividing $|A|$. Let $C_q = A_q(A \cap B)$. Then C_q and B are mutually permutable by Lemma 4.1.21 (3). Suppose that C_q is a proper subgroup of A for all primes q. Since, by Proposition 4.1.16 (3), C_q and B are mutually permutable, the minimal choice of (G, A, B) implies that C_q is normalised by $B^{\mathfrak{N}}$. Let $\{q_1, q_2, \dots, q_n\}$ be the set of primes dividing $|A|$. Then $B^{\mathfrak{N}} \leq \mathrm{N}_G(\langle C_{q_1}, C_{q_2}, \dots, C_{q_n} \rangle) = \mathrm{N}_G(A)$, contrary to supposition. Therefore there exists a prime q dividing $|A|$ such that $A = A_q(A \cap B)$.

7. $O^p(B) \leq N_G(A \cap B)$.

Let $\pi = \mathbb{P} \setminus \{p\}$. We apply Lemma 4.4.1 and conclude that

$$O^p(B) \leq N_G(O^\pi(A \cap B)) = N_G(A \cap B).$$

8. *A is not a p-group. In particular, $p \neq q$.*

Suppose that A is a p-group. If N is a minimal normal subgroup of G contained in $F(G)$, we have that AN is a p-group. Hence $A(N \cap Z)$ is a p-group and so A is a subnormal subgroup of $A(N \cap Z)$. By Statement 1, A is a subnormal subgroup of $Z = AB^{\mathfrak{N}}$. By Lemma 4.4.1 applied to $Y = A$, $X = (A \cap B)B^{\mathfrak{N}}$, and $\pi = \mathbb{P} \setminus \{p\}$, we obtain that $O^p((A \cap B)B^{\mathfrak{N}}) \leq N_G(O^\pi(A)) = N_G(A)$. Consequently $O^p(B^{\mathfrak{N}}) \leq N_G(A)$. Now if $O^p(B)(A \cap B)$ is a proper subgroup of B, by minimality of (G, A, B), we conclude that $(O^p(B))^{\mathfrak{N}} \leq (O^p(B)(A \cap B))^{\mathfrak{N}} \leq N_G(A)$. By Lemma 4.4.4, $B^{\mathfrak{N}} = O^p(B^{\mathfrak{N}}) \cdot (O^p(B))^{\mathfrak{N}} \leq N_G(A)$, against the assumption. Thus $B = O^p(B)(A \cap B)$. By Statement 7, $A \cap B$ is a normal subgroup of B. Hence $\langle (A \cap B)^G \rangle = \langle (A \cap B)^A \rangle$ is a non-trivial normal subgroup of G contained in A. This contradicts Statement 2. Consequently, A is not a p-group.

9. *$A \cap B$ is a normal subgroup of A.*

By Statement 6, $A = A_q(A \cap B)$ and by Statement 8, $p \neq q$. Statements 4 and 5 yield that $A \cap B$ is a subnormal Sylow p-subgroup of A. Hence $A \cap B$ is normal in A.

10. *$A \cap B$ is not a normal subgroup of B.*

Suppose that $A \cap B$ is a normal subgroup of B. Then $A \cap B$ is a non-trivial normal subgroup of $AB = G$ by Statements 5 and 9. This contradicts Statement 2.

11. *Let N be a minimal normal subgroup of G contained in* $F(G)$. *Then A is not subnormal in AN.*

Suppose that A is a subnormal subgroup of AN. Since $|AN : A|$ divides $|N|$, we obtain that $|AN : A| = p^\alpha$ for a suitable integer α by Statement 5. Then $O^p(AN) = O^p(A)$ by Lemma 1.1.11. Consequently $O^p(A)$ is a characteristic subgroup of AN, which is in turn a normal subgroup of ZN by minimality of (G, A, B). Thus $O^p(A)$ is normal in ZN. In particular, $O^p(A)$ is normalised by $B^{\mathfrak{N}}$. Moreover, $B^{\mathfrak{N}} \leq O^p(B) \leq N_G(A \cap B)$ by Statement 7. Hence $A = O^p(A)(A \cap B)$ is normalised by $B^{\mathfrak{N}}$, contrary to assumption.

12. *B is soluble and hence G is soluble.*

Suppose that $B^{\mathfrak{S}} \neq 1$. Then $B^{\mathfrak{S}} \leq C_G(A)$ by Statement 2 and Theorem 4.3.5 and $B^{\mathfrak{S}}$ is a normal subgroup of G by Corollary 4.3.6. Hence if N is a normal subgroup of G contained in $B^{\mathfrak{S}}$, then A is a normal subgroup of AN. Applying Statement 11, N is non-abelian. In this case, A is the soluble radical of AN. The normality of A in ZN follows now form the normality of AN in ZN, a contradiction which shows that $B^{\mathfrak{S}} = 1$, and B is soluble. The solubility of G follows from Statement 3 and Theorem 4.1.14.

13. *If N is a minimal normal subgroup of G, then N is cyclic.*

Assume that N is not cyclic. By Statement 2, $N \not\leq A$. By Lemma 4.3.3 (4), we have that $N \cap A = 1$. By Lemma 4.3.9, $N \cap B \neq 1$ and so N is contained in B. Applying Lemma 4.3.3 (5), $N \leq C_G(A)$ and A is a normal subgroup of AN, contradicting again Statement 11.

14. *q divides $p - 1$.*

Applying Lemmas 4.3.3 (5) and 4.3.9, $A \cap B \leq C_G(N)$. Since A is not a normal subgroup of AN by 11, and $A/C_A(N)$ is both an abelian group of exponent dividing $p-1$ by Statement 12 and a q-group by Statement 6, we conclude that q divides $p-1$.

15. *$O^{p'}(B)$ is a normal subgroup of G.*

By Theorem 4.4.2 applied to $\pi = \{p\}$ and $\rho = \{q\}$ together with Statement 14, we conclude that $1 \neq O^{p'}(B)$ is subnormal in $BO^{q'}(A)$, which coincides with G by Statement 6. Let $Y = O^{p'}(B)$. By Statements 4 and 5, $A \cap B \leq Y$ and Y is subnormal in G. Lemma 4.4.1 applied to $\pi = \mathbb{P} \setminus \{p\}$ yields that $O^{p'}(Y) = Y$ is normalised by $O^p(A)$. But $A \cap B \leq N_G(Y)$ and so $A = O^p(A)(A \cap B) \leq N_G(Y)$. Hence Y is a normal subgroup of G.

16. *$O^{p'}(B) = B$ and so B is a normal subgroup of G.*

Suppose that $Y = O^{p'}(B)$ is a proper subgroup of B. Then A is a normal subgroup of $AY^{\mathfrak{N}}$ by minimality of (G, A, B). By Statement 15, Y is a normal subgroup of G. If $Y^{\mathfrak{N}} \neq 1$, then there exists a minimal normal subgroup N of G such that $N \leq Y^{\mathfrak{N}}$. In this case, A is a normal subgroup of AN and this contradicts Statement 11. Therefore $Y^{\mathfrak{N}} = 1$, that is, Y is nilpotent and so Y is contained in the Fitting subgroup $F(G)$. Thus $B = F(G)B_{p'}$, where $B_{p'}$ denotes a Hall p'-subgroup of G. In particular, $B = F(G)O^p(B)$. By [47, 6.5.26], $B^{\mathfrak{N}} = (O^p(B))^{\mathfrak{N}}$. Since B does not normalise $A \cap B$ by Statement 10, it follows by Statement 7 that $O^p(B)(A \cap B)$ is a proper subgroup of B. Hence by minimality of (G, A, B) and [47, 6.5.26], $(O^p(B)(A \cap B))^{\mathfrak{N}} = (O^p(B))^{\mathfrak{N}} = B^{\mathfrak{N}}$ normalises A. This contradiction shows that $O^{p'}(B) = B$ is a normal subgroup of G.

17. *The final contradiction.*

Suppose that for a prime r, $O^r(B)$ is a proper subgroup of B. Suppose that $r \neq p$. Then $B/O^r(B)$ is a p'-group and so $B = O^{p'}(B) \leq O^r(B)$. This contradiction shows that $r = p$. Moreover, $O^p(B)$ is a normal subgroup of G. If $(O^p(B))^{\mathfrak{N}} \neq 1$, then there exists a minimal normal subgroup of G such that $N \leq (O^p(B))^{\mathfrak{N}}$. Since $O^p(B)$ is a proper subgroup of B by Statement 7, we conclude that N normalises A and A is a normal subgroup of AN, against Statement 11. Consequently $O^p(B)$ is nilpotent.

Since $\mathrm{O}^p(B) \leq \mathrm{F}(G)$, $\mathrm{O}^p(B)$ is a p-group. Hence B is a p-group and so $B^{\mathfrak{N}} = 1$. This contradicts the fact that $B^{\mathfrak{N}}$ does not normalise A and completes the proof of the theorem. □

Corollary 4.4.6. *Let the group $G = G_1 G_2 \cdots G_r$ be the product of the pairwise mutually permutable subgroups $G_1, G_2, \dots, G_r$. Then $\prod_{j \in \{1,\dots,r\}\setminus\{i\}} G_j$ is normalised by $G_i^{\mathfrak{N}}$.*

Proof. The result follows from Theorem 4.2.7 taking into account that if $i \neq j$, $G_i^{\mathfrak{N}}$ normalises G_j because G_i and G_j are mutually permutable. □

Let $\mathfrak{F}$ be a formation containing the class of all nilpotent groups. Then $G^{\mathfrak{F}}$ is contained in $G^{\mathfrak{N}}$ for all groups G. Therefore we have:

Corollary 4.4.7. *Let the group $G = G_1 G_2 \cdots G_r$ be the product of the pairwise mutually permutable subgroups $G_1, G_2, \dots, G_r$ and let $\mathfrak{F}$ be a formation containing the class of all nilpotent groups. Then $\prod_{j \in \{1,\dots,r\}\setminus\{i\}} G_j$ is normalised by $G_i^{\mathfrak{F}}$ and $G_i^{\mathfrak{F}} G_j^{\mathfrak{F}}$ is a subnormal subgroup of G for all i, j. In particular, $\prod_{j \in \{1,\dots,r\}} G_j^{\mathfrak{F}}$ is subnormal in G.*

Proof. Let $i, j \in \{1, \dots, r\}$. Since G_j is normalised by $G_i^{\mathfrak{F}}$ and $G_j^{\mathfrak{F}}$ is a characteristic subgroup of G_j, we conclude that $G_i^{\mathfrak{F}}$ normalises $G_j^{\mathfrak{F}}$. Hence $G_i^{\mathfrak{F}} G_j^{\mathfrak{F}}$ is a subgroup of G. Applying Corollary 4.1.26, G_i' is subnormal in G for all i. Since $G_i^{\mathfrak{F}}$ is normal in G_i', we have that $G_i^{\mathfrak{F}}$ is subnormal in G. Consequently $G_i^{\mathfrak{F}} G_j^{\mathfrak{F}}$ is a subnormal subgroup of G by [119, A, 14.4]. The same argument leads to the subnormality of $\prod_{j \in \{1,\dots,r\}} G_j^{\mathfrak{F}}$. □

If $\mathfrak{F}$ does not contain the class of nilpotent groups, the permutability of $\mathfrak{F}$-residuals of mutually permutable subgroups is not guaranteed as the following example due to Bochtler [74, 5.1.3] shows.

Example 4.4.8. Consider the class $\mathfrak{F}$ composed of all groups whose 3-chief factors are not central. Applying [119, A, 3.2], $\mathfrak{F}$ is a formation. It is clear that $\mathfrak{F}$ does not contain the class of all nilpotent groups.

Let P be the extraspecial group of order 27 and exponent 3 with the presentation

$$P = \langle x, y : x^3 = y^3 = 1, [x, y]^3 = [x, y, x] = [x, y, y] = 1 \rangle.$$

Then P possesses two automorphisms α given by $x^\alpha = x$ and $y^\alpha = y^{-1}$ and β given by $x^\beta = x^{-1}$ and $y^\beta = y$. It is easy to see that $\langle \alpha, \beta \rangle \leq \mathrm{Aut}(P)$ is an elementary abelian group of order 4. Let $G = [P]\langle \alpha, \beta \rangle$ be the corresponding semidirect product. Consider now the subgroups $A = \langle \alpha, \mathrm{Z}(P), x \rangle$ and $B = \langle \beta, \mathrm{Z}(P), y \rangle$. Then $A = \langle \alpha \rangle \mathrm{Z}(P) \times \langle x \rangle$ and $B = \langle \beta \rangle \mathrm{Z}(P) \times \langle y \rangle$. We obtain that $G = AB$. Note that neither A nor B is a subnormal subgroup of G.

Next we prove that A and B are mutually permutable. By Lemma 4.1.12, it is enough to show that $A/\mathrm{Z}(P)$ and $B/\mathrm{Z}(P)$ are mutually permutable. Since the unique non-trivial proper subgroups of $A/\mathrm{Z}(P) = (\langle\alpha\rangle\mathrm{Z}(P)/\mathrm{Z}(P)) \times (\langle x\rangle\mathrm{Z}(P)/\mathrm{Z}(P))$ are $\langle\alpha\rangle\mathrm{Z}(P)/\mathrm{Z}(P)$ and $\langle x\rangle\mathrm{Z}(P)/\mathrm{Z}(P)$, the permutability of all subgroups of $A/\mathrm{Z}(P)$ with $B/\mathrm{Z}(P)$ is clear. A symmetric argument shows that all subgroups of $B/\mathrm{Z}(P)$ permute with $A/\mathrm{Z}(P)$ and so A and B are mutually permutable.

Since the minimal normal 3-subgroup $\langle x\rangle$ of A is central in A, we obtain that $A^{\mathfrak{F}} \neq 1$. Since the automorphism α inverts $\mathrm{Z}(P)$, we obtain that $\langle\alpha\rangle\mathrm{Z}(P)$ is a non-abelian group of order 6 and $A/\langle x\rangle \cong \langle\alpha\rangle\mathrm{Z}(P) \in \mathfrak{F}$. It follows that $1 \neq A^{\mathfrak{F}} \leq \langle x\rangle$ and so $A^{\mathfrak{F}} = \langle x\rangle$. Analogously, $B^{\mathfrak{F}} = \langle y\rangle$. Since $\langle x\rangle$ does not permute with $\langle y\rangle$, we obtain that the product $A^{\mathfrak{F}}B^{\mathfrak{F}}$ is not a subgroup of G.

Note that in this example $\langle x\rangle = \mathrm{Z}(A)$ and $\langle y\rangle = \mathrm{Z}(B)$. It follows that characteristic subgroups of the factors do not necessarily permute.

If $\mathfrak{F}$ does not contain the class $\mathfrak{N}$ of all nilpotent groups, we cannot conclude in general that $\langle A^{\mathfrak{F}}, B^{\mathfrak{F}}\rangle$ is a subnormal subgroup of G. It is enough to consider the class $\mathfrak{F} = \mathfrak{S}_3$ of all 3-groups and the symmetric group $G = \mathrm{Sym}(3)$ of degree 3, which is a totally permutable product the of its Sylow 3-subgroup and a Sylow 2-subgroup.

Our second substantive result of this section is due to Bochtler [74] and shows that the residuals of mutually permutable subgroups associated to Fitting formations permute. We need a preparatory structural result [75].

Theorem 4.4.9. *Let the group $G = AB$ be the product of the mutually permutable subgroups A and B. Then there exist subgroups L and M with the following properties:*

1. $A' \leq L \leq A$, $B' \leq M \leq B$,
2. $A \cap B \leq L \cap M$,
3. *L and M are subnormal in G,*
4. $G' \leq LB \cap MA = LM$.

Proof. Clearly we may assume that $G' \neq 1$. We consider part of a chief series $1 = N_r \trianglelefteq N_{r-1} \trianglelefteq \cdots \trianglelefteq N_1 = G'$ of G passing through G'. By Corollary 4.3.4, both A and B cover or avoid each of these chief factors.

Let N_j/N_{j+1} be one of these chief factors and suppose that it is avoided, but not centralised, by A. By Lemma 4.3.3 and Lemma 4.3.9, we have that N_j/N_{j+1} is cyclic and centralised by B. Since N_j/N_{j+1} has an abelian automorphism group, it is also centralised by G'.

We set

$$J = \{j \in \{1, \dots, r-1\} : N_j/N_{j+1} \text{ is avoided but not centralised by } A\}$$

and

$$K = \bigcap_{j \in J} C_G(N_j/N_{j+1}), \quad L = K \cap A.$$

If $J = \emptyset$, we define $K = G$. Thus $BG' \leq K$ and $A' \leq L \leq A$. We also note that by definition of K, L centralises all chief factors in the above series that are avoided by A. Moreover, since $B \leq K$, it follows that $A \cap B \leq L$.

For the following, define $[K,_1 L] = [K, L]$ and $[K,_i L] = [[K,_{i-1} L], L]$ for $i > 1$. We prove by induction on i that $[K,_i L] \leq LN_i$. The case $i = 1$ is trivial because $[K, L] \leq G'$. Suppose that $[K,_i L] \leq LN_i$. If N_i/N_{i+1} is avoided by A, then, as mentioned before, L centralises N_i/N_{i+1}. It follows that

$$[K,_{i+1} L] \leq [LN_i, L] \leq LN_{i+1}.$$

Otherwise, A covers N_i/N_{i+1}. Then, by [157, III, 1.10 (a)], we have

$$[K,_{i+1} L] \leq [LN_i, L] \leq [AN_i, L] = [AN_{i+1}, L] = [A, L][N_{i+1}, L] \leq LN_{i+1}.$$

This completes the induction argument. It follows that $[K,_r, L] \leq L$. Hence L is subnormal in K and therefore also in G.

Define $M \leq B$ analogously. Then $B' \leq M \leq B$ and Conditions 1, 2, and 3 are satisfied. Moreover, since $B \leq K$, we have that $K = K \cap AB = (K \cap A)B = LB$. By symmetry,

$$G' \leq LB \cap MA = L(B \cap MA) = L(B \cap A)M = LM$$

and Condition 4 holds. □

Theorem 4.4.10. *Let the group $G = AB$ be the product of the mutually permutable subgroups A and B and let $\mathfrak{F}$ be a Fitting formation. Then $A^{\mathfrak{F}}B^{\mathfrak{F}}$ is a subgroup of G.*

Proof. We suppose that the result is false and derive a contradiction. Let $G = AB$ be a counterexample with $|G| + |A| + |B|$ as small as possible. For this group we have:

1. *Let N_1 and N_2 be normal subgroups of A with $A = N_1N_2$. Then $A = N_1(A \cap B)$ or $A = N_2(A \cap B)$. The same statement holds for B.*

The subgroups $N_i(A \cap B)$ and B are mutually permutable by Lemma 4.1.21 (3). Suppose that $N_i(A \cap B)$ is a proper subgroup of A for both $i \in \{1, 2\}$. Then the assumption about G gives that $(N_i(A \cap B))^{\mathfrak{F}}B^{\mathfrak{F}}$ is a subgroup of G. Moreover $N_i(A \cap B)$ is the product of two subnormal subgroups of A by Proposition 4.1.16 (4). Hence $(N_i(A \cap B))^{\mathfrak{F}} = (N_i)^{\mathfrak{F}}(A \cap B)^{\mathfrak{F}}$ by [47, 6.5.5]. Since $(A \cap B)^{\mathfrak{F}} \leq B^{\mathfrak{F}}$, it follows that

$$(N_i)^{\mathfrak{F}}B^{\mathfrak{F}} = (N_i)^{\mathfrak{F}}(A \cap B)^{\mathfrak{F}}B^{\mathfrak{F}} = (N_i(A \cap B))^{\mathfrak{F}}B^{\mathfrak{F}} \leq G$$

for both $i \in \{1, 2\}$. Applying again [47, 6.5.5], $A^{\mathfrak{F}} = (N_1)^{\mathfrak{F}}(N_2)^{\mathfrak{F}}$, and so

$$A^{\mathfrak{F}}B^{\mathfrak{F}} = (N_1)^{\mathfrak{F}}(N_2)^{\mathfrak{F}}B^{\mathfrak{F}} = (N_1)^{\mathfrak{F}}B^{\mathfrak{F}}(N_2)^{\mathfrak{F}} = B^{\mathfrak{F}}(N_1)^{\mathfrak{F}}(N_2)^{\mathfrak{F}} = B^{\mathfrak{F}}A^{\mathfrak{F}},$$

which implies that $A^{\mathfrak{F}}B^{\mathfrak{F}}$ is a subgroup of G. This contradicts the choice of G. Therefore $A = N_1(A \cap B)$ or $A = N_2(A \cap B)$, and an analogous statement follows for B.

By Theorem 4.4.9, there exist subgroups L and M with the following properties: $A' \leq L \leq A$ and $B' \leq M \leq B$, $A \cap B = L \cap M$, L and M subnormal in G, and $G' \leq LB \cap AM = LM$.

2. *Either A or B is subnormal in G.*

Suppose that neither A nor B is a subnormal subgroup of G. Since A/L is then non-trivial and abelian, there exists a prime number p such that $\mathrm{O}^p(A/L) = \mathrm{O}^p(A)L/L$ is a proper subgroup of A/L and so $\mathrm{O}^p(A)(A \cap B) \leq \mathrm{O}^p(A)L$ is a proper subgroup of A. By Statement 1, it follows that $A = \mathrm{O}^{p'}(A)(A \cap B)$. The same argument yields the existence of a prime number q such that $B = \mathrm{O}^{q'}(B)(A \cap B)$. Assume without loss of generality that $p \geq q$. Then by Theorem 4.4.2 we obtain that $\mathrm{O}^{p'}(A)$ is a subnormal subgroup of $A\mathrm{O}^{q'}(B) = AB = G$. Since $A \cap B$ is also subnormal in G by Proposition 4.1.16 (4), we conclude that $A = \mathrm{O}^{p'}(A)(A \cap B)$ is a subnormal subgroup of G. This contradicts our initial assumption. Therefore either A or B is subnormal in G.

3. *The final contradiction.*

By Statement 2, we can assume without loss of generality that A is a subnormal subgroup of G. Thus A and M are subnormal in AM and therefore $(AM)^{\mathfrak{F}} = A^{\mathfrak{F}}M^{\mathfrak{F}}$ by [47, 6.5.5]. Furthermore $(AM)^{\mathfrak{F}}$ is a characteristic subgroup of the normal subgroup AM of G. It follows that $A^{\mathfrak{F}}M^{\mathfrak{F}} = (AM)^{\mathfrak{F}}$ is a normal subgroup of G. Since $M^{\mathfrak{F}} \leq B^{\mathfrak{F}}$, we have that $A^{\mathfrak{F}}B^{\mathfrak{F}} = A^{\mathfrak{F}}M^{\mathfrak{F}}B^{\mathfrak{F}} \leq G$. This is the desired contradiction. □

In view of Theorem 4.4.10 we now get the following.

Corollary 4.4.11. *Let the group $G = G_1G_2\cdots G_r$ be the product of the pairwise mutually permutable subgroups $G_1, G_2, \dots, G_r$ and let $\mathfrak{F}$ be a Fitting formation. Then $\prod_{j\in\{1,\dots,r\}} G_j^{\mathfrak{F}}$ is a subgroup of G which is a pairwise permutable product of the subgroups $G_j^{\mathfrak{F}}$, $j \in \{1,\dots,r\}$.*

The immediate purpose of the example that we are about to describe is to give a negative answer to the following question: is the product $A^{\mathfrak{F}}B^{\mathfrak{F}}$, where $\mathfrak{F}$ is a Fitting formation and A and B are mutually permutable subgroups, mutually permutable?

Example 4.4.12 ([74]). Let $p \geq 5$ be a prime number, and let

$$\mathrm{Sym}(3) = \langle x, y : x^3 = y^2 = 1, x^y = x^{-1}\rangle$$

be the symmetric group of degree 3. Applying [119, B, 11.7], there exists a faithful and irreducible Sym(3)-module V of dimension 2 over the field GF(p) of p elements such that V, considered as a module over $\langle y\rangle$, contains the trivial $\langle y\rangle$-module both as a submodule and as a quotient. Let $G = [V]\mathrm{Sym}(3)$ be the corresponding semidirect product and consider the subgroups $A = \langle x\rangle V$ and $B = \langle y\rangle V$. Clearly $G = AB$.

Since $\langle x\rangle$ and $\langle y\rangle$ are mutually permutable subgroups of Sym(3), we get that A/V and B/V are mutually permutable and so are A and B by Lemma 4.1.12.

Clearly, $B^{\mathfrak{N}}$ is a non-trivial proper subgroup of V. The irreducibility of V as a Sym(3)-module shows that $B^{\mathfrak{N}}$ cannot be normal in G.

Now we see that $\mathrm{O}^p(A) = A$ and $\mathrm{O}^p(B)$ are not mutually permutable. Suppose the contrary, in particular, that $\mathrm{O}^p(B)$ permutes with the Sylow 3-subgroup $\langle x\rangle$ of A. The subgroup $\langle x\rangle \mathrm{O}^p(B)$ has then a normal Sylow p-subgroup, as G, while $B^{\mathfrak{N}}$ is the unique Sylow p-subgroup of $\mathrm{O}^p(B)$. Since $|\langle x\rangle\mathrm{O}^p(B) : \mathrm{O}^p(B)| = 3 \neq p$, $B^{\mathfrak{N}}$ is also the unique Sylow p-subgroup of $\langle x\rangle\mathrm{O}^p(B)$. It follows that $B^{\mathfrak{N}}$ is a normal subgroup of $\langle x\rangle\mathrm{O}^p(B)$ and so $B^{\mathfrak{N}}$ is a normal subgroup of $\langle x\rangle B = AB = G$, contradicting the previous observation. It follows that A and $\mathrm{O}^p(B)$ are not mutually permutable.

The class $\mathfrak{F} = \mathfrak{S}_p$ of all p-groups is a subgroup-closed saturated Fitting formation such that if $G = AB$ is the product of the mutually permutable subgroups A and B; $A^{\mathfrak{F}} = \mathrm{O}^p(A)$ and $B^{\mathfrak{F}} = \mathrm{O}^p(B)$ permute by Theorem 4.4.10, but they are not mutually permutable.

Theorem 4.4.2 can be generalised to finitely many pairwise mutually permutable products (see [74]).

Corollary 4.4.13. *Let $G_1, G_2, \dots, G_r$ be pairwise mutually permutable subgroups of a group G and π and ρ two sets of prime numbers such that p does not divide $q-1$ for every $p \in \pi$ and every $q \in \pi \cup \rho$. Then $\mathrm{O}^{\pi'}(G_i)$ is subnormal in $G_i(\prod_{j=1}^r \mathrm{O}^{\rho'}(G_j))$ for every $i \in \{1, 2, \dots, r\}$. If, furthermore, $\pi \cap \rho = \emptyset$, then $\mathrm{O}^{\pi'}(G_i)$ is a normal subgroup of $G_i(\prod_{j=1}^r \mathrm{O}^{\rho'}(G_j))$.*

Proof. Let $i \in \{1, 2, \dots, r\}$. By Theorem 4.4.10, $\mathrm{O}^{\rho'}(G_j)\mathrm{O}^{\rho'}(G_k)$ is a subgroup of G for all $j, k \in \{1, 2, \dots, r\}$, and by the mutual permutability of G_i with G_j and G_k we obtain that $G_i\mathrm{O}^{\rho'}(G_j)$ permutes with $G_i\mathrm{O}^{\rho'}(G_k)$. By the choice of the sets of primes π and ρ, we obtain that $\mathrm{O}^{\pi'}(G_i)$ is subnormal in $G_i\mathrm{O}^{\rho'}(G_j)$ by Theorem 4.4.2. By Corollary 1.1.8, we conclude that $\mathrm{O}^{\pi'}(G_i)$ is subnormal in $\prod_{j=1}^r G_i\mathrm{O}^{\rho'}(G_j) = G_i(\prod_{j=1}^r \mathrm{O}^{\rho'}(G_j))$.

In the case $\pi \cap \rho = \emptyset$, by Theorem 4.4.2 it follows that $\mathrm{O}^{\pi'}(G_i)$ is a normal subgroup of $G_i\mathrm{O}^{\rho'}(G_j)$ and thus actually $\mathrm{O}^{\pi'}(G_i)$ is a normal subgroup of $G_i(\prod_{j=1}^r \mathrm{O}^{\rho'}(G_j))$. □

We now touch on the question of permutability of residuals of mutually permutable products associated to saturated formations. We already know that, by Corollary 4.4.7, the residuals of pairwise mutually permutable products associated to saturated formations of full characteristic permute. This is due to the fact that these formations always contain the class of all nilpotent groups. We shall prove that full characteristic is unnecessary to guarantee permutability. The key point is the following result proved by Bochtler [74].

Theorem 4.4.14. *Let the group $G = AB$ be the product of the mutually permutable subgroups A and B and let $\mathfrak{F}$ be a formation of characteristic π contained in $\mathfrak{E}_\pi$, the class of all π-groups, and containing $\mathfrak{N}_\pi$, the class of all nilpotent π-groups for a set of primes π. Then $A^{\mathfrak{F}}$ and $B^{\mathfrak{F}}$ permute.*

Proof. Since $A/A^{\mathfrak{F}}$ and $B/B^{\mathfrak{F}}$ are π-groups, we have that $\mathrm{O}^{\pi}(A^{\mathfrak{F}}) = \mathrm{O}^{\pi}(A)$ and $\mathrm{O}^{\pi}(B^{\mathfrak{F}}) = \mathrm{O}^{\pi}(B)$ by Lemma 1.1.11. By Theorem 4.4.10, $\mathrm{O}^{\pi}(A)$ and $\mathrm{O}^{\pi}(B)$ permute, that is,

$$\mathrm{O}^{\pi}(A^{\mathfrak{F}})\mathrm{O}^{\pi}(B^{\mathfrak{F}}) = \mathrm{O}^{\pi}(B^{\mathfrak{F}})\mathrm{O}^{\pi}(A^{\mathfrak{F}}). \tag{4.3}$$

Since $\mathfrak{N}_\pi \subseteq \mathfrak{F}$, we obtain that $\mathrm{O}^{\pi'}(A^{\mathfrak{F}}) \leq \mathrm{O}^{\pi'}(A^{\mathfrak{N}_\pi}) \leq A^{\mathfrak{N}}$. By Theorem 4.4.5, $\mathrm{O}^{\pi'}(A^{\mathfrak{F}})$ normalises B and thus the characteristic subgroups $\mathrm{O}^{\pi}(B^{\mathfrak{F}})$ and $\mathrm{O}^{\pi'}(B^{\mathfrak{F}})$. A similar argument follows by exchanging the roles of A and B. This means that

$$\begin{aligned} \mathrm{O}^{\pi'}(A^{\mathfrak{F}}) &\leq \mathrm{N}_G(\mathrm{O}^{\pi}(B^{\mathfrak{F}})) \cap \mathrm{N}_G(\mathrm{O}^{\pi'}(B^{\mathfrak{F}})), \\ \mathrm{O}^{\pi'}(B^{\mathfrak{F}}) &\leq \mathrm{N}_G(\mathrm{O}^{\pi}(A^{\mathfrak{F}})) \cap \mathrm{N}_G(\mathrm{O}^{\pi'}(A^{\mathfrak{F}})). \end{aligned} \tag{4.4}$$

Now from (4.3) and (4.4) we get

$$\begin{aligned} A^{\mathfrak{F}}B^{\mathfrak{F}} &= \mathrm{O}^{\pi'}(A^{\mathfrak{F}})\mathrm{O}^{\pi}(A^{\mathfrak{F}})\mathrm{O}^{\pi}(B^{\mathfrak{F}})\mathrm{O}^{\pi'}(B^{\mathfrak{F}}) \\ &= \mathrm{O}^{\pi'}(A^{\mathfrak{F}})\mathrm{O}^{\pi}(B^{\mathfrak{F}})\mathrm{O}^{\pi}(A^{\mathfrak{F}})\mathrm{O}^{\pi'}(B^{\mathfrak{F}}) \\ &= \mathrm{O}^{\pi'}(B^{\mathfrak{F}})\mathrm{O}^{\pi}(B^{\mathfrak{F}})\mathrm{O}^{\pi}(A^{\mathfrak{F}})\mathrm{O}^{\pi'}(A^{\mathfrak{F}}) \\ &= B^{\mathfrak{F}}A^{\mathfrak{F}}. \end{aligned}$$

This completes the proof of the theorem. □

Corollary 4.4.15. *Let the group $G = AB$ be the product of the mutually permutable subgroups A and B and let $\mathfrak{F}$ be a saturated formation. Then $A^{\mathfrak{F}}B^{\mathfrak{F}}$ is a subgroup of G.*

Proof. If $\mathfrak{F}$ is a saturated formation with characteristic π, then the hypothesis $\mathfrak{N}_\pi \subseteq \mathfrak{F} \subseteq \mathfrak{E}_\pi$ of Theorem 4.4.14 is satisfied (see [119, IV, 4.3]). □

Corollary 4.4.16. *Let the group $G = G_1G_2\cdots G_r$ be the product of the pairwise mutually permutable subgroups $G_1, G_2, \ldots, G_r$ and let $\mathfrak{F}$ be a saturated formation. Then $\prod_{j\in\{1,\ldots,r\}} G_j^{\mathfrak{F}}$ is a subgroup of G which is a pairwise permutable product of the subgroups $G_j^{\mathfrak{F}}$, $j \in \{1, \ldots, r\}$.*

To show that groups which are products of finitely many pairwise mutually permutable have prefactorised residuals associated to saturated formations of soluble groups is the purpose of the following result.

Theorem 4.4.17 (Ballester-Bolinches et al. [34]). *Let the group $G = G_1 G_2 \cdots G_r$ be the product of the pairwise mutually permutable subgroups $G_1, G_2, \ldots, G_r$. If $\mathfrak{F}$ is a saturated formation of soluble groups, then the $\mathfrak{F}$-residual $G^{\mathfrak{F}}$ of G is prefactorised.*

Proof. Denote by $N = G^{\mathfrak{F}}$. Assume that the result is not true and let G be a counterexample with least order. By Remark 4.1.47, we may assume that N is a minimal normal subgroup of G and $N \cap G_i = 1$ for all $i \in \{1, 2, \ldots, r\}$. Since G/N is soluble, we have that $G_i \cong G_i N/N$ is soluble for all $i \in \{1, 2, \ldots, r\}$ and so G is soluble by Theorem 4.1.15. Then N is an elementary abelian p-group for some prime p. Let F denote the canonical local definition of $\mathfrak{F}$, that is, the unique determined function defining $\mathfrak{F}$ which is integrated and full, that is, $F(q) \subseteq \mathfrak{F}$ and $F(q) = \mathfrak{S}_q F(q)$ for all primes q (see [119, IV, 3.9]). Clearly $G^{\mathfrak{F}}$ is contained in $L = G^{F(p)}$, the $F(p)$-residual of G, and L/N is p-nilpotent by [119, IV, 3.2 (b)]. Denote by M/N the normal Hall p'-subgroup of L/N. By Proposition 4.1.45 (2), there exists a prefactorised Hall p'-subgroup H of G. Then $H = (H \cap G_1)(H \cap G_2) \cdots (H \cap G_r)$ is a product of pairwise mutually permutable subgroups, $G_i H$ is a subgroup of G, and the Sylow p-subgroups of G_i are Sylow p-subgroups of $G_i H$. Now $G_i H \cap N$ is a normal p-subgroup of $G_i H$ and so it is contained in every Sylow p-subgroup of $G_i H$. In particular, it is contained in every Sylow p-subgroup of G_i. Consequently $HG_i \cap N = 1$. Now $M \cap H$ is a Hall p'-subgroup of M. Since M/N is a p'-group, it follows that $M = (M \cap H)N$. This implies that $M \cap H$ is normalised by all G_i and therefore also by G. Therefore $M = (M \cap H) \times N$ and $M \leq C_G(N)$. Since N is a minimal normal p-subgroup of G, we have that $G/C_G(N)$ does not possess non-trivial normal p-subgroups by [119, A, 13.6 (b)]. Consequently $LC_G(N)/C_G(N) = 1$, that is, $L \leq C_G(N)$. We have that G/N is an $\mathfrak{F}$-group and $G/C_G(N) \in F(p)$. Therefore applying [119, IV, 3.2], $G \in \mathfrak{F}$ and $N = 1$. This is a contradiction. □

Unfortunately, $\mathfrak{F}$-residuals, even for subgroup-closed saturated formations, are not necessarily factorised subgroups in the group, as the following example shows.

Example 4.4.18. Consider the group G of Example 4.1.46. Then it is clear that the nilpotent residual of G is $G^{\mathfrak{N}} = \langle x \rangle$, which is clearly prefactorised in G. However, recall that $A \cap B = \langle b \rangle$, which is not contained in $G^{\mathfrak{N}}$.

We now turn our attention to the behaviour of radicals associated to Fitting classes in mutually permutable products. According to Theorem 4.4.10 and Corollary 4.4.15, if $\mathfrak{F}$ is either a Fitting formation or a saturated formation, the $\mathfrak{F}$-residuals $A^{\mathfrak{F}}$ and $B^{\mathfrak{F}}$ of the mutually permutable subgroups A and B permute, but they are not in general mutually permutable (Example 4.4.12). However, if the formation $\mathfrak{F}$ contains all the nilpotent groups, then $A^{\mathfrak{F}} B^{\mathfrak{F}}$ is a normal product by Corollary 4.4.7. If $\mathfrak{F}$ is a Fitting class, the $\mathfrak{F}$-radicals of mutually permutable subgroups are mutually permutable, as Bochtler proved in [73, 74].

Theorem 4.4.19. *Let the group G be the product of the mutually permutable subgroups A and B and let $\mathfrak{F}$ be a Fitting class. Then $A_{\mathfrak{F}}$ and $B_{\mathfrak{F}}$ are mutually permutable subgroups of G. In particular, $A_{\mathfrak{F}}B_{\mathfrak{F}}$ is a subgroup of G.*

Proof. We show first that $A_{\mathfrak{F}}B_{\mathfrak{F}}$ is a subgroup of G. To this end, we assume that there exist mutually permutable products $G = AB$ in which $A_{\mathfrak{F}}B_{\mathfrak{F}}$ is not a subgroup of G and choose a counterexample with least $|G| + |A| + |B|$. Then the following statements hold:

1. $A = A_{\mathfrak{F}}(A \cap B)$ *and* $B = B_{\mathfrak{F}}(A \cap B)$.

The subgroups $A_{\mathfrak{F}}(A \cap B) = A \cap A_{\mathfrak{F}}B$ and B are mutually permutable by virtue of Lemma 4.1.21 (3). Assume that $A_{\mathfrak{F}}(A \cap B)$ is a proper subgroup of A. In this case $(A_{\mathfrak{F}}(A \cap B))_{\mathfrak{F}}B_{\mathfrak{F}}$ is a subgroup of G by the choice of G. Since $A_{\mathfrak{F}}(A \cap B)$ is subnormal in A by Corollary 4.1.17, we have

$$(A_{\mathfrak{F}}(A \cap B))_{\mathfrak{F}} = A_{\mathfrak{F}}(A \cap B) \cap A_{\mathfrak{F}} = A_{\mathfrak{F}}.$$

Consequently

$$A_{\mathfrak{F}}B_{\mathfrak{F}} = (A_{\mathfrak{F}}(A \cap B))_{\mathfrak{F}}B_{\mathfrak{F}}$$

is a subgroup of G, which contradicts the choice of G. Therefore $A = A_{\mathfrak{F}}(A \cap B)$ and $B = B_{\mathfrak{F}}(A \cap B)$ by symmetry.

2. $G = \langle A_{\mathfrak{F}}, B_{\mathfrak{F}} \rangle$.

Let J denote $\langle A_{\mathfrak{F}}, B_{\mathfrak{F}} \rangle$. By Lemma 4.1.21 (1), the subgroups $J \cap A$ and $J \cap B$ are mutually permutable. Hence if J is a proper subgroup of G, then $(J \cap A)_{\mathfrak{F}}(J \cap B)_{\mathfrak{F}}$ is a subgroup of G by the choice of G. Since $\langle J, A \cap B \rangle = G$ by Statement 1, and $A \cap B$ normalises both $A_{\mathfrak{F}}$ and $B_{\mathfrak{F}}$, we have that J is normal in $G = \langle J, A \cap B \rangle$. Then $(J \cap A)_{\mathfrak{F}} = J \cap A_{\mathfrak{F}} = A_{\mathfrak{F}}$ and $(J \cap B)_{\mathfrak{F}} = B_{\mathfrak{F}}$. This contradiction yields $J = G = \langle A_{\mathfrak{F}}, B_{\mathfrak{F}} \rangle$.

Applying Theorem 4.4.9, there exist subgroups L and M such that $A' \leq L \leq A$ and $B' \leq M \leq B$, $A \cap B = L \cap M$, L and M subnormal in G, and $G' \leq LB \cap AM = LM$.

3. $L_{\mathfrak{F}}M_{\mathfrak{F}} = (LM)_{\mathfrak{F}}$ *is a normal subgroup of G.*

By Statement 1,

$$L = L \cap A_{\mathfrak{F}}(A \cap B) = (L \cap A_{\mathfrak{F}})(A \cap B) = L_{\mathfrak{F}}(A \cap B),$$

where the last equality holds since L is a normal subgroup of A. Bearing in mind that L and M are subnormal subgroups of G, we deduce that

$$\begin{aligned} L_{\mathfrak{F}}M_{\mathfrak{F}} &= L_{\mathfrak{F}}(M \cap (LM)_{\mathfrak{F}}) \\ &= L_{\mathfrak{F}}(A \cap B)M \cap (LM)_{\mathfrak{F}} = LM \cap (LM)_{\mathfrak{F}} = (LM)_{\mathfrak{F}}, \end{aligned}$$

which is characteristic in LM and LM is normal in G. Thus $L_{\mathfrak{F}}M_{\mathfrak{F}} = (LM)_{\mathfrak{F}}$ is a normal subgroup of G.

4. *$G/L_{\mathfrak{F}}M_{\mathfrak{F}}$ is nilpotent.*

Note first that $A_{\mathfrak{F}}M_{\mathfrak{F}} = A_{\mathfrak{F}}L_{\mathfrak{F}}M_{\mathfrak{F}}$ is a subgroup of G because $L_{\mathfrak{F}}M_{\mathfrak{F}}$ is a normal subgroup of G. In addition, $M = M_{\mathfrak{F}}(A \cap B)$ as in Statement 3 and we obtain that $A_{\mathfrak{F}}M_{\mathfrak{F}}$ is a normal subgroup of $A_{\mathfrak{F}}(A \cap B)M_{\mathfrak{F}}$, which coincides with AM by Statement 1. Hence

$$\begin{aligned} [LM, A_{\mathfrak{F}}M_{\mathfrak{F}}] &\le LM \cap A_{\mathfrak{F}}M_{\mathfrak{F}} = (LM \cap A_{\mathfrak{F}})M_{\mathfrak{F}} \\ &= (LM \cap A \cap A_{\mathfrak{F}})M_{\mathfrak{F}} = (L(M \cap A) \cap A_{\mathfrak{F}})M_{\mathfrak{F}} \\ &= (L(A \cap B) \cap A_{\mathfrak{F}})M_{\mathfrak{F}} = (L \cap A_{\mathfrak{F}})M_{\mathfrak{F}} = L_{\mathfrak{F}}M_{\mathfrak{F}}. \end{aligned}$$

Consequently, $A_{\mathfrak{F}}$ centralises $LM/L_{\mathfrak{F}}M_{\mathfrak{F}}$, and the same is true for $B_{\mathfrak{F}}$ by symmetry. Since $G = \langle A_{\mathfrak{F}}, B_{\mathfrak{F}} \rangle$ by Statement 2, we have that $[G, LM] \le L_{\mathfrak{F}}M_{\mathfrak{F}}$ and, of course, $G' \le LM$. This implies that $G/L_{\mathfrak{F}}M_{\mathfrak{F}}$ is nilpotent.

5. *$G/L_{\mathfrak{F}}M_{\mathfrak{F}}$ is a p-group for a prime p.*

For every prime q, we set $G(q) = G_qL_{\mathfrak{F}}M_{\mathfrak{F}} \le G$, where G_q denotes a Sylow q-subgroup of G. Note that the definition of $G(q)$ does not depend on the special choice of the Sylow q-subgroup G_q, because $G/L_{\mathfrak{F}}M_{\mathfrak{F}}$ is nilpotent. Suppose that $G(q)$ is a proper subgroup of G for every prime q. Then by Lemma 4.1.21 (1), the product $(G(q) \cap A)(G(q) \cap B)$ is mutually permutable. It follows that $(G(q) \cap A)_{\mathfrak{F}}(G(q) \cap B)_{\mathfrak{F}}$ is a subgroup of $G(q)$ by the choice of G. Since $G/L_{\mathfrak{F}}M_{\mathfrak{F}}$ is nilpotent, we have that $G(q)$ is a normal subgroup of G. Thus $(G(q) \cap A)_{\mathfrak{F}} = G(q) \cap A_{\mathfrak{F}}$ and $(G(q) \cap B)_{\mathfrak{F}} = G(q) \cap B_{\mathfrak{F}}$. Consequently $(G(q) \cap A_{\mathfrak{F}})(G(q) \cap B_{\mathfrak{F}})$ is a subgroup of G for every prime q.

We note that $(G(q) \cap A_{\mathfrak{F}})(G(q) \cap B_{\mathfrak{F}})$ permutes with $(G(p) \cap A_{\mathfrak{F}})(G(p) \cap B_{\mathfrak{F}})$ for all primes $q \neq p$ because the quotients of these products modulo $L_{\mathfrak{F}}M_{\mathfrak{F}}$ are a q-group and a p-group, respectively, so they centralise each other due to the nilpotency of $G/L_{\mathfrak{F}}M_{\mathfrak{F}}$. Now we prove that $G(q) \cap A_{\mathfrak{F}}$ permutes with $G(p) \cap B_{\mathfrak{F}}$ for all primes $q \neq p$. Clearly $(G(q) \cap A_{\mathfrak{F}})M_{\mathfrak{F}}$ permutes with $L_{\mathfrak{F}}(G(p) \cap B_{\mathfrak{F}})$. Therefore

$$\begin{aligned} (G(q) \cap A_{\mathfrak{F}})M_{\mathfrak{F}}L_{\mathfrak{F}}(G(p) \cap B_{\mathfrak{F}}) &= (G(q) \cap A_{\mathfrak{F}})L_{\mathfrak{F}}M_{\mathfrak{F}}(G(p) \cap B_{\mathfrak{F}}) \\ &= (G(q) \cap A_{\mathfrak{F}})(G(p) \cap B_{\mathfrak{F}}) \end{aligned}$$

is a subgroup of G. Moreover, from the definition and the normality of $G(q)$, we obtain that

$$A_{\mathfrak{F}} = \prod_{q \in \mathbb{P}} (G(q) \cap A_{\mathfrak{F}}) \quad \text{and} \quad B_{\mathfrak{F}} = \prod_{q \in \mathbb{P}} (G(q) \cap B_{\mathfrak{F}}).$$

We conclude that

$$\prod_{q \in \mathbb{P}} (G(q) \cap A_{\mathfrak{F}})(G(q) \cap B_{\mathfrak{F}}) = \prod_{q \in \mathbb{P}} (G(q) \cap A_{\mathfrak{F}}) \prod_{q \in \mathbb{P}} (G(q) \cap B_{\mathfrak{F}}) = A_{\mathfrak{F}} B_{\mathfrak{F}}$$

is a subgroup of G. This contradicts the choice of G. Therefore there exists a prime p such that $G(p) = G = G_p L_{\mathfrak{F}} M_{\mathfrak{F}}$ and $G/L_{\mathfrak{F}} M_{\mathfrak{F}}$ is a p-group.

6. $A = \mathrm{O}^{p'}(A)(A \cap B)$ *and* $B = \mathrm{O}^{p'}(B)(A \cap B)$.

We know that $A/(L_{\mathfrak{F}} M_{\mathfrak{F}} \cap A)$ is a p-group by Statement 5. Taking into account that $A \cap B$ is subnormal in G by Corollary 4.1.17, we obtain

$$L_{\mathfrak{F}} M_{\mathfrak{F}} \cap A = L_{\mathfrak{F}}(M_{\mathfrak{F}} \cap A) = L_{\mathfrak{F}}(M_{\mathfrak{F}} \cap A \cap B) = L_{\mathfrak{F}}(A \cap B)_{\mathfrak{F}} = L_{\mathfrak{F}},$$

so we have $A = \mathrm{O}^{p'}(A) L_{\mathfrak{F}}$ and $A_{\mathfrak{F}} = \mathrm{O}^{p'}(A_{\mathfrak{F}}) L_{\mathfrak{F}}$. Now $\mathrm{O}^{p'}(A_{\mathfrak{F}})$ is a normal $\mathfrak{F}$-subgroup of $\mathrm{O}^{p'}(A)(A \cap B)$, and so $\mathrm{O}^{p'}(A_{\mathfrak{F}}) \leq (\mathrm{O}^{p'}(A)(A \cap B))_{\mathfrak{F}}$. Since $A \cap B$ is subnormal in G, we obtain that $(\mathrm{O}^{p'}(A)(A \cap B))_{\mathfrak{F}}$ is a subnormal $\mathfrak{F}$-subgroup of A and hence it is a subgroup of $A_{\mathfrak{F}}$. Thus $A_{\mathfrak{F}} = (\mathrm{O}^{p'}(A)(A \cap B))_{\mathfrak{F}} L_{\mathfrak{F}}$. Now we suppose that $\mathrm{O}^{p'}(A)(A \cap B)$ is a proper subgroup of A. The subgroups $\mathrm{O}^{p'}(A)(A \cap B)$ and B are mutually permutable by Lemma 4.1.21 (3), so $(\mathrm{O}^{p'}(A)(A \cap B))_{\mathfrak{F}} B_{\mathfrak{F}}$ is a subgroup of G by the choice of G. Since $L_{\mathfrak{F}} M_{\mathfrak{F}}$ is a normal subgroup of G, we have that

$$A_{\mathfrak{F}} B_{\mathfrak{F}} = (\mathrm{O}^{p'}(A)(A \cap B))_{\mathfrak{F}} L_{\mathfrak{F}} B_{\mathfrak{F}} = (\mathrm{O}^{p'}(A)(A \cap B))_{\mathfrak{F}} (L_{\mathfrak{F}} M_{\mathfrak{F}}) B_{\mathfrak{F}}$$

is a subgroup of G. This contradicts the choice of (G, A, B). Therefore

$$A = \mathrm{O}^{p'}(A)(A \cap B).$$

By symmetry,

$$B = \mathrm{O}^{p'}(B)(A \cap B).$$

7. *The counterexample does not exist.*

By Theorem 4.4.2, we have that $\mathrm{O}^{p'}(A)$ is subnormal in $A\mathrm{O}^{p'}(B)$. By Statement 6,

$$A\mathrm{O}^{p'}(B) = A(A \cap B)\mathrm{O}^{p'}(B) = AB = G.$$

Since, by Corollary 4.1.17, $A \cap B$ is subnormal in G, it follows that A is subnormal in G by [119, A, 14.4]. Hence $A_{\mathfrak{F}}$ is subnormal in G. Analogously, B and $B_{\mathfrak{F}}$ are subnormal in G. Hence $G \in \mathfrak{F}$ since, by Statement 2, $G = \langle A_{\mathfrak{F}}, B_{\mathfrak{F}} \rangle$ is generated by two subnormal $\mathfrak{F}$-subgroups. Since A and B are subnormal subgroups of G, we obtain that A and B belong to $\mathfrak{F}$, which yields the final contradiction $A_{\mathfrak{F}} B_{\mathfrak{F}} = AB = G$.

We conclude that $A_{\mathfrak{F}} B_{\mathfrak{F}}$ is a subgroup of G. Now we will show that $A_{\mathfrak{F}}$ and $B_{\mathfrak{F}}$ are mutually permutable. Since $A_{\mathfrak{F}} B_{\mathfrak{F}}$ is a subgroup of G, we can apply Lemma 4.1.21 (1) to deduce that $A_{\mathfrak{F}} B_{\mathfrak{F}} \cap A$ and $A_{\mathfrak{F}} B_{\mathfrak{F}} \cap B$ are mutually permutable. By the subnormality of $A \cap B$ in G, we have that

$$A_{\mathfrak{F}} B_{\mathfrak{F}} \cap A = A_{\mathfrak{F}}(B_{\mathfrak{F}} \cap A) = A_{\mathfrak{F}}(B_{\mathfrak{F}} \cap A \cap B) = A_{\mathfrak{F}}(A \cap B)_{\mathfrak{F}} = A_{\mathfrak{F}}$$

and $A_{\mathfrak{F}} B_{\mathfrak{F}} \cap B = B_{\mathfrak{F}}$. This completes the proof of the theorem. □

Corollary 4.4.20. *Let the group $G = G_1 G_2 \cdots G_r$ be the product of the pairwise mutually permutable subgroups $G_1, G_2, \ldots, G_r$ and let $\mathfrak{F}$ be a Fitting class. Then $(G_i)_{\mathfrak{F}}$ and $(G_j)_{\mathfrak{F}}$ are mutually permutable subgroups of G, for all i, j. In particular, $\prod_{j \in \{1,\ldots,r\}} (G_j)_{\mathfrak{F}}$ is the product of the pairwise mutually permutable subgroups $(G_j)_{\mathfrak{F}}$, $j \in \{1, 2, \ldots, r\}$.*

According to Corollary 4.4.7, if $\mathfrak{F}$ is a formation containing all nilpotent groups, the product of the $\mathfrak{F}$-residuals of the factors of a mutually permutable product is subnormal. To show that the corresponding result for Fitting classes is not true is the purpose of the next example, with which we close this section.

Example 4.4.21. Let S be the symmetric group of degree 3. Clearly, S is the mutually permutable product of its Sylow 3-subgroup A_1 and a Sylow 2-subgroup B_1. By [119, B, 11.7], S has an irreducible and faithful module V over the field of 5 elements. Let $G = [V]S$ be the corresponding semidirect product. Applying Lemma 4.1.12, G is the mutually permutable product of $A = VA_1$ and $B = VB_1$. Let $\mathfrak{F}$ be the Fitting class of all groups X such that $\mathrm{O}^2(X)$ is nilpotent. It is clear that $\mathfrak{F}$ contains all nilpotent groups. We have that $A_{\mathfrak{F}} = V$ and $B_{\mathfrak{F}} = B$. However, $B = A_{\mathfrak{F}} B_{\mathfrak{F}}$ is not subnormal in G.

4.5 Mutually permutable products: the core-free intersection case

It is abundantly clear that the intersection of the factors plays a central part in the study of mutually permutable products. The case when the intersection of two mutually permutable subgroups is core-free in the group is of special interest and may repay further study. It is our impression that, in the present state of knowledge, core-free intersection is the property which marks a sharp dividing line between amenable and intractable behaviour of mutually permutable products.

Our first goal in this section is to describe a configuration, due to Bochtler [74], often encountered in the study of mutually permutable products with core-free intersection. It considers the effect on the structure of a mutually permutable product when the core-free property is not inherited by a given quotient of the group. This result will find many applications in the rest of the section, where the behaviour of mutually permutable products with core-free intersection with respect to some families of formations is analysed.

Theorem 4.5.1. *Let the group $G = AB$ be the mutually permutable product of the subgroups A and B such that $(A \cap B)_G = 1$. Let N be a minimal normal subgroup of G and let $T = (AN \cap BN)_G$. If $T \neq N$, then*

1. *T is an elementary abelian p-group for a prime p.*

If, moreover, N is contained in B and $C = \mathrm{C}_G(T)$, then we have:

2. *BC/C is a p-group,*
3. *$T/(A \cap T)$ ($\cong N$) has order p,*
4. *the derived subgroup of AC/C is a p-group,*
5. *all primes dividing $|AC/C|$ different from p divide $p-1$,*
6. *G/C has a normal p-subgroup.*

Proof. 1. We begin by proving that there exists a prime p such that N is an elementary abelian p-group. By Corollary 4.3.4, A and B either cover or avoid each factor of G. If A and B cover N, then N is contained in $(A \cap B)_G = 1$, which is impossible. If both A and B avoid N, then N has prime order by Lemma 4.3.9. Hence we are left with the case that N is covered by one of the factors A or B and avoided by the other one. Without loss of generality we can assume that $A \cap N = 1$ and $N \leq B$. Since $T \leq AN \cap B \leq B$, it follows that $A \cap T = A \cap B \cap T$ and so $A \cap T$ is subnormal in T by Proposition 4.1.16 (5). In addition, $A \cap B$ is nilpotent by Lemma 4.1.18. Hence $A \cap T$ is nilpotent. This gives that $A \cap T$ is contained in the Fitting subgroup $\mathrm{F}(T)$ of T. Since $\mathrm{F}(T)$ is a characteristic subgroup of the normal subgroup T of G, it follows that either $N \cap \mathrm{F}(T) = 1$ or $N \leq \mathrm{F}(T)$. Assume that $N \cap \mathrm{F}(T) = 1$. Then from $T = AN \cap T = (A \cap T)N$ and $A \cap T \leq \mathrm{F}(T)$ we obtain the equality

$$\mathrm{F}(T) = T \cap \mathrm{F}(T) = (A \cap T)N \cap \mathrm{F}(T) = (A \cap T)(N \cap \mathrm{F}(T)) = A \cap T.$$

Consequently $\mathrm{F}(T) = A \cap T$ is a normal subgroup of G contained in $A \cap B$. Hence $A \cap T = 1$ and

$$T = (A \cap T)N = N.$$

This contradicts the assumption $T \neq N$. Therefore $N \leq \mathrm{F}(T)$ and so N is nilpotent. Since N is a minimal normal subgroup of G, N is an elementary abelian group for a prime p.

Now we prove that T is an elementary abelian p-group. First of all, $T = AN \cap T = (A \cap T)N$ and, by symmetry, $T = (B \cap T)N$. Then $|T : A \cap T|$ and $|T : B \cap T|$ are powers of p. Further $(A \cap T)(B \cap T)$ is a subgroup of G by Lemma 4.1.21 (1). From there we obtain that $|A \cap T : A \cap B \cap T| = |(A \cap T)(B \cap T) : B \cap T|$, which is a divisor of $|T : B \cap T|$, is likewise a power of p. This means, according to Lemma 1.1.11, that $\mathrm{O}^p(A \cap B \cap T) = \mathrm{O}^p(T)$ is a characteristic subgroup of the normal subgroup T of G so that $\mathrm{O}^p(T)$ is normal in G which is contained in $A \cap B$. Hence $\mathrm{O}^p(T)$ is trivial and so T is a p-group. In particular, $N \leq \mathrm{Z}(T)$ by [119, A, 13.6 (b)]. From this, $A \cap T$ is a normal subgroup of $(A \cap T)N = T$ and $B \cap T$ is a normal subgroup of $(B \cap T)N = T$. Hence $T/(A \cap T)$ and $T/(B \cap T)$ are isomorphic to quotients of N, and they are elementary abelian. Therefore $T'T^p \leq A \cap B \cap T$ and $T'T^p$ is a characteristic subgroup of the normal subgroup T of G. It follows that $T'T^p = 1$ and T is actually an elementary abelian p-group.

For the rest of the proof, we also assume that $N \leq B$. Then $T = (AN \cap B)_G$ is obviously contained in B. Moreover, we have that $A \cap N = 1$, since otherwise we would have $N \leq A \cap B$ and that would contradict the hypothesis $(A \cap B)_G = 1$.

2. *BC/C is a p-group (for $C = \mathrm{C}_G(T)$).*

From the mutual permutability of A and B, it follows that $A \cap B$ is a permutable subgroup of B by Corollary 4.1.17. By Corollary 1.5.6, $(A \cap B)/(A \cap B)_B$ is contained in the hypercentre $\mathrm{Z}_\infty(B/(A \cap B)_B)$ of $B/(A \cap B)_B$. In particular, every p-subgroup of $A \cap B$ is centralised by all p'-elements of B (and hence by the subgroup $\mathrm{O}^p(B)$ they generate) modulo $(A \cap B)_B$ by Lemma 1.2.22. Since $A \cap T$ is a p-group by Statement 1 and is contained in $A \cap B$, we also have

$$[A \cap T, \mathrm{O}^p(B)] \leq (A \cap B)_B \cap T = A_B \cap B \cap T = A_{AB} \cap T = A_G \cap T.$$

The subgroup $A_G \cap T$ is a normal subgroup of G which is contained in $A \cap B$ and is therefore trivial. It follows that $A \cap T$ is centralised by $\mathrm{O}^p(B)$. Consequently $\langle (A \cap T)^B \rangle = \langle (A \cap T)^{AB} \rangle = \langle (A \cap T)^G \rangle$ is also centralised by $\mathrm{O}^p(B)$. Moreover either $N \cap \langle (A \cap T)^G \rangle = 1$ or $N \leq \langle (A \cap T)^G \rangle$. Assume that $N \cap \langle (A \cap T)^G \rangle = 1$. Then since $T = AN \cap T = (A \cap T)N$, we would obtain the equality

$$\begin{aligned}\langle (A \cap T)^G \rangle &= T \cap \langle (A \cap T)^G \rangle = (A \cap T)N \cap \langle (A \cap T)^G \rangle \\ &= (A \cap T)(N \cap \langle (A \cap T)^G \rangle) = A \cap T.\end{aligned}$$

Hence $A \cap T$ would be a normal subgroup of G contained in $A \cap B$ and so $A \cap T$ would be trivial. Thus $T = (A \cap T)N = N$, contradicting the assumption $T \neq N$. Therefore $N \leq \langle (A \cap T)^G \rangle$, and since $T = (A \cap T)N \leq \langle (A \cap T)^G \rangle \leq T$, it follows that

$$\mathrm{O}^p(B) \leq \mathrm{C}_G(\langle (A \cap T)^G \rangle) = \mathrm{C}_G(T) = C.$$

Therefore BC/C is a p-group.

3. *$T/(A \cap T)$, which is isomorphic to N, has order p.*

It is clear that $A \cap T$ is normal in T since T is abelian by Statement 1. The equalities $T = (A \cap T)N$ and $A \cap N = 1$ yield

$$T/(A \cap T) = (A \cap T)N/(A \cap T) \cong N/(A \cap T \cap N) \cong N.$$

Assume that N is not cyclic. Then $A \leq \mathrm{C}_G(N)$ by Lemma 4.3.3 (5). Moreover, by Statement 2, $\mathrm{O}^p(B) \leq \mathrm{C}_G(T) \leq \mathrm{C}_G(N)$. Therefore $G/\mathrm{C}_G(N)$ is a p-group. It follows that the minimal normal p-subgroup N is central in G and, in particular, cyclic. This contradiction yields $|N| = p$.

4. *The derived subgroup $(AC/C)'$ of AC/C is a p-group.*

The group G/C is the product of the mutually permutable subgroups AC/C and BC/C. Therefore the product $(AC/C)'(BC/C)$ is a subgroup of G/C and the subgroup $(AC/C)'$ is a subnormal subgroup of G/C by Corollary 4.1.25, whose index in $(AC/C)'(BC/C)$ is a power of p by virtue of Statement 2. By Lemma 1.1.11, $\mathrm{O}^p((AC/C)') = \mathrm{O}^p((AC/C)'(BC/C))$ is a normal subgroup of $(AC/C)'(BC/C)$. This implies that $\mathrm{O}^p((AC/C)') = \mathrm{O}^p(A')C/C$ is normalised by BC/C and, obviously, by AC/C. Consequently $\mathrm{O}^p(A')C$ is a normal subgroup of $G = AB$. Hence $[\mathrm{O}^p(A'), T] = [\mathrm{O}^p(A')C, T]$ is a normal subgroup of G. By Statement 3, $T/(A \cap T)$ is cyclic and so the automorphism group induced by A on $T/(A \cap T)$ by conjugation is abelian. Thus A' and, in particular, $\mathrm{O}^p(A')$ act trivially on $T/(A \cap T)$. This gives $[\mathrm{O}^p(A'), T] \leq A \cap T$. Furthermore, $[\mathrm{O}^p(A'), T]$ is a normal subgroup of G contained in $A \cap B$. Then $[\mathrm{O}^p(A'), T] = 1$ and $\mathrm{O}^p(A') \leq \mathrm{C}_G(T) = C$. Hence $(AC/C)'$ is indeed a p-group.

5. *All prime divisors of $|AC/C|$ different from p divide $p - 1$.*

Suppose that there exists a prime q dividing $|AC/C|$, $q \neq p$, such that q does not divide $p - 1$. The sets of primes $\pi = \{q\}$ and $\rho = \{p\}$ satisfy the hypothesis of Theorem 4.4.2, with the additional condition $\pi \cap \rho = \emptyset$. Since, by Lemma 4.1.10, AC/C and BC/C are mutually permutable subgroups of G/C, we obtain that $\mathrm{O}^{q'}(AC/C)$ is a normal subgroup of $(AC/C)\mathrm{O}^{p'}(BC/C)$. From Statement 2 we also obtain $\mathrm{O}^{p'}(BC/C) = BC/C$, and so $\mathrm{O}^{q'}(A)C/C = \mathrm{O}^{q'}(AC/C)$ is a normal subgroup of $(AC/C)(BC/C) = G/C$. Hence $[\mathrm{O}^{q'}(A), T] = [\mathrm{O}^{q'}(A)C, T]$ is a normal subgroup of G. The order of the automorphism group induced by A on $T/(A \cap T)$ divides $p - 1$ by Statement 3. Since q does not divide $p - 1$, we conclude that $[\mathrm{O}^{q'}(A), T] \leq A \cap T$. Consequently $[\mathrm{O}^{q'}(A), T]$ is a normal subgroup of G which is contained in $A \cap B$. Thus $[\mathrm{O}^{q'}(A), T] = 1$ and so $\mathrm{O}^{q'}(A) \leq \mathrm{C}_G(T) = C$. Therefore AC/C is a q'-group and q does not divide $|AC/C|$. This contradiction yields that all prime divisors of A different from p actually divide $p - 1$.

6. *G/C has a normal Sylow p-subgroup.*

The set of prime numbers

$$\pi = \{q \in \mathbb{P} : q \text{ divides } |AC/C| \text{ and } q \neq p\}$$

is p-special (see Definition 4.1.27): every prime q from π divides $p-1$ by Statement 5 and so for all $q \in \pi$ we obtain that p cannot be a divisor of $q(q-1)$.

From Statement 4 and the definition of π we obtain that AC/C belongs to the class product $\mathfrak{S}_p\mathfrak{E}_\pi$. The same happens for the p-group BC/C by Statement 2. From Theorem 4.1.28, we obtain that G/C, which is the product of the mutually permutable subgroups AC/C and BC/C, also belongs to $\mathfrak{S}_p\mathfrak{E}_\pi$. Since p does not belong to π, we obtain that G/C has a normal Sylow p-subgroup. This completes the proof of the theorem. □

We shall draw two conclusions from the preceding theorem.

Corollary 4.5.2. *Let the group $G = AB$ be the product of the mutually permutable subgroups A and B such that $(A \cap B)_G = 1$. Let N be a minimal normal subgroup of G contained in B. Denote by $T = (AN \cap B)_G$. If $T \neq N$, then B centralises all chief factors of G below T.*

Proof. A chief factor H/K of G with $1 \leq K < H \leq T$ is a p-group by Theorem 4.5.1. Let now $X/C = \mathrm{O}_p(G/C)$, where $C = \mathrm{C}_G(T)$. Note that C is contained in $\mathrm{C}_G(H/K)$. It follows that $X\mathrm{C}_G(H/K)/\mathrm{C}_G(H/K)$ is a normal p-subgroup of $G/\mathrm{C}_G(H/K)$. Since H/K is a p-chief factor, $X\mathrm{C}_G(H/K)/\mathrm{C}_G(H/K)$ is trivial by [119, A, 13.6 (b)]. Furthermore G/C has a normal Sylow p-subgroup and BC/C is a p-subgroup by Theorem 4.5.1. This yields $BC \leq X \leq \mathrm{C}_G(H/K)$ and hence that H/K is centralised by B. □

Assume that $T = N$ in the above result. Then $A \leq \mathrm{C}_G(T)$ or $B \leq \mathrm{C}_G(T)$ by Lemma 4.3.3 (5). This observation, together with Corollary 4.5.2, leads to the following corollary.

Corollary 4.5.3. *Let the group $G = AB$ be the product of the mutually permutable subgroups A and B such that $(A \cap B)_G = 1$. Let N be a minimal normal subgroup of G contained in B. Denote by $T = (AN \cap B)_G$. Then A or B centralises all chief factors of G below T.*

As we have already proved in Corollary 4.4.15, the residuals of two mutually permutable subgroups associated to saturated formations permute. Unfortunately this subgroup is not contained in the residual of the whole group in general. However if $G = AB$ is a mutually permutable product of the subgroups A and B with core-free intersection and $\mathfrak{F}$ is a saturated formation, then $A^{\mathfrak{F}}B^{\mathfrak{F}}$ is a subgroup of $G^{\mathfrak{F}}$. We make this explicit in the next theorem proved by Bochtler [74], and then go on to show that for saturated formations containing all supersoluble groups, the equality holds ([74, 6.3.1]).

Theorem 4.5.4. *Let the group $G = AB$ be the product of the mutually permutable subgroups A and B and let $\mathfrak{F}$ be a saturated formation. If $(A \cap B)_G = 1$, then $A^{\mathfrak{F}}B^{\mathfrak{F}}$ is contained in $G^{\mathfrak{F}}$.*

Proof. We proceed by induction on the order of G. By Theorem 4.3.11, there exists a minimal normal subgroup N of G contained in A or B. Without loss of generality, assume that $N \leq B$. Let $T = (AN \cap B)_G$. Since AT/T and B/T are mutually permutable by Lemma 4.1.10, and their intersection $(AT/T) \cap (B/T) = (AT \cap B)/T = (A \cap B)T/T$ is core-free in G/T, we obtain that the hypotheses of the theorem hold in G/T. Since $|G/T| < |G|$, we can conclude by induction that

$$(A^{\mathfrak{F}}T/T)(B^{\mathfrak{F}}T/T) = (AT/T)^{\mathfrak{F}}(B/T)^{\mathfrak{F}} \leq (G/T)^{\mathfrak{F}} = G^{\mathfrak{F}}T/T,$$

and so

$$A^{\mathfrak{F}}B^{\mathfrak{F}} \leq G^{\mathfrak{F}}T. \tag{4.5}$$

Since $TG^{\mathfrak{F}}/G^{\mathfrak{F}}$ is G-isomorphic to $T/(T \cap G^{\mathfrak{F}})$, by Corollary 4.5.3 we can deduce that all G-chief factors H/K of G with $G^{\mathfrak{F}} \leq K < H \leq TG^{\mathfrak{F}}$ are centralised by A or centralised by B.

Assume that all such G-chief factors are centralised by A. Let H/K be one of these chief factors. Since A centralises H/K, we have that it is a chief factor of $BG^{\mathfrak{F}}/G^{\mathfrak{F}}$ and G induces the same automorphism group by conjugation as $BG^{\mathfrak{F}}/G^{\mathfrak{F}}$. In particular, since H/K is an $\mathfrak{F}$-central chief factor of G, it is an $\mathfrak{F}$-central chief factor of $BG^{\mathfrak{F}}$. Therefore $G^{\mathfrak{F}}T/G^{\mathfrak{F}} \leq \mathrm{Z}_{\mathfrak{F}}(BG^{\mathfrak{F}}/G^{\mathfrak{F}})$. Since $B^{\mathfrak{F}} \leq G^{\mathfrak{F}}T$ by (4.5), we obtain that $(BG^{\mathfrak{F}}/G^{\mathfrak{F}})^{\mathfrak{F}} = B^{\mathfrak{F}}G^{\mathfrak{F}}/G^{\mathfrak{F}} \leq TG^{\mathfrak{F}}/G^{\mathfrak{F}}$ and so

$$BG^{\mathfrak{F}}/TG^{\mathfrak{F}} \cong (BG^{\mathfrak{F}}/G^{\mathfrak{F}})/(TG^{\mathfrak{F}}/G^{\mathfrak{F}}) \in \mathfrak{F}.$$

Consequently $BG^{\mathfrak{F}}/G^{\mathfrak{F}} \in \mathfrak{F}$ and we conclude that $B^{\mathfrak{F}} \leq G^{\mathfrak{F}}$.

By Lemma 4.3.3 (4), $NG^{\mathfrak{F}}/G^{\mathfrak{F}}$ is either covered or avoided by A. Suppose now that $NG^{\mathfrak{F}}/G^{\mathfrak{F}}$ is covered by A. In this case, $NG^{\mathfrak{F}} \leq AG^{\mathfrak{F}}$ and so $T = (AN \cap B)_G \leq AG^{\mathfrak{F}}$. Consider a chief factor H/K of G with $G^{\mathfrak{F}} \leq K < H \leq TG^{\mathfrak{F}}$. Then H/K is a chief factor of the $\mathfrak{F}$-group $G/G^{\mathfrak{F}}$. Thus the primes dividing $|H/K|$ belong to the characteristic of $\mathfrak{F}$ by [119, IV, 4.2]. It follows that H/K is $\mathfrak{F}$-hypercentral in $AG^{\mathfrak{F}}/G^{\mathfrak{F}}$. Therefore $TG^{\mathfrak{F}}/G^{\mathfrak{F}} \leq \mathrm{Z}_{\mathfrak{F}}(AG^{\mathfrak{F}}/G^{\mathfrak{F}})$. By (4.5), $A^{\mathfrak{F}} \leq TG^{\mathfrak{F}}$. We obtain that $AG^{\mathfrak{F}}/TG^{\mathfrak{F}} \in \mathfrak{F}$ and so $AG^{\mathfrak{F}}/G^{\mathfrak{F}} \in \mathfrak{F}$. Hence $A^{\mathfrak{F}} \leq G^{\mathfrak{F}}$.

Suppose now that $NG^{\mathfrak{F}}/G^{\mathfrak{F}}$ is avoided by A. We now argue with the chief factors H/K of G such that $G^{\mathfrak{F}}N \leq H < K \leq G^{\mathfrak{F}}T$ like in the previous paragraph to conclude that $TG^{\mathfrak{F}}/NG^{\mathfrak{F}} \leq \mathrm{Z}_{\mathfrak{F}}(ANG^{\mathfrak{F}}/NG^{\mathfrak{F}})$ and, bearing in mind that $A^{\mathfrak{F}} \leq TG^{\mathfrak{F}}$ by (4.5), we obtain that $ANG^{\mathfrak{F}}/TG^{\mathfrak{F}} \in \mathfrak{F}$ and so $ANG^{\mathfrak{F}}/NG^{\mathfrak{F}} \in \mathfrak{F}$. Therefore $A^{\mathfrak{F}} \leq NG^{\mathfrak{F}} \cap A = G^{\mathfrak{F}}$, because $NG^{\mathfrak{F}}/G^{\mathfrak{F}}$ is avoided by A.

Assume that the chief factors H/K of G such that $G^{\mathfrak{F}} \leq K < H \leq TG^{\mathfrak{F}}$ are centralised by B. An argument like the one used in the previous paragraphs shows that $G^{\mathfrak{F}}T/G^{\mathfrak{F}} \leq Z_{\mathfrak{F}}(BG^{\mathfrak{F}}/G^{\mathfrak{F}})$ and, since $BG^{\mathfrak{F}}/G^{\mathfrak{F}}T \in \mathfrak{F}$, $BG^{\mathfrak{F}}/G^{\mathfrak{F}} \in \mathfrak{F}$ and so $B^{\mathfrak{F}} \leq G^{\mathfrak{F}}$. Note that G induces on H/K the same group of automorphisms as A by conjugation. Hence A acts $\mathfrak{F}$-centrally on H/K. As in the previous paragraphs, we conclude that $A^{\mathfrak{F}} \leq G^{\mathfrak{F}}$.

Therefore, in all cases, $A^{\mathfrak{F}}B^{\mathfrak{F}} \leq G^{\mathfrak{F}}$, which yields the desired result. □

An immediate consequence of Theorem 4.5.4 is:

Corollary 4.5.5. *Let the group $G = AB$ be the product of the mutually permutable subgroups A and B and let $\mathfrak{F}$ be a saturated formation. If $(A \cap B)_G = 1$ and $G \in \mathfrak{F}$, then $A \in \mathfrak{F}$ and $B \in \mathfrak{F}$.*

Example 4.5.6. Theorem 4.5.4 and Corollary 4.5.5 are false in general when the formation $\mathfrak{F}$ is not saturated. For example, consider the formation $\mathfrak{F}$ of all groups whose 3-chief factors are not central like in Example 4.4.8 and G the direct product of two copies of the symmetric group Sym(3) of degree 3. Then G is the mutually permutable product of two cyclic groups A and B of order 6, which have trivial intersection. Then G belongs to $\mathfrak{F}$, but the cyclic factors A and B do not belong to $\mathfrak{F}$.

Example 4.5.7. The conclusion of Corollary 4.5.5 (and hence obviously the inclusion $A^{\mathfrak{F}}B^{\mathfrak{F}} \leq G^{\mathfrak{F}}$ of Theorem 4.5.4) does not hold when $(A \cap B)_G \neq 1$. A counterexample can be found for every saturated formation $\mathfrak{F}$ which is not closed under taking normal subgroups (for example, the product of the class $\mathfrak{N}$ of all nilpotent groups and the formation $\mathfrak{F}$ of Example 4.5.6). There exists an $\mathfrak{F}$-group G with a normal subgroup A which does not belong to $\mathfrak{F}$. This group can be expressed as the mutually permutable product of A and G.

Theorem 4.5.8. *Let the group $G = AB$ be the product of the mutually permutable subgroups A and B and let $\mathfrak{F}$ be a saturated formation containing the class $\mathfrak{U}$ of all supersoluble groups. If $(A \cap B)_G = 1$, then $G^{\mathfrak{F}} = A^{\mathfrak{F}}B^{\mathfrak{F}}$.*

Proof. We suppose that the theorem is false and derive a contradiction. Let (G, A, B) be a counterexample with G of minimal order. Then:

1. *Let N be a minimal normal subgroup of G and let $T = (AN \cap BN)_G$. Then $G^{\mathfrak{F}} = H(T \cap G^{\mathfrak{F}})$, where $H = A^{\mathfrak{F}}B^{\mathfrak{F}}$.*

Applying Lemma 4.1.10, G/T is the mutually permutable product of the subgroups AT/T and BT/T and

$$((AT/T) \cap (BT/T))_{G/T} = 1.$$

The minimality of G yields $G^{\mathfrak{F}}T = A^{\mathfrak{F}}B^{\mathfrak{F}}T$. Since $A^{\mathfrak{F}}B^{\mathfrak{F}}$ is a subgroup of $G^{\mathfrak{F}}$ by Theorem 4.5.4, it follows that

$$G^{\mathfrak{F}} = G^{\mathfrak{F}}T \cap G^{\mathfrak{F}} = A^{\mathfrak{F}}B^{\mathfrak{F}}T \cap G^{\mathfrak{F}} = A^{\mathfrak{F}}B^{\mathfrak{F}}(T \cap G^{\mathfrak{F}}).$$

2. *$A^{\mathfrak{F}}$ and $B^{\mathfrak{F}}$, and hence $H = A^{\mathfrak{F}}B^{\mathfrak{F}}$, are subnormal subgroups of G.*

The hypothesis $\mathfrak{U} \subseteq \mathfrak{F}$ implies that $A^{\mathfrak{F}}$ is a normal subgroup of A', and A' is a subnormal subgroup of G by Corollary 4.1.25. Therefore $A^{\mathfrak{F}}$ is a subnormal subgroup of G. A similar argument shows that $B^{\mathfrak{F}}$ is a subnormal subgroup of G. It follows that $A^{\mathfrak{F}}B^{\mathfrak{F}}$ is a subnormal subgroup of G by [119, A, 14.4].

3. *$H = A^{\mathfrak{F}}B^{\mathfrak{F}}$ is a normal subgroup of $G^{\mathfrak{F}}$.*

By a result of Wielandt [119, A, 14.3], $A^{\mathfrak{F}}$ and $B^{\mathfrak{F}}$ are normalised by N. Since $A \cap T$ obviously normalises $A^{\mathfrak{F}}$, we obtain that $T = AN \cap T = (A \cap T)N \leq \mathrm{N}_G(A^{\mathfrak{F}})$. With an analogous argument we conclude that $T \leq \mathrm{N}_G(B^{\mathfrak{F}})$. Therefore H is normalised by T and so H is a normal subgroup of $H(T \cap G^{\mathfrak{F}}) = G^{\mathfrak{F}}$.

4. *All minimal normal subgroups of G are abelian.*

Assume that this is false and that N is a non-abelian minimal normal subgroup of G. Then $T = (AN \cap BN)_G = N$ by Theorem 4.5.1. In this case, $T \cap G^{\mathfrak{F}} = N \cap G^{\mathfrak{F}}$ cannot be contained in H because (G, A, B) is a counterexample. Hence $N \cap G^{\mathfrak{F}} = N$. Since $H \cap N$ is a subnormal subgroup of N, there exists a complement N^* of $H \cap N$ in N such that $N = (H \cap N) \times N^*$. Moreover, since N is not contained in H, $H \cap N \neq N$ and so N^* is not trivial. Then $G^{\mathfrak{F}} = HN = H(H \cap N)N^*$ and $H \cap N^* = H \cap N \cap N^* = 1$. Thus N^* centralises H.

Suppose that there exists another minimal normal subgroup M of G. If M is non-abelian, then $G^{\mathfrak{F}} = HM$ by the same argument just given. On the other hand, $M \cap N = 1$ and so N centralises M. Since N^* centralises both H and M, we obtain that $N^* \leq \mathrm{Z}(G^{\mathfrak{F}})$. Assume that M is abelian. Let us denote by $S = (AN \cap BN)_G$. By Statement 1, $H(S \cap G^{\mathfrak{F}}) = G^{\mathfrak{F}}$ and, by Theorem 4.5.1, S is elementary abelian. Hence $S \cap N = 1$. It follows that N^* centralises both H and S. Therefore $N^* \leq \mathrm{Z}(G^{\mathfrak{F}})$. In any case we have a contradiction.

Consequently, N is the unique minimal normal subgroup of G and $\mathrm{C}_G(N) = 1$. Now Lemma 4.3.3 (2), either $N \leq B$ or $N \leq A$. Assume that $N \leq B$. In this case, $N \cap A = 1$ because $(A \cap B)_G = 1$ and by Lemma 4.3.3 (5), $A \leq \mathrm{C}_G(N) = 1$ since N is not cyclic. Consequently $G = B$ and so $G^{\mathfrak{F}} = B^{\mathfrak{F}}$. This contradicts our assumption. Therefore G does not possess non-abelian minimal normal subgroups.

5. *$(G^{\mathfrak{F}})' \leq A^{\mathfrak{F}}B^{\mathfrak{F}}$.*

Let N be a minimal normal subgroup of G. Then N is abelian by Statement 4. If $T = (AN \cap BN)_G$, we have that T is elementary abelian by Theorem 4.5.1. By Statement 1, it follows that

$$G^{\mathfrak{F}}/A^{\mathfrak{F}}B^{\mathfrak{F}} = A^{\mathfrak{F}}B^{\mathfrak{F}}(T \cap G^{\mathfrak{F}})/A^{\mathfrak{F}}B^{\mathfrak{F}} \cong (T \cap G^{\mathfrak{F}})/(T \cap A^{\mathfrak{F}}B^{\mathfrak{F}})$$

and so $G^{\mathfrak{F}}/A^{\mathfrak{F}}B^{\mathfrak{F}}$ is abelian. This implies that $(G^{\mathfrak{F}})'$ is contained in $A^{\mathfrak{F}}B^{\mathfrak{F}}$, as claimed.

6. *The final contradiction.*

Let $1 = K_r \trianglelefteq K_{r-1} \trianglelefteq \cdots \trianglelefteq K_1 = T \cap G^{\mathfrak{F}}$ be a series such that all factors K_i/K_{i+1} are chief factors of G. Our aim is to show that for all i the inclusion $K_i \leq A^{\mathfrak{F}}B^{\mathfrak{F}}K_{i-1}$ holds. By Statement 5, there is nothing to prove if K_i/K_{i+1} is covered by $(G^{\mathfrak{F}})'$: in this case, $K_i \leq (G^{\mathfrak{F}})'K_{i+1}$ and hence $K_i \leq (G^{\mathfrak{F}})'K_{i+1} \leq A^{\mathfrak{F}}B^{\mathfrak{F}}K_{i+1}$. In the rest of the proof we can assume that K_i/K_{i+1} is avoided by $(G^{\mathfrak{F}})'$. In this case, $K_i(G^{\mathfrak{F}})'/K_{i+1}(G^{\mathfrak{F}})'$ is G-isomorphic to K_i/K_{i+1}, and so $K_i(G^{\mathfrak{F}})'/K_{i+1}(G^{\mathfrak{F}})'$ is a G-chief factor between $(G^{\mathfrak{F}})'$ and $G^{\mathfrak{F}}$. As the $\mathfrak{F}$-residual of $G/(G^{\mathfrak{F}})'$ is abelian, it does not contain $\mathfrak{F}$-central chief factors by [119, 4.2.19]. Therefore $K_i(G^{\mathfrak{F}})'/K_{i+1}(G^{\mathfrak{F}})'$ and K_i/K_{i+1} are both $\mathfrak{F}$-eccentric in G. Since $\mathfrak{U} \subseteq \mathfrak{F}$, K_i/K_{i+1} cannot be cyclic. Moreover, there exists $X \in \{A, B\}$ such that X centralises K_i/K_{i+1} by Corollary 4.5.3. These properties of K_i/K_{i+1} can happen at the same time only when the subgroup $Y \in \{A, B\} \setminus \{X\}$ covers this chief factor, by Lemma 4.1.10, Lemma 4.3.3, and Lemma 4.3.9. Since $X \leq C_G(K_i/K_{i+1})$, G and Y induce the same automorphism group on K_i/K_{i+1} by conjugation. In particular, since K_i/K_{i+1} is a minimal normal of G/K_{i+1} it is also a minimal normal subgroup of YK_{i+1}/K_{i+1} and hence it is covered or avoided by $Y^{\mathfrak{F}}$. If $Y^{\mathfrak{F}}$ avoids K_i/K_{i+1}, then $Y^{\mathfrak{F}}K_{i+1}/K_{i+1}$ centralises K_i/K_{i+1} and therefore the automorphism group induced by Y on K_i/K_{i+1} belongs to $\mathfrak{F}$. This contradicts the $\mathfrak{F}$-eccentricity of K_i/K_{i+1} in G. Therefore K_i/K_{i+1} cannot be avoided by $Y^{\mathfrak{F}}$. Thus K_i/K_{i+1} is covered by $Y^{\mathfrak{F}}$. In particular, $K_i \leq A^{\mathfrak{F}}B^{\mathfrak{F}}K_{i+1}$. Hence we also have the equality $A^{\mathfrak{F}}B^{\mathfrak{F}}K_i = A^{\mathfrak{F}}B^{\mathfrak{F}}K_{i+1}$ for all i. By considering the representation $G^{\mathfrak{F}} = A^{\mathfrak{F}}B^{\mathfrak{F}}(T \cap G^{\mathfrak{F}})$ of $G^{\mathfrak{F}}$ given in Statement 1, we obtain successively

$$\begin{aligned} G^{\mathfrak{F}} = A^{\mathfrak{F}}B^{\mathfrak{F}}(T \cap G^{\mathfrak{F}}) &= A^{\mathfrak{F}}B^{\mathfrak{F}}K_1 \\ &= A^{\mathfrak{F}}B^{\mathfrak{F}}K_2 = \cdots = A^{\mathfrak{F}}B^{\mathfrak{F}}K_r = A^{\mathfrak{F}}B^{\mathfrak{F}}. \end{aligned}$$

This is the desired contradiction. □

Corollary 4.5.9. *Let the group $G = AB$ be the product of the mutually permutable subgroups A and B and let $\mathfrak{F}$ be a saturated formation containing the class $\mathfrak{U}$ of all supersoluble groups. If $(A \cap B)_G = 1$, then $G \in \mathfrak{F}$ if and only if $A \in \mathfrak{F}$ and $B \in \mathfrak{F}$.*

The result of Alejandre, Ballester-Bolinches, and Cossey [3, Theorem 1] and the corresponding local result proved by Beidleman and Heineken [65, Theorem 6] can be read off by specialising to the case where $\mathfrak{F}$ is either the class of all supersoluble groups or the class of all p-supersoluble groups. The theorems about mutually permutable products of SC-groups with core-free intersection which were proved by Ballester-Bolinches, Cossey, and Pedraza-Aguilera [41] and Beidleman and Heineken [66] in the general case, also follow as a consequence of the above study, despite the fact that the formation of all SC-groups is not saturated (see Section 1.6).

Theorem 4.5.10. *Let the group $G = AB$ be the product of the mutually permutable subgroups A and B and assume that $(A \cap B)_G = 1$. Then if A and B are* SC*-groups, then G is an* SC*-group.*

Proof. We proceed by induction on the order of G. By Theorem 4.3.11, there exists a minimal normal subgroup of G contained in either A or B. We assume without loss of generality that $N \leq B$. Let $T = (AN \cap B)_G$. Since, by Lemma 4.1.10, $G/T = (AN/T)(B/T)$ is the product of the mutually permutable subgroups A/T and B/T and $(AN/T) \cap (B/T)$ contains no non-trivial normal subgroups of G/T, we conclude that G/T is an SC-group. Assume that $T \neq N$. Then, by Theorem 4.5.1, T is abelian and, by Corollary 4.5.2, B centralises all chief factors of G below T. Assume there exists a chief factor H/K of G below T such that H/K is not cyclic. Then A covers it by Corollary 4.3.4, Lemma 4.1.10, and Lemma 4.3.9. This means that $(H \cap A)/(K \cap A)$ is chief factor of A which is isomorphic to H/K. Since A is an SC-group, we conclude that H/K is simple. This contradiction shows that every chief factor of G below T factor is simple. Assume now that $N = T$. Then N is avoided by A by Corollary 4.3.4 and so, if N were not cyclic, N would be centralised by A by Lemma 4.3.3 (5). Hence N would be a minimal normal subgroup of the SC-group B, contradicting our assumption. Consequently N must be cyclic. In both cases, we have that G is an SC-group. This completes the induction argument. □

The converse does not require the restriction on the intersection and it holds for finitely many pairwise mutually permutable products.

Theorem 4.5.11. *Let the group $G = G_1G_2 \cdots G_r$ be the product of the pairwise mutually permutable subgroups $G_1, G_2, \ldots, G_r$. If G is an* SC*-group, then G_i is an SC-group for every $i \in \{1, 2, \ldots, r\}$.*

Proof. Suppose that G is an SC-group. We prove that G_i is an SC-group by induction on the order of G. Let N be a minimal normal subgroup of G. Then G/N is an SC-group which is the product of the pairwise mutually permutable subgroups $G_1N/N, G_2N/N, \ldots, G_nN/N$. By induction, G_iN/N is an SC-group for every $i \in \{1, 2, \ldots, r\}$. If N is abelian, then N is cyclic of prime order and if N is non-abelian, either $N \cap G_i = 1$ or $N \leq G_i$ for every $i \in \{1, 2, \ldots, r\}$ by Corollary 4.3.8. In both cases, we have that G_i is an SC-group for every $i \in \{1, 2, \ldots, r\}$. □

In the case of totally permutable products, we obtain the natural extension of Asaad and Shaalan's result (Corollary 4.1.38) from supersoluble groups to SC-groups (see Ballester-Bolinches and Cossey [37] for the 2-case).

Theorem 4.5.12. *Let the group* $G = G_1 G_2 \cdots G_r$ *be the product of the pairwise totally permutable subgroups* G_1, $G_2, \ldots, G_r$*. Then* G *is an* SC*-group if and only if* G_i *is an* SC*-group for every* $i \in \{1, 2, \ldots, r\}$*.*

Proof. By Theorem 4.5.11, we only must prove that if all G_i are SC-groups for $i \in \{1, 2, \ldots, r\}$, then G is an SC-group. In this case, $G_i Z_{\mathfrak{U}}(G)/Z_{\mathfrak{U}}(G)$ is an SC-group. By Corollary 4.2.13,

$$G/Z_{\mathfrak{U}}(G) = (G_1 Z_{\mathfrak{U}}(G)/Z_{\mathfrak{U}}(G)) \times (G_2 Z_{\mathfrak{U}}(G)/Z_{\mathfrak{U}}(G)) \times \cdots \times (G_r Z_{\mathfrak{U}}(G)/Z_{\mathfrak{U}}(G)).$$

Hence $G/Z_{\mathfrak{U}}(G)$ is an SC-group. Since all chief factors of G below $Z_{\mathfrak{U}}(G)$ are cyclic, we obtain that G is an SC-group. □

Chapter 5

Totally and mutually permutable products and classes of groups

5.1 Totally and mutually permutable products of nilpotent groups

Our theme in this section is totally and mutually permutable products of nilpotent groups. The totally permutable products were first investigated by Shunkov [228], and later by Ballester-Bolinches, Cossey, and Esteban-Romero [39], while the mutually permutable products were studied by Ballester-Bolinches, Beidleman, Cossey, Heineken, and Pedraza-Aguilera [30], and Ballester-Bolinches, Beidleman, Cossey, and Heineken [29]. In the following, $G^2 = \langle g^2 : g \in G\rangle$.

Lemma 5.1.1 (Ballester-Bolinches et al. [30]). *Let the group $G = AB$ be the product of the mutually permutable nilpotent subgroups A and B. Assume that $G^{\mathfrak{N}}$ is a p-group and that $G^{\mathfrak{N}} \le A$. Then G satisfies one of the following statements:*

1. *G^2 is nilpotent, or*

2. *$G^{\mathfrak{N}}$ is abelian.*

Proof. Assume that the result is false and G is a minimal counterexample. Hence G is not nilpotent, $G^{\mathfrak{N}} \neq 1$, and p is a prime divisor of G. By Corollary 4.1.39, G is supersoluble. Moreover, $p > 2$. We begin with two statements and collect the consequences afterwards.

1. $O_{p'}(G) = 1$.

The conditions of the lemma are inherited by quotient groups. Assume that $O_{p'}(G) = M \neq 1$. Then $M \cap G^{\mathfrak{N}} = 1$ and M is contained in the hypercentre of G because M and $MG^{\mathfrak{N}}/G^{\mathfrak{N}}$ are G-isomorphic. Now the nilpotency of $(G/M)^2$ leads to the nilpotency of G^2M/M and G^2. On the other hand, if $(G/M)^{\mathfrak{N}} = (G^{\mathfrak{N}} \times M)/M$ is abelian, so is $G^{\mathfrak{N}}$. Thus the statement is true by minimality of G.

In particular, the Fitting subgroup $F(G) = P$ of G is a Sylow p-subgroup of G, and p is the biggest prime dividing the order of G.

2. $O^p(G) = G$.

Let W be a normal subgroup of G. By Lemma 4.1.21 (1 and 2), we have that $(W \cap A)(W \cap B)$ is a normal subgroup of G and it is the product of the two mutually permutable subgroups $W \cap A$ and $W \cap B$. Furthermore, $|G : (W \cap A)(W \cap B)|$ is a divisor of $|G : W|^2$ by Theorem 1.1.4. If we apply this to the smallest normal subgroup of p-power index $K = O^p(G)$, we have $K = (K \cap A)(K \cap B)$. If $K \neq G$, then the lemma is true for K. The minimality of G implies that K^2 is nilpotent or $K^{\mathfrak{N}}$ is abelian. Since $|G : K|$ is a power of p, we have that $G = PK$.

Since P is a normal subgroup of G of odd order, it follows that $P^2 = P$. Hence $P^2K^2 = PK^2 \leq G^2$. Let $x \in P$, $k \in K$, then $(xk)^2 = xx^{k^{-1}}k^2 \in PK^2$ and so $G^2 = PK^2$. Now the quotient group $G^2/(K^2)^{\mathfrak{N}}$ is the product of the nilpotent normal subgroups $P(K^2)^{\mathfrak{N}}/(K^2)^{\mathfrak{N}}$ and $K^2/(K^2)^{\mathfrak{N}}$. Consequently it is nilpotent and $(G^2)^{\mathfrak{N}} = (K^2)^{\mathfrak{N}}$. Therefore G^2 is nilpotent if and only if K^2 is nilpotent. Moreover $K^{\mathfrak{N}}$ is a normal subgroup of G. Thus $G/K^{\mathfrak{N}}$ is the product of the two normal nilpotent subgroups $PK^{\mathfrak{N}}/K^{\mathfrak{N}}$ and $K/K^{\mathfrak{N}}$. Consequently $G/K^{\mathfrak{N}}$ is nilpotent and so $G^{\mathfrak{N}} = K^{\mathfrak{N}}$. Hence $G^{\mathfrak{N}}$ is abelian if and only if $K^{\mathfrak{N}}$ is abelian. Therefore we can suppose that $O^p(G) = G$.

Let L be a Hall p'-subgroup of G. The subgroup $LG^{\mathfrak{N}}/G^{\mathfrak{N}}$ is a Hall p'-subgroup of the nilpotent group $G/G^{\mathfrak{N}}$. It follows that $LG^{\mathfrak{N}}$ is a normal subgroup of G. Moreover the index $|G : LG^{\mathfrak{N}}|$ is a power of the prime p. Hence $LG^{\mathfrak{N}} = G$ and so $G^{\mathfrak{N}} = P$.

Now we collect the consequences: $F(G)$ is the Sylow p-subgroup P of G by Statement 1, furthermore $P = G^{\mathfrak{N}}$ by Statement 2. By Lemma 2.1.6, G' is a p-group. Therefore $F(G) = G^{\mathfrak{N}} = G' = P$. Since $C_G(F(G)) \leq F(G)$ and $G^{\mathfrak{N}} \leq A$, we obtain that A does not possess elements of order prime to p, so $A = G^{\mathfrak{N}} = P$ and $G = AH$, where H is the Hall p'-subgroup of B. Moreover, $G/A \cong H$ is abelian of exponent dividing $p - 1$.

3. $[A, H] = A$.

Since A is nilpotent, $A' \leq \Phi(A)$. Moreover, $A = G^{\mathfrak{N}}$. It follows that $A = [G, A] = [H, A]A' \leq [H, A]\Phi(A)$ and hence $[H, A] = A$.

4. $A \cap B \leq A'$.

By [119, A, 12.5] and Statement 3, $A/A' = ([A, H]A'/A') \times C_{A/A'}(H) = (A/A') \times C_{A/A'}(H)$. Since $B \cap A$ is a p-subgroup of the nilpotent group B, it is centralised by H and so $(B \cap A)A'/A' \leq C_{A/A'}(H) = 1$. This proves Statement 4.

We denote by T the subgroup $\gamma_3(A) = [A', A]$. We have the following strengthening of Statement 4.

5. $(A \cap B)T = A'$.

First, $(A \cap B)T$ is normalised by A because $[A \cap B, A] \leq [A', A] = T$ and T is normal in G, and by H, because both A and B are normalised by H. Let U be a subgroup of

A with $(A \cap B)T \leq U \leq A$. Then $UB \cap A = U(B \cap A) = U$ is normal in UB and so B normalises U. Hence H induces a group of power automorphisms on $A/(A \cap B)T$. By Theorem 1.3.3, $[A, [A, H]] \leq (A \cap B)T$. Now by Statement 3 $[A, [A, H]] = A'$ and $A \cap B \leq A'$ by Statement 4. It follows that $(A \cap B)T = A'$.

6. *If A is non-abelian, then G^2 is nilpotent.*

Note that $G/A' = (A/A')(BA'/A')$ is a mutually permutable product of A/A' and BA'/A' and their intersection is trivial. By Lemma 4.1.37, BA/A' induces a group of power automorphisms on the normal subgroup A/A'. Let $1 \neq x$ be a generator of H. It follows that if $A' \leq D \leq A$, then D is normalised by x. Hence there is a number s such that $u^x A' = u^s A'$ for all $u \in A$. Let $u_1, u_2 \in A$ and $z_1, z_2 \in A'$ such that $u_i^x = z_i u^s$ for $i \in \{1, 2\}$. Then $[u_1, u_2]^x T = [z_1 u_1^s, z_2 u_2^s]T = [u_1^s, u_2^s]T = [u_1, u_2]^{s^2} T$ (recall that $[ab, c] = [a, b]^c [b, c]$ and $[z_1, z_2 u_2^s]$, $[z_2, u_1^s] \in T$, and A/T has nilpotency class at most 2). But x centralises A'/T because x centralises $A \cap B$ and $A' = (A \cap B)T$. Consequently $s^2 \equiv 1 \pmod{\exp(A'/T)}$ and $u^x A' = u^{-1} A'$ for all $u \in A$. Now $[x^2, A] \leq A'$. Since $\langle x^2, A\rangle/A'$ is abelian and A is nilpotent, it follows that $\langle x^2, A\rangle$ is nilpotent. In particular, the p'-element x^2 centralises A. Consequently $x^2 = 1$ and $H^2 = 1$. Therefore $G^2 = AH^2 = A$ is nilpotent.

Therefore there is no minimal counterexample and the proof is complete. □

Lemma 5.1.2. *Under the conditions of Lemma* 5.1.1, *we have that $G^{\mathfrak{N}} \cap B = 1$ if and only if $G^{\mathfrak{N}}$ is abelian.*

Proof. Suppose that G is a group under the hypotheses of Lemma 5.1.1 of least order such that $G^{\mathfrak{N}} \cap B = 1$, but $G^{\mathfrak{N}}$ is not abelian. The arguments of Lemma 5.1.1 show that $G^2 = A = G^{\mathfrak{N}}$ is nilpotent. Since A is the Sylow p-subgroup of G, we have that $B = H$ and $B^2 = 1$. Hence B is a cyclic group of order 2. Since B induces a group of power automorphisms on A by Lemma 4.1.37, $a^b = a^{-1}$ for $B = \langle b\rangle$ and all $a \in A$. This implies that A is abelian. This contradiction proves that if $G^{\mathfrak{N}} \cap B = 1$, then $G^{\mathfrak{N}}$ is abelian.

Conversely, suppose that G is a group under the hypotheses of Lemma 5.1.1 of smallest order such that $G^{\mathfrak{N}}$ is abelian, but $B \cap G^{\mathfrak{N}} \neq 1$. Let $R = \langle (B \cap G^{\mathfrak{N}})^G \rangle$, which is contained in $G^{\mathfrak{N}}$, and consider a normal subgroup N of G such that $N \leq R$ and $|R : N| = p$. We have that G/N satisfies the hypothesis of Lemma 5.1.1. Moreover, $(BN/N) \cap (G^{\mathfrak{N}}/N) = (B \cap G^{\mathfrak{N}})N/N \neq 1$ (otherwise, $R = \langle (B \cap G^{\mathfrak{N}})^G \rangle \leq N$). Since $(B \cap G^{\mathfrak{N}})N/N \leq R/N$, we have that $R = (B \cap G^{\mathfrak{N}})N$. By minimality, we can suppose that $N = 1$. Hence $R = B \cap G^{\mathfrak{N}}$ is a minimal normal subgroup of G. Moreover, R is a minimal normal subgroup of A and B. In particular, R is centralised by A and B and so $R \leq Z(G)$. Then $G^{\mathfrak{N}}$ is abelian and contains a central minimal normal subgroup of G. This contradicts [47, 4.2.19]. Hence $B \cap G^{\mathfrak{N}}$ must be trivial if $G^{\mathfrak{N}}$ is abelian. □

Theorem 5.1.3. *Let the group $G = AB$ be the product of the mutually permutable nilpotent subgroups A and B. Then $(G^2)^{\mathfrak{N}}$ is abelian. In particular, if G has odd order, then $G^{\mathfrak{N}}$ is abelian.*

Proof. Assume that the result is false and G is a minimal counterexample. By Corollary 4.1.39, G is supersoluble. By Lemma 2.1.6, we have that $G^{\mathfrak{N}} \leq G' \leq \mathrm{F}(G)$ and $P \leq \mathrm{F}(G)$ if P is the Sylow p-subgroup of G with p the largest prime divisor of the order of G. Clearly, every quotient group G/N of G inherits the hypothesis. The proof will follow as a consequence of the following steps.

1. *G possesses only one minimal normal subgroup.*

Assume that N_1 and N_2 are two different minimal normal subgroups of G. Then $((G^2)^{\mathfrak{N}})' \leq N_i$ for $i \in \{1, 2\}$ and so $((G^2)^{\mathfrak{N}})' \leq N_1 \cap N_2 = 1$.

By Statement 1, we have now that $P = \mathrm{F}(G)$ and therefore $G^{\mathfrak{N}} \leq G' \leq P = P^2 < G^2 \leq G$.

By Theorem 1.1.19, $P = (P \cap A)(P \cap B)$ and the product of the two Hall p'-subgroups of A and B, respectively, gives rise to a Hall p'-subgroup Q of G, with $(Q \cap A)(Q \cap B) = Q$, and $G = PQ$. The quotient group $G/P \cong Q$ is abelian because G is supersoluble.

We study now the subgroup BP. Our intention is to prove that BP satisfies the hypothesis of the Lemma 5.1.1. We consider the intersections of A and B with P and Q, and obtain

2. *$B \cap Q$ normalises $A \cap P$.*

The subgroup $A \cap P$ is the Sylow p-subgroup of

$$(B \cap Q)A = (B \cap Q)(A \cap Q)(A \cap P) = Q(A \cap P)$$

and Statement 2 follows because the Sylow p-subgroup is normal.

3. *$BP = (B \cap Q)P = (A \cap P)B$ satisfies the conditions of Lemma 5.1.1 with $A \cap P$ and B instead of A and B, respectively.*

Note that $[B \cap Q, P]$ is a normal subgroup of $(B \cap Q)P = BP$ by [157, III, 1.6 (b)]. Now $[B \cap Q, P] = [B \cap Q, (A \cap P)(B \cap P)] = [B \cap Q, A \cap P] \leq A \cap P$ by Statement 2. But $B \cap Q$ acts on P and then $P = [B \cap Q, P]\mathrm{C}_P(B \cap Q)$ by [119, A, 12.5]. It follows that $BP/[B \cap Q, P]$ is nilpotent. Consequently, $(BP)^{\mathfrak{N}} \leq [B \cap Q, P] \leq A \cap P$. This shows Statement 2.

By symmetry, we also have

4. *AP satisfies the hypothesis of Lemma 5.1.1, with $B \cap P$ and A instead of A and B.*

5. *AP and BP are normal subgroups of G.*

This follows because G/P is abelian.

Now we will exclude all possibilities for the counterexample G.

6. *AP cannot be nilpotent.*

Suppose that AP is nilpotent. Then $AP \leq \mathrm{F}(G) = P$ and $P = A(B \cap P)$. In this case $BP = G$ and either G^2 is nilpotent $G^{\mathfrak{N}} = (BP)^{\mathfrak{N}}$ is abelian by Statement 2 and Lemma 5.1.1, which contradicts the choice of G.

7. *$(AP)^{\mathfrak{N}}$ cannot be abelian.*

Suppose now that $W = (AP)^{\mathfrak{N}}$ is non-trivial and abelian. If $(BP)^2$ is nilpotent, then $1 \leq W \leq (AP)(BP)^2 \leq G$ is a sequence of normal subgroups of G with W abelian, $(AP)(BP)^2/W$ nilpotent, and $G/(AP)(BP)^2$ of exponent 2. It follows that $(G^2)^{\mathfrak{N}}$ is abelian. If $(BP)^{\mathfrak{N}}$ is non-trivial and abelian, we have on one hand by Lemma 5.1.2 and Statement 3 that $W \cap A = 1$ and $W \leq B \cap P$; on the other hand, by Lemma 5.1.2 and Statement 2, $(BP)^{\mathfrak{N}} \cap B = 1$ and $(BP)^{\mathfrak{N}} \leq A \cap P$. Now $(BP)^{\mathfrak{N}} \cap W = 1$, contrary to the existence of a unique minimal normal subgroup.

Now we prove the conclusion.

8. *There is no counterexample to the theorem.*

By Statement 6, we know that the only possibility left is $(AP)^2$ and $(BP)^2$ are nilpotent. But then $(AP)^2(BP)^2 = G^2$, which is the product of two nilpotent normal subgroups, is nilpotent. This finishes the proof of the theorem. □

The nilpotent residual of G need not be abelian in a product of two nilpotent subgroups. In the following we give two examples:

Example 5.1.4. Consider the group

$$B = \langle a, b : a^3 = [b,a,a], b^3 = [b,a,a]^2, [b,a]^3 = [b,a,a]^3 = [b,a,b] = 1 \rangle.$$

Then B has order 3^4 and nilpotency class 3.

1. All elements of $B \setminus B'$ have order 9. Note that $(ab^k)^3 = a^3 b^{3k} [b,a,a]^k = [b,a,a]^m$ with $m = 1 + 2k + k \equiv 1 \pmod{\mathrm{o}([b,a,a])}$. Thus $B^3 = \langle [b,a,a] \rangle$.

2. $B' = \langle [a,b], [b,a,a] \rangle$.

3. Let α be the automorphism of B given by $a^\alpha = a^2$, $b^\alpha = b^2[b,a,a]^2 = b^5$. Note that α defines an automorphism of the group B by realising that $[a,b]^\alpha = [a^2, b^5] = [a^{-1}, b^{-1}]$ is conjugate to $[a,b]$.

4. Let $d = [b,a][b^{-1}, a^{-1}]$. It is clear by the above remark that $d^\alpha = d$. Therefore we can consider $A = \langle d, \alpha \rangle = \langle d \rangle \times \langle \alpha \rangle$.

Let $G = [B]\langle \alpha \rangle$ be the corresponding semidirect product. The group G is a mutually permutable product of A and B. Moreover, $(A \cap B)_G = 1$ and $G^{\mathfrak{N}} = B$ is not abelian.

Example 5.1.5. Let p be and odd prime; denote by P an extraspecial group of order p^{2t+1} with $t \geq 1$ and of exponent p. It is clear that the subgroup $\Phi(P) = P' = \mathrm{Z}(P)$ has order p. Moreover P has nilpotency class 2 and the quotient $P/\mathrm{Z}(P)$ is elementary abelian. Now by [119, A, 20.8 (b)], there exists an automorphism α of order 2 of P such that a fixes the elements of $\mathrm{Z}(P)$ and induces the inversion on $P/\mathrm{Z}(P)$.

Consider now the corresponding semidirect product $G = [P]\langle\alpha\rangle$. Denote $B = P$ and $A = \langle\alpha, \mathrm{Z}(P)\rangle$. Then G is the mutually permutable product of A and B, both nilpotent. Moreover $A \cap B = \mathrm{Z}(P)$ and $G^{\mathfrak{N}} = P$ is not abelian.

The following results were proved by Ballester-Bolinches, Beidleman, Cossey, and Heineken in [29]. We begin with two lemmas.

Lemma 5.1.6. *Let p be a prime and let A be a normal p-subgroup of a p-soluble group G such that $A \leq \mathrm{Z}(L)$, where $L = G^{\mathfrak{N}}$ is the nilpotent residual of G. Let Q be a Hall p'-subgroup of G. Then $A = [A, Q] \times \mathrm{C}_A(Q)$ and the factors $[A, Q]$ and $\mathrm{C}_A(Q)$ are both normal in G.*

Proof. The equality $A = [A, Q] \times \mathrm{C}_A(Q)$ follows from [119, A, 12.5]. What has to be proved is that both factors are normal in G. Let $C = \mathrm{C}_G(A)$. Since L centralises A, we obtain that G/C is nilpotent and so $G/C = (PC/C) \times (QC/C)$, where P is a Sylow p-subgroup of G. Therefore it is enough to verify that P and Q normalise $[A, Q]$ and $\mathrm{C}_A(Q)$. Note that $[P, Q] \leq C$.

Let $a \in A$, $y \in Q$, and $x \in P$. Then $[a, y]^x = (a^{-1}a^y)^x = (a^x)^{-1}a^{yx} = (a^x)^{-1}a^{[y^{-1},x^{-1}]xy} = (a^x)^{-1}(a^x)^y \in [A, Q]$ because $[y^{-1}, x^{-1}] \in [P, Q]$ centralises A and $a^x \in A$. Let now $a \in A$, $y \in Q$, and $z \in Q$. Then $[a, y]^z = (a^{-1}a^y)^z = (a^z)^{-1}a^{yz} = [a, z]^{-1}[a, yz] \in [A, Q]$. It follows that $[A, Q]$ is normal in G.

Let $a \in \mathrm{C}_A(Q)$, then $a^y = a$ for all $y \in Q$. Let $x \in P$. Then $(a^x)^y = a^{xy} = a^{[x^{-1},y^{-1}]yx} = a^{yx} = a^x$ and so $a^x \in \mathrm{C}_A(Q)$. This yields that $\mathrm{C}_A(Q)$ is a normal subgroup of G. □

The following situation appears in the proof of Theorem 5.1.11.

Hypothesis 5.1.7. Let $H = UV$ be a group with U a normal p-subgroup of class at most 3, p an odd prime, and V a nilpotent subgroup of H for which $PV = VP$ for all subgroups P of U. Assume that U^* is a normal abelian subgroup of H such that $U' \leq U^*$ and $[U^*, U] = \gamma_3(U)$. Suppose further that $|V_{p'}/\mathrm{C}_{V_{p'}}(U)| = 2$, where $V_{p'}$ denotes the Hall p'-subgroup of V, and if $v \in V_{p'} \setminus \mathrm{C}_{V_{p'}}(U)$, then $u^v U^* = u^{-1}U^*$ for all $u \in U$ while $U^*/\gamma_3(U)$ is centralised by v.

Note that v inverts the elements of $\gamma_3(U)$. Since U^* is abelian, we have $U^* = [U^*, v] \times \mathrm{C}_{U^*}(v)$ by [119, A, 12.5].

Let us denote by K the subgroup $C_{U^*}(v)$. Since v inverts $\gamma_3(U)$, we obtain that $U^* = \gamma_3(U) \times K$. We also note that UV_p is a normal subgroup of H, where V_p is a Sylow p-subgroup of V.

Lemma 5.1.8. *Under Hypothesis 5.1.7, if $u \in U \setminus U^*$, there is an element $w \in U$ such that $uU^* = wU^*$ and $\langle w \rangle$ is normalised by V. Moreover, if $u \in U$ satisfies $u^v = u^{-1}$, then $\langle u \rangle$ is normalised by V.*

Proof. We suppose first that U has class 2. In this case, $U^* \leq Z(U)$. Then $U^*\langle u \rangle$ is abelian because

$$[U^*\langle u \rangle, U^*\langle u \rangle] = [U^*, U^*][U^*, \langle u \rangle][\langle u \rangle, \langle u \rangle] = 1.$$

We also claim that $U^*\langle u \rangle$ is normalised by V. To see this, we observe that $V_{p'}$ normalises $U^*\langle u \rangle / U^*$ by our hypothesis. Moreover, by [119, A, 12.5], we have $U^*\langle u \rangle = [U^*\langle u \rangle, V_{p'}] \times C_{U^*\langle u \rangle}(V_{p'})$. Since $C_{U^*\langle u \rangle}(V_{p'}) \leq U^*$ by hypothesis, we have that $J = [U^*\langle u \rangle, V_{p'}]$ is cyclic, normalised by $V_{p'}$, and for some $w \in J$ we have $wU^* = uU^*$. Now let $z \in V_p$. Then, since $JV = VJ$, we have that $JV_p = V_pJ$ by Theorem 1.1.19 and so $wz = xw^t$ for some $x \in V_p$ and some integer t. Then $z^{-1}wz = z^{-1}xw^t = yw^t$ with $y = z^{-1}x \in V_p$. Moreover, w^z, $w^t \in U$ and so $y \in V_p \cap U$, which must be contained in U^* because $v \in V_{p'}$ inverts all elements of U/U^*. It follows that V_p normalises $U^*\langle u \rangle$. By Lemma 5.1.6 applied to $(U^*\langle u \rangle)V$, we see that J is normalised by V. If $u^v = u^{-1}$, then $[u, v] = u^{-2} \in J$ and this implies that $u \in J$ because p is odd. Hence $\langle u \rangle = J$ is normalised by V.

Suppose now that U has class 3 and suppose that we have chosen w to satisfy the conclusion of the lemma modulo $\gamma_3(U)$. Note that $\gamma_3(U)\langle w \rangle$ is abelian and normalised by V. By Lemma 5.1.6, we have that $\gamma_3(U)\langle w \rangle = [\gamma_3(U)\langle w \rangle, V_{p'}] \times C_{\gamma_3(U)\langle w \rangle}(V_{p'})$ and both subgroups are normalised by V. Since $\gamma_3(U) \leq [\gamma_3(U)\langle w \rangle, V_{p'}]$, it follows that $C_{\gamma_3(U)\langle w \rangle}(V_{p'})$ is cyclic. Let z be a generator of this subgroup, then $w\gamma_3(U) = z^j\gamma_3(U)$ for some integer j and $\langle z \rangle$ is normalised by V. Moreover $uU^* = wU^* = z^jU^*$.

If $u^v = u^{-1}$, then $\gamma_3(U)\langle u \rangle$ is normalised by V because the conclusion holds for $H/\gamma_3(U)$ and so $u = u_1u_2$ with $u_1 \in [\gamma_3(U)\langle u \rangle, V_{p'}]$ and $u_2 \in C_{\gamma_3(U)\langle u \rangle}(V_{p'})$, which implies that $u^v = u_1^tu_2 = u^{-1} = u_1^{-1}u_2^{-1}$, and so $u_2 = 1$, that is, $\langle u \rangle$ is normalised by $V_{p'}$. This completes the proof of the result. □

The proof of the next lemma relies on the concept of regular p-group.

Definition 5.1.9. Let p be a prime. A p-group G is said to be *regular* if for every x, y in G, there exist elements $d_i \in \langle x, y \rangle'$, $1 \leq i \leq r$, such that $x^py^p = (xy)^p \prod_{i=1}^{r} d_i^p$.

The main properties of regular p-groups appear in Chapter 3, Section 10 of Huppert's book [157] or in Chapter 7 of the book of Berkovich [70].

Lemma 5.1.10. *Under Hypothesis* 5.1.7, *suppose further that* $K \leq V$. *Suppose also that* U *has class* 3 *and let* $M \leq U'$ *be the subgroup of* H *generated by all normal subgroups* R *of* H *such that* $R \leq U'$ *and* $R \cap \gamma_3(U) = 1$. *Then* $\gamma_3(U)$ *has at most two generators and if* $\gamma_3(U)$ *is not cyclic or* $p > 3$, *then* $U'/(M\gamma_3(U))$ *is cyclic.*

Proof. Since M is normal in H and $M \cap \gamma_3(U) = 1$, we see that H/M satisfies the hypothesis of the lemma and it will be enough to prove the result for H/M. Thus we may assume that $M = 1$.

By Lemma 5.1.8, we can choose a set of generators $u_1, u_2, \ldots, u_n$ for U such that $\langle u_i \rangle$ is normalised by V, $1 \leq i \leq n$. We then have $U^* = \gamma_3(U) \times K$ and so $\gamma_3(U) = [U^*, U] = [\gamma_3(U) \times K, U] = [K, U]$. It follows that

$$\gamma_3(U) = \langle [u_i, z] : i = 1, 2, \ldots, n; z \in K \rangle$$

and so $\gamma_3(U)$ is generated by powers of the generators u_i since $[u_i, z] \in \langle u_i \rangle$ for all $i \in \{1, 2, \ldots, n\}$ and $z \in K$.

For each u_i, set $C_i = C_K(u_i)$. Then $\bigcap_{i=1}^n C_i = 1$. Suppose that K is not cyclic. Then we can find i and j such that $C_i \neq C_i \cap C_j \neq C_j$. We may suppose that $\{i, j\} = \{1, 2\}$ and then choose $x_i \in C_i \setminus (C_1 \cap C_2)$, $i \in \{1, 2\}$. Suppose first that $\langle u_1 \rangle \cap \langle u_2 \rangle = 1$. Note that $u_1u_2u_1$ is inverted by v and hence $\langle u_1u_2u_1 \rangle$ is normalised by V by Lemma 5.1.8. We then have that $1 \neq [u_1u_2u_1, x_1] = [u_2, x_1] \in \langle u_2 \rangle$, while $1 \neq [u_1u_2u_1, x_2] = [u_1, x_2]^2 \in \langle u_1 \rangle$. But both these commutators must lie in $\langle u_1u_2u_1 \rangle$, a contradiction. Thus if $\langle u_i \rangle \cap \langle u_j \rangle = 1$, we must have $C_i \leq C_j$ or vice versa.

Now we can renumber the u_i so that C_1 has order as small as possible. Then we choose $u_2, \ldots, u_m$ so that if $\langle u_i \rangle \cap \operatorname{Soc}(\gamma_3(U)) = \langle z_i \rangle$, then $\{z_1, z_2, \ldots, z_m\}$ is a minimal generating set for $\operatorname{Soc}(\gamma_3(U))$. Note that $\langle u_1 \rangle \cap \langle u_i \rangle = 1$ for $i \in \{2, \ldots, m\}$ and so $C_1 \leq C_i$ for each $i \in \{1, 2, \ldots, m\}$.

Suppose that $m \geq 2$, that is, $\gamma_3(U)$ is not cyclic. Now let $j \notin \{1, 2, \ldots, m\}$. Assume that C_1 is not contained in C_j. Then $\langle u_j \rangle \cap \langle u_1 \rangle \neq 1$. Denote $w_j = u_2u_ju_2$ and note that $[w_j, v] = [u_2, v]^2[u_j, v] \notin \langle u_1 \rangle$. Further, if $C_j^* = C_K(w_j)$, then $D = C_2 \cap C_j \leq C_j^*$ and K/D is a subdirect product of two cyclic groups. Thus K/C_j^* is a cyclic quotient of the subdirect product of two cyclic groups and so it must have order at most the largest of the orders of the factors K/C_2 and K/C_j. Since both of these are at most the order of K/C_1, we have $|C_1| \leq |C_j^*|$. Since $\langle u_1 \rangle \cap \langle w_j \rangle = 1$, we have $C_1 \leq C_j^*$. Let z be a generator of C_1, then $z \in C_j^*$ and $z \in C_2$ and so $1 = [w_j, z] = [u_2, z]^2[u_j, z] = [u_j, z]$, which implies that $z \in C_j$ and so $C_1 \leq C_j$. This contradiction shows that C_1 must be contained in all C_j and, since $\bigcap_{i=1}^n C_i = 1$, we obtain that $C_1 = 1$. Therefore K is isomorphic to a group of automorphisms of $\langle u_i \rangle$, which is cyclic by [157, I, 13.19]. This contradicts our assumption about K. Therefore we have proved that $K \cong U^*/\gamma_3(U)$ is cyclic (and thus $U'/\gamma_3(U)$ is also cyclic) if $\gamma_3(U)$ is not cyclic.

Since K is cyclic, we can choose u_i and u_j such that $K\gamma_3(U)/\gamma_3(U)$ is generated by $[u_i, u_j]\gamma_3(U)$. Now consider u_k with $k \notin \{i, j\}$ ($i, j, k \in \{1, 2, \ldots, n\}$). Then by the Witt identity ([157, III, 1.4]) and the fact that $\gamma_3(U)$ is central in U we have that

$$[u_i, u_j, u_k][u_k, u_i, u_j][u_j, u_k, u_i] = 1.$$

But $[u_k, u_i] \equiv [u_i, u_j]^t \pmod{\gamma_3(U)}$ and $[u_j, u_k] \equiv [u_i, u_j]^s \pmod{\gamma_3(U)}$, s and t integers. Hence we have $[u_i, u_j, u_k] = [u_i, u_j, u_i]^{-t}[u_i, u_j, u_j]^{-s}$. Since $\gamma_3(U)$ is generated by the $[u_i, u_j, u_k]$, $k \in \{1, 2 \ldots, n\}$, it follows that $\gamma_3(U)$ is generated by $[u_i, u_j, u_i]$ and $[u_i, u_j, u_j]$, as desired. Therefore $\gamma_3(U)$ has two generators if $\gamma_3(U)$ is not cyclic

It only remains to show that the conclusion holds when $\gamma_3(U)$ is cyclic and $p > 3$. In this case, U is a regular p-group by [157, III, 10.2 (a)]. Choose a generator $[u_i, u_j, u_k]$ for $\gamma_3(U)$, where u_i, u_j, u_k are some generators of U chosen as before. We also have that $\gamma_3(U) = \langle u_k^{p^r} \rangle$ for some r because $U^* = \gamma_3(U) \times K$ and so $[u_i, u_j] = gz$ with $g \in \gamma_3(U)$ and $z \in K$, which implies that $[u_i, u_j, u_k] = [gz, u_k] = [z, u_k] \in \langle u_k \rangle$ because $z \in K \le V$ normalises $\langle u_k \rangle$. We choose k so that r is as large as possible and then renumber if necessary to make $k = 1$. Suppose that for some generator u_m we have $\langle u_m \rangle \cap \gamma_3(U) = \langle u_m^{p^s} \rangle \ne 1$ with $s > r$. Since u_1 and u_m are inverted by v, $u_1 u_m u_1$ is also inverted by v and by Lemma 5.1.8, $\langle u_1 u_m u_1 \rangle$ is also normalised by V. Therefore we can expand $u = [u_i, u_j, u_m u_1 u_m]$ to obtain

$$u = [u_i, u_j, u_m][u_i, u_j, u_1][u_i, u_j, u_m] = u_m^{ap^s} u_1^{bp^r} u_m^{ap^s} = (u_m u_1 u_m)^{cp^t}$$

with b and c prime to p. By the choice of r, we obtain that $t \le r$. Note that $(u_m u_1 u_m)^{p^t} \in \langle u_1^{p^r} \rangle$. This implies that $[(u_m u_1 u_m)^{p^t}, u_1] = 1$. By [157, III, 10.6 (b)], this is equivalent to $[u_m u_1 u_m, u_1]^{p^t} = 1$, that is, $[u_m, u_1]^{2p^t} = 1$, or, in other words, $[u_m, u_1]^{p^t} = 1$. This is also equivalent to $[u_m, u_1^{p^t}] = 1$. It follows that $(\langle u_1, u_m \rangle')^{p^t} = 1$ (the subgroup generated by all p^tth powers) and so $(u_m u_1 u_m)^{cp^t} = u_m^{cp^t} u_1^{cp^t} u_m^{cp^t}$ by [157, III, 10.8 (g)]. Let $u_m^{p^s} = u_1^{dp^r}$. Then $u = u_1^{adp^r} u_1^{bp^r} u_1^{adp^r} = u_m^{cp^t} u_1^{cp^t} u_m^{cp^t}$, that is, $u_m^{2cp^t} = u_1^{sp^t}$ for $s = (2ad + b)p^{r-t} - c$. Consequently $u_m^{2cp^r} = u_1^{sp^r}$ and so $s \le r$. This contradiction shows that $s \le r$ if $\langle u_m \rangle \cap \gamma_3(U) \ne 1$.

For $m > 1$ we set $u_m^{p^s} = u_1^{dp^r} = u_1^{-2e(m)p^s}$. From the regularity of U and [157, III, 10.6 (a)], $(u_m u_1^{2e(m)})^{p^s} = 1$, and if we conjugate by $u_1^{-e(m)}$ we get the equality $(u_1^{e(m)} u_m u_1^{e(m)})^{p^s} = 1$. Call $v_1 = u_1$ $v_m = u_1^{e(m)} u_m u_1^{e(m)}$, note that v_m is inverted by v. Suppose that $\langle v_m \rangle \cap \gamma_3(U) \ne 1$, then for some integers β and x with $0 \le x < s$, the equality $v_m^{p^x} = u_1^{\beta p^r}$ holds. Hence $(v_m u_1^{-\beta p^{r-x}})^{p^x} = 1$ by [157, III, 10.6 (b)]. Conjugation by $u_1^{e(m)}$ yields $(u_m u_1^{-\beta p^{r-x} + 2e(m)})^{p^x} = 1$. This is equivalent to $u_m^{p^x} = u_1^{\beta p^r - 2e(m)p^x}$ and so $u_m^{p^s} = u_1^{\beta p^{r-s+x} - 2e(m)p^s}$. Hence $u_1^{\beta p^{r-s+x}} = 1$. Since $s - x > 0$, this contradicts the fact $u_1^{p^r} \neq 1$. Consequently $\langle v_m \rangle \cap \gamma_3(U) = 1$ for $m \geq 2$.

We now have $\gamma_3(U) = \langle [v_i, v_j, v_1] \rangle$ for some i and j. By the Witt identity [157, III, 1.4] and bearing in mind that $\gamma_3(U) \leq \mathrm{Z}(U)$, we have

$$1 = [v_i, v_j, v_1][v_1, v_i, v_j][v_j, v_1, v_i].$$

The factors $[v_i, v_j, v_1]$, $[v_1, v_i, v_j]$, and $[v_j, v_1, v_i]$ belong to $\langle v_1 \rangle$, $\langle v_j \rangle$, and $\langle v_i \rangle$, respectively, and so one of $[v_1, v_i, v_j]$ and $[v_j, v_1, v_i]$ must be non-trivial. By our choice of generators, we must have one of i, j equal to 1. We renumber the indices if necessary to make the other v_2. Let $i > 2$ and $[v_1, v_i, v_1] = [v_1, v_2, v_1]^{-2\delta(i)}$, then we can replace v_i by $v_2^{\delta(i)} v_i v_2^{\delta(i)}$ if necessary in order to guarantee that $[v_1, v_i, v_1] = 1$ for $i > 2$. Since $[v_i, v_j, v_k] \in \langle v_k \rangle$ we obtain that $[v_i, v_j, v_k] = 1$ if $k \neq 1$. If $k = 1$, another application of the Witt identity shows that $i = 1$ or $j = 1$ or $[v_i, v_j, v_1] = 1$. If $i = 1$, then $[v_1, v_j, v_1] = 1$ unless $j = 2$, and if $j = 1$, then $[v_i, v_1, v_1] = [v_1, v_i, v_1]^{-1} = 1$ unless $i = 2$. Therefore $[v_i, v_j, v_k] = 1$ unless $\{i, j\} = \{1, 2\}$ and $k = 1$. Now consider $[v_i, v_j]$. Since v inverts v_i, we obtain that $[v_i, v_j]\gamma_3(U)$ is centralised by v. Since v_i and v_j are normalised by V_p, we also obtain that $[v_i, v_j]^w = [v_i^m, v_j^n] = [v_i, v_j]^{mn} z$ for some integers m and n, with $z \in \gamma_3(U)$. This shows that $Z = \langle [v_i, v_j], \gamma_3(U) \rangle = \langle [v_i, v_j] \rangle \gamma_3(U)$ is a normal subgroup of H contained in U'. Under the action of v, $Z = [Z, v] \times \mathrm{C}_Z(v)$ by [119, A, 12.5]. Note that $[Z, v] \leq \gamma_3(U)$, because if $g \in \gamma_3(U)$ and $c = [v_i, v_j]$, we have that $c^v = cd$, with $d \in \gamma_3(U)$, and so $[gc, v] = c^{-1}g^{-1}(gc)^v = c^{-1}g^{-1}g^{-1}cd$ is an element of $\gamma_3(U)$. Assume in addition that $\{i, j\} \neq \{1, 2\}$. We see that $[v_i, v_j] \in \mathrm{Z}(U)$ and so $Z \leq \mathrm{Z}(U)$. Now $\gamma_3(U) = [Z, v]\mathrm{C}_Z(v) \cap \gamma_3(U) = [Z, v](\mathrm{C}_Z(v) \cap \gamma_3(U)) = [Z, v]$. By Lemma 3.3.39, we see that $\mathrm{C}_Z(v)$ is normalised by V. Therefore $\mathrm{C}_Z(v)$ is normal in H. Since $\mathrm{C}_Z(v) \cap \gamma_3(U) = 1$, by hypothesis we have that $\mathrm{C}_Z(v) = 1$. Hence $\langle [v_i, v_j] \rangle \gamma_3(U) = \gamma_3(U)$ and $[v_i, v_j] \in \gamma_3(U)$. This shows that $U'/\gamma_3(U) = \langle [v_1, v_2]\gamma_3(U) \rangle$ is cyclic, as desired.

Since $Z/\gamma_3(U)$ is cyclic, we see that $\mathrm{C}_Z(v)$ is cyclic. This completes the proof of the lemma. □

The next theorem provides some insight into the structure of the nilpotent residual of a mutually permutable product of two nilpotent groups. It has been proved by Ballester-Bolinches, Beidleman, Cossey, and Heineken [29].

Theorem 5.1.11. *Let the group $G = AB$ be the mutually permutable product of the nilpotent groups A and B. Let $L = G^{\mathfrak{N}}$. Then L is nilpotent. Assume further than p is a prime dividing $|L|$. Then $L_p = A_0 B_0$ and L_p has class at most 3, where L_p is the Sylow p-subgroup of L, $A_0 = L_p \cap A$ and $B_0 = L_p \cap B$. Moreover:*

1. *If L_p is abelian, then $A_0 \cap B_0 = 1$, and A_0 and B_0 are normal in G. Further, $A_{p'}$ acts on B_0 as a group of power automorphisms and $B_{p'}$ acts on A_0 as a group of power automorphisms.*

2. *If L_p has class 2, then $G_{p'}/C_{G_{p'}}(L_p)$ is an elementary abelian 2-group of order at most 4. Further, A_0 and B_0 are normal subgroups of G, L_p is a central product of A_0 and B_0, $L'_p = A'_0 B'_0$ is hypercentral in G, and $L'_p = L_p \cap A \cap B \leq A \cap B$, and so if $(A \cap B \cap L_p)_G = 1$, then L_p cannot have class 2.*

3. *If L_p has class 3, then $\gamma_3(L_p) = \gamma_3(A_0)\gamma_3(B_0)$, $[A'_0, B_0] = [B'_0, A_0] = 1$, and $\gamma_3(A_0) \cap \gamma_3(B_0) = 1$. If A_0 has class 3, then A_0 is normal in G, $\gamma_3(A_0)$ has at most 2 generators and if $\gamma_3(A_0)$ is non-cyclic or $p > 3$, then $A'_0 Z(A_0)/Z(A_0)$ is cyclic. The symmetric statement obtained by replacing A_0 by B_0 also holds.*

Proof. Since G is supersoluble, we know that L is nilpotent. We must have $L_p \leq O^p(G)$. But $(O^p(G) \cap A)(O^p(G) \cap B)$ is a normal subgroup of G contained in $O^p(G)$ by Lemma 4.1.21 (1) and has p-power index because by Theorem 1.1.4 the index $|G : (O^p(G) \cap A)(O^p(G) \cap B)|$ divides $|G : O^p(G)|^2$. Therefore

$$O^p(G) = (O^p(G) \cap A)(O^p(G) \cap B).$$

Then since $L_p = O_p(O^p(G))$, we have

$$L_p = (L_p \cap (O^p(G) \cap A))(L_p \cap (O^p(G) \cap B)) = (L_p \cap A)(L_p \cap B)$$

by Theorem 3.2.25. We fix a prime p and write $A_0 = L_p \cap A$, $B_0 = L_p \cap B$ for the remainder of the proof. We also assume that $L_p \neq 1$ and then since at least one of A_0, B_0 is non-trivial, we assume that $A_0 \neq 1$. We will defer the proof that L_p has class at most 3 to the end. We can choose a Hall p'-subgroup Q of G such that $Q = (Q \cap A)(Q \cap B)$ by Theorem 1.1.19. We denote by A_p and B_p the Sylow p-subgroups of A and B, respectively, and by $A_{p'}$ and $B_{p'}$ the Hall p'-subgroups of A and B, respectively.

Proof of part 1. Suppose that L_p is abelian. Note that $L_p = [L_p, Q] \times C_{L_p}(Q)$, with $[L_p, Q]$ and $C_{L_p}(Q)$ both normal in G, by Lemma 5.1.6. Since L is the nilpotent residual of G, we must have $C_{L_p}(Q) = 1$.

Now we have that $L_p = A_0 B_0$ and observe that $A_{p'}$ centralises A_0 and $B_{p'}$ centralises B_0. Moreover, A permutes with $B_{p'}$ because AB is a mutually permutable product. Note that a Sylow p-subgroup A_p of A is a Sylow p-subgroup of $AB_{p'}$. Since $AB_{p'} \cap L_p$ is a normal p-subgroup of $AB_{p'}$, it is contained in all Sylow p-subgroups of A. Hence $AB_{p'} \cap L_p \leq A_p \cap L_p = A_0$. Since $A_0 \leq AB_{p'}$, we obtain that $AB_{p'} \cap L_p = A_0$ and hence $B_{p'}$ normalises A_0. Therefore, by [119, A, 12.5], $A_0 = [A_0, B_{p'}] \times C_{A_0}(B_{p'})$. Let $x \in C_{A_0}(B_{p'})$, then x centralises $A_{p'}B_{p'} = Q$ and so $x = 1$. It follows that $A_0 = [A_0, B_{p'}]$. By symmetry, $B_0 = [B_0, A_{p'}]$. Further, $A_0 \cap B_0 = 1$, because this intersection is centralised by both $A_{p'}$ and $B_{p'}$ and $C_{L_p}(Q) = 1$.

Let $x \in B_{p'}$ and $a \in A_0$. Then $\langle a\rangle B$ is a subgroup of G and $\langle a\rangle B \cap L_p = \langle a\rangle(B \cap L_p) = \langle a\rangle B_0$ is normal in $\langle a\rangle B$. But now we have $a^x \in A_0 \cap \langle a\rangle B_0 = \langle a\rangle(A_0 \cap B_0) = \langle a\rangle$. It follows that x acts as a power automorphism on A_0. A similar argument applies to the action of $A_{p'}$ on B_0.

Finally, we prove that A_0 and B_0 are normal subgroups of G. Since A_0 is normalised by A and by $B_{p'}$, it suffices to show that A_0 is normalised by B_p, where B_p is the Sylow p-subgroup of B. We argue by induction on $|G| + |A| + |B|$. Let $T = \mathrm{O}_{p'}(G)$ and suppose that $T \neq 1$. Then $G/T = (AT/T)(BT/T)$ satisfies the hypothesis, its nilpotent residual $(G/T)^{\mathfrak{N}}$ of G/T coincides with LT/T, and the Sylow p-subgroup of $(G/T)^{\mathfrak{N}}$ coincides with L_pT/T. By induction, B_pT/T normalises $(L_pT/T) \cap (AT/T)$. Let $x \in L_p \cap A$ and $b \in B_p$, then $x^b \in L_p$ because L_p is normal in G and $A\langle b\rangle$ is a subgroup of G because A and B are mutually permutable. Moreover, $x^b \in AT$ by induction. Hence $x^b \in AT \cap A\langle b\rangle \cap L_p = A(T \cap A\langle b\rangle) \cap L_p$. Since $T \cap A\langle b\rangle$ is a p'-group and $A(T \cap A\langle b\rangle) \cap L_p$ is a normal p-subgroup of $A(T \cap A\langle b\rangle)$, it is contained in all Sylow p-subgroups of $A(T \cap A\langle b\rangle)$. Since A_p is one of them, we have that $x^b \in A_p \cap L_p = A_0$. Hence A_0 is normalised by B_p. Since A_0 is normalised by $B_{p'}$, it follows that A_0 is normal in G.

Therefore we can assume that $\mathrm{O}_{p'}(G) = 1$. In particular, G has a normal Sylow p-subgroup, p is the largest prime dividing $|G|$ and G' is a p-group. Since $A \cap B$ is a nilpotent subnormal subgroup of G by Corollary 4.1.17, $A \cap B$ is a p-group. Now assume that there exists a minimal normal subgroup N of G contained in A_0. Then $G/N = (A/N)(BN/N)$ satisfies the hypothesis and so $(G/N)^{\mathfrak{N}} \cap (A/N)$ is a normal subgroup of G/N, that is, $G^{\mathfrak{N}} \cap A$ is a normal subgroup of G. Therefore we can assume that $(A_0)_G = 1$.

Assume now that A is not a p-group. Since $A \cap B$ is contained in A_p, we have by Lemma 4.1.21 (3) that $X = A_pB$ satisfies the hypothesis, and by induction $X^{\mathfrak{N}} \cap A$ is normalised by B_p. But X is a normal subgroup of G because it contains the Sylow p-subgroup of G and hence G'. Moreover, $X^{\mathfrak{N}} \cap A_p \leq L \cap A$ is normalised by $B_{p'}$. It follows that $X^{\mathfrak{N}} \cap A_p$ is a normal subgroup of G. Hence $X^{\mathfrak{N}} \cap A_p = 1$. In this case, on the one hand, $(B_{p'}X^{\mathfrak{N}}/X^{\mathfrak{N}})$ centralises $(A_0X^{\mathfrak{N}}/X^{\mathfrak{N}})$ and, on the other hand, $B_{p'}$ normalises A_0. It follows that $[A_0, B_{p'}] \leq X^{\mathfrak{N}} \cap A_0 = 1$ and so $B_{p'}$ centralises A_0. In this case, $B_p \leq \mathrm{C}_{A_0B_0}(Q) = 1$, that is, $B = B_{p'}$ is a p'-group and B normalises A_0. Therefore we can assume that A is a p-group. But in this case, $Q = B_{p'}$ and $B_0 \leq \mathrm{C}_{A_0B_0}(Q) = 1$. It follows that $L_p = A_0$ is a normal subgroup of G. This completes the proof of this part.

Proof of part 2. Suppose now that L_p has class 2. Then the conclusion of part 1 holds for G/L_p'. For the ease of reading, we break the argument into separately-stated steps.

1. $L_p' = A_0'B_0'[A_0, B_0]$.

Assume that $B_0 = 1$. Then $L_p = A_0$ and $L_p' = A_0'$. If $B_0 \neq 1$, then A_0L_p' and B_0L_p' are normal subgroups of G by part 1. Note that $(A_0L_p')' = A_0'$ and $(B_0L_p')' = B_0'$

since $L'_p \leq Z(L_p)$, the centre of L_p. Further, $L'_p = A'_0 B'_0 [A_0, B_0]$ and $A_0 \cap B_0 \leq L'_p$ by part 1.

2. *Q acts as a group of power automorphisms on each of A'_0, B'_0, and $[A_0, B_0]$.*

Since $B_{p'}$ acts as a group of power automorphisms on $A_0 L'_p / L'_p$ by part 1, if $g \in B_{p'}$ and $a \in A_0$, we have that $a^g L_{p'} = a^{\alpha} L_{p'}$ for some integer α prime to p. But now if $a, a^* \in A_0$, we have $[a, a^*]^g = [a, a^*]^{\alpha^2}$ and so $B_{p'}$ acts as a group of power automorphisms on A'_0. Similarly, if $h \in A_{p'}$ and $b, b^* \in B_0$, then $b^h L_{p'} = b^{\beta} L_{p'}$ for some integer β coprime to p. Now we have $[b, b^*]^h = [b, b^*]^{\beta^2}$ and $[a, b]^g = [a, b]^{\alpha}$ and $[a, b]^h = [a, b]^{\beta}$. Since $A_{p'}$ centralises A_0 and $B_{p'}$ centralises B_0, Q acts as a group of power automorphisms on each of A'_0, B'_0, and $[A_0, B_0]$.

3. $A'_0 B'_0 \cap [A_0, B_0] = 1$.

Suppose that $A'_0 B'_0 \cap [A_0, B_0] \neq 1$ and $1 \neq ab \in A'_0 B'_0 \cap [A_0, B_0]$ with $a \in A'_0$ and $b \in B'_0$. If $B_{p'}$ centralises $A_0 L'_p / L'_p$ and A'_p centralises $B_0 L'_p / L'_p$, then Q centralises L_p / L'_p, but this contradicts that $C_{L_p / L'_p}(Q) = 1$. Therefore we can assume that $B_{p'}$ does not centralise $A_0 L'_p / L'_p$ and consider $g \in B_{p'}$ such that $z^g L'_p = z^{\alpha} L'_p \neq z L'_p$ for all $z \in A_0$ and an integer α. We note that $\alpha \not\equiv 1 \pmod{p}$ because otherwise g would centralise all factors of a chief series of G between L'_p and $\langle z \rangle L'_p$ and by [119, A, 12.3], the p'-element g acts on $\langle z \rangle L'_p / L'_p$ as a p-group, which implies that g centralises $\langle z \rangle L'_p / L'_p$. Hence $(ab)^g = (ab)^{\alpha} = a^{\alpha^2} b$. Since L'_p is abelian, we obtain that $a^{\alpha} b^{\alpha} = a^{\alpha^2} b$, that is, $a^{\alpha - \alpha^2} = b^{1-\alpha}$. Let μ be the inverse of $1 - \alpha$ modulo the order of ab. Then we get $a^{\alpha} = b$, which is an element of $A'_0 \cap B'_0$. Therefore $ab \in A'_0 \cap B'_0$ is centralised by $B_{p'}$, which contradicts that $\alpha \not\equiv 1 \pmod{p}$. Consequently $A'_0 B'_0 \cap [A_0, B_0] = 1$.

4. *If $A'_0 \cap B'_0 \neq 1$, then A'_0 and B'_0 are hypercentral in G.*

Now assume that $A_{0'} \cap B_{0'} \neq 1$ and consider $1 \neq [a, a^*] = [b, b^*] \in A_{0'} \cap B_{0'}$. Let $g \in B'_p$, then $[a, a^*]^g = [a, a^*]^{\alpha^2} = [b, b^*]^g = [b, b^*] = [a, a^*]$, which implies that g centralises $[a, a^*]$. A similar argument shows that if $h \in A_{p'}$, then h centralises $[a, a^*]$. By Lemma 1.2.22, we have that $A'_0 \cap B'_0 = 1$ unless both terms are hypercentral.

5. *L_p is a central product of A_0 and B_0. Therefore we can assume that A_0 is non-abelian.*

Now suppose that $[A_0, B_0] \neq 1$ and let $1 \neq x \in [A_0, B_0]$. Since $L_p = A_0 B_0$, we can write $x = ab$ with $a \in A_0$ and $b \in B_0$. Then since $(A_0 L'_p / L'_p) \cap (B_0 L'_p / L'_p) = 1$, we must have $a \in L'_p$, $b \in L'_p$. If both a and b belong to $A'_0 B'_0$, then $x \in A'_0 B'_0 \cap [A_0, B_0] = 1$, against our supposition. Therefore we can assume that one of a or b does not belong to $A'_0 B'_0$, say b. But now $b = uv$ with $u \in A'_0 B'_0$ and $v \in [A_0, B_0]$. Consider $w \in B_{p'}$. Then $uv = b^w = u^w v^w$ and so $(u^w)^{-1} u = v^w v^{-1} \in A'_0 B'_0 \cap [A_0, B_0] = 1$. Then $v^w = v$. Since $v \neq 1$ and $B_{p'}$ acts as a group of power automorphisms on $[A_0, B_0]$, $B_{p'}$ centralises $[A_0, B_0]$. But the action of $B_{p'}$ on A_0/A_0' coincides with the action of $B_{p'}$ on $[A_0, B_0]$. Hence A_0/A_0' is centralised by $B_{p'}$ and so is A_0'. Since B is nilpotent, $B_{p'}$ centralises B_0. By applying [119, A, 12.3], we conclude that $B_{p'}$ centralises L_p. Thus $B_{p'}$ centralises $L_p/L_p' = (A_0L_p'/L_p') \times (B_0L_p'/L_p')$. But $A_{p'}$ centralises A_0L_p'/L_p'. It follows that Q centralises A_0L_p'/L_p' and so $L/B_0L_p'L_{p'}$ is contained in the hypercentre of $G/B_0L_p'L_{p'}$. Since L is the nilpotent residual of G, it follows that $L = B_0L_p'L_{p'}$, that is, $L_p = B_0L_p'$. Since $L_p' \leq \Phi(L_p)$, $L_p = B_0$. Consequently $A_0 = A_0 \cap B_0 \leq L_p'$. Since $L_p' \leq Z(L_p)$, $[A_0, B_0] = 1$. This contradicts our initial assumption. Thus L_p is a central product of A_0 and B_0. We may assume that A_0 is non-abelian.

6. *$B_{p'}$ normalises A_0.*

Let $a \in A_0$ and $g \in B_{p'}$. Since $B_{p'}$ acts as a group of power automorphisms on A_0L_p'/L_p', we obtain that $a^g \in \langle a\rangle L_p'$. Moreover, since A and B are mutually permutable, it follows that $a^g \in A\langle g\rangle$. Therefore $a^g \in \langle a\rangle L_p' \cap A\langle g\rangle = \langle a\rangle(L_p' \cap A\langle g\rangle)$. Since $L_p' \cap A\langle g\rangle$ is a normal p-subgroup of $A\langle g\rangle$, it is contained in all its Sylow p-subgroups, in particular, in the Sylow p-subgroup A_p of A. Therefore $L_p' \cap A\langle g\rangle = L_p' \cap A_p$ is contained in A_0 and so $a^g \in A_0$.

7. *$A_0/A_0' = (A_1/A_0') \times ((A_0 \cap L_p')/A_0')$, where $C_{A_0/A_0'}(B_{p'}) = (A_0 \cap L_p')/A_0'$, $A_1/A_0' = [A_0/A_0', B_{p'}]$, and A_1' coincides with A_0'.*

Since A_0L_p'/L_p' is $B_{p'}$-isomorphic to $A_0/(A_0 \cap L_p')$, we have that $A_0/(A_0 \cap L_p')$ has no fixed points under the action of $B_{p'}$. Also $A_0/A_0' = [A_0/A_0', B_{p'}] \times C_{A_0/A_0'}(B_{p'})$ by [119, A, 12.5]. Now we have that $A_0 \cap L_p' = A_0 \cap (A_0'B_0') = A_0'(A_0 \cap B_0')$. Note that $A_0 \cap B_0'$ is centralised by $B_{p'}$. We now have that $(A_0 \cap L_p')/A_0' = A_0'(A_0 \cap B_0')/A_0' \cong (A_0 \cap B_0')/(A_0' \cap B_0')$ as $B_{p'}$-groups and so $(A_0 \cap L_p')/A_0' \leq C_{A_0/A_0'}(B_{p'})$. Then since $A_0/(A_0 \cap L_p')$ has no fixed points under the action of $B_{p'}$, we have $C_{A_0/A_0'}(B_{p'}) = (A_0 \cap L_p')/A_0'$. Let $A_1/A_0' = [A_0/A_0', B_{p'}]$. Note that A_1 is normal in G by virtue of Lemma 5.1.6. We then have $A_1' = A_0'$ because $A_0 = A_1(A_0 \cap L_p')$ and $A_0 \cap L_p' \leq Z(L_p)$.

8. *$B \cap A_1$ is non-trivial and $B_{p'}$ centralises A_1'.*

Note that since A_0 is non-abelian, then A_1 is non-abelian, too. By Lemma 1.3.4, $B_{p'}$ cannot induce a power automorphism group on A_1. Hence for some $x \in A_1$ and some $y \in B_{p'}$, we have $\langle x^y\rangle \neq \langle x\rangle$. Since $\langle x^y\rangle \leq (\langle x\rangle B) \cap A_1 = \langle x\rangle(B \cap A_1)$, we must have $B \cap A_1 \neq 1$. Note that $B \cap A_1$ is centralised by $B_{p'}$ and so, since A_1/A_1' has no fixed points under the action of $B_{p'}$, we must have $B \cap A_1 \leq A_1'$. Since $B_{p'}$ acts as a group of power automorphisms on A_1', we obtain that $B_{p'}$ centralises A_1'.

9. *$C_{B_{p'}}(A_1)$ has index 2 in $B_{p'}$.*

If $x \in B_{p'}$ and $u, v \in A_1$, then $u^x = u^\alpha s$ and $v^x = v^\alpha t$ with $s, t \in A_1 \cap L_p'$. Since $A_1 \cap (A_0 \cap L_p') = A_0'$, we obtain that $s, t \in A_0'$. We then have $[u, v]^x =$

$[u, v]^{\alpha^2} = [u, v]$. It follows that $\alpha^2 \equiv 1 \pmod p$ and hence $C_{B_{p'}}(A_1)$ has index 2 in $B_{p'}$ (because $B_{p'}$ cannot centralise A_1: otherwise, it would centralise $A_0/A_0' = (A_1/A_0') \times C_{A_0/A_0'}(B_{p'})$ and hence A_0L_p'/L_p'). Moreover for some $x \in B_{p'}$, we then have $u^x = u^{-1}s$ with $s \in A_0'$. Note that $B \cap A_1$ is a normal subgroup of B because A_1 is normal in G.

10. *If B_0 is also non-abelian, then L_p' is hypercentral in G.*

Now suppose that $B_0' \neq 1$, too. We have by the same argument as above that $A_{p'}$ centralises B_0' and $C_{A_{p'}}(B_0)$ has index 2 in $A_{p'}$. Then Q centralises both A_0' and B_0' and hence $A_0'B_0' = L_p'$.

11. *A_0' is contained in B.*

Next we suppose that A_0' is not contained in B. We choose $u \in A_0' \setminus B$ and then $v \in A_1 \setminus A_0'$ of maximal order, p^β say. Since $u \in A_0'$, we have $u \in C_{A_0'}(B_{p'})$. We also have that since $B_{p'}$ acts as a group of power automorphisms on A_1L_p'/L_p', we have for some $x \in B_{p'}$ that $v^x = v^{-1}s$ with $s \in A_1 \cap L_p' = A_1 \cap (A_0 \cap L_p') = A_0'$. Now $A_0'\langle v\rangle = Y$ is abelian and so by Lemma 5.1.6 we have $Y = C_Y(\langle x\rangle) \times [Y, \langle x\rangle] = A_0' \times \langle w\rangle$ with $w^x = w^{-1}$ and w of order p^β, because $C_Y(\langle x\rangle) = A_0'$. Then no non-trivial power of w can lie in L_p'. Since A_0' is generated by commutators of elements of A_1, all of which have order at most p^β modulo L_p', A_0' has exponent at most p^β. In particular, u has order at most p^β. Now consider $\langle\langle uw\rangle^{B_{p'}}\rangle$. We have $(uw)^x = u(w^x) = uw^{-1}$ and so $(uw)^x(uw) = uw^{-1}uw = u^2 \in \langle uw\rangle$. Since $p \neq 2$, we have that $u \in \langle\langle uw\rangle^{B_{p'}}\rangle$.

Thus $\langle u\rangle \leq B\langle uw\rangle \cap (A \cap L_p) = \langle uw\rangle(B \cap A \cap L_p)$. From the part 1, we have $B \cap A \cap L_p \leq L_p'$ and so $\langle u\rangle \leq L_p' \cap \langle uw\rangle(B \cap A \cap L_p') = (\langle uw\rangle \cap L_p')(B \cap A \cap L_p')$. Assume that a non-trivial power $(uw)^s$ of uw is centralised by $B_{p'}$. Then $u^s w^s$ is centralised by $B_{p'}$ and, since $u^s \in A_0'$ is centralised by $B_{p'}$, then w^s is centralised by $B_{p'}$. Since uw has order p^β, this cannot happen, because $\langle w\rangle \cap A_0' = C_{B_{p'}}(w) = 1$. Recall that $L_p' = A_0'B_0'$ is centralised by $B_{p'}$. Since no non-trivial power of uw can be centralised by $B_{p'}$, it follows that $\langle uw\rangle \cap L_p' = 1$, giving $u \in B$, a contradiction. Hence we have $A_0' \leq B$.

12. *$L_p' = A_0 \cap B_0$ is hypercentral in G and hence both A_0 and B_0 are normal subgroups of G.*

We now have $A_0' \leq B \cap L_p = B_0$. If $B_0' = 1$, then we have $L_p' = A_0'$. If $B_0' \neq 1$, then we have $B_0' \leq A_0$ by symmetry and so $L_p' = A_0'B_0' \leq A_0 \cap B_0$. On the other hand, $A_0 \cap B_0 \leq L_p'$ by part 1. Hence $L_p' = A_0 \cap B_0$. Since L_p' is centralised by Q, we obtain that L_p' is hypercentral. Moreover, $A_0 = A_0(A_0 \cap B_0) = A_0L_p'$ is a normal subgroup of G, and, by symmetry, B_0 is a normal subgroup of G.

13. *$Q/C_Q(L_p)$ is an elementary abelian group of order dividing* 4.

Since $A_0 = A_1(A_0 \cap L_p')$, $|B_{p'}/C_{B_{p'}}(A_1)| = 2$, and L_p' is centralised by $B_{p'}$, we conclude that $|B_{p'}/C_{B_{p'}}(A_0)| = 2$. Assume first that $B_0' \neq 1$. By symmetry, we have that $|A_{p'}/C_{A_{p'}}(B_0)| = 2$. If B_0 is abelian, we know that $|A_{p'}/C_{A_{p'}}(B_0)| = 1$. In any case, we have that $|B_{p'}/C_{B_{p'}}(A_0)| = 2$ and $|A_{p'}/C_{A_{p'}}(B_0)| \leq 2$. It follows that $|Q/C_{A_{p'}}(B_0)B_{p'}| \leq 2$ and $|Q/A_{p'}C_{B_{p'}}(A_0)| \leq 2$. Since

$$C_{A_{p'}}(B_0)B_{p'} \cap A_{p'}C_{B_{p'}}(A_0) = C_{A_{p'}}(B_0)(B_{p'} \cap A_{p'})C_{B_{p'}}(A_0)$$
$$= C_{A_{p'}}(B_0)C_{B_{p'}}(A_0) \leq C_Q(L_p),$$

it follows that $Q/C_Q(L_p)$ is elementary abelian of order dividing 4.

Proof of part 3. Assume that L_p has class 3. In this case, $G/\gamma_3(L_p)$ satisfies the conditions of part 2. We also break the proof into steps for the ease of reading.

1. $[A_0', B_0] = [B_0', A_0] = 1$ *and* $\gamma_3(L_p) = \gamma_3(A_0)\gamma_3(B_0)$.

Since $L_p = A_0B_0$, it follows that $\gamma_3(L_p)$ is generated by commutators of the form $[x, y, z]$ with each of x, y, z in either A_0 or B_0 and it is enough to show that such a commutator with just one entry from A_0 or from B_0 is trivial. Suppose that we have $x \in A_0$, $y \in B_0$. By part 2, we have that $[x, y] \in \gamma_3(L_p)$ and so $[x, y, z] = 1$ for any $z \in L_p$. Hence we suppose that x, $y \in A_0$ and $z \in B_0$. Now by the Witt identity (see [157, III, 1.4]), we have that $[x, y, z][z, x, y][y, x, z] = 1$ and so since $[z, x, y] = [y, z, x] = 1$, we also have $[x, y, z] = 1$ and so $[A_0', B_0] = 1$. By a symmetrical argument, $[B_0', A_0] = 1$. Thus the only non-trivial commutators must have all three entries from the same factor and so $\gamma_3(L_p) = \gamma_3(A_0)\gamma_3(B_0)$.

Note in particular that at least one of A_0 and B_0 must have class 3. Suppose that A_0 has class 3.

2. $B_{p'}$ *acts on* $\gamma_3(A_0)$ *as a group of power automorphisms and* $A_{p'}$ *acts on* $\gamma_3(B_0)$ *as a group of power automorphisms. More precisely, if* $g \in B_{p'}$ *satisfies* $a^g L_{p'} = a^r L_{p'}$ *for all* $a \in A_0$, *then* $z^g = z^{r^3}$ *for all* $z \in \gamma_3(A_0)$.

This follows by a similar argument to the one used in the proof of part 2 to show that $B_{p'}$ induces power automorphisms on A_0' when L_p has class 2.

3. $\gamma_3(A_0) \cap \gamma_3(B_0) = 1$.

Assume that $g \in B_{p'}$ induces an automorphism of order 2 on L_p/L_p'. Then for every $a \in A_0$, we have $a^g L_p = a^{-1} L_p$, since the inversion is the unique power automorphism of order 2 of an abelian p-group with p odd and so g inverts all elements of $\gamma_3(A_0)$. Since $B_{p'}$ centralises $\gamma_3(B_0)$, we obtain that $\gamma_3(A_0) \cap \gamma_3(B_0) = 1$.

4. A_0 *is a normal subgroup of* G.

We know by part 2 that $A_0\gamma_3(L_p)$ is a normal subgroup of G. Since $\gamma_3(L_p) = \gamma_3(A_0)\gamma_3(B_0)$, we obtain that $A_0\gamma_3(B_0)$ is a normal subgroup of G. In order to prove that A_0 is normal in G, we argue by induction on $|G| + |A| + |B|$. Let $T = O_{p'}(G)$. First of all, suppose that $T \neq 1$. Then $G/T = (AT/T)(BT/T)$ satisfies the hypothesis and so $AT/T \cap L_pT/T$ is normalised by the Sylow p-subgroup

$(BT/T)_p = B_pT/T$ of BT/T. Let $a \in A \cap L_p$ and $b \in B_p$, then $a^b \in L_p \cap AT \cap A\langle b\rangle = L_p \cap A(T \cap A\langle b\rangle)$. Now this is a normal p-subgroup of $A(T \cap A\langle b\rangle)$ and so it is contained in A_p, which is a Sylow p-subgroup of $A(T \cap A\langle b\rangle)$. Hence $a^b \in L_p \cap A_p = A_0$. It follows that we can assume $T = 1$. In this case, $A \cap B$ is a subnormal nilpotent subgroup of G and so $A \cap B$ is contained in the Fitting subgroup $\mathrm{F}(G)$ of G, which coincides with $\mathrm{O}_p(G)$ and is the Sylow p-subgroup of G. Moreover, since G is supersoluble by Corollary 4.1.39, we have that Q is abelian of exponent dividing $p-1$ and that G' is a p-group.

Assume that N is a minimal normal subgroup of G contained in A_0. Then $G/N = (A/N)(BN/N)$ satisfies the hypothesis of the theorem and so $(L_pN/N) \cap (A/N)$ is normalised by B. Consequently $L_p \cap A$ is a normal subgroup of G. Hence we can assume that $(A_0)_G = 1$.

Suppose that A is not a p-group. Let $X = A_pB$, which is a normal subgroup of G because X contains G'. By induction, we have that $X^{\mathfrak{N}} \cap A_p$ is a normal subgroup of X. But $X^{\mathfrak{N}}$ is a normal subgroup of G. It follows that $X^{\mathfrak{N}} \cap A_p$ is a normal subgroup of G because it is normalised by X and centralised by $A_{p'}$. Since $(A_0)_G = 1$, it follows that $X^{\mathfrak{N}} \cap A_p = 1$. Furthermore, the nilpotency of $X/X^{\mathfrak{N}}$ yields $[A_0X^{\mathfrak{N}}/X^{\mathfrak{N}}, B_{p'}X^{\mathfrak{N}}/X^{\mathfrak{N}}] = 1$. We deduce that $[A_0, B_{p'}] \leq X^{\mathfrak{N}} \cap AB_{p'} = X^{\mathfrak{N}} \cap A_p = 1$ because $X^{\mathfrak{N}} \cap AB_{p'}$ a normal p-subgroup of $AB_{p'}$ and hence it is contained in the Sylow p-subgroup A_p of $AB_{p'}$. Therefore $B_{p'}$ centralises A_0. Since $A_{p'}$ also centralises A_0, it follows that Q centralises $A_0L'_p/L'_p$. Consequently $A_0L'_p/L'_p \leq \mathrm{C}_{L_p/L'_p}(Q) = 1$, that is, $A_0 \leq L'_p$. But since L_p has class 3, we conclude that L'_p is abelian by [157, III, 2.12]. Hence A_0 is abelian, which contradicts that A_0 has class 3.

Therefore A is a p-group. Hence $A_{p'} = 1$ centralises B_p and so does $B_{p'}$. The same argument as above shows that B_0 is contained in L'_p and so $L_p = A_0B_0 = A_0L'_p$. Recall that L'_p is contained in $\Phi(L_p)$ because L is nilpotent. Consequently $L_p = A_0$ is a normal subgroup of G.

5. $[L'_p \cap A_0, A_0] = \gamma_3(A_0)$.

Since $L'_p = A'_0B'_0\gamma_3(L_p) = A'_0B'_0\gamma_3(A_0)\gamma_3(B_0) = A'_0B'_0$, it follows that $L'_p \cap A_0 = A'_0(A_0 \cap B'_0)$. Moreover $A_0 \cap B_0 \leq \gamma_3(L_p)$ because $L_p/\gamma_3(L_p)$ is a central product of $A_0\gamma_3(L_p)/\gamma_3(L_p)$ and $B_0\gamma_3(L_p)/\gamma_3(L_p)$. Hence

$$[L'_p \cap A_0, A_0] = [A'_0(A_0 \cap B'_0), A_0] = [A'_0, A_0][A_0 \cap B'_0, A_0] = \gamma_3(A_0)$$

by Step 1.

6. *$B_{p'}$ centralises $(L'_p \cap A_0)/\gamma_3(A_0)$.*

Let $g \in B_{p'}$ such that $a^gL'_p = a^{-1}L'_p$ for all $a \in A_0$. Let $a_1, a_2 \in A_0$, and suppose that $a_i^g = a_i^{-1}x_i$ with $x_i \in L'_p \cap A_0$ for $i \in \{1,2\}$. Then $[a_1,a_2]^g\gamma_3(A_0) = [a_1^{-1}x_1, a_2^{-1}x_2]\gamma_3(A_0) = [a_1^{-1}, a_2^{-1}]\gamma_3(A_0) = [a_1,a_2]\gamma_3(A_0)$ bearing in mind that $A_0/\gamma_3(A_0)$ has class 2. It follows that $A'_0/\gamma_3(A_0)$ is centralised by $B_{p'}$. Since $A_0 \cap B'_0$

is also centralised by $B_{p'}$, it follows that $(L'_p \cap A_0)/\gamma_3(A_0) = A'_0(A_0 \cap B'_0)/\gamma_3(A_0)$ is centralised by $B_{p'}$ as well.

7. *If* $v \in B_{p'} \setminus \mathrm{C}_{B_{p'}}(A_0 L'_p/L'_p)$, *then* $\mathrm{C}_{L'_p \cap A_0}(v) = A_0 \cap B_0$.

Since

$$L'_p/\gamma_3(L_p) = (A_0\gamma_3(L_p)/\gamma_3(L_p)) \cap (B_0\gamma_3(L_p)/\gamma_3(L_p))$$

and $\gamma_3(L_p) = \gamma_3(A_0)\gamma_3(B_0)$, we obtain that $L'_p = A_0\gamma_3(B_0) \cap B_0\gamma_3(A_0)$ and so $L'_p \cap A_0 = A_0 \cap B_0\gamma_3(A_0) = (A_0 \cap B_0)\gamma_3(A_0)$. We have that $A_0 \cap B_0 \leq \mathrm{C}_{L'_p \cap A_0}(B_{p'})$. Consequently $\mathrm{C}_{L'_p \cap A_0}(v) = (A_0 \cap B_0)\gamma_3(A_0) \cap \mathrm{C}_{L'_p \cap A_0}(v) = (A_0 \cap B_0)(\gamma_3(A_0) \cap \mathrm{C}_{L'_p \cap A_0}(v)) = A_0 \cap B_0$, because v inverts all elements of $\gamma_3(A_0)$.

8. *Let* $M \leq A'_0$ *be the subgroup of* $H = A_0 B$ *generated by all normal subgroups* R *of* H *such that* $R \leq A'_0$ *and* $R \cap \gamma_3(A_0) = 1$. *Then* $\gamma_3(A_0)$ *has at most two generators and if* $\gamma_3(A_0)$ *is non-cyclic or* $p > 3$, *then* $A'_0/(M\gamma_3(A_0))$ *is cyclic.*

This follows by applying Lemma 5.1.10 with $U = A_0$, $U^* = L'_p \cap A_0$, $V = B$, $K = \mathrm{C}_{L'_p \cap A_0}(v) = A_0 \cap B_0 \leq B$.

9. *If* $\gamma_3(A_0)$ *is non-cyclic or* $p > 3$, *then* $A'_0 \mathrm{Z}(A_0)/\mathrm{Z}(A_0)$ *is cyclic.*

If R is a normal subgroup of H such that $R \leq A'_0$ and $R \cap \gamma_3(A_0) = 1$, then $[R, A_0] \leq [A'_0, A_0] = \gamma_3(A_0)$ and $[R, A_0] \leq R$ because A_0 normalises R. It follows that $[R, A_0] = 1$ and so $R \leq \mathrm{Z}(A_0)$. Hence $M \leq \mathrm{Z}(A_0)$ and so $M\gamma_3(A_0) \leq \mathrm{Z}(A_0) \cap A'_0$. Since $A'_0/M\gamma_3(A_0)$ is cyclic, so is $A'_0/(A'_0 \cap \mathrm{Z}(A_0))$, which is isomorphic to $A_0\mathrm{Z}(A_0)/\mathrm{Z}(A_0)$. This completes the proof for L_p of class 3.

Proof that L_p *has class at most* 3. Suppose that L_p has class 4. It will be enough to show that commutators w of the form $w = [x_1, x_2, x_3, x_4]$ with each x_i in A_0 or B_0 are trivial. If x_1 and x_2 come from different factors, say $x_1 \in A_0$ and $x_2 \in B_0$, then $[x_1, x_2] \in \gamma_3(L_p)$ by part 2 and so $w \in \gamma_5(L_p) = 1$. Thus we may assume that x_1 and x_2 come from the same factor, A_0 say. If $x_3 \in B_0$, then $[x_1, x_2, x_3] \in \gamma_4(L_p)$ by part 3 and so again $w \in \gamma_5(L_p) = 1$. Hence we may suppose also that $x_3 \in A_0$. Put $c = [x_1, x_2]$ and suppose that $x_4 \in B_0$. Then by the Witt identity [157, III, 1.4] we have $[c, x_3, x_4][x_4, c, x_3][x_3, x_4, c] = 1$. But $[x_4, c] = [c, x_4]^{-1} = [x_1, x_2, x_4]^{-1} \in \gamma_4(L_p)$ by part 3. Also $[x_3, x_4] \in \gamma_3(L_p)$ by part 2. It now follows that $[x_4, c, x_3]$ and $[x_3, x_4, c]$ are both trivial and hence $[x_1, x_2, x_3, x_4] = 1$. We now have, as in the proof of part 3, that $\gamma_4(L_p)$ is generated by commutators all of whose entries come from either A_0 or B_0 so that $\gamma_4(L_p) = \gamma_4(A_0)\gamma_4(B_0)$. In particular, one of A_0 and B_0 must have class 4. We assume that A_0 has class 4. Note that also since $A_0 L'_p$ is normal in G by part 1, $\gamma_4(A_0) = \gamma_4(A_0 L'_p)$ is normal in G. By taking $G/\gamma_4(A_0)^p$, we may also suppose that $\gamma_4(A_0)$ has exponent p. It then follows that $\gamma_3(A_0)^p \leq \mathrm{Z}(A_0)$ because if $g \in \gamma_3(A_0)$ and $a \in A_0$, then $[g^p, a] = [g, a]^p = 1$.

Note that by Lemma 5.1.8, we can choose generators for A_0, say $a_1, a_2, \dots, a_n$, such that $\langle a_i \gamma_4(A_0)\rangle$ are normalised by B. Suppose first that $\gamma_3(A_0)/\gamma_4(A_0)$ is cyclic, generated by $c\gamma_4(A_0)$ say. Observe that if $\gamma_3(A_0)/\gamma_4(A_0)$ is a proper subgroup of $\langle a_i \gamma_4(A_0)\rangle$ for some i, we can assume that $\gamma_3(A_0)/\gamma_4(A_0)$ is a proper subgroup of $\langle a_j \gamma_4(A_0)\rangle$ for some j by replacing a_j by $a_i a_j a_i$ if necessary (note that $\langle a_i a_j a_i\rangle$ is also normalised by B and that if a_j has order smaller than a_i, then $a_i a_j a_i$ has the same order as a_i). Now consider $c = [x_1, x_2, x_3, x_4]$ with $x_i \in \{a_1, a_2, \dots, a_n\}$. If $[x_1, x_2, x_3] \in \gamma_4(A_0)$, then $c = 1$. Hence we may suppose that $[x_1, x_2, x_3] = d$ generates $\gamma_3(A_0)$ modulo $\gamma_4(A_0)$. But then $d \in \langle x_4\rangle\gamma_4(A_0)$ and so $[d, x_4] = 1$. It now follows that A_0 has class 3, which contradicts the fact that A_0 has class 4.

Now suppose that $\gamma_3(A_0)/\gamma_4(A_0)$ is non-cyclic. As in the proof of Lemma 5.1.10, we can choose $x, y \in A_0$ such that $\langle x\gamma_4(A_0)\rangle$ and $\langle y\gamma_4(A_0)\rangle$ are normalised by B, $[x, y]$ generates A_0' modulo $\gamma_3(A_0)$, and $[x, y, x]$ and $[x, y, y]$ generate $\gamma_3(A_0)$ modulo $\gamma_4(A_0)$. Thus $\gamma_4(A_0)$ is generated by the commutators $[x, y, x, u]$ and $[x, y, y, v]$ for $u, v \in A_0$. Suppose that $[x, y, x, u] \neq 1$ for some u. By the Witt identity [157, III, 1.4], we have

$$[[x, y], x, u][u, [x, y], x][x, u, [x, y]] = 1.$$

But now $[x, u] = [x, y]^{\alpha} w$ with $w \in \gamma_3(A_0)$ and so $[x, u, [x, y]] = 1$. Also

$$[u, [x, y]] = [x, y, x]^{\beta} [x, y, y]^{\delta} z$$

with $z \in \gamma_4(A_0)$ and so to show $[x, y, x, u] = 1$ it will be enough to show that $[x, y, x, x] = [x, y, y, x] = 1$. Note that conjugation by $[x, y]$ is a power automorphism of $\langle x\rangle$ modulo $\gamma_4(A_0)$ as in the proof of Lemma 5.1.10 and so $[x, y, x]$ is a power of x modulo $\gamma_4(A_0)$, giving $[x, y, x, x] = 1$. Similarly $[x, y, y, y] = 1$ and $[x, y, xyx, xyx] = 1$ (note in this case that $xyx\gamma_4(A_0)$ is inverted by $v \in B_{p'} \setminus \mathrm{C}_{B_{p'}}(A_0)$ and by Lemma 5.1.8, $\langle xyx\gamma_4(A_0)\rangle$ is normalised by B). But

$$[x, y, xyx, xyx] = [x, y, y, x]^2 [x, y, x, y]^2.$$

A 2-generator group of class 4 is metabelian by [157, III, 2.12]. Moreover, by the Witt identity [157, III, 1.4],

$$[[x, y], y, x][y, x, [x, y]][x, [x, y], y] = 1,$$

since A_0 is metabelian, $[y, x, [x, y]] = 1$, and since A_0 has class 4 we have

$$[x, [x, y], y] = [[x, y], x, y]^{-1}$$

we obtain that $[x, y, x, y]^4 = 1$ and, since $p \neq 2$, we have $[x, y, x, y] = 1$. A similar argument shows that $[x, y, y, v] = 1$ and hence A_0 has class 3, a contradiction which completes the proof. □

Example 5.1.12. Let p be an odd prime and let F be the n generator (relatively) free group of class 2 and exponent p ($n \geq 2$). Then every map t inverting a set of free generators of F extends to an automorphism of F of order 2 (which inverts all elements of F/F' and centralises F'). Set $G = F\langle t\rangle$, $A = F$, and $B = F'\langle t\rangle$. It is easy to check that G is the mutually permutable product of A and B. Note that any subgroup N of $F' = A \cap B$ is normal in G and so G/N also provides an example. For details about relatively free groups and their automorphisms, we refer the reader to Sections 1.3, 1.4, and 4.1 of [206].

Example 5.1.13. Let $p > 3$ be a prime and set $q = p^n$ for some $n > 0$. Let $C = \langle c\rangle \times \langle d\rangle \times \langle e\rangle$ be isomorphic to $C_q \times C_q \times C_q$ and construct an extension $S = \langle C, a\rangle$ with $a^q = e$, $c^a = cd$, $d^a = d$, $e^a = e$, and the extension $T_0 = \langle S, b\rangle$ with $b^q = c$, $a^b = ac$, $c^b = ce$, $d^b = d$, $e^b = e$. Two applications of [223, 9.7.1] show that T_0 is a group of order q^5. Moreover T_0 has class 3 and $\mathrm{Z}(T_0) = \langle d, e\rangle$. Let t be an automorphism of T_0 defined by $a^t = a^{-1}$, $b^t = b^{-1}$; this automorphism fixes $cc^t = [a,b][a^{-1},b^{-1}]$. Note that $c^t = aba^{-1}b^{-1} = abcb^{-1}a^{-1} = ce^{q-1}d^{q-1}$. We then form the semidirect product of T_0 and $\langle t\rangle$ and denote it by T. Note that $(cc^t)^{a^m b^n} = c^{a^m b^n} c^{a^{-m}b^{-n}t} = cc^t d^{2m} e^{2n}$. If $p > 3$, then T_0 is regular by [157, III, 10.2 (a)] and this yields $d^{2m}e^{2n} \in \langle a^m b^n\rangle$. Hence all cyclic subgroups of T_0 permute with $\langle cc^t\rangle$. We then have that T is the mutually permutable product of $A = T_0$ and $B = \langle cc^t, t\rangle$ with A the nilpotent residual of T. Note that $A \cap B = \langle cc^t\rangle$ is core-free since $cc^t = c^2 e^{q-1} d^{q-1}$ intersects trivially $\mathrm{Z}(T_0)$.

We now take isomorphic copies T_1 and T_2 of T with A_1, B_1 and A_2, B_2 the corresponding factors, respectively, so that $W = U_1 \times U_2$ is the mutually permutable product of $A_3 = A_1 \times B_2$ and $B_3 = B_1 \times A_2$. The nilpotent residual of W has class 3 and both $\gamma_3(A_3)$ and $\gamma_3(B_3)$ are two generator groups. This shows that the result of part 3 of Theorem 5.1.11 cannot be improved.

This construction is not valid for $p = 3$: in this case, we have that B does not permute with $\langle ab\rangle$. But a slight modification of T_0 gives a corresponding example for $p = 3$. With $q = 3^n$, we choose $\langle c\rangle \times \langle d\rangle \times \langle e\rangle \cong C_q \times C_q \times C_q$ and consider a and b such that $a^{3q} = d$, $b^{3q} = e$, $[a,b] = c$, $c^a = cd$, $c^b = ce$, $d^a = d$, $d^b = d$, $e^a = e$, $e^b = e$. The corresponding extension T_0 gives a group of order $9q^5$. We consider again t such that $a^t = a^{-1}$, $b^t = b^{-1}$ and the semidirect product $T = [T_0]\langle t\rangle$. In a similar construction with $B = \langle cc^t, t\rangle$ and T_0, we see that $(xy)^{3q} = x^{3q}y^{3q}$ for any $x, y \in T_0$ and we obtain that T is the mutually permutable product of T_0 and B.

Example 5.1.14. Let S be the set of 4×4 matrices with entries from $\mathrm{GF}(3^3)$ defined as follows:

$$S = \{(m_{i,j}) : m_{i+1,j+1} = m_{i,j}^3,\ m_{1,1} = 1,\ m_{i,1} = 0,\ 2 \leq i \leq 4\}.$$

This is a group of upper unitriangular matrices of order 3^9, exponent 3^2, and class 3. Let t be the diagonal matrix $\mathrm{diag}(1,-1,1,-1)$. Note that conjugation by t induces an automorphism of order 2 on S. Let L be a subgroup of order 3 not containing 1 of the additive subgroup of $\mathrm{GF}(3^3)$ and set $Z = \{(m_{i,j}) \in S : m_{1,2} = m_{1,3} = 0, m_{1,4} \in L\}$. Note that $Z \le \mathrm{Z}(S)$ and if $(m_{i,j})$ is a matrix in Z, then $(m_{i,j})^t = (m_{i,j})^{-1}$ and so Z is a normal subgroup of $S\langle t\rangle$. Easy calculations show that t inverts $\mathrm{Z}(S) = \{(m_{i,j}) \in S : m_{1,2} = m_{1,3} = 0\}$ and that $C = \mathrm{C}_S(t) = \{(m_{i,j}) : m_{1,2} = m_{1,4} = 0\}$.

Let $T = S\langle t\rangle/Z$ and set $A = S/Z$ and $B = CZ\langle t\rangle/Z$. Then T is the mutually permutable product of A and B. Note that $S/Z = T^{\mathfrak{N}}$ and $\gamma_3(S/Z)$ is cyclic. Further $(S/Z)'\mathrm{Z}(S/Z)/\mathrm{Z}(S/Z)$ is a 3 generator group.

Note that this product is not totally permutable, even though $(A \cap B)_T = 1$.

Totally permutable products of two nilpotent groups have been studied by Ballester-Bolinches, Cossey, and Esteban-Romero in [39].

Theorem 5.1.15. *Let the group $G = AB$ be the product of the totally permutable nilpotent subgroups A and B. Then:*

1. *The nilpotent residual $L = G^{\mathfrak{N}}$ of G is abelian.*
2. *If p is a prime dividing $|L|$, A_p is the Sylow p-subgroup of A and B_p is the Sylow p-subgroup of B, and $A_p \cap L \neq 1$, then $A_p \le L$. In particular, the Sylow p-subgroup L_p of L must be either A_p, B_p, or $A_p \times B_p$.*

Proof. 1. Assume that the result is false and that G is a counterexample of least order.

First of all, note that G is supersoluble by Corollary 4.1.39. Therefore the nilpotent residual $L = G^{\mathfrak{N}}$ of G is nilpotent. Since the class of abelian-by-nilpotent groups is a formation, we have that G has a unique minimal normal subgroup N and that $N = L'$. In particular, L is a p-group for the largest prime p dividing $|G|$. Moreover, G has a normal Sylow p-subgroup P. Since L is nilpotent, N is contained in the centre $\mathrm{Z}(L)$ and so L has class 2. According to Theorem 5.1.11 (2), $A_0 = L \cap A$ and $B_0 = L \cap B$ are normal subgroups of G, $N = L' = L \cap A \cap B = A_0'B_0'$, and L is a central product of A_0 and B_0. Since L is non-abelian, one of the factors A_0 or B_0 is non-abelian. We assume without loss of generality that A_0 is non-abelian. By Lemma 4.1.37, we have that $B_{p'}$ acts on A_0 as a group of power automorphisms. By Lemma 1.3.4, $B_{p'}$ centralises A_0. Let Q be a Hall p'-subgroup of G factorising as $Q = A_{p'}B_{p'}$, where $A_{p'}$ and $B_{p'}$ are the Hall p'-subgroups of A and B, respectively. Since A_0/N is centralised by both $B_{p'}N/N$ and $A_{p'}N/N$, it follows that A_0/N is centralised by Q. Assume that $A_0/N \neq 1$. Then there exists a minimal normal subgroup C/N of G/N contained in A_0/N. Since C/N is centralised by Q and is central in P/N, it follows that C/N is central in G/N. But L/N is abelian and by [47, 4.2.19], L/N cannot contain central G-chief factors. Therefore $A_0 = N$ and so A_0 is abelian, the final contradiction which proves the result. 2. We argue by induction on $|G|$. Let $A_0 = A \cap L_p$ and $B_0 = B \cap L_p$. Assume that $A_0 \neq 1$. Let q be the largest prime dividing $|G|$ and suppose that $p < q$. Since G is supersoluble, G has a normal Sylow q-subgroup Q. We have that $(G/Q)^{\mathfrak{N}} = LQ/Q$ has L_pQ/Q as a Sylow p-subgroup. Moreover $AQ \cap L_pQ = AQ \cap A_0B_0Q = A_0(AQ \cap B_0Q) = A_0(AQ \cap B_0)Q = A_0Q$ because $AQ \cap B_0$ is contained in the Sylow p-subgroup A_p of AQ and so $AQ \cap B_0 \leq A \cap B_0 = A_0 \cap B_0 = 1$ by Theorem 5.1.11 (1). By induction, if $A_0 \neq 1$, then $A_0Q/Q \neq 1$ and so $A_0Q/Q = A_pQ/Q$. Since $A_0 \leq A_p$, by order considerations we obtain $A_0 = A_p$. An analogous statement holds for B_p. Therefore we can assume that $p = q$. Let P be the normal Sylow p-subgroup of G. Since A_0 is normal in G, we have that G/A_0 is the totally permutable product of A/A_0 and BA_0/A_0 and $(G/A_0)^{\mathfrak{N}} = L/A_0$. By induction, we have that $L = A_0$, $L = A_0B_p$, $L = A_p$, or $L = A_pB_p$.

Since $A_pB_{p'}$ is the totally permutable product of A_p and $B_{p'}$, we obtain that $B_{p'}$ acts as a group of power automorphisms on A_p by Lemma 4.1.37. Since L is abelian, L cannot have central G-chief factors by [47, 4.2.19]. This implies that there exists a p'-element g such that g does not centralise A_0, and since $A_{p'}$ centralises A_0, we can assume $g \in B_{p'}$. By Lemma 1.3.4, we obtain that A_p is abelian. □

We record as consequence of Theorem 5.1.15 a nice result of Shunkov [228] (see [95, Theorem 3.18]).

Theorem 5.1.16. *A group $G = G_1G_2 \cdots G_r$ is a product of the pairwise totally permutable subgroups of prime power order G_i, $1 \leq i \leq r$, if and only if G is a semidirect product $G = [A]B$ of an abelian group A and a nilpotent group B such that if p divides $|A|$ and A_p is the Sylow p-subgroup of A, then A_p can be decomposed as a direct product $A_p = A_{p,1} \times \cdots \times A_{p,n_p}$ and G induces a group of power automorphisms on $A_{p,j}$ for every $j \in \{1, 2, \ldots, n_p\}$.*

Proof. It is clear that if G has the structure described in the statement of the theorem, then G is the pairwise totally permutable product of the Sylow subgroups of B and the $A_{p,j}$ for each p dividing $|A|$ and every $j \in \{1, 2, \ldots, n_p\}$.

Conversely, suppose that the group $G = G_1G_2 \cdots G_r$ is the product of the pairwise totally permutable non-trivial subgroups $G_1, G_2, \ldots, G_r$ of prime power order. What has to be proved is that G is a semidirect product of an abelian subgroup A and a nilpotent subgroup B and A has the properties described in the statement of the theorem. Let p_i be the prime dividing $|G_i|$ for $1 \leq i \leq r$. By Theorem 5.1.15, it follows that the nilpotent residual $(G_iG_j)^{\mathfrak{N}}$ of G_iG_j for $i \neq j$ is either 1, or G_i, or G_j. If $p_i = p_j$, then G_iG_j is nilpotent. Assume that $p_i > p_j$. Then, since G_iG_j is supersoluble by Theorem 4.1.38, G_i is normal in G_iG_j by [119, VII, 2.1]. Thus $(G_iG_j)^{\mathfrak{N}}$ is either 1 or G_i. Let

$$I = \{i \in \{1, 2, \ldots, r\} : (G_iG_j)^{\mathfrak{N}} = G_i \text{ for some } j \in \{1, 2, \ldots, r\}\}.$$

The conclusion will hold after proving the following series of claims.

1. *If $l \in I$, then G_l is abelian.*

Consider $m \in I$ such that $(G_l G_m)^{\mathfrak{N}} = G_l$. Applying Theorem 5.1.15, G_l is abelian.

2. *If $l \in I$, then every subgroup of G_l is normal in G.*

Choose $m \in I$ such that $(G_l G_m)^{\mathfrak{N}} = G_l$ and $n \in \{1, 2, \dots, r\}$. Then $p_l > p_m$, and if $g_l \in G_l$, there exists $g_m \in G_m$ such that $\langle g_l \rangle \langle g_m \rangle$ is not abelian (note that G_m induces a non-trivial group of power automorphisms on G_l by Lemma 4.1.37). Let $g_n \in G_n$ and consider the subgroup $\langle g_n \rangle \langle g_l \rangle \langle g_m \rangle$. If $\langle g_n \rangle \langle g_l \rangle$ is not abelian, then g_n normalises $\langle g_l \rangle$ by Lemma 3.1.11. If $\langle g_n \rangle \langle g_l \rangle$ is abelian, then g_n normalises $\langle g_l \rangle$. Since this can be done for all $n \in \{1, 2, \dots, r\}$, it follows that $\langle g_l \rangle$ is normal in $\langle G_n : 1 \leq n \leq r \rangle = G$.

3. *Let $l \in I$ and $n \in \{1, 2, \dots, r\}$. If $p_l < p_n$ or $n \in I$, then $G_l \leq \mathrm{C}_G(G_n)$.*

Let $m \in I$ such that $(G_l G_m)^{\mathfrak{N}} = G_l$. Then $p_l > p_m$. Let $g_l \in G_l$. As in Statement 2, we choose and $g_m \in G_m$ such that $\langle g_l \rangle \langle g_m \rangle$ is not abelian. Let $g_n \in G_n$. Suppose first that $p_l < p_n$, or that $p_l = p_n$ and $\langle g_l \rangle \cap \langle g_l \rangle \neq 1$. We apply Lemma 3.1.11 to the group $\langle g_n \rangle \langle g_l \rangle \langle g_m \rangle$. We conclude that $\langle g_n \rangle \langle g_l \rangle$ must be abelian. Now assume that $p_l = p_n$, $n \in I$, and $\langle g_n \rangle \cap \langle g_l \rangle = 1$. In this case, $\langle g_n \rangle$ and $\langle g_l \rangle$ are normal subgroups of G by Statement 2. Therefore $\langle g_n \rangle$ centralises $\langle g_l \rangle$. In both cases, g_n centralises g_l. Since this can be done for all $g_l \in G_l$ and all $g_n \in G_n$, we conclude that $G_l \leq \mathrm{C}_G(G_n)$.

These considerations demonstrate:

4. *$A = \langle G_l : l \in I \rangle$ is abelian (if $I = \emptyset$, it is understood $A = 1$).*

5. *A satisfies the conditions of the theorem.*

By Theorem 1.3.2, p_l'-elements of G induce universal power automorphisms on the abelian group G_l. By Lemma 2.1.3, all chief factors of G below G_l are cyclic and isomorphic as G-modules by conjugation. Define on the set $\{G_j : j \in I, \ p_j = p_l\}$ the following equivalence relation: $G_i \sim G_j$ if and only if all chief factors of G below G_i and all chief factors of G below G_j are G-isomorphic. Let $C_{p_l,1}, \dots, C_{p_l,n_{p_l}}$ be the corresponding equivalence classes and let $A_{p_l,j} = \langle G_n : G_n \in C_{p_l,j} \rangle$, $1 \leq j \leq n_{p_l}$. Then $A_{p_l,j}$ is normal in G since it is a product of normal subgroups of G and all chief factors of G below $A_{p_l,j}$ are G-isomorphic. By Lemma 2.1.3, all subgroups of $A_{p_l,j}$ are normal in G. Now we prove that A is the direct product of the $A_{p_l,j}$ for $l \in I$ and $1 \leq j \leq n_{p_l}$. It is clear that A is the product of all these factors. In order to prove that the product is direct, it will be enough to check that if $a_1, a_2, \dots, a_k$ are elements of some of these factors and $a_1 a_2 \cdots a_k = 1$, then all a_i are trivial. Assume that this is false and that $a_1 a_2 \cdots a_k = 1$ with least k and a_i non-trivial elements of some of these factors. Since A is abelian, we can assume that all a_i are p-elements for a fixed prime p. Since the chief factors of G below $\langle a_1 \rangle$ and $\langle a_2 \rangle$ are not G-isomorphic, we can determine a p'-element g of G such that $a_1^g = a_1^{r_1}$ and $a_2^g = a_2^{r_2}$ for some integers $r_1 \geq 0$, $r_2 \geq 0$ such that $r_1 \neq r_2$. Let $a_i^g = a_i^{r_i}$ with $r_i \geq 0$ for $1 \leq i \leq k$. Then $1 = (a_1 a_2 \cdots a_k)^g (a_1 a_2 \cdots a_n)^{-r_1} = a_2^{r_2 - r_1} \cdots a_n^{r_n - r_1}$ and $a_2^{r_2 - r_1} \neq 1$. This contradicts the minimality of k. Therefore A is the direct product of these factors.

6. *Let $B = \langle G_i : i \in \{1, 2, \ldots, r\} \setminus I \rangle$. Then B is nilpotent.*

Let B_{p_j} be the subgroup of G generated by all G_i with $i \in \{1, 2, \ldots, r\} \setminus I$ and $p_i = p_j$. It is clear that B_{p_j} is a p_j-group for all j and that all B_{p_j} generate B. Now assume that $p_i \neq p_j$, with $i, j \notin I$, and consider the nilpotent residual $(G_i G_j)^{\mathfrak{N}}$ of $G_i G_j$. Clearly $(G_i G_j)^{\mathfrak{N}} \neq G_i$ and $(G_i G_j)^{\mathfrak{N}} \neq G_j$. Therefore $(G_i G_j)^{\mathfrak{N}} = 1$ and so $[G_i, G_j] = 1$. It follows that $[B_{p_i}, B_{p_j}] = 1$. Consequently B is nilpotent.

7. *$G = [A]B$ is the semidirect product of A and B.*

Obviously, G is the product of A and B and A is normal in G. Assume that $A \cap B \neq 1$. Consider $b = a_1 a_2 \cdots a_k$ an element of prime order p of $A \cap B$. Assume that each of the a_i belong to one of the $A_{p,j}$, $1 \leq j \leq n_p$, and that these factors are all pairwise different. By definition of B, there exists a p'-element $g \in B$ such that $a_i^g = a_i^{r_i}$, $i \in \{1, \ldots, k\}$, and $a_1^{r_1} \neq a_1$. Since b and g have coprime orders and B is nilpotent, we have that

$$1 = b^g b^{-1} = a_1^{r_1 - 1} a_2^{r_2 - 1} \cdots a_n^{r_n - 1},$$

where $a_1^{r_1 - 1} \neq 1$. This contradiction shows that $A \cap B = 1$ and the theorem is proved. □

Since nilpotent groups are direct products of their Sylow subgroups, it follows that totally permutable products of nilpotent groups are totally permutable products of subgroups of prime-power order. Hence we have:

Corollary 5.1.17. *The group $G = G_1 G_2 \cdots G_r$ is a product of the pairwise totally permutable nilpotent subgroups G_i, $1 \leq i \leq r$, if and only if G is a semidirect product $G = [A]B$ of an abelian group A and a nilpotent group B such that if p divides $|A|$ and A_p is the Sylow p-subgroup of A, then A_p can be decomposed as a direct product $A_p = A_{p,1} \times \cdots \times A_{p,n_p}$ and G induces a group of power automorphisms on $A_{p,j}$ for every $j \in \{1, 2, \ldots, n_p\}$.*

We bring this section to a close with some beautiful characterisations of soluble PST-groups by means of Sylow systems.

Theorem 5.1.18 ([95, Proposition 3.2]). *The group $G = G_1 G_2 \cdots G_r$ is the product of the totally permutable Sylow subgroups $G_1, G_2, \ldots, G_r$ of G if and only if G is a soluble* PST-*group.*

Proof. If $G = G_1 G_2 \cdots G_r$ is the product of its totally permutable Sylow subgroups $G_1, G_2, \dots, G_r$, then we can argue like in Theorem 5.1.16 to show that if

$$I = \{i \in \{1,2,\dots,r\} : (G_i G_j)^{\mathfrak{N}} = G_i \text{ for some } j \in \{1,2,\dots,r\}\},$$

then $A = \langle G_i : i \in I\rangle$ and $B = \langle G_i : i \notin I\rangle$ satisfy that A is abelian, B is nilpotent, and $G = [A]B$. Moreover, if p_i is the prime dividing $|G_i|, i \in I$, p_i'-elements of G act on G_i as power automorphisms. Furthermore, A and B are of coprime order. Suppose that G_i is a Sylow 2-subgroup of G and that G_j is another Sylow subgroup of G. Since $G_i G_j$ is supersoluble, we have that G_j is normal in $G_i G_j$ by [119, VII, 2.1]. Thus $(G_i G_j)^{\mathfrak{N}}$ is contained in G_j, and $(G_i G_j)^{\mathfrak{N}} \neq G_i$. This means that A has odd order. Hence by Theorem 2.1.8, G is a soluble PST-group.

Conversely, assume that G is a soluble PST-group. Applying Theorem 2.1.8, the nilpotent residual A of G is an abelian Hall subgroup of odd order acted on by conjugation by G as a group of power automorphisms. Let B a complement of A in G. Then G is the pairwise totally permutable product of the Sylow subgroups of A and B. □

In [67] Beidleman, Heineken, and Ragland introduced the notion of strong Sylow system of a group G: if $\pi(G) = \{p_1, p_2, \dots, p_k\}$ is the set of prime divisors of G, we say that $\Sigma = \{P_1, P_2, \dots, P_k\}$ is a *strong Sylow system* if Σ consists of Sylow subgroups of G, P_i is a Sylow p_i-subgroup of G for $1 \leq i \leq k$, and for each $P_i \in \Sigma$ with $p_i \in \pi(G)$, we have that $P_i P_j$ is a mutually permutable product for all $P_j \in \Sigma$. Note that in that case the product is totally permutable by Lemma 4.1.21 (3). With this terminology, Theorem 5.1.18 tells us that a group G is a soluble PST-group if and only if G has a strong Sylow system.

As an application of Theorem 5.1.18 and Theorem 3 of the paper of Ballester-Bolinches, Cossey, and Soler-Escrivà [42], we offer the following nice characterisation of soluble PST-groups proved by Beidleman, Heineken, and Ragland in [67].

Corollary 5.1.19. *Let G be a soluble group and D a system normaliser of G. Then G is a soluble* PST-*group if and only if D is a Hall subgroup of G with a Hall complement L for which every subgroup of L permutes with D.*

Proof. If G is a soluble PST-group, then by Theorem 2.1.8 and [47, 4.2.17], $G = LD$ is a totally permutable product of D and L, where L is the nilpotent residual of G. Conversely, assume that D is a Hall subgroup of G complemented by L, and every subgroup of L permutes with D. Let $g \in G$. Then $LD^g = G$ so that L permutes with every system normaliser of G by [119, I, 5.2]. Applying [42, Theorem 3], L is subnormal in G. But then L is a subnormal Hall subgroup of G and hence normal in G. It implies that every system normaliser of G induces a power automorphism on L and so every subgroup of L permutes with every system normaliser. Consequently, L is nilpotent and so G is a pairwise totally permutable product of the Sylow subgroups of D and L. By Theorem 5.1.18, G is a soluble PST-group. □

5.2 Totally and mutually permutable products and formation subgroups

The topic of this section is the interplay between formation subgroups and totally and mutually permutable products. Before we start, it is as well to remember that saturated formations containing all supersoluble groups are closed under the formation of totally permutable products (Theorem 4.2.10). Hence one would expect that subgroups associated to these type of formations such as residuals, projectors, and normalisers would have a noticeable influence on the structure of such groups. Saturated formations containing all supersoluble groups are not closed under the formation of mutually permutable products in general and so it is a much harder problem to understand the role of the above formation subgroups in the structure of mutually permutable products.

Our first result is concerned with the widest classes of groups under consideration, formations containing all supersoluble groups. It extends Theorem 4.2.10 and it was proved by Ballester-Bolinches and Pérez-Ramos [55, Theorem] in the two factors case and by Ballester-Bolinches, Pedraza-Aguilera, and Pérez-Ramos [54, Theorem 1] in the general one.

Theorem 5.2.1. *Let the group $G = G_1G_2 \cdots G_r$ be the product of the pairwise totally permutable subgroups $G_1, G_2, \ldots, G_r$. Let $\mathfrak{F}$ be a formation containing the class $\mathfrak{U}$ of all supersoluble groups. If $G_i \in \mathfrak{F}$ for all $i \in \{1, 2, \ldots, r\}$, then G belongs to $\mathfrak{F}$.*

Proof. Assume that the result is false and consider a counterexample $(G, G_1, \ldots, G_r)$ with $|G| + |G_1| + \cdots + |G_r|$ as small as possible. By Theorem 4.2.10, there exists $i \in \{1, 2, \ldots, r\}$ such that G_i is not supersoluble. Without loss of generality, we suppose $i = 1$. Then $G_1^{\mathfrak{U}} \neq 1$. By Corollary 4.2.9, the subgroup $K = G_2 \cdots G_r$ centralises $G_1^{\mathfrak{U}}$. Hence $G_1^{\mathfrak{U}}$ is a normal subgroup of G. Moreover, $G_1^{\mathfrak{U}}$ centralises the normal closure $\langle K^G \rangle$ of K in G. On the other hand, $G_1 = G_1^{\mathfrak{U}} X$ for a $\mathfrak{U}$-projector X of G_1. Consider now $Z = [\langle K^G \rangle]G_1$, the semidirect product of $\langle K^G \rangle$ and G_1 with respect to the action by conjugation. By Lemma 1.1.5, G is isomorphic to a quotient of Z. We prove that Z is an $\mathfrak{F}$-group. It is clear that $G_1^{\mathfrak{U}}$ is a normal subgroup of Z because it is centralised by $\langle K^G \rangle$. Moreover $Z/G_1^{\mathfrak{U}}$ is isomorphic to $[\langle K^G \rangle](X/(X \cap G_1^{\mathfrak{U}}))$, which is a quotient of $[\langle K^G \rangle]X$. Next we see that $[\langle K^G \rangle]X \in \mathfrak{F}$. From the fact that $G_1^{\mathfrak{U}}$ centralises K, one can easily deduce that $\langle K^G \rangle = \langle K^X \rangle$. A new application of Lemma 1.1.5 yields the existence of an epimorphism ψ from $[\langle K^X \rangle]X$ onto KX. Hence $[\langle K^X \rangle]X/\operatorname{Ker} \psi$ is isomorphic to KX, which is a product of pairwise totally permutable $\mathfrak{F}$-subgroups. Note that $|KX| + |X| + |G_2| + \cdots + |G_r| < |G| + |G_1| + |G_2| + \cdots + |G_r|$. It follows that $KX \in \mathfrak{F}$ by the minimal choice of $(G, G_1, \ldots, G_r)$. Now, since $[\langle K^X \rangle]X/\langle K^X \rangle$ is isomorphic to $X \in \mathfrak{U} \subseteq \mathfrak{F}$, we have that $[\langle K^X \rangle]X/(\operatorname{Ker} \psi \cap \langle K^X \rangle) \in \mathrm{R}_0 \mathfrak{F} = \mathfrak{F}$. But $(\operatorname{Ker} \psi) \cap \langle K^X \rangle = 1$ by Lemma 1.1.5. Consequently $[\langle K^X \rangle]X \in \mathfrak{F}$ and then $Z/G_1^{\mathfrak{U}} \in \mathfrak{F}$.

Moreover $Z/\langle K^G\rangle$ is isomorphic to $G_1 \in \mathfrak{F}$. Therefore $Z \cong Z/(\langle K^G\rangle \cap G_1^{\mathfrak{U}}) \in \mathrm{R}_0\,\mathfrak{F} = \mathfrak{F}$ and $G \in \mathfrak{F}$, the final contradiction. □

For saturated formations containing the class of all supersoluble groups, the converse of the above theorem holds.

Theorem 5.2.2 (Ballester-Bolinches et al. [54]). *Let $\mathfrak{F}$ be a saturated formation containing the class $\mathfrak{U}$ of all supersoluble groups. Let the group $G = G_1G_2\cdots G_r$ be the product of the pairwise totally permutable subgroups $G_1, G_2, \dots, G_r$. If $G \in \mathfrak{F}$, then $G_i \in \mathfrak{F}$ for all $i \in \{1, 2, \dots, r\}$.*

Proof. We argue by induction on the order of $|G|$. By Lemma 4.1.20, we can assume that G has a unique minimal normal subgroup N and $G_iN/N \in \mathfrak{F}$ for all $i \in \{1, 2, \dots, r\}$. By Corollary 4.2.13, N is contained in the supersoluble hypercentre of G and so N is cyclic. In particular, either $G_i \cap N = 1$ and $G_i \in \mathfrak{F}$, or N is contained in G_i and $G_i/C_{G_i}(N)$ is cyclic. In the second case, N is $\mathfrak{U}$-central in G_i. Applying [119, IV, 3.5 (c)], $G_i \in \mathfrak{F}$. Therefore $G_i \in \mathfrak{F}$ for all $i \in \{1, 2, \dots, r\}$. □

In attaining our next objective, which is to prove that the converse of Theorem 5.2.1 remains true in the case of pairwise totally permutable subgroups and formations which are soluble, the following result, proved by Ballester-Bolinches, Pedraza-Aguilera, and Pérez-Ramos [54] and by Ballester-Bolinches and Pérez-Ramos [55] in the two factor case, will be helpful.

Lemma 5.2.3. *Let $\mathfrak{F}$ be a formation containing the class $\mathfrak{U}$ of all supersoluble groups and let the group $G = G_1G_2\cdots G_r$ be the product of the pairwise totally permutable subgroups $G_1, G_2, \dots, G_r$. If $G_2, G_3, \dots, G_r$, and G are $\mathfrak{F}$-groups, then $G_1 \in \mathfrak{F}$.*

Proof. It is clear that we can assume $G_1^{\mathfrak{U}} \neq 1$ and $G_1 = G_1^{\mathfrak{U}}U$ for a supersoluble projector U of G_1. By Lemma 1.1.5, G_1 is an epimorphic image of $C = [G_1^{\mathfrak{U}}]U$, the semidirect product of $G_1^{\mathfrak{U}}$ and U with respect to the action by conjugation. Hence it is enough to prove that $C \in \mathfrak{F}$. Note that by Corollary 4.2.9, $G_1^{\mathfrak{U}}$ is a normal subgroup of G. Moreover $G = G_1^{\mathfrak{U}}(UK)$, where $K = G_2G_3\cdots G_r$. Also by Lemma 1.1.5, there exists an epimorphism $\alpha\colon X \to G$ with $(\operatorname{Ker}\alpha) \cap G_1^{\mathfrak{U}} = 1$, where $X = [G_1^{\mathfrak{U}}](UK)$. In particular, $X/(\operatorname{Ker}\alpha) \cong G \in \mathfrak{F}$. On the other hand, $X/G_1^{\mathfrak{U}} \cong UK$. Note now that UK is a product of pairwise totally permutable $\mathfrak{F}$-subgroups. Consequently, by Theorem 5.2.1, $UK \in \mathfrak{F}$. Now $X/G_1^{\mathfrak{U}} \in \mathfrak{F}$ and from here $X/(G_1^{\mathfrak{U}} \cap \operatorname{Ker}\alpha) \cong X \in \mathfrak{F}$. By Corollary 4.2.9, we know that $\langle K^X\rangle = \langle K^U\rangle$, the normal closure of K in KU. In particular, $\langle K^X\rangle$ is contained in KU. This implies that $\langle K^X\rangle \cap G_1^{\mathfrak{U}} = 1$. Moreover $X/\langle K^X\rangle \cong [G_1^{\mathfrak{U}}](U/(\langle K^X\rangle \cap U))$ and so $[G_1^{\mathfrak{U}}](U/(\langle K^X\rangle \cap U))$ belongs to $\mathfrak{F}$. This means that $[G_1^{\mathfrak{U}}](U/(\langle K^X\rangle \cap G_1^{\mathfrak{U}} \cap U)) \cong [G_1^{\mathfrak{U}}]U \in \mathrm{R}_0\,\mathfrak{F} = \mathfrak{F}$. Consequently, $C = [G_1^{\mathfrak{U}}]U \in \mathfrak{F}$ and the lemma is proved. □

Theorem 5.2.4 (Ballester-Bolinches *et al.* [54]). *Let $\mathfrak{F}$ be a formation of soluble groups containing the formation $\mathfrak{U}$ of all supersoluble groups. Let $G = G_1G_2 \cdots G_r$ be the product of the pairwise totally permutable subgroups $G_1, G_2, \dots, G_r$. If $G \in \mathfrak{F}$, then $G_i \in \mathfrak{F}$ for every $i \in \{1, 2, \dots, r\}$.*

Proof. We argue by induction on $|G| + |G_1| + |G_2| + \cdots + |G_r|$. We can assume that there exists $i \in \{1, 2, \dots, r\}$ such that G_i does not belong to $\mathfrak{U}$. Suppose without loss of generality that $i = r$. Since $\mathfrak{U}$ is a saturated formation, we have that $G_r^{\mathfrak{U}}$ is not contained in $\Phi(G_r)$, the Frattini subgroup of G_i. Moreover, since $G \in \mathfrak{F}$, we have that G is soluble. In particular, G_r is soluble and then the Fitting subgroup of $G_r^{\mathfrak{U}}/(G_r^{\mathfrak{U}} \cap \Phi(G_r))$, denoted by $F/(G_r^{\mathfrak{U}} \cap \Phi(G_r))$, is not trivial. Since $F/(G_r^{\mathfrak{U}} \cap \Phi(G_r))$ is a characteristic subgroup of $G_r^{\mathfrak{U}}/(G_r^{\mathfrak{U}} \cap \Phi(G_r)) = (G_r/(G_r^{\mathfrak{U}} \cap \Phi(G_r)))^{\mathfrak{U}}$, we conclude that F is a normal subgroup of G_r. Moreover $G_r^{\mathfrak{U}} \cap \Phi(G_r) = F \cap \Phi(G_r)$ and so $F\Phi(G_r)/\Phi(G_r) \cong F/(F \cap \Phi(G_r))$ is nilpotent. By [157, III, 3.5], we conclude that $F\Phi(G_r)$ is nilpotent and so F is nilpotent, too. Note also that F is normal in $G_r^{\mathfrak{U}}$ and $G_r^{\mathfrak{U}}$ is a normal subgroup of G by Corollary 4.2.9. Hence F is a subnormal subgroup of G. It follows that F is contained in $\mathrm{F}(G)$. On the other hand, F is a normal subgroup of G_r which is not contained in $\Phi(G_r)$. Hence there exists a maximal subgroup M of G_r such that F is not contained in M. This implies that $G_r = FM$ and $G = \mathrm{F}(MG_1G_2 \cdots G_{r-1}) = \mathrm{F}(G)(MG_1G_2 \cdots G_{r-1})$. Thus $MG_1G_2 \cdots G_{r-1}$ is a subgroup of G supplementing a nilpotent normal subgroup of G. By [119, IV, 1.14], we have that $J = MG_1G_2 \cdots G_{r-1} \in \mathfrak{F}$. Moreover J is a product of pairwise totally permutable subgroups such that $|J| + |M| + |G_1| + \cdots + |G_{r-1}| < |G| + |G_1| + \cdots + |G_r|$. This implies by induction that $G_1, G_2, \dots, G_{r-1}$ are $\mathfrak{F}$-groups. By Lemma 5.2.3, we conclude that $G_r \in \mathfrak{F}$. □

The following construction of Doerk and Hawkes [120, Proposition 2.3] is used to show that Theorems 5.2.2 and 5.2.4 are not true without assuming that $\mathfrak{F}$ is a saturated formation or $\mathfrak{F}$ is contained in the class $\mathfrak{S}$ of all soluble groups.

Theorem 5.2.5. *Let $\mathfrak{F}$ be the class of groups G satisfying the following properties:*

1. *$\mathrm{Z}(G)$ is an elementary abelian 2-group,*
2. *$G/\mathrm{Z}(G) \cong X_1 \times \cdots \times X_n$ with each $X_i \cong \mathrm{Alt}(5)$,*
3. *if x_i is a non-trivial element of a Sylow 2-subgroup of X_i and $x_i = y_i\mathrm{Z}(G)$, then $y_1^2 y_2^2 \cdots y_n^2 = 1$.*

Then the class $\mathfrak{F}$ is a formation.

Proof. First we verify that $\mathfrak{F}$ is closed under taking quotients. Let K be a normal subgroup of $G \in \mathfrak{F}$. Then $\mathrm{Z}(G)K/\mathrm{Z}(G)$ is a normal subgroup of $G/\mathrm{Z}(G)$, which is by hypothesis the direct product of non-abelian simple groups $X_1, X_2, \dots, X_n$. By [119, I, 4.13], $\mathrm{Z}(G)K/\mathrm{Z}(G)$ is the direct product of a suitable subset, say $\mathrm{Z}(G)K/\mathrm{Z}(G) =$

$X_{r+1} \times \cdots \times X_n$ after suitably renumbering the components. Let $W/K = \mathrm{Z}(G/K)$. Then clearly $\mathrm{Z}(G)K/K \le W/K$. Since

$$G/\mathrm{Z}(G)K \cong (G/\mathrm{Z}(G))/(\mathrm{Z}(G)K/\mathrm{Z}(G)) \cong X_1 \times \cdots \times X_r,$$

the centre of $G/\mathrm{Z}(G)K$ is trivial. Therefore the central factor $W/\mathrm{Z}(G)K$ is trivial and we have $W = K\mathrm{Z}(G)$. Hence $\mathrm{Z}(G/K) \cong \mathrm{Z}(G)/(\mathrm{Z}(G) \cap K)$ is an elementary abelian 2-group and $(G/K)/\mathrm{Z}(G/K) \cong G/\mathrm{Z}(G)K \in \mathrm{D}_0(1, \mathrm{Alt}(5))$. It remains to verify requirement 3 for G/K. Let $y_i\mathrm{Z}(G)$ be an involution of X_i. It is not difficult to see that $(y_iK)(\mathrm{Z}(G)K/K)$ is an involution of the ith direct component of

$$(G/K)/\mathrm{Z}(G/K) \cong X_1 \times \cdots \times X_r, \quad 1 \le i \le r.$$

If $r+1 \le j \le n$, we have $y_i \in K\mathrm{Z}(G)$, because $K\mathrm{Z}(G)/\mathrm{Z}(G) = X_{r+1} \times \cdots \times X_n$. Let $y_j = kz$, with $k \in K$ and $z \in \mathrm{Z}(G)$. Then $y_j^2 = k^2z^2 = k^2 \in K$ because $\mathrm{Z}(G)$ is elementary. Thus $(y_iK)^2 \cdots (y_rK)^2 = y_1^2 \cdots y_r^2K = y_1^2 \cdots y_n^2K = K$ and condition 3 is fulfilled by G/K. Therefore $\mathfrak{F}$ is closed under taking epimorphic images.

Now we shall prove that $\mathfrak{F} = \mathrm{R}_0\,\mathfrak{F}$. Assume that G has normal subgroups N_1 and N_2 such that $G/N_1, G/N_2 \in \mathfrak{F}$ and $N_1 \cap N_2 = 1$. For $i \in \{1,2\}$, let $Z_i/N_i = \mathrm{Z}(G/N_i)$ and let $Z = Z_1 \cap Z_2$. First we show that $Z = \mathrm{Z}(G)$. Clearly

$$[Z, G] \le [Z_1, G] \cap [Z_2, G] \le N_1 \cap N_2 = 1$$

and so $Z \le \mathrm{Z}(G)$. On the other hand, $\mathrm{Z}(G)N_i/N_i \le Z_i/N_i$ and therefore $\mathrm{Z}(G) \le Z$. Hence $Z = \mathrm{Z}(G)$. Since $ZN_i/N_i \le Z_i/N_i$ and Z_i/N_i is an elementary abelian 2-group by hypothesis, it follows that each of $Z/(Z \cap N_1)$, $Z/(Z \cap N_2)$, and therefore Z itself, is an elementary abelian 2-group. Moreover

$$G/\mathrm{Z}(G) = G/Z = G/(Z_1 \cap Z_2) \in \mathrm{R}_0\,\mathrm{D}_0(1, \mathrm{Alt}(5)),$$

which coincides with $\mathrm{D}_0(1, \mathrm{Alt}(5))$ (see [119, II, 2.13]). Thus G fulfils the requirements 1 and 2 of a group in $\mathfrak{F}$. It remains to verify requirement 3. Since N_1Z/Z and N_2Z/Z are normal subgroups of G/Z, which is a direct product of non-abelian simple groups, and since $N_1Z \cap N_2Z \le Z_1 \cap Z_2 = Z$, by [119, A, 4.6] we may write

$$N_1Z/Z = (A_1/Z) \times \cdots \times (A_r/Z),$$
$$N_2Z/Z = (B_1/Z) \times \cdots \times (B_s/Z), \quad \text{and}$$
$$G/Z = (N_1Z/Z) \times (N_2Z/Z) \times (C_1/Z) \times \cdots \times (C_t/Z),$$

where the A_i/Z, B_j/Z and C_k/Z are isomorphic to $\mathrm{Alt}(5)$. Since G/N_1Z has trivial centre, we conclude that $N_1Z = Z_1$. Let a_iZ, b_jZ, c_kZ denote elements of order 2 in the direct factors A_i/Z, B_j/Z, and C_k/Z, respectively, $1 \le i \le r$, $1 \le j \le s$, and $1 \le k \le t$. Since $G/N_1 \in \mathfrak{F}$ and

$$(G/N_1)/\mathrm{Z}(G/N_1) \cong G/N_1Z = \Big(\bigtimes_{j=1}^{s} (N_1B_j/N_1Z)\Big) \times \Big(\bigtimes_{k=1}^{t} (N_1C_k/N_1Z)\Big),$$

from the definition of $\mathfrak{F}$ we have

$$\prod_{j=1}^{s} b_j^2 \prod_{k=1}^{t} c_k^2 \in N_1. \tag{5.1}$$

But for $1 \le j \le s$ we have $b_j \in N_2Z$ and hence $b_j^2 \in N_2 \cap Z$. Therefore

$$\prod_{k=1}^{t} c_k^2 \in N_1\Big(\prod_{j=1}^{s} b_j^2\Big)^{-1} \subseteq N_1N_2 = N_1 \times N_2.$$

Let $\prod_{k=1}^{t} c_k^2 = n_1n_2$ with (unique) $n_i \in N_i$, $i = 1, 2$. Then from (5.1) we conclude that $n_2 = (\prod_{i=1}^{t} b_j^2)^{-1}$. Similarly $n_1 = (\prod_{i=1}^{r} a_i^2)^{-1}$. Therefore

$$\prod_{i=1}^{r} a_i^2 \prod_{j=1}^{s} b_j^2 \prod_{k=1}^{t} c_k^2 = 1,$$

as required for 3. Thus $\mathfrak{F}$ is R_0-closed and hence a formation. □

Example 5.2.6. Theorems 5.2.2 and 5.2.4 are not true without assuming that $\mathfrak{F}$ is a saturated formation or $\mathfrak{F}$ is contained in the class $\mathfrak{S}$ of all soluble groups. The group $H = \mathrm{SL}_2(5)$ has a normal subgroup of order 2, say $\mathrm{Z}(H) = \mathrm{F}(H) = \Phi(H)$, with $H/\mathrm{Z}(H) \cong \mathrm{Alt}(5)$. Take $G = X \times Y$ with $X \cong H \cong Y$ and denote $D = \langle xy \rangle$, where x and y are central involutions of X and Y, respectively. Consider the group $T = G/D$, which is a central product of X and Y (see [119, A, 19] for further details). Let $\mathfrak{G} = \mathrm{Q}\,\mathrm{R}_0(T)$ be the formation generated by the group T and $\mathfrak{G} \circ \mathfrak{U}$ the formation product of $\mathfrak{G}$ and $\mathfrak{U}$. Recall that $\mathfrak{U} \subseteq \mathfrak{G} \circ \mathfrak{U}$ (see [119, IV, 1.7] or [47, 2.2.10] for the definition of formation product). Note that T is a perfect group. Hence $T^{\mathfrak{G}\circ\mathfrak{U}} = T^{\mathfrak{G}} = 1$ and $T \in \mathfrak{G} \circ \mathfrak{U}$. However H does not belong to $\mathfrak{G} \circ \mathfrak{U}$: Note that T belongs to the formation $\mathfrak{F}$ constructed in Theorem 5.2.5, but H does not, because it does not fulfil condition 3. Thus T is an $\mathfrak{G} \circ \mathfrak{U}$-group which is a totally permutable product with factors X and Y not in $\mathfrak{G} \circ \mathfrak{U}$.

The next theorem was proved by Ballester-Bolinches, Pedraza-Aguilera, and Pérez-Ramos in [54, Theorem 4].

Theorem 5.2.7. *Let $\mathfrak{F}$ be a formation containing $\mathfrak{U}$ such that $\mathfrak{F}$ is saturated or $\mathfrak{F}$ consists or soluble groups. Let the group $G = G_1G_2 \cdots G_r$ be the product of the pairwise totally permutable subgroups G_i for $i \in \{1, 2, \dots, r\}$. Then $G^{\mathfrak{F}} = G_1^{\mathfrak{F}} G_2^{\mathfrak{F}} \cdots G_r^{\mathfrak{F}}$.*

Proof. Consider $G/G^{\mathfrak{F}} = (G_1G^{\mathfrak{F}}/G^{\mathfrak{F}})(G_2G^{\mathfrak{F}}/G^{\mathfrak{F}})\cdots(G_rG^{\mathfrak{F}}/G^{\mathfrak{F}})$, which is the product of the pairwise totally permutable subgroups $G_iG^{\mathfrak{F}}/G^{\mathfrak{F}}$ by Lemma 4.1.20. Since $G/G^{\mathfrak{F}} \in \mathfrak{F}$, by Theorem 5.2.2 or 5.2.4, we obtain that $G_iG^{\mathfrak{F}}/G^{\mathfrak{F}}$ is an $\mathfrak{F}$-group for every $i \in \{1, 2, \dots, r\}$. Since $G_iG^{\mathfrak{F}}/G^{\mathfrak{F}} \cong G_i/(G^{\mathfrak{F}} \cap G_i)$ is an $\mathfrak{F}$-group, $G_i^{\mathfrak{F}} \leq G^{\mathfrak{F}} \cap G_i$. Therefore $G_1^{\mathfrak{F}}G_2^{\mathfrak{F}}\cdots G_r^{\mathfrak{F}} \leq G^{\mathfrak{F}}$. Now consider $K = G_1^{\mathfrak{F}}G_2^{\mathfrak{F}}\cdots G_r^{\mathfrak{F}}$. Since $G_iK/K \in \mathfrak{F}$ for every $i \in \{1, 2, \dots, r\}$, by Theorem 5.2.1, $G/K \in \mathfrak{F}$. It follows that $G^{\mathfrak{F}} \leq K$ and so $K = G^{\mathfrak{F}}$, as desired. □

Obviously Theorem 5.2.7 is not valid for formations which do not contain the class $\mathfrak{U}$ of all supersoluble groups as the saturated formation $\mathfrak{N}$ of all nilpotent groups and the symmetric group of degree 3 shows.

Remark 5.2.8. Example 5.2.6 shows that the conditions on $\mathfrak{F}$ in Theorem 5.2.7 are essential to have the result: Since H does not belong to $\mathfrak{G} \circ \mathfrak{U}$ and $H \in \mathrm{Q}(G)$, we have that G is not an $(\mathfrak{G} \circ \mathfrak{U})$-group. On the other hand, $G/D \in \mathfrak{G} \circ \mathfrak{U}$. Moreover D is a minimal normal subgroup of G. Then $G^{\mathfrak{G}\circ\mathfrak{U}} = D$. Consider $H/\mathrm{Z}(H) \cong \mathrm{Alt}(5) \in \mathrm{Q}(G/D) \subseteq \mathfrak{G} \circ \mathfrak{U}$. Also $\mathrm{Z}(H)$ is a minimal normal subgroup of H. Thus $H^{\mathfrak{G}\circ\mathfrak{U}} = \mathrm{Z}(H)$. Hence $X^{\mathfrak{G}\circ\mathfrak{U}} \times Y^{\mathfrak{G}\circ\mathfrak{U}} = \langle x\rangle \times \langle y\rangle \neq \langle xy\rangle = D = G^{\mathfrak{G}\circ\mathfrak{U}}$.

We now describe the connection between $\mathfrak{F}$-projectors and $\mathfrak{F}$-normalisers of pairwise totally permutable subgroups and those of their product when $\mathfrak{F}$ is a saturated formation containing all supersoluble groups ([54]).

Theorem 5.2.9. *Let $\mathfrak{F}$ be a saturated formation containing the class $\mathfrak{U}$ of all supersoluble groups. Let the group $G = G_1G_2\cdots G_r$ be the product of the pairwise totally permutable subgroups G_i for $i \in \{1, 2, \dots, r\}$. If $A_1, A_2, \dots, A_r$ are $\mathfrak{F}$-projectors of $G_1, G_2, \dots, G_r$, respectively, then the product $A_1A_2\cdots A_r$ is an $\mathfrak{F}$-projector of G.*

Proof. Assume that the result is not true and let G be a counterexample of minimal order. By Lemma 4.1.20, there exists $i \in \{1, 2, \dots, r\}$ such that G_i contains a minimal normal subgroup of G, N say. Assume without loss of generality that $i = 1$. Now the minimality of G and Lemma 4.1.10 yields that $(A_1N/N)(A_2N/N)\cdots(A_rN/N) = (A_1A_2\cdots A_r)N/N$ is an $\mathfrak{F}$-projector of G/N. Denote $C = (A_1A_2\cdots A_r)N$ and assume that C is a proper subgroup of G. We have then that C as the product of the pairwise totally permutable subgroups $A_1N, A_2, \dots, A_r$. Moreover, by [47, 2.3.16], we have that A_1 is an $\mathfrak{F}$-projector of A_1N. Therefore, by the minimality of G, $A_1A_2\cdots A_r$ is an $\mathfrak{F}$-projector of C. Applying now [119, III, 3.7], we have that $A_1A_2\cdots A_r$ is an $\mathfrak{F}$-projector of G, a contradiction. Consequently $G = (A_1A_2\cdots A_r)N$. By Theorem 5.2.1, $A_1A_2\cdots A_r \in \mathfrak{F}$. Hence $G^{\mathfrak{F}} \leq N$ and $G^{\mathfrak{F}} = 1$ or $G^{\mathfrak{F}} = N$. If $G^{\mathfrak{F}} = 1$, then the result follows from Theorem 5.2.2 because $G_i \subset \mathfrak{F}$ for $i \in \{1, 2, \dots, r\}$ and so $A_i = G_i$ for $i \in \{1, 2, \dots, r\}$. Hence we may assume $G^{\mathfrak{F}} = N$. Consider now an $\mathfrak{F}$-maximal subgroup U of G containing $A_1A_2\cdots A_r$. By [47, 2.3.16], U is an $\mathfrak{F}$-projector of G. Moreover U is the product of the pairwise totally permutable subgroups $A_1(U \cap N), A_2, \dots, A_r$. By Lemma 5.2.3, we have that $A_1(U \cap N)$ belongs to $\mathfrak{F}$. But A_1 is an $\mathfrak{F}$-maximal subgroup of G_1. This implies that $U \cap N \leq A_1$ and $U = A_1A_2 \cdots A_r$ is an $\mathfrak{F}$-projector of G. This concludes the proof of the theorem. □

Because we need conjugation and other "soluble properties" of the $\mathfrak{F}$-normalisers, we make the blanket hypothesis that the $\mathfrak{F}$-residual is soluble in the next result.

Theorem 5.2.10. *Let the group $G = G_1G_2 \cdots G_r$ be a group which is the product of the pairwise totally permutable subgroups $G_1, G_2, \dots, G_r$. Let $\mathfrak{F}$ be a saturated formation containing the class $\mathfrak{U}$ of all supersoluble groups. Assume that the $\mathfrak{F}$-residual of G is soluble. If $A_1, A_2, \dots, A_r$ are $\mathfrak{F}$-normalisers of $G_1, G_2, \dots, G_r$, respectively, then $A_1A_2 \cdots A_r$ is an $\mathfrak{F}$-normaliser of G.*

Proof. We argue by induction on $|G| + |G_1| + \cdots + |G_r|$. By Theorem 5.2.1, we can assume that there exists $i \in \{1, 2, \dots, r\}$ such that G_i does not belong to $\mathfrak{F}$. Suppose that $i = 1$. Since $\mathfrak{F}$ is a saturated formation, $G_1^{\mathfrak{F}}$ is not contained in the Frattini subgroup $\Phi(G_1)$ of G_1. Hence $T/(G_1^{\mathfrak{F}} \cap \Phi(G_1))$, the Fitting subgroup of $G_1^{\mathfrak{F}}/(G_1^{\mathfrak{F}} \cap \Phi(G_1))$, is non-trivial because $G_1^{\mathfrak{F}}$ is soluble (note that, by Theorem 5.2.7, $G_i^{\mathfrak{F}} \leq G^{\mathfrak{F}}$ for all i). Now T is a normal subgroup of G_1 and $G_1^{\mathfrak{F}} \cap \Phi(G_1) = T \cap \Phi(G_1)$. Thus $T/(G_1^{\mathfrak{F}} \cap \Phi(G_1)) = T/(T \cap \Phi(G_1))$ is isomorphic to $T\Phi(G_1)/\Phi(G_1)$ and so $T\Phi(G_1)$ is nilpotent by [157, III, 3.5]. In particular, T is nilpotent. Note also that $G_1^{\mathfrak{F}}$ is a normal subgroup of G by Corollary 4.2.9. Hence T is a nilpotent normal subgroup of G not contained in $\Phi(G_1)$. This yields the existence of a maximal subgroup M of G_1 such that T is not contained in M. Therefore $G_1 = TM = G_1^{\mathfrak{F}}M = \mathrm{F}(G_1)M$ and M is $\mathfrak{F}$-critical in G_1. By [47, 2.2.8], $M^{\mathfrak{F}}$ is contained in $G_1^{\mathfrak{F}}$ and so it is soluble. Now, applying [47, 4.2.2 and 4.2.15], every $\mathfrak{F}$-normaliser of M is an $\mathfrak{F}$-normaliser of G and they are conjugate in G. Therefore we can assume that $A_1 \leq M$ by taking a suitable conjugate of M. Then $G = T(MG_2G_3 \cdots G_r) = \mathrm{F}(G)(MG_2G_3 \cdots G_r) = G^{\mathfrak{F}}(MG_2G_3 \cdots G_r)$. If we assume $G = MG_2 \cdots G_r$, then $M, G_2, \dots, G_r$ are pairwise totally permutable subgroups of G and $|G| + |M| + |G_2| + \cdots + |G_r| < |G| + |G_1| + |G_2| + \cdots + |G_r|$. By induction, we have that $A_1A_2 \cdots A_r$ is an $\mathfrak{F}$-normaliser of G and the result is true. Hence we can assume that $MG_2G_3 \cdots G_r$ is a proper subgroup of G. Then $MG_2G_3 \cdots G_r$ is an $\mathfrak{F}$-critical maximal subgroup of G and each $\mathfrak{F}$-normaliser of M is an $\mathfrak{F}$-normaliser of G. Since the $\mathfrak{F}$-residual of $MG_2G_3 \cdots G_r$ is soluble, we have that $A_1A_2 \cdots A_r$ is an $\mathfrak{F}$-normaliser of $MG_2G_3 \cdots G_r$. Hence $A_1A_2 \cdots A_r$ is an $\mathfrak{F}$-normaliser of G. □

Our next goal is a theorem about factorisations of some type of subgroups which applies to projectors and normalisers associated to saturated formations containing the class of all supersoluble groups. It was proved in [34].

Theorem 5.2.11. *Let the soluble group $G = G_1G_2 \cdots G_r$ be the product of the pairwise totally permutable subgroups $G_1, G_2, \dots, G_r$. If $\mathfrak{F}$ is a saturated formation containing the class $\mathfrak{U}$ of all supersoluble groups, then every sub-$\mathfrak{F}$-abnormal subgroup of G is factorised (see Definition* 3.2.15*).*

Proof. We proceed by induction on the index $|G : S|$ of S in G. Suppose that S is a maximal $\mathfrak{F}$-abnormal subgroup of G. If $S_G = 1$, then, by Lemma 4.2.15, G is supersoluble or there exists $i \in \{1, 2, \dots, r\}$ such that $\prod_{i \neq j} G_j = 1$ and $G = G_i$. It is clear that in this case S is a factorised subgroup of G. Assume that $D = S_G \neq 1$. As G/D does not belong to $\mathfrak{F}$, it cannot be supersoluble. Applying now Lemma 4.1.10 and Lemma 4.2.15 we obtain that there exists $i \in \{1, 2, \dots, r\}$ (we can assume without loss of generality that $i = 1$) such that $G_2G_3 \cdots G_r$ is contained in D. Then $S = G_2G_3 \cdots G_r(S \cap G_1) = (S \cap G_2)(S \cap G_3) \cdots (S \cap G_r)(S \cap G_1)$ and $G_i \cap \prod_{j \neq i} G_j \leq S$ for all $i \in \{1, 2, \dots, r\}$. Hence S is factorised. Assume now that S is not a maximal subgroup of G and let S_1 be an $\mathfrak{F}$-abnormal maximal subgroup of G containing S such that S is sub-$\mathfrak{F}$-abnormal in S_1. Then S_1 is a factorised subgroup of G and $|S_1 : S| < |G : S|$. The induction hypothesis implies that S is a factorised subgroup of S_1. Then S is factorised in G by Lemma 4.1.42 (2). □

Remark 5.2.12. Solubility is needed in Theorem 5.2.11: Let G be the direct product of two copies of the alternating group Alt(5) of degree 5. Then G possesses a diagonal subgroup M which is maximal in G and obviously a $\mathfrak{U}$-abnormal maximal subgroup of G, but M is not even prefactorised.

As it has been pointed out in Section 3.2, $\mathfrak{F}$-normalisers of soluble groups associated to saturated formations are typical examples of sub-$\mathfrak{F}$-abnormal subgroups ([47, Chapter 4]). Hence we have:

Corollary 5.2.13. *Let the soluble group $G = G_1G_2 \cdots G_r$ be the product of the pairwise totally permutable subgroups $G_1, G_2, \dots, G_r$. If $\mathfrak{F}$ is a saturated formation containing the class $\mathfrak{U}$ of all supersoluble groups, then the $\mathfrak{F}$-normalisers of G are factorised subgroups.*

Applying [47, 2.3.15 and 4.2.1], $\mathfrak{F}$-projectors of groups with soluble $\mathfrak{F}$-residual are also sub-$\mathfrak{F}$-abnormal subgroups, in which case we derive the following result.

Corollary 5.2.14. *Let the soluble group $G = G_1G_2 \cdots G_r$ be the product of the pairwise totally permutable subgroups $G_1, G_2, \dots, G_r$. If $\mathfrak{F}$ is a saturated formation containing the class $\mathfrak{U}$ of all supersoluble groups, then the $\mathfrak{F}$-projectors of G are factorised subgroups.*

The corresponding results for dinilpotent groups were proved in Theorem 3.2.17 and Corollary 3.2.18.

For the saturated formation $\mathfrak{N}$ of all nilpotent groups we have much weaker statements. It also appears in [34].

Theorem 5.2.15. *Let the soluble group $G = G_1G_2 \cdots G_r$ be the product of the pairwise totally permutable subgroups $G_1, G_2, \dots, G_r$. Then there exists a prefactorised Carter subgroup of G.*

Proof. Assume that the result is not true and let G be a counterexample with $|G| + |G_1| + |G_2| + \cdots + |G_r|$ as small as possible. If G is a nilpotent group, then G is its own Carter subgroup and the conclusion follows. Assume G is not nilpotent and let C denote a Carter subgroup of G. Then there exists a non-normal maximal subgroup M of G with $C \leq M$. Write $K = M_G$. If $K = 1$, then G is a primitive group. By Lemma 4.2.15, there exists $i \in \{1, 2, \dots, r\}$ such that $\prod_{j \neq i} G_j = 1$ and $G = G_i$ or G is supersoluble. In the first case, it is clear that C is prefactorised. If G is supersoluble, then $G = NM$ where $N = \mathrm{C}_G(N) = G^{\mathfrak{N}}$ is the unique minimal normal subgroup of G, $|N| = p$, p a prime, and M an abelian complement of N with trivial core. By [119, IV, 5.18], the Carter subgroups of G are the complements of N in G. Applying Lemma 4.2.16, C is prefactorised. This contradiction yields $K \neq 1$. Then $G/K = (N/K)(M/K)$ is a primitive group, N/K is the unique minimal normal subgroup of G/K, $\mathrm{C}_{G/K}(N/K) = N/K$, and M/K is a maximal subgroup of G/K with trivial core. By Lemma 4.1.10, G/K is the product of the pairwise totally permutable subgroups $G_1K/K, G_2K/K, \dots, G_rK/K$. Applying Lemma 4.2.15, G/K is supersoluble or there exists $i \in \{1, 2, \dots, r\}$, suppose $i = 1$ without loss of generality, such that $(G_2K/K)(G_3K/K) \cdots (G_rK/K) = 1$. In the last case we have $G_2G_3 \cdots G_r \leq K \leq M$. Thus $M = G_2G_3 \cdots G_r(M \cap G_1)$ is a pairwise totally permutable product. By the choice of G, M has a prefactorised Carter subgroup. But C is a Carter subgroup of M. Therefore there exists $m \in M$ such that C^m is prefactorised in M. By Lemma 4.1.42 (2), C^m is prefactorised in G. Consequently we may assume that G/K is supersoluble. In particular, $|N/K| = p$ and M/K is abelian with exponent dividing $p - 1$.

On the other hand, applying Corollary 4.1.22 (2), $W = (K \cap G_1)(K \cap G_2) \cdots (K \cap G_r)$ is a normal subgroup of G. If $W \neq 1$, then the choice of G implies that G/W has a prefactorised Carter subgroup, C^gW/W, for some $g \in G$. Since W is prefactorised, it follows that C^gW is prefactorised in G by Lemma 4.1.42 (1), that is, $C^gW = (C^gW \cap G_1)(C^gW \cap G_2) \cdots (C^gW \cap G_r)$. Now C^gW is a proper subgroup of G because it is contained in M^g. By the minimal choice of G, there exists $t \in C^gW$ such that C^{gt} is prefactorised in C^gW. Lemma 4.1.42 (2) implies that C^{gt} is prefactorised in G. Consequently we may assume that $W = 1$, that is, $K \cap G_i = 1$ for all $i \in \{1, 2, \dots, r\}$. It now follows that G_i is supersoluble so that G is also supersoluble by Corollary 4.1.38.

It is clear that p is the largest prime dividing the order of G/K. Assume that p is not the largest prime dividing $|G|$ and let $Q \neq 1$ denote a Sylow q-subgroup of G with q the largest prime divisor of $|G|$, $q \neq p$. Then Q is a normal subgroup of G with $Q \leq K$. Moreover, by Proposition 4.1.45 (1), Q is prefactorised in G. Arguing as above we obtain $Q = 1$. This contradicts the choice of Q. Thus we may suppose that p is the largest prime dividing $|G|$ and G has a normal Sylow p-subgroup, P say, and $PK/K = N/K$.

Now we apply Lemma 4.2.16 to G/K and analyse all the cases appearing there. Assume we are in the hypothesis of Lemma 4.2.16 (2). Then $G_1K/K = G_2K/K = \cdots = G_nK/K = N/K$ for some $n \in \{1, 2, \dots, r\}$ and $(G_{n+1}G_{n+2}\cdots G_r)K/K$ is a complement of N/K in G/K. Since $K \cap G_i = 1$ for all $i \in \{1, 2, \dots, r\}$, we have $|G_i| = p$ for $i \in \{1, 2, \dots, n\}$ and G_j is an abelian p'-group for $j > n$. Further $P = G_1G_2\cdots G_n$ is the normal Sylow p-subgroup of G and $G_{n+1}G_{n+2}\cdots G_r$ is a Hall p'-subgroup of G. In particular, P is a prefactorised subgroup of G. The above arguments imply that $G = C^gP = CP$. Hence the nilpotent residual $G^{\mathfrak{N}}$ of G is contained in P.

On the other hand, $P = G_1G_2\cdots G_n$ with $|G_i| = p$ for all $i \in \{1, 2, \dots, n\}$. Assume that $G_i = G_j$ for $i \neq j$, $i, j \in \{1, 2, \dots, n\}$. Then G can be regarded as a totally permutable product of less than r factors. The choice of G implies that there is a conjugate of C which is prefactorised with respect to this new factorisation. This clearly implies that this conjugate is actually prefactorised with respect to $G = G_1G_2\cdots G_r$. Hence we may assume that $G_i \neq G_j$ for $i \neq j$, $i, j \in \{1, 2, \dots, n\}$. Then $[G_i, G_j] = 1$ for all $i, j \in \{1, 2, \dots, n\}$, $i \neq j$, that is, P is elementary abelian. Next we prove that $P = G_1 \times G_2 \times \cdots \times G_n$. Assume that $G_1 \cap G_2G_3\cdots G_n \neq 1$. Then G_1 is contained in $G_2G_3\cdots G_n$ and then $G = G_2G_3\cdots G_r$. By the choice of G, there exists a Carter subgroup T of G which is prefactorised with respect to this new factorisation. This clearly implies that T is prefactorised in $G = G_1G_2\cdots G_r$. Hence we have that $G_1 \cap (G_2G_3\cdots G_n) = 1$. Repeating the argument with the other factors, we have that P is the direct product of $G_1, G_2, \dots, G_n$. In particular $G^{\mathfrak{N}}$ is abelian and so it is complemented in G (see [119, IV, 5.18]). If $P = G^{\mathfrak{N}}$, then C would be a Hall p'-subgroup of G and by Proposition 4.1.45 (2), C would have a prefactorised conjugate, contrary to hypothesis. Hence $G^{\mathfrak{N}}$ is a proper subgroup of P. By Theorem 4.4.17, $G^{\mathfrak{N}} = (G^{\mathfrak{N}} \cap G_1)(G^{\mathfrak{N}} \cap G_2)\cdots(G^{\mathfrak{N}} \cap G_n)$. Since all the factors G_i have order p, it follows that $G^{\mathfrak{N}}$ is the product of all factors G_j which are contained in $G^{\mathfrak{N}}$. Without loss of generality, we may assume that $G^{\mathfrak{N}} = G_1G_2\cdots G_h$ for some $h < n$. Let $T = (G_{h+1}G_{h+2}\cdots G_n)(G_{n+1}G_{n+2}\cdots G_r)$. Thus $G = G^{\mathfrak{N}}T$ and $G^{\mathfrak{N}} \cap T = 1$. Furthermore T is a Carter subgroup of G which is clearly prefactorised, against supposition.

Now suppose G/K satisfies Lemma 4.2.16 (1). If $G_1K/K = G_2K/K = \cdots = G_rK/K = N/K$, then $G/K = N/K$ is abelian. Hence $M/K = 1$, that is, $M = K$, contrary to assumption. Therefore there exists $i \in \{1, 2, \dots, r\}$, we suppose $i = 1$, with $G/K = G_1K/K$ and $G_iK/K = N/K$ for all $i \neq 1$. Since $K \cap G_i = 1$ for all i, we have $|G_i| = p$ for all $i \neq 1$, $G = G_1K$ and $M = K(M \cap G_1)$. Thus $M \cap G_1$ is a p'-group. Moreover $N = G_2G_3\cdots G_r(N \cap G_1)$. Now $(N \cap G_1)K/K$ is a subgroup of N/K which is of order p. Hence either $(N \cap G_1)K/K = 1$ or $N = (N \cap G_1)K$. If $(N \cap G_1)K/K = 1$, then $N \cap G_1 \leq K \cap G_1 = 1$. Therefore $N = G_2G_3\cdots G_r$ and $G = G_1N$. Furthermore G/N is abelian. Thus G_1 is also abelian. But then G/K

is abelian and $M = K$, contrary to supposition. Consequently we may assume that $N = (N \cap G_1)K$ and $|N \cap G_1| = p$. In particular, N and K are both p-groups. Now $G = NM = NK(M \cap G_1) = N(M \cap G_1)$. Thus N is the Sylow p-subgroup of G and $M \cap G_1$ is a Hall p'-subgroup of G. Also $N \cap G_1$ is the Sylow p-subgroup of G_1 and $G_1 = (N \cap G_1)(M \cap G_1)$ Arguing as in the above case, we obtain that N is elementary abelian. We prove that $N = (N \cap G_1) \times G_2 \times G_3 \times \cdots \times G_r$. Suppose that $G_2 \leq (N \cap G_1)G_3 \cdots G_r$, then $G = G_1G_3 \cdots G_r$ and, by the choice of G, there exists a Carter subgroup prefactorised with respect to $G = G_1G_3 \cdots G_r$. Now a standard argument shows that this Carter subgroup is factorised with respect to $G = G_1G_2 \cdots G_r$. This is a contradiction. If $N \cap G_1 \leq G_2G_3 \cdots G_r$, then $G = (M \cap G_1)G_2G_3 \cdots G_r$. By the choice of G, there exists a Carter subgroup T of G with $T = (T \cap M \cap G_1)(T \cap G_2) \cdots (T \cap G_r) \leq (T \cap G_1)(T \cap G_2) \cdots (T \cap G_r) \leq T$, and T is prefactorised, against supposition. Note that this argument could be used with every factor of the decomposition of N. Consequently $N = (N \cap G_1) \times G_2 \times G_3 \times \cdots \times G_r$. In particular, $G^{\mathfrak{N}} \leq N$ is abelian and so it is complemented in G by C (see [119, IV, 5.18]). If $N = G^{\mathfrak{N}}$, then C would be a Hall p'-subgroup of G and, by Proposition 4.1.45 (2), C would have a prefactorised conjugate, contrary to hypothesis. Thus $G^{\mathfrak{N}}$ is a proper subgroup of N. By Theorem 4.4.17,

$$\begin{aligned} G^{\mathfrak{N}} &= (G^{\mathfrak{N}} \cap G_1)(G^{\mathfrak{N}} \cap G_2) \cdots (G^{\mathfrak{N}} \cap G_r) \\ &= (G^{\mathfrak{N}} \cap N \cap G_1)(G^{\mathfrak{N}} \cap G_2) \cdots (G^{\mathfrak{N}} \cap G_r). \end{aligned}$$

Assume first that $G^{\mathfrak{N}} \cap G_1 = 1$, $G_i \leq G^{\mathfrak{N}}$ for $i \in \{1, 2, \dots, h\}$ and $G_j \cap G^{\mathfrak{N}} = 1$ for $i \in \{h+1, \dots, r\}$. Then

$$\begin{aligned} G &= G^{\mathfrak{N}}(G_{h+1} \cdots G_r)(N \cap G_1)(M \cap G_1) \\ &= (G_2 \cdots G_h)(G_{h+1} \cdots G_r)(N \cap G_1)(M \cap G_1). \end{aligned}$$

Now let $T = (G_{h+1} \cdots G_r)(N \cap G_1)(M \cap G_1)$. Arguing as in the above case, $G^{\mathfrak{N}} \cap T = 1$ (recall that N is a direct product of the subgroups $N \cap G_1, G_2, \dots, G_r$). Therefore T is a Carter subgroup of G and by construction it is prefactorised. On the other hand, if $N \cap G_1 \leq G^{\mathfrak{N}}$, then a similar argument yields $G^{\mathfrak{N}} = (N \cap G_1)G_2 \cdots G_h$ and $G = G^{\mathfrak{N}}T$ where $T = G_{h+1} \cdots G_r(M \cap G_1)$ is a Carter subgroup of G and it is clearly prefactorised in G. We reach a contradiction in both cases.

Assume now that we are in case 3 of Lemma 4.2.16. Note that if $G_jK/K = 1$ for $j \geq n+1$, then as $G_j \cap K = 1$, we obtain $G_j = 1$ for all $j \geq n+1$. Therefore we would be in case 1.

Finally, if we are in the hypothesis of Statement 4 of Lemma 4.2.16, we have $G/K = G_iK/K$ for some i and $G_jK/K = 1$ for all $j \neq i$. As before, this means that $G_j = 1$ for all $j \neq i$, therefore $G = G_i$ and a Carter subgroup of G is clearly prefactorised, the final contradiction. □

Theorem 5.2.16. *Let the soluble group $G = G_1 G_2 \cdots G_r$ be the product of the pairwise totally permutable subgroups $G_1, G_2, \ldots, G_r$. Then G possesses a prefactorised system normaliser.*

Proof. Let k denote the nilpotent length of the soluble group G. Then $G \in \mathfrak{N}^k$, the class of soluble groups with nilpotent length at most k. If $k = 2$, then by [119, V, 4.2] the system normalisers of G coincide with the Carter subgroups of G. Therefore the conclusion follows by Theorem 5.2.15. Thus we may assume that $k > 2$. Set $H_0 = G$ and denote by H_i an $\mathfrak{N}^{k-1-i}\mathfrak{N}$-projector of H_{i-1} for $i \in \{1, 2, \ldots, k-1\}$. By [119, V, 4.3 (b)], H_{k-1} is a system normaliser of G. We have the chain:

$$H_k \leq H_{k-1} \leq \cdots \leq H_1 \leq H_0 = G.$$

Since the class $\mathfrak{U}$ of all supersoluble groups is contained in $\mathfrak{N}^2 \subseteq \mathfrak{N}^k$, we can apply Corollary 5.2.13 and Lemma 4.1.42 (2) to conclude that H_{k-2} is factorised in G. Now H_{k-1} is a Carter subgroup of H_{k-2}. Therefore, by Theorem 5.2.15, there exists $h \in H_{k-2}$ such that H_{k-1}^h is prefactorised in H_{k-2}. By Lemma 4.1.42 (2), H_{k-1}^h is prefactorised in G. The proof of the theorem is now complete. □

The following example shows that a Carter subgroup of a group need not be factorised.

Example 5.2.17. Consider the group

$$G = \langle a, b, c : a^3 = b^7 = c^7 = [b, c] = 1,\ a^2ba = b^2,\ a^2ca = c^2 \rangle,$$

which is the product of the totally permutable subgroups $A = G$ and $B = \langle b \rangle$. The Carter subgroups of G (which coincide with the system normalisers of G) are the conjugates of $\langle a \rangle$, which are not factorised in G.

The remainder of this section is devoted to connections between the structure of a mutually permutable product on the one hand and properties of formation subgroups of its factors on the other. We saw in the Example 4.1.32 that a mutually permutable product of supersoluble groups is not supersoluble in general and so in this setting the theory is less satisfactory. However, in Section 4.5, we have seen that mutually permutable products of two factors with core-free intersection have an excellent behaviour with respect Fitting formations and saturated formations containing the class of all supersoluble groups. These results have extended our knowledge of the theory's scope and limitations. The following result, proved by Alejandre, Ballester-Bolinches, and Cossey [3, Theorem 2], illustrates this claim.

Theorem 5.2.18. *Let the group G be the product of the mutually permutable supersoluble subgroups A and B. Then $G/\mathrm{F}(G)$ is supersoluble and metabelian.*

Proof. Let $M = G^{\mathfrak{U}}$ denote the supersoluble residual of G. By Corollary 4.5.9, $G/(A \cap B)_G$ is supersoluble. Therefore M is contained in $(A \cap B)_G$. In particular, M is supersoluble.

Let $\mathrm{F}(M)$ be the Fitting subgroup of M. Since A and B are supersoluble, the derived subgroups of A and B are nilpotent and we have that $[M, A] \leq \mathrm{F}(A) \cap M \leq \mathrm{F}(M)$ and $[M, B] \leq \mathrm{F}(B) \cap M \leq \mathrm{F}(M)$. Consequently $[M, G]$ is contained in $\mathrm{F}(M)$. Note now that the chief factors of G between $\mathrm{F}(M)$ and M are cyclic, and recall that G/M is supersoluble. Therefore we have that $G/\mathrm{F}(M)$ is supersoluble. This implies that $M = \mathrm{F}(M)$ and thus M is nilpotent. Consequently, $G/\mathrm{F}(G)$ is supersoluble.

We will show now that $G/\mathrm{F}(G)$ is metabelian. We prove first that A' and B' both centralise every chief factor of G. Let H/K be a chief factor of G. If H/K is cyclic, then as G' centralises H/K, so do A' and B'. Hence we may assume that H/K is a non-cyclic p-chief factor of G for some prime p. Note that we may assume that H is contained in M because G/M is supersoluble and H/K is not cyclic. To simplify the notation, we will assume that $K = 1$. Since $\mathrm{F}(G)$ centralises H (see [119, A, 13.8 (b)]), we have that $G/\mathrm{C}_G(H)$ is supersoluble. Let $A_{p'}$ be a Hall p'-subgroup of A. By Maschke's theorem [119, A, 11.5], H is a completely reducible $A_{p'}$-module and $HA_{p'}$ is supersoluble because H is contained in A. Therefore $A_{p'}/\mathrm{C}_{A_{p'}}(H)$ is abelian of exponent dividing $p-1$. This implies that the primes dividing $|A/\mathrm{C}_A(H)|$ can only be p or divisors of $p-1$. The same is true for $|B/\mathrm{C}_B(H)|$. This implies that if p divides $|G/\mathrm{C}_G(H)|$, then p is the largest prime dividing $|G/\mathrm{C}_G(H)|$. But since $\mathrm{O}_p(G/\mathrm{C}_G(H)) = 1$ and $G/\mathrm{C}_G(H)$ is supersoluble, it follows that $G/\mathrm{C}_G(H)$ must be a p'-group. Consider H as A-module over $\mathrm{GF}(p)$. Since $A\mathrm{C}_G(H)/\mathrm{C}_G(H)$ is a p'-group, we have that H is a completely reducible A-module and every irreducible A-submodule of H is cyclic. Consequently A' centralises H, and the same is true for B'.

Let now U/V be a chief factor of G. Then $G/\mathrm{C}_G(U/V)$ is the product of the abelian subgroups $A\mathrm{C}_G(U/V)/\mathrm{C}_G(U/V)$ and $B\mathrm{C}_G(U/V)/\mathrm{C}_G(U/V)$. By Theorem 3.1.7, we have that $G/\mathrm{C}_G(U/V)$ is metabelian. Since $\mathrm{F}(G)$ is the intersection of the centralisers of all chief factors ([119, A, 13.8 (b)]), we can conclude that $G/\mathrm{F}(G)$ is metabelian. □

Open question 5.2.19. *What can be said about the structure of a product of finitely many pairwise mutually permutable product of supersoluble groups?*

We now direct our attention towards possible extensions of the classical results of Asaad and Shaalan [24]. We recall once more that they showed that a group G factorised by two supersoluble mutually permutable subgroups is supersoluble provided G' is nilpotent or at least one of the factors is nilpotent. Unfortunately, the second result does not hold in general for saturated formations containing all supersoluble groups, as the following example shows.

Example 5.2.20. For the saturated formation $\mathfrak{F} = \mathfrak{N}^2 \supseteq \mathfrak{U}$, there exists a mutually permutable product $G = AB$ of two subgroups A and B such that $A \in \mathfrak{F}$ and $B \in \mathfrak{N}$, but G does not belong to $\mathfrak{F}$. Take for G the symmetric group Sym(4) of degree 4, $A = \mathrm{Alt}(4)$ the corresponding alternating group, and B a Sylow 2-subgroup of G, which is isomorphic to the dihedral group of order 8. It is clear that $G = AB$ and the subgroups A and B are mutually permutable. In this group, A belongs to $\mathfrak{F}$, B belongs to $\mathfrak{N}$, but G has nilpotent length 3 and so G does not belong to $\mathfrak{F}$.

However mutually permutable products with nilpotent commutator subgroup would seem to offer a better prospect of developing new results. It takes as its starting point the following theorem.

Theorem 5.2.21. *Let the group $G = G_1G_2 \cdots G_r$ be the product of the pairwise mutually permutable subgroups $G_1, G_2, \ldots, G_r$. Let $\mathfrak{F}$ be a saturated formation containing the class $\mathfrak{U}$ of all supersoluble groups. Assume that G_i belongs to $\mathfrak{F}$ for all $i \in \{1, 2, \ldots, r\}$. If G' is nilpotent, then G belongs to $\mathfrak{F}$.*

Proof. Assume that the result is not true and let G be a counterexample with $|G|$ as small as possible. Clearly $G \neq G_i$ for all i. By minimality of G and Lemma 4.1.10, we know that every proper quotient of G belongs to $\mathfrak{F}$. Consequently, G is a soluble primitive group because $\mathfrak{F}$ is a saturated formation. Then $N = \mathrm{F}(G) = \mathrm{Soc}(G)$ is the unique minimal normal subgroup of G and if p is the prime dividing its order, we have that $\mathrm{F}(G) = G'$ is the unique Sylow p-subgroup of G. For every $i \in \{1, \ldots, r\}$, let H_i be a Hall p'- subgroup H_i of G_i. Since $N \cap G_i$ is the Sylow p-subgroup of G_i, we have that $G_i = (N \cap G_i)H_i$. Let $j \in \{1, \ldots, r\}$. Then G_jH_i is a subgroup of G and, since $|G_j : N \cap G_j|$ and $|G_jH_i : G_j|$ are not divisible by p, we have $N \cap G_jH_i = N \cap G_j$. Hence, $N \cap G_j$ is normalised by $N\langle H_1, \ldots, H_r\rangle = G_1G_2 \cdots G_r = G$. Since N is a minimal normal subgroup of G, we conclude that either $N = N \cap G_j$ or $N \cap G_j = 1$. Therefore either N is contained in G_j or G_j is an abelian p'-group.

Since G is not a p'-group, there exists $i \in \{1, 2, \ldots, r\}$ such that p divides the order of G_i. Without loss of generality we may assume that $i = 1$. Suppose that G_j is a p'-group for some $j \in \{1, 2, \ldots, r\}$. Let X be a subgroup of N. Then since the product G_1G_j is mutually permutable, it follows that XG_j is a subgroup of G. Therefore $X = N \cap XG_j$ is normal in XG_j and G_j normalises each subgroup of N. Set $A = G_1$ and $B = G_2 \cdots G_r$. Then A is normal in G and either B is normal in G or it normalises every subgroup of N.

Consider N regarded as an A-module over $\mathrm{GF}(p)$. By Clifford's theorem [119, B, 7.3], we know that N is a completely reducible A-module, that is, $N = N_1 \times \cdots \times N_r$, where N_i is a minimal normal subgroup of A for all i. Now $A \in \mathfrak{F}$. This means that $A/\mathrm{C}_A(N_i) \in F(p)$ for all i (here F denotes the canonical local definition of $\mathfrak{F}$). Since $F(p)$ is a formation, we have that $A/\mathrm{C}_A(N) \in F(p)$. Hence $A/N \in F(p)$. Suppose that B is not normal in G. Then B normalises each subgroup of N. Then $B/\mathrm{C}_B(N) = B$ is cyclic of order dividing $p-1$. Applying [119, IV, 3.11], $B \in F(p)$. Assume that B contains N. Then, arguing as above, B is normal in G and so $B/N \in F(p)$. In both cases, $B/N \in F(p)$. Now the group G/N is an abelian group which is the product of the subgroups A/N and B/N. Since formations are closed under central products, it follows that $G/N \in F(p)$. This implies that $G \in \mathfrak{S}_p F(p) = F(p) \subseteq \mathfrak{F}$, the final contradiction. □

The two factor case was proved by Ballester-Bolinches and Pedraza-Aguilera [51].

The converse of the above theorem is valid for factorised groups with nilpotent commutator subgroup.

Lemma 5.2.22. *Let the group $G = G_1 G_2 \cdots G_r$ be the product of the pairwise mutually permutable subgroups $G_1, G_2, \dots, G_r$. Let $\mathfrak{F}$ be a saturated formation containing the class $\mathfrak{U}$ of all supersoluble groups. Assume that G' is nilpotent. If G belongs to $\mathfrak{F}$, then G_i belongs to $\mathfrak{F}$ for all $i \in \{1, 2, \dots, r\}$.*

Proof. Assume that the result is false and let G be a counterexample of least order. Note that G is a soluble group. By minimality of G and Lemma 4.1.10, we can assume that G has a unique minimal normal subgroup, N say. Let p be the prime dividing the order of N. Since G' is nilpotent, we have that G' is contained in $\mathrm{F}(G) = \mathrm{O}_p(G)$ and then G' is a p-group. In particular, G has a unique Sylow p-subgroup P and $P = \mathrm{F}(G)$. For every $i \in \{1, 2, \dots, r\}$ let P_i be the unique Sylow p-subgroup of G_i. If $i \neq j$, the Sylow p-subgroup of $G_i G_j$ is $P_i P_j$ by Theorem 1.1.19. Thus the subgroups $P_1, P_2, \dots, P_r$ pairwise permute and $P = P_1 P_2 \cdots P_r$. Let $i \in \{1, 2, \dots, r\}$. Then there exists a Hall p'-subgroup H of G such that $H \cap G_i$ is a Hall p'-subgroup of G_i. We conclude that $G = PH$ and $P \cap H = 1$. Since $G \in \mathfrak{F}$, it follows that $G/\mathrm{O}_{p',p}(G)$ belongs to $F(p)$, where F is the canonical local definition of $\mathfrak{F}$. Since $\mathrm{O}_{p'}(G) = 1$, we have that $\mathrm{O}_{p',p}(G) = \mathrm{O}_p(G) = P$. Therefore H belongs to $F(p)$. Now G' is contained in P. This implies that H is an abelian group. We can now apply [119, IV, 1.14] to conclude that $H \cap G_i$ belongs to $F(p)$. Then $G_i = P_i(H \cap G_i)$ belongs to $\mathfrak{S}_p F(p) = F(p) \subseteq \mathfrak{F}$, and the proof of the lemma is complete. □

The next result, proved by Ballester-Bolinches and Pedraza-Aguilera [51] for mutually permutable products of two factors, shows that the residuals associated to saturated formations containing all supersoluble groups behave well in pairwise mutually permutable products with nilpotent derived subgroup.

Theorem 5.2.23. *Let the group $G = G_1 G_2 \cdots G_r$ be the product of the pairwise mutually permutable subgroups $G_1, G_2, \dots, G_r$. Let $\mathfrak{F}$ be a saturated formation containing the class $\mathfrak{U}$ of all supersoluble groups. Assume that G' is nilpotent. Then $G^{\mathfrak{F}} = G_1^{\mathfrak{F}} G_2^{\mathfrak{F}} \cdots G_r^{\mathfrak{F}}$.*

Proof. Suppose that the result is not true and let G be a counterexample with $|G| + |G_1| + |G_2| + \cdots + |G_r|$ as small as possible. Denote by $H = G_1^{\mathfrak{F}} G_2^{\mathfrak{F}} \cdots G_r^{\mathfrak{F}}$. Applying Corollary 4.4.7, H is a subnormal subgroup of G. If N is a minimal normal subgroup of G, then G/N is the product of the mutually permutable subgroups G_1N/N, $G_2N/N, \dots, G_rN/N$ by Lemma 4.1.10. Moreover $(G/N)'$ is nilpotent. Hence, by the minimal choice of G, we have that $G^{\mathfrak{F}}N = HN$. Consider now $G/G^{\mathfrak{F}} = (G_1G^{\mathfrak{F}}/G^{\mathfrak{F}})(G_2G^{\mathfrak{F}}/G^{\mathfrak{F}})\cdots(G_rG^{\mathfrak{F}}/G^{\mathfrak{F}})$, which is the product of the pairwise totally permutable subgroups $G_iG^{\mathfrak{F}}/G^{\mathfrak{F}}$, $i \in \{1, 2, \ldots, r\}$, by Lemma 4.1.20. By Lemma 5.2.22, we have that $G_i/(G_i \cap G^{\mathfrak{F}}) \in \mathfrak{F}$ and so $G_i^{\mathfrak{F}} \leq G^{\mathfrak{F}}$ for all i. Thus H is contained in $G^{\mathfrak{F}}$. Hence $G^{\mathfrak{F}} = H(G^{\mathfrak{F}} \cap N)$. Therefore if $G^{\mathfrak{F}} \cap N = 1$, it follows that $G^{\mathfrak{F}} = H$. This contradiction forces the conclusion that $\mathrm{Soc}(G)$ is contained in $G^{\mathfrak{F}}$ and $G^{\mathfrak{F}} = HN$ for every minimal normal subgroup N of G. In particular, $H_G = 1$. Moreover, by [119, A, 14.3], $\mathrm{Soc}(G)$ normalises H. Hence H is a normal subgroup of $G^{\mathfrak{F}}$ (note that the subnormality of H in G can be also deduced from the nilpotency of G'). Let p be the prime dividing the order of a minimal normal subgroup N of G. Then $G^{\mathfrak{F}}/H$ is an abelian p-group and so $\mathrm{O}^p(G^{\mathfrak{F}})(G^{\mathfrak{F}})'$ is a normal subgroup of G contained in H. Thus $\mathrm{O}^p(G^{\mathfrak{F}})(G^{\mathfrak{F}})' = 1$ and $G^{\mathfrak{F}}$ is an abelian p-group. This means that $\mathrm{F}(G)$ is a p-group and then G' is also a p-group because G' is nilpotent. Consequently, $\mathrm{F}(G)$ is the unique Sylow p-subgroup of G.

Let $A = G_1$ and $B = G_2 \cdots G_r$. The choice of G implies that $B^{\mathfrak{F}} = G_2^{\mathfrak{F}} \cdots G_r^{\mathfrak{F}}$. Furthermore, $A_p = A \cap \mathrm{F}(G)$ and $B_p = B \cap \mathrm{F}(G)$ are the Sylow p-subgroups of A and B, respectively. Let $A_{p'}$ and $B_{p'}$ be Hall p'-subgroups of A and B, respectively. It follows that $BA_{p'}$ is a subgroup of G. Suppose that $G = BA_{p'}$. Then $\mathrm{F}(G)$ is contained in B and so B is a normal subgroup of G. Hence $B^{\mathfrak{F}}$ is a normal subgroup of G. Since $H_G = 1$, it follows that $B^{\mathfrak{F}} = 1$ and so $G_i \in \mathfrak{F}$ for all $i \in \{2, \ldots, r\}$ by Lemma 5.2.22. Assume that $BA_{p'}$ is a proper subgroup of G. Let M be a maximal subgroup of G containing $BA_{p'}$. Then $M = (M \cap A)B$ and $G = A_pM = \mathrm{F}(G)M$. By Lemma 4.1.21 (3), M is the pairwise mutually permutable product of the subgroups $(M \cap A)$, $G_2, \ldots, G_n$. The minimal choice of G yields $M^{\mathfrak{F}} = (M \cap A)^{\mathfrak{F}}B^{\mathfrak{F}}$. Moreover, $A = A_p(M \cap A)$ and so, by [47, 2.2.8], $(M \cap A)^{\mathfrak{F}}$ is contained in $A^{\mathfrak{F}}$. Therefore $M^{\mathfrak{F}}$ is contained in H. Suppose that $G^{\mathfrak{F}}$ is not contained in M. Then $G = MG^{\mathfrak{F}}$. Since $G^{\mathfrak{F}}$ is abelian and M normalises $M^{\mathfrak{F}}$, we have that $M^{\mathfrak{F}}$ is a normal subgroup of G. In particular, $M^{\mathfrak{F}} \leq H_G = 1$ and so, by Lemma 5.2.22, $B \in \mathfrak{F}$. Assume that $G^{\mathfrak{F}}$ is contained in M. Let N be a minimal normal subgroup of G. Then $N \leq G^{\mathfrak{F}} \leq M$. Since $G = M\mathrm{F}(G)$ and $\mathrm{F}(G)$ centralises N, it follows that N is a minimal normal subgroup of M. If $N \cap M^{\mathfrak{F}} = N$, then $N \leq H$, which contradicts our supposition. Hence $N \cap M^{\mathfrak{F}} = 1$ and $NM^{\mathfrak{F}}/M^{\mathfrak{F}}$ is a minimal normal subgroup of $M/M^{\mathfrak{F}} \in \mathfrak{F}$. In particular, N is $\mathfrak{F}$-central in M and so N is $\mathfrak{F}$-central in G, which contradicts [47, 4.2.19]. Consequently, $G^{\mathfrak{F}}$ is not contained in M and $B \in \mathfrak{F}$. In both cases, we have $G_i \in \mathfrak{F}$ for all $i \in \{2, \ldots, r\}$.

If we denote by $C = G_2$ and $D = G_1G_3 \cdots G_r$ ($D = G_1$ if $r = 1$), we can use the arguments just given above to conclude that $G_1 \in \mathfrak{F}$. By Theorem 5.2.21, $G \in \mathfrak{F}$, yielding the final contradiction. □

One might wonder, as in the totally permutable case, whether there exists some $\mathfrak{F}$-projector (respectively $\mathfrak{F}$-normaliser) of a mutually permutable product which is a product of $\mathfrak{F}$-projectors (respectively $\mathfrak{F}$-normalisers) of the factors. The answer is "no" in general, even in groups with nilpotent derived subgroup, as the following example of Ballester-Bolinches and Pedraza-Aguilera [51] shows.

Example 5.2.24. Let G be the direct product of a cyclic group $\langle a \rangle$ of order 3 with an alternating group Alt(4) of degree 4. Let V be the Klein 4-group of Alt(4). Then G is the mutually permutable product of $H = \langle a \rangle \times V$ and $K = A_4$. Moreover $G' = V$ is nilpotent. Note that H is the supersoluble projector of H and a Sylow 3-subgroup B of Alt(4) is a supersoluble projector of K. However $HB = G$ is not supersoluble. This implies that Theorem 5.2.9 cannot be extended to mutually permutable products.

Since $G \in \mathfrak{N}\mathfrak{U}$, by [47, 4.1.18] we see that the $\mathfrak{U}$-normalisers of G coincide with the $\mathfrak{U}$-projectors of G. Consequently Theorem 5.2.10 cannot be extended to mutually permutable products.

In the final part of this section we show, following [34], that $\mathfrak{F}$-projectors and $\mathfrak{F}$-normalisers of soluble mutually permutable products are prefactorised.

Theorem 5.2.25. *Let the soluble group $G = AB$ be the product of the mutually permutable subgroups A and B. If $\mathfrak{F}$ is a saturated formation containing the class $\mathfrak{U}$ of all supersoluble groups, then every sub-$\mathfrak{F}$-abnormal subgroup of G is prefactorised.*

Proof. Assume that the result is false and let G be a counterexample of least order. Then there exists a sub-$\mathfrak{F}$-abnormal subgroup S of G which is not prefactorised. Let us take S of maximal order. If S is not a maximal subgroup of G, then there exists an $\mathfrak{F}$-abnormal maximal subgroup M of G with $S \leq M$ and S is sub-$\mathfrak{F}$-abnormal in M. By the choice of S, $M = (M \cap A)(M \cap B)$. Now the minimality of G implies that S is prefactorised in M and therefore in G by Lemma 4.1.42 (2). Consequently, we may suppose that S is an $\mathfrak{F}$-abnormal maximal subgroup of G.

Then G satisfies the following properties:

1. $(A \cap B)_G = 1$.

Assume that there exists a minimal normal subgroup N of G such that $N \leq A \cap B$. Suppose that $N \leq S$. Now S/N is an $\mathfrak{F}$-abnormal subgroup of G/N. By induction, S/N is prefactorised in G/N. But N is prefactorised in G. By Lemma 4.1.42 (1), S is prefactorised, contrary to supposition. Therefore $G = SN$, $S \cap N = 1$, $A = N(S \cap A)$ and $B = N(S \cap B)$. Consequently $G = NS = N(S \cap A)(S \cap B)$. Therefore $S = (S \cap A)(S \cap B)$, which contradicts the choice of G.

2. *Every minimal normal subgroup of G is contained in S.*

Assume that there exists a minimal normal subgroup M of G which is not contained in S. Then $G = SM$. Moreover, by Corollary 4.1.22 (2), either $M \cap A = M \cap B = 1$ or $M = (M \cap A)(M \cap B)$.

Assume first that $M \cap A = M \cap B = 1$. Then by Lemma 4.3.9, $|M| = p$, a prime number. Therefore $G/C_G(M)$ is an $\mathfrak{F}$-group and $G^{\mathfrak{F}} \leq C_G(M)$. Applying [119, IV, 1.17 (b)], $S^{\mathfrak{F}} \leq G^{\mathfrak{F}}$ and so $S^{\mathfrak{F}}$ is normal in G. If $S^{\mathfrak{F}} = 1$, then $S \in \mathfrak{F}$. In this case, $G = SM$ would be the totally permutable product of two $\mathfrak{F}$-subgroups. Therefore, by Theorem 5.2.2, $G \in \mathfrak{F}$, contrary to the fact that S is $\mathfrak{F}$-abnormal in G. Hence $S^{\mathfrak{F}} \neq 1$. On the other hand, $G/S^{\mathfrak{F}}$ is the product of the totally permutable subgroups $S/S^{\mathfrak{F}}$ and $MS^{\mathfrak{F}}/S^{\mathfrak{F}}$, and both are $\mathfrak{F}$-groups. A new application of Theorem 5.2.2 yields $G/S^{\mathfrak{F}}$ is an $\mathfrak{F}$-group and so $G^{\mathfrak{F}} = S^{\mathfrak{F}}$. The minimality of G implies that $S/S^{\mathfrak{F}}$ is prefactorised in $G/S^{\mathfrak{F}}$ and, by Theorem 4.4.17, $S^{\mathfrak{F}}$ is prefactorised in G. By Lemma 4.1.42 (2), S is prefactorised in G, against supposition.

Suppose now that $M = (M \cap A)(M \cap B)$. We know by Statement 1 that M is not contained in $A \cap B$. Assume that $M \leq A$ and $M \cap B = 1$ (see Lemma 4.3.3). Now $A = M(S \cap A)$. We claim that $S \cap A \neq 1$. Suppose, arguing by contradiction, that $S \cap A = 1$. Then $A = M$, $G = AB = MB = MS$ and $A \cap B = M \cap B = 1$. Let X be a subgroup of A of prime order. Then $X = XB \cap M$ is normalised by B. Consequently X is normal in G and $X = M$, that is, $|M| = p$, where p is a prime. Now consider the quotient $G/S_G = (MS_G/S_G)(S/S_G)$. It is a primitive group and MS_G/S_G has order p. Thus G/S_G is supersoluble, contradicting the fact that S is $\mathfrak{F}$-abnormal. Hence $S \cap A \neq 1$. On the other hand, applying Lemma 4.3.3 (5), M is cyclic of prime order or $M \leq C_G(B)$. We prove that $S \cap A$ is a maximal subgroup of A. If M has prime order, then it is clear that $S \cap A$ is a maximal subgroup of A. Assume that $M \leq C_G(B)$. Then M is a minimal normal subgroup of A. If $A = S \cap A$, then S is prefactorised by Proposition 1.1.16, against the choice of G. Consequently $S \cap A$ is a proper subgroup of A. Thus $S \cap A$ is a maximal subgroup of A. Suppose that $A \cap B$ is not contained in $S \cap A$. Then $A = (S \cap A)(A \cap B)$, since $A \cap B$ is a permutable subgroup of A by Corollary 4.1.17. Thus $G = (S \cap A)(A \cap B)B = (S \cap A)B$ and $S = (S \cap A)(S \cap B)$, contrary to the choice of S. Consequently we may assume that $A \cap B$ is contained in S. On the other hand, by Corollary 4.1.17, $A \cap B$ is a subnormal subgroup of G. Therefore M normalises $A \cap B$ by [119, A, 14.3]. Then $T = \langle (A \cap B)^G \rangle = \langle (A \cap B)^S \rangle \leq S$. If $A \cap B = 1$, then the product is totally permutable by Proposition 4.1.16 (3) and we have that S is factorised by Theorem 5.2.11. This contradiction yields $A \cap B \neq 1$. Since $A \cap B$ is contained in $(T \cap A)(T \cap B)$, it follows that $(T \cap A)(T \cap B)$ is a non-trivial factorised normal subgroup of G contained in S (see Lemma 4.1.21 (2)). By induction, $S/(T \cap A)(T \cap B)$ is prefactorised. Applying Lemma 4.1.42 (1), S is prefactorised. This contradiction proves Statement 2.

3. *The final contradiction.*

Let N be a minimal normal subgroup of G. Then $N \leq S$. If $N = (N \cap A)(N \cap B)$, the minimal choice of G implies that S/N is prefactorised in G/N. By Lemma 4.1.42 (1), the same is true for S. This contradiction yields $N \cap A = N \cap B = 1$ for all minimal normal subgroups of G. This is to say that $A_G B_G = 1$, which contradicts Theorem 4.3.11. □

Corollary 5.2.26. *Let the soluble group $G = AB$ be the product of the mutually permutable subgroups A and B. If $\mathfrak{F}$ is a saturated formation containing the class $\mathfrak{U}$ of all supersoluble groups, then the $\mathfrak{F}$-projectors and $\mathfrak{F}$-normalisers of G are prefactorised subgroups of G.*

$\mathfrak{F}$-projectors and $\mathfrak{F}$-normalisers of soluble mutually permutable products are not factorised in general.

Example 5.2.27. Let $G = \mathrm{Sym}(4)$ be the symmetric group of degree 4. Then $G = AB$ is the product of the mutually permutable subgroups A and B, where A denotes the alternating group of degree 4 and B is a Sylow 2-subgroup of G. Now consider a subgroup H isomorphic to the symmetric group of degree 3. Then H is a $\mathfrak{U}$-projector (and also a $\mathfrak{U}$-normaliser of G) which is prefactorised in G, but not factorised.

Open questions 5.2.28.

1. *Can Theorems* 5.2.15 *and* 5.2.16 *be extended to pairwise mutually permutable products?*
2. *Is Theorem* 5.2.25 *true for mutually permutable products with more than two factors?*

5.3 Totally permutable products and Fitting classes

The theory developed in this section is largely the work of Hauck, Martínez-Pastor, and Pérez-Ramos and grew out of an attempt to study the behaviour of totally permutable products with respect to Fitting classes. The challenging unsolved problem in this theory is whether the Fitting classes containing all supersoluble groups are closed under the formation of totally permutable products. A partial response to this general question is to impose extra conditions on the Fitting classes under consideration; for example, since the nilpotent residuals of totally permutable subgroups have an excellent behaviour in the product, one can try initially to answer the question for Fischer classes, which are Fitting classes closed under the formation of subgroups with subnormal nilpotent residual. In fact, according to Corollary 4.1.25, if $G = AB$ is a mutually permutable product of the subgroups A and B, then A' and B' are subnormal subgroups of G. Hence, since the nilpotent residuals of A and B are subnormal subgroups of A' and B', respectively, they are in fact subnormal subgroups of G. Consequently, if $\mathfrak{F}$ is a Fischer class and $G \in \mathfrak{F}$, then A and B belong to $\mathfrak{F}$. The converse is true for totally permutable products and Fischer classes containing all supersoluble groups. This result, proved in [144], is probably one of the most satisfying ones in this theory (Theorem 5.3.3). It follows after a description of a minimal configuration often encountered in the study of totally permutable products and Fitting classes. The analysis leading up to the proof of this result has enormously increased our understanding of the behaviour of these products with respect to Fitting classes.

Lemma 5.3.1 (Hauck et al. [144]). *Let $\mathfrak{H}$ be a subgroup-closed saturated formation and let*

$$\mathfrak{T}_{\mathfrak{H}} = (G = AB : A \text{ and } B \text{ totally permutable}, A \cap B \le \mathrm{Z}_{\mathfrak{H}}(G)).$$

Let $\mathfrak{F}$ be a Fitting class containing $\mathfrak{U}$. Suppose that one of the following cases holds:

Case 1: *there exists $G = AB \in \mathfrak{T}_{\mathfrak{H}}$ with $A, B \in \mathfrak{F}$, but $G \notin \mathfrak{F}$, and that among all these groups in $\mathfrak{T}_{\mathfrak{H}}$ we choose one with least $|G| + |A|\,|B|$, or*

Case 2: *there exists $G = AB \in \mathfrak{T}_{\mathfrak{H}}$ with $G \in \mathfrak{F}$, but not both of A and B in $\mathfrak{F}$, and that among all these groups in $\mathfrak{T}_{\mathfrak{H}}$ we choose a group with least $|G|+|A|\,|B|$.*

Then, after interchanging the roles of A and B if necessary, the following holds:

1. *$A/A^{\mathfrak{N}}$ is a cyclic p-group for a prime p and $A^{\mathfrak{N}}$ is a non-trivial normal subgroup of G.*
2. *B is a normal abelian p'-subgroup and p divides $q-1$ for all prime divisors q of $|B|$.*
3. *$B = [A, B]$.*
4. *$A \cap B = A^{\mathfrak{N}} \cap B \le \mathrm{Z}_{\mathfrak{U}}(G)$ but $A \cap B \nleq \mathrm{Z}_{\infty}(G)$.*
5. *A acts as a group of power automorphisms on B.*

If $\mathfrak{U} \subseteq \mathfrak{H}$, then condition 2 can be replaced by

2′. *B is a normal cyclic q-subgroup, where q is a prime such that p divides $q-1$.*

Proof. Note first that, since $\mathfrak{H}$ is subgroup-closed, we obtain that $\mathrm{Z}_{\mathfrak{H}}(G) \cap M \le \mathrm{Z}_{\mathfrak{H}}(M)$ for every subgroup M of G. Thus if $A_1 \le A$ and $B_1 \le B$, it is clear that $M = A_1B_1 \in \mathfrak{T}_{\mathfrak{H}}$.

We split the proof into the following steps.

1. *We may assume that B is a nilpotent group and that A is not nilpotent. Moreover, $[B, A^{\mathfrak{N}}] = 1$.*

Note that, by Theorem 4.2.7, $[B, A^{\mathfrak{N}}] = 1$ and $[A, B^{\mathfrak{N}}] = 1$.

It is clear that A and B cannot be both nilpotent groups, because if we are in Case 1, then G would be supersoluble by Corollary 4.1.38, and if we are in Case 2, that would contradict the choice of (G, A, B).

Suppose that neither A nor B is nilpotent. Then $A^{\mathfrak{N}} \ne 1$ and $B^{\mathfrak{N}} \ne 1$. Note also that $A^{\mathfrak{N}}$ cannot be central in A and $B^{\mathfrak{N}}$ cannot be central in B. Then $B \le \mathrm{C}_G(A^{\mathfrak{N}}) < G$ and $A \le \mathrm{C}_G(B^{\mathfrak{N}}) < G$. Hence $\mathrm{C}_G(A^{\mathfrak{N}}) = B(A \cap \mathrm{C}_G(A^{\mathfrak{N}}))$ is a product of two totally permutable subgroups. Assume that Case 1 holds. Then $\mathrm{C}_G(A^{\mathfrak{N}})$ is a normal subgroup of G and so $\mathrm{C}_G(A^{\mathfrak{N}}) \in \mathfrak{F}$ by the choice of (G, A, B).

Analogously $C_G(B^{\mathfrak{N}}) \in \mathfrak{F}$ and so $G = C_G(A^{\mathfrak{N}})C_G(B^{\mathfrak{N}}) \in \mathfrak{F}$ because it is the product of two normal $\mathfrak{F}$-subgroups. Assume now that Case 2 holds. Then $C_G(A^{\mathfrak{N}})$ is an $\mathfrak{F}$-group because it is a normal subgroup of G. By the choice of (G, A, B), we obtain that $B \in \mathfrak{F}$. An analogous argument yields $A \in \mathfrak{F}$, contradicting the choice of (G, A, B).

Hence we can assume that B is nilpotent and A is not nilpotent.

2. $G = \langle A^G \rangle = A[A, B]$, *$\langle B^G \rangle \cap A$ is nilpotent, and* $\langle B^G \rangle \in \mathfrak{U} \subseteq \mathfrak{F}$.

Note first that $[B, A^{\mathfrak{N}}] = 1$ implies that $[\langle B^G \rangle, A^{\mathfrak{N}}] = 1$. Now

$$(\langle B^G \rangle \cap A)/(\langle B^G \rangle \cap A^{\mathfrak{N}}) \cong (\langle B^G \rangle \cap A)A^{\mathfrak{N}}/A^{\mathfrak{N}} \in \mathfrak{N}$$

and

$$\langle B^G \rangle \cap A^{\mathfrak{N}} \leq Z(\langle B^G \rangle \cap A),$$

so $\langle B^G \rangle \cap A$ is nilpotent. Consequently $\langle B^G \rangle = B(\langle B^G \rangle \cap A)$ is a product of two totally permutable nilpotent groups. Then $\langle B^G \rangle \in \mathfrak{U} \subseteq \mathfrak{F}$ by Corollary 4.1.38.

Assume now that $\langle A^G \rangle$ is a proper subgroup of G. Then $\langle A^G \rangle = A(\langle A^G \rangle \cap B)$ is the product of two totally permutable subgroups. Assume that Case 1 holds. Then $\langle A^G \rangle \cap B$ is a normal subgroup of B and hence an $\mathfrak{F}$-group. By minimality of (G, A, B), $\langle A^G \rangle \in \mathfrak{F}$ and hence $G = \langle A^G \rangle \langle B^G \rangle \in \mathrm{N}_0 \mathfrak{F} = \mathfrak{F}$, against the choice of G. Assume that Case 2 holds. Then $\langle A^G \rangle \in \mathfrak{F}$ because $\langle A^G \rangle$ is a normal subgroup of G and by the choice of (G, A, B), we deduce that $A \in \mathfrak{F}$, against assumption.

3. *There exists a prime number p such that $G = A^{\mathfrak{N}} A_p \langle B^G \rangle$, where A_p is a Sylow p-subgroup of A.*

Assume that $A^{\mathfrak{N}} A_q \langle B^G \rangle$ is a proper subgroup of G for all primes q, where H_q denotes a Sylow q-subgroup of H. Then

$$A^{\mathfrak{N}} A_q \langle B^G \rangle = A^{\mathfrak{N}} A_q \langle B^G \rangle \cap AB = A^{\mathfrak{N}} A_q (\langle B^G \rangle \cap AB) = A^{\mathfrak{N}} A_q (\langle B^G \rangle \cap A)B.$$

Note also that $A^{\mathfrak{N}} A_q \langle B^G \rangle$ is a normal subgroup of G. The subgroups $X_q = A^{\mathfrak{N}} A_q (\langle B^G \rangle \cap A)$ and B are totally permutable. Assume that Case 1 holds. Then $X_q \in \mathfrak{F}$ because X_q is a normal subgroup of A and, by minimality of (G, A, B), $X_q B \in \mathfrak{F}$. Hence $G = \prod_{q \in \mathbb{P}} X_q B \in \mathrm{N}_0 \mathfrak{F} = \mathfrak{F}$, against assumption. Assume now that Case 2 holds. Then $X_q B$ is a normal subgroup of G and so $X_q B \in \mathfrak{F}$. By the minimal choice of (G, A, B), $X_q \in \mathfrak{F}$ for all primes q. It follows that $A = \prod_{q \in \mathbb{P}} X_q \in \mathrm{N}_0 \mathfrak{F} = \mathfrak{F}$, against supposition.

4. *Assume that Case 1 holds. Then for all primes $q \neq p$, $A^{\mathfrak{N}} A_q [A, B]$ is a normal $\mathfrak{F}$-subgroup of G, where A_q is a Sylow q-subgroup of A, but $A^{\mathfrak{N}} A_p [A_p, B] \notin \mathfrak{F}$.*

Since $G = \langle A^G \rangle = \langle A^B \rangle = A[A, B]$, $[A, B]$ is a normal subgroup of G, and $A^{\mathfrak{N}} A_q$ is a normal subgroup of A, we obtain that $A^{\mathfrak{N}} A_q [A, B]$ is a normal subgroup of $G = AB$. Since $A^{\mathfrak{N}} \in \mathfrak{F}$ and $\langle B^G \rangle \in \mathfrak{U} \subseteq \mathfrak{F}$, we conclude that $A^{\mathfrak{N}} \langle B^G \rangle \in$

$N_0 \mathfrak{F} = \mathfrak{F}$. Since $G/A^{\mathfrak{N}}\langle B^G\rangle$ is a p-group, it follows that $A^{\mathfrak{N}} A_q[A, B]$ is contained in $A^{\mathfrak{N}}\langle B^G\rangle$. Hence $A^{\mathfrak{N}} A_q[A, B]$ is an $\mathfrak{F}$-group.

Consider now the normal subgroup $C = \langle (A^{\mathfrak{N}} A_p)^G\rangle$ of G. Since $[B, A^{\mathfrak{N}}] = 1$, we obtain that $C = \langle (A^{\mathfrak{N}} A_p)^B\rangle = A^{\mathfrak{N}} A_p[A^{\mathfrak{N}} A_p, B] = A^{\mathfrak{N}} A_p[A_p, B]$. Suppose that $C \in \mathfrak{F}$. Then by Statement 2, it follows that

$$G = \langle A^G\rangle = A[A, B] = C\Big(\prod_{q\neq p} A^{\mathfrak{N}} A_q[A, B]\Big) \in N_0 \mathfrak{F} = \mathfrak{F},$$

contradicting the choice of (G, A, B).

5. *Assume that Case 2 holds. Then for all primes $q \neq p$, $A^{\mathfrak{N}} A_q \in \mathfrak{F}$, where A_q is a Sylow q-subgroup of A. Moreover $A^{\mathfrak{N}} A_p \notin \mathfrak{F}$.*

Note that $X_q = A^{\mathfrak{N}} A_q \leq A^{\mathfrak{N}}\langle B^G\rangle = X_q\langle B^G\rangle$ by Statement 3. Since $A^{\mathfrak{N}} \in \mathfrak{F}$ and $\langle B^G\rangle \in \mathfrak{U} \subseteq \mathfrak{F}$, we conclude that $A^{\mathfrak{N}}\langle B^G\rangle \in N_0 \mathfrak{F} = \mathfrak{F}$. Since $X_q\langle B^G\rangle$ is a proper subgroup of G which is the totally permutable product of the subgroups $A \cap X_q\langle B^G\rangle$ and B, it follows that $A \cap X_q\langle B^G\rangle \in \mathfrak{F}$ by the choice of (G, A, B). Then $X_q \in \mathfrak{F}$ because X_q is normal in $A \cap X_q\langle B^G\rangle$. Finally, if $X_p \in \mathfrak{F}$, then $A = \prod_{q\in\mathbb{P}} X_q \in N_0 \mathfrak{F} = \mathfrak{F}$, against assumption.

6. $G = A^{\mathfrak{N}} A_p[A_p, B] = A^{\mathfrak{N}} A_p B$.

Let $C = A^{\mathfrak{N}} A_p[A_p, B] = \langle (A^{\mathfrak{N}} A_p)^G\rangle$. Assume that C is a proper subgroup of G. Then $C = C \cap (A^{\mathfrak{N}} A_p B) = A^{\mathfrak{N}} A_p(C \cap B)$. If Case 1 holds, then C is the product of two totally permutable subgroups in $\mathfrak{F}$ and $C \in \mathfrak{F}$ by the choice of (G, A, B), against Statement 4. Assume that Case 2 holds. Then $C \in \mathfrak{F}$ and C is the totally permutable product of the subgroups $A^{\mathfrak{N}} A_p$ and $C \cap B$ and so $A^{\mathfrak{N}} A_p \in \mathfrak{F}$ by the choice of G, against Statement 5. It follows that $C = G$ and since $C \leq A^{\mathfrak{N}} A_p B$, we also obtain that $G = A^{\mathfrak{N}} A_p B$.

7. *$A/A^{\mathfrak{N}}$ is a p-group. In particular, $A = A^{\mathfrak{N}} A_p$.*

By Statement 6 and the minimal choice of (G, A, B), we have that $A^{\mathfrak{N}} A_p = A$ and so $A/A^{\mathfrak{N}}$ is a p-group.

8. *p divides $q - 1$ for all prime divisors q of $|B|$ different from p. Moreover, $G = A^{\mathfrak{N}} A_p O_p(B) O_{p'}(B)$ and $O_{p'}(B)$ is a normal subgroup of G.*

We first claim that there exists a prime q dividing $|B|$ with $p < q$. Suppose that $p \geq q$ for all primes q dividing $|B|$. Since $A_p B$ is a supersoluble group by Corollary 4.1.38, it follows that $A_p O_p(B)$ is a normal subgroup of $A_p B$. Therefore $A = A^{\mathfrak{N}} A_p$ is a subnormal subgroup of $G = A^{\mathfrak{N}} A_p B$. If we are in Case 1, this implies that $G = A\langle B^G\rangle \in N_0 \mathfrak{F} = \mathfrak{F}$, against the choice of (G, A, B). If we are in Case 2, then A is a subnormal subgroup of $G \in \mathfrak{F}$ and so $A \in \mathfrak{F}$, which contradicts the choice of (G, A, B).

Let $\pi(B) \cup \{p\} = \{p_1, p_2, \dots, p_t = p, p_{t+1}, \dots, p_n\}$ with $p_1 < p_2 < \dots < p_t = p < p_{t+1} < \dots < p_n$. We denote by $\pi = \{p, p_{t+1}, \dots, p_n\}$ and $\pi' = (\pi(B) \cup \{p\}) \setminus \pi$. Since $A_p B$ is a supersoluble group, B normalises $A_p \mathrm{O}_\pi(B)$. Then $[A_p, B] = [A_p, B_{\pi'}\mathrm{O}_\pi(B)] \leq A_p[A_p, \mathrm{O}_\pi(B)] \leq A_p\mathrm{O}_\pi(B)$ and, by Statement 6, $G = A^{\mathfrak{N}} A_p[A_p, B] = A^{\mathfrak{N}} A_p[A_p, \mathrm{O}_\pi(B)]$ is contained in $A^{\mathfrak{N}} A_p \mathrm{O}_\pi(B)$ and so $G = A^{\mathfrak{N}} A_p \mathrm{O}_\pi(B)$. Now, by the choice of (G, A, B), it follows that $B = \mathrm{O}_\pi(B)$ and so $p \leq q$ for all primes $q \in \pi(B)$.

Finally, the facts that p divides $q - 1$ for all primes $q \in \pi(B) \setminus \{p\}$ and that $\mathrm{O}_{p'}(B)$ is a normal subgroup of $G = A^{\mathfrak{N}} A_p \mathrm{O}_p(B)\mathrm{O}_{p'}(B)$ follow again because $A_p B$ is a supersoluble group and $B \leq [A_p, B]$.

9. *B is a normal p'-subgroup of G. Moreover, $B = [A, B]$.*

Suppose that $A\mathrm{O}_{p'}(B) < G$. If we are in Case 1, $A\mathrm{O}_{p'}(B) \in \mathfrak{F}$ by minimality of (G, A, B). By Statement 8, $A^{\mathfrak{N}}\mathrm{O}_{p'}(B)$ is a normal subgroup of G and $G/A^{\mathfrak{N}}\mathrm{O}_{p'}(B)$ is a p-group. Hence $A\mathrm{O}_{p'}(B)$ is subnormal in G. This means that

$$G = (A\mathrm{O}_{p'}(B))\langle B^G \rangle \in \mathfrak{F},$$

against the choice of (G, A, B). Assume that we are in Case 2. Then $A\mathrm{O}_{p'}(B)$ is a subnormal subgroup of $G \in \mathfrak{F}$. We obtain that $A\mathrm{O}_{p'}(B) \in \mathfrak{F}$. By minimality of (G, A, B), $A \in \mathfrak{F}$, against assumption. Therefore $G = A\mathrm{O}_{p'}(B)$ and $B = \mathrm{O}_{p'}(B)$ is a normal p'-subgroup of G by the choice of (G, A, B).

Since $[A, B]$ is a normal subgroup of G contained in B and $G = \langle A^G \rangle = A[A, B]$, if $[A, B]$ is a proper subgroup of B, by minimality of (G, A, B) we obtain that $G \in \mathfrak{F}$ in Case 1 and $A \in \mathfrak{F}$ in Case 2. In both cases, we have a contradiction. Therefore $[A, B] = B$.

10. *A acts on B by conjugation as a group of power automorphisms and B is an abelian group.*

By Lemma 4.1.37 applied to the totally permutable product $A_p B$, we obtain that A_p acts as a group of power automorphisms on B. Since $[A^{\mathfrak{N}}, B] = 1$ and $A = A_p A^{\mathfrak{N}}$ by Statement 7, we conclude that A normalises each subgroup of B.

Since $B = [A^{\mathfrak{N}} A_p, B] = [A_p, B]$ is nilpotent, we obtain that $\mathrm{O}_q(B)$ cannot be centralised by A for each prime q dividing $|B|$. Consequently $\mathrm{O}_q(B)$ is abelian for all q by virtue of Lemma 1.3.4. It follows that B is an abelian group.

11. *$A \cap B = A^{\mathfrak{N}} \cap B \leq \mathrm{Z}_{\mathfrak{U}}(G)$, but $A \cap B \nleq \mathrm{Z}_\infty(G)$.*

Since $A \cap B$ is a p'-group and $A = A_p A^{\mathfrak{N}}$ by Statement 7 and Statement 9, we have that $A \cap B = A^{\mathfrak{N}} \cap B$. Moreover, $A \cap B \leq \mathrm{Z}_{\mathfrak{U}}(G)$ by Lemma 4.2.12. Assume that $A \cap B \leq \mathrm{Z}_\infty(G)$. Since by Statements 1, 9, and 10, $B = [B, A] = [B, A_p]$ is an abelian p'-group and $A \cap B$ is A_p-invariant, we deduce from [119, A, 12.6] that $A \cap B = [A \cap B, A_p]$. But then $A \cap B \leq \mathrm{Z}_\infty(G)$ implies $A \cap B = 1$. By Statement 9, we can consider $G/A^{\mathfrak{N}}B$, which is a nilpotent group, and $G/A^{\mathfrak{N}}$, which is supersoluble because it is a product of two totally permutable nilpotent groups (Corollary 4.1.38). Assume that we are in Case 1. Then $G/B \cong A$ is an $\mathfrak{F}$-group. Applying the quasi-R_0 lemma [119, IX, 1.13], we deduce that $G \in \mathfrak{F}$, against the choice of (G, A, B). Now assume that we are in Case 2. Then by the quasi-R_0 lemma, $A \cong G/B \in \mathfrak{F}$, against the choice of (G, A, B).

12. *$A/A^{\mathfrak{N}}$ is a cyclic p-group.*

Assume that this is false. Let L and M two different maximal subgroups of A containing $A^{\mathfrak{N}}$. Assume that we are in Case 1. By the choice of (G, A, B), we have that BL and BM belong to $\mathfrak{F}$. Consequently $G = (BL)(BM) \in \mathrm{N}_0\mathfrak{F} = \mathfrak{F}$, against the choice of (G, A, B). Assume that Case 2 holds. Then BL and BM are subnormal subgroups of $G \in \mathfrak{F}$ and we obtain that $BL \in \mathfrak{F}$ and $BM \in \mathfrak{F}$. Hence by the choice of (G, A, B), L and M belong to $\mathfrak{F}$. Consequently $A = \mathrm{N}_0\mathfrak{F} = \mathfrak{F}$, which contradicts our assumption.

The statements 1–5 are now completely proved by means of the previous steps.

In the sequel, we also assume that $\mathfrak{U} \subseteq \mathfrak{H}$. By Lemma 4.2.12, we see that the assertion $A \cap B \leq \mathrm{Z}_{\mathfrak{H}}(G)$ is always satisfied whenever $G = AB$ is the product of the totally permutable subgroups A and B. In this case, we have:

13. *B is a q-group for a prime $q \neq p$.*

Let q be a prime dividing $|B|$ and assume that the result is not true. We can consider a prime q such that $A \cap \mathrm{O}_{q'}(B) \neq 1$ by Statement 11. From Statement 10, G is the product of the two totally permutable subgroups $A\mathrm{O}_{q'}(B)$ and $\mathrm{O}_q(B)$. Assume that Case 1 holds. Then $A\mathrm{O}_{q'}(B)$ is an $\mathfrak{F}$-group by the choice of (G, A, B) and $\mathrm{O}_q(B)$ is an $\mathfrak{F}$-group. Since $|G| + |A\mathrm{O}_{q'}(B)||\mathrm{O}_q(B)| < |G| + |A||B|$, by our choice of (G, A, B) we obtain that $G \in \mathfrak{F}$. Now assume that Case 2 holds. By the choice of (G, A, B), we obtain that $A\mathrm{O}_{q'}(B)$ is an $\mathfrak{F}$-group. Then, again by the choice of (G, A, B), we would obtain that $A \in \mathfrak{F}$, against the hypothesis.

14. *B is a cyclic q-group.*

Assume that B is not cyclic. Since B is an abelian q-group and $A \cap B \neq 1$ by Statement 11, there is a direct decomposition $B = M \times N$ with N a cyclic group such that $A \cap N \neq 1$. Now we have that $G = (AN)M$ is a product of the totally permutable subgroups AN and M by Statement 10. Assume that we are in Case 1. Then, by the choice of (G, A, B), it follows that $AN \in \mathfrak{F}$. Now by the choice of (G, A, B), it follows that $G \in \mathfrak{F}$, against the hypothesis. Assume that Case 2 holds. Then $HN \in \mathfrak{F}$ by the choice of (G, A, B), and again by the choice of (G, A, B), we deduce that $A \in \mathfrak{F}$. This contradicts the hypothesis. □

As a consequence of Lemma 5.3.1, we have that Fitting classes containing $\mathfrak{U}$ are closed under the formation of totally permutable products whose intersection is hypercentral.

Theorem 5.3.2. *Let $\mathfrak{F}$ be a Fitting class containing $\mathfrak{U}$. Let the group $G = AB$ be the product of the totally permutable subgroups A and B. Assume that $A \cap B \leq \mathrm{Z}_\infty(G)$. Then G belongs to $\mathfrak{F}$ if and only if A and B belong to $\mathfrak{F}$.*

We have now arrived at our promised objective, namely Hauck, Martínez-Pastor, and Pérez-Ramos' result about Fischer classes and totally permutable products.

Theorem 5.3.3. *Let $\mathfrak{F}$ be a Fischer class containing $\mathfrak{U}$. Let the group $G = AB$ be the product of the totally permutable subgroups A and B. Then G belongs to $\mathfrak{F}$ if and only if A and B belong to $\mathfrak{F}$.*

Proof. If $G \in \mathfrak{F}$, then clearly $A \in \mathfrak{F}$ and $B \in \mathfrak{F}$. Let us prove that if A and B belong to $\mathfrak{F}$, then G belongs to $\mathfrak{F}$. Assume that the result is false and let $G = AB$ be a minimal counterexample as in Lemma 5.3.1 with $\mathfrak{H} = \mathfrak{E}$, the class of all finite groups.

Let us denote $C = \mathrm{C}_A(B)$ and $L = A \cap B$. Recall that $A^{\mathfrak{N}} \leq C$ by Theorem 4.2.7. Then, by Lemma 5.3.1, (1) and (2′), we can consider elements $b \in B$, $l \in L$ and $a \in A/C$ such that $B = \langle b \rangle$, $L = \langle l \rangle$, and $A/C = \langle a \rangle$. Let $|B/L| = s$ and $|A/C| = m$. Since B is a normal subgroup of G, a acts on B by taking every element of B to the ith power for a suitable integer i. Then a^{-1} acts on B by taking every element of B to the rth power for a positive integer r such that $ir \equiv 1 \pmod{|B|}$. Let $[B](A/C)$ be the semidirect product of B with A/C corresponding to this action. By [119, A, 18.9], there exists a monomorphism $\mu\colon [B](A/C) \to L \wr_{\mathrm{reg}} [B/L](A/C)$ in such a way that:

1. $\mu(l) = (\underbrace{l, \dots, l}_{(s)}, \underbrace{l^r, \dots, l^r}_{(s)}, \dots, \underbrace{l^{r^{m-1}}, \dots, l^{r^{m-1}}}_{(s)})$.

Note that $L^\natural = \mu(L) \times (L_2 \times \cdots \times L_{sm})$, where $L^\natural = L_1 \times \cdots \times L_{sm}$ denotes the base group of the considered regular wreath product and every L_i denotes a copy of L for $i \in \{1, 2, \dots, sm\}$.

2. $\mu(A/C) = A/C$, in fact, $\mu(a) = a$.
3. $L^\natural(B/L) = L^\natural\mu(B)$; in particular, $L^\natural([B/L](A/C)) = L^\natural\mu(B)(A/C)$.
4. $\mu(b)^a = \mu(b)^{\mu(a)} = \mu(b^i) = \mu(b)^i$.
5. $\mu(b) = l^*(Lb)$ for some $l^* \in L^\natural$.

Consider a cyclic group $\langle \gamma \rangle$ isomorphic to A/C acting on each component of $L^\natural$ by $l^\gamma = l^r$. Note that γ commutes with each element of $[B/L](A/C)$ when acting on $L^\natural$. This means that the group $([B/L](A/C)) \times \langle \gamma \rangle$ acts on $L^\natural$. Now consider the corresponding semidirect product $[L^\natural]((B/L)\langle (a, \gamma) \rangle)$ of $L^\natural$ with the subgroup $(B/L)\langle (a, \gamma) \rangle$ of this direct product. We remark that, in this group, $\mu(l)^{(a,\gamma)} = \mu(l)^{ir} = \mu(l)$.

Consider now a group A_1 isomorphic to A acting on each component of $L^\natural$ by means of an action equivalent to the action of A on L by conjugation. We note that each element of A_1 commutes with each element of $(B/L)\langle(a,\gamma)\rangle$ when acting on $L^\natural$. Then $((B/L)\langle(a,\gamma)\rangle) \times A_1$ acts on $L^\natural$ and we can construct the semidirect product $X = [L^\natural]((B/L)\langle(a,\gamma)\rangle \times A_1)$. Let $L_1 = \langle l_1\rangle$ be the corresponding subgroup of A_1 isomorphic to $L \leq A$. Let $D = \{\mu(l)^t l_1^t : t \in \mathbb{Z}\}$, which is a subgroup of X. Moreover, D is normal in X because $(B/L)\langle(a,\gamma)\rangle$ centralises $\mu(L)$ and L_1 by Statement 1 and the previous remark, and besides $\mu(L)$ and L_1 are A_1-isomorphic. Hence we can consider the group $\bar{X} = X/D$. We will use the "bar" convention to denote the images in this factor group.

We claim that $\bar{X} \in \mathfrak{F}$. We note that $\overline{L^\natural(B/L)\langle(a,\gamma)\rangle}$ is supersoluble and so it is an $\mathfrak{F}$-group. This is because obviously $[B](A/C)$ is supersoluble, which implies in particular that $|A/C|$ divides $q-1$ for the prime divisor q of $|B|$. Now we are going to see that $\overline{L^\natural A_1} \in \mathfrak{F}$. Then it will be clear that $\bar{X} \in \mathrm{N}_0\mathfrak{F} = \mathfrak{F}$. We have that $\overline{L^\natural A_1} = \overline{(L_2 \times \cdots \times L_{sm})}\,\overline{A_1}$ with $\overline{(L_2 \times \cdots \times L_{sm})} \cap \overline{A_1} = \bar{1}$ and $\overline{A_1} \cong A_1$. If C_1 is the corresponding subgroup of A_1 isomorphic to the subgroup C of A, it is clear that $\overline{L^\natural A_1}/\overline{C_1}$ is an $\mathfrak{F}$-group because it is supersoluble. Obviously $\overline{L^\natural A_1}/\overline{(L_2 \times \cdots \times L_{sm})}$ is isomorphic to $A \in \mathfrak{F}$ and $\overline{A_1}/\overline{C_1}$ is cyclic. Then it follows by the quasi-R_0 lemma [119, IX, 1.13] that $\overline{L^\natural A_1} \in \mathfrak{F}$.

Let $a_1 \in A_1$ such that $A_1/C_1 = \langle C_1 a_1\rangle$ and consider the subgroup $\bar{Y}$ of $\bar{X}$, where Y is the semidirect product of $L^\natural$ with a subdirect product of $((B/L)\langle(a,\gamma)\rangle) \times A_1$ with amalgamated factor group $\langle(a,\gamma)\rangle \cong A_1/C_1$ (see [119, A, 19.1 and 19.2]). Then $\bar{Y} \in \mathfrak{F}$ because $\bar{Y}$ is a normal subgroup of $\bar{X}$. We claim now that the group G is embedded in $\bar{Y}$. To see this, we construct Y in such a way $\bar{Y}$ contains the following subgroups:

$$A_2 = \overline{C_1\langle(a,\gamma,a_1)\rangle} \cong \overline{A_1} \cong A_1 \cong A,$$

$$B_2 = \overline{\mu(B)} \cong \mu(B) \cong B.$$

We have that $A_2 \cap B_2 = \overline{\mu(L)} = \overline{L_1}$. Moreover

$$\begin{aligned}\mu(b)^{(a,\gamma,a_1)} &= (l^*(Lb))^{(a,\gamma,a_1)} = (l^*)^{(a,\gamma,a_1)}(Lb)^{(a,\gamma,a_1)}\\ &= (l^*)^a(Lb)^a = (l^*(Lb))^a = \mu(b)^a = \mu(b)^i.\end{aligned}$$

From these remarks, it follows that $A_2B_2 \cong AB = G$. Finally we note that $(A_2B_2)^{\mathfrak{N}}$ is a normal subgroup of $B_2\overline{C_1} = \overline{\mu(B)\,C_1}$, this is subnormal in $\overline{L^\natural(B/L)\,C_1}$, and this is a normal subgroup of $\bar{Y}$. In particular, $(A_2B_2)^{\mathfrak{N}}$ is subnormal in $\bar{Y}$. Since $\bar{Y} \in \mathfrak{F}$ and $\mathfrak{F}$ is a Fischer class, we have that $G \cong A_2B_2 \in \mathfrak{F}$, which provides the final contradiction. □

The next theorem appears in [144, Theorem 3] and shows in particular that every Fitting class which is closed under taking epimorphic images is also closed under forming totally permutable products.

Theorem 5.3.4. *Let $\mathfrak{F}$ be a Fitting class containing the class $\mathfrak{U}$ of all supersoluble groups. Assume that whenever $G \in \mathfrak{F}$ and $N \le Z_{\mathfrak{U}}(G)$, then $G/N \in \mathfrak{F}$ (in particular, this holds for a* Q*-closed Fitting class). Let the group $G = AB$ be the product of the totally permutable subgroups A and B. If A and B belong to $\mathfrak{F}$, then G belongs to $\mathfrak{F}$.*

Proof. Assume that the result is false and take $G = AB$ a minimal counterexample as in Lemma 5.3.1 with $\mathfrak{H} = \mathfrak{E}$, the class of all finite groups.

Since B is a normal subgroup of G, A acts by conjugation on B and we can construct the semidirect product $C = [B]A$ with respect to this action. By Lemma 1.1.5, there exists an epimorphism $\alpha\colon C \to G$ such that $\operatorname{Ker}\alpha \cap B = 1$ given by $(b,a)\alpha = ba$ for all $a \in A$, $b \in B$. Note that $\operatorname{Ker}\alpha \le [A \cap B](A \cap B)$, which is a direct product because B is cyclic by Lemma 5.3.1 (2′).

We have that $C/A^{\mathfrak{N}} \cong [B](A_p/(A_p \cap A^{\mathfrak{N}}))$ is a product of two supersoluble totally permutable subgroups by Lemma 5.3.1 (1) and (2′) and so $C/A^{\mathfrak{N}} \in \mathfrak{U} \subseteq \mathfrak{F}$ by Corollary 4.1.38. Moreover, $C/B \cong A \in \mathfrak{F}$. Since $C/([B]A^{\mathfrak{N}})$ is a p-group, we deduce that $C \in \mathfrak{F}$ by the quasi-R_0 lemma [119, IX, 1.13]. By Lemma 4.2.12, we can deduce that $A \cap B \le Z_{\mathfrak{U}}(G)$. Since B is cyclic and G acts on B as a group of power automorphisms by Lemma 5.3.1 (2′) and (5), we conclude that $[A \cap B](A \cap B) \le Z_{\mathfrak{U}}(C)$. This means that $C/\operatorname{Ker}\alpha \cong G \in \mathfrak{F}$, which provides the final contradiction. □

Next we show that the converse of this result also holds for R_0-closed Fitting classes [144, Theorem 4].

Theorem 5.3.5. *Let $\mathfrak{F}$ be an R_0-closed Fitting class containing the class $\mathfrak{U}$ of all supersoluble groups. Let the group $G = AB$ be the product of the totally permutable subgroups A and B. If $G \in \mathfrak{F}$, then $A \in \mathfrak{F}$ and $B \in \mathfrak{F}$.*

Proof. Assume that the result is false and take a minimal counterexample $G = AB$ as in Lemma 5.3.1 with $\mathfrak{H} = \mathfrak{E}$. In particular, B is a nilpotent normal p'-subgroup of G and $A/A^{\mathfrak{N}}$ is a p-group for a prime p.

Since $A^{\mathfrak{N}}$ is a normal subgroup of G, by Lemma 1.1.5 there exists an epimorphism $\alpha\colon X \to G$ such that $\operatorname{Ker}\alpha \cap A^{\mathfrak{N}} = 1$ for the corresponding semidirect product $X = [A^{\mathfrak{N}}]A_pB$. We show first that X is an $\mathfrak{F}$-group. Note that $X/A^{\mathfrak{N}} \cong A_pB \in \mathfrak{F}$ because it is supersoluble by Corollary 4.1.38. On the other hand, $X/\operatorname{Ker}\alpha \cong G \in \mathfrak{F}$. Hence $X \cong X/(A^{\mathfrak{N}} \cap \operatorname{Ker}\alpha) \in \mathrm{R}_0\mathfrak{F} = \mathfrak{F}$.

Now, since $X/([A^{\mathfrak{N}}]B)$ is a p-group and $A^{\mathfrak{N}} \cap B = 1$, we deduce that $X/B \in \mathfrak{F}$ by the quasi-R_0-lemma [119, IX, 1.13]. This means that $C = [A^{\mathfrak{N}}]A_p \in \mathfrak{F}$.

Again by Lemma 1.1.5, there exists an epimorphism $\varphi\colon C \to A$ such that $\operatorname{Ker}\varphi \cap A^{\mathfrak{N}} = 1$. Note that $C/A^{\mathfrak{N}}$ and $C/(A^{\mathfrak{N}}\operatorname{Ker}\varphi)$ are nilpotent groups. Since $C \in \mathfrak{F}$, another application of the quasi-R_0 lemma [119, IX, 1.13] yields $C/\operatorname{Ker}\varphi \in \mathfrak{F}$. This means that $A \in \mathfrak{F}$, the final contradiction which proves the theorem. □

The classes of the type $\mathfrak{N} \diamond \mathfrak{F}$ with $\mathfrak{F}$ a Fitting class such that $\mathfrak{N} \subseteq \mathfrak{F}$ (see [47, 2.4.5] or [119, IX, 1.10] for the definition of the Fitting product) also behave nicely with respect to totally permutable products.

Theorem 5.3.6. *Let $\mathfrak{F}$ be a Fitting class containing the class $\mathfrak{N}$ of all nilpotent groups. Consider the Fitting product $\mathfrak{N} \diamond \mathfrak{F}$. Let the group $G = AB$ be the product of the totally permutable subgroups A and B. Then $G \in \mathfrak{N} \diamond \mathfrak{F}$ if and only if $A \in \mathfrak{N} \diamond \mathfrak{F}$ and $B \in \mathfrak{N} \diamond \mathfrak{F}$.*

Proof. For both implications we can assume, from Lemma 5.3.1, that B is a nilpotent group and $G = A\mathrm{F}(G)$.

Assume first that $A \in \mathfrak{N} \diamond \mathfrak{F}$, that is, $A/\mathrm{F}(A) \in \mathfrak{F}$. Since $A/A^{\mathfrak{N}} \in \mathfrak{N}$, we deduce that $A/(A^{\mathfrak{N}} \cap \mathrm{F}(A)) \in \mathfrak{F}$ by the quasi-R_0 lemma [119, IX, 1.13]. But $A^{\mathfrak{N}} \cap \mathrm{F}(A) = A^{\mathfrak{N}} \cap \mathrm{F}(G)$ since $A^{\mathfrak{N}} \cap \mathrm{F}(A)$ is a normal nilpotent subgroup of G by Theorem 4.2.7. Now, by the quasi-R_0-lemma [119, IX, 1.13] applied to the group $A/(A^{\mathfrak{N}} \cap \mathrm{F}(G)) \in \mathfrak{F}$ and the normal subgroups $(A \cap \mathrm{F}(G))/(A^{\mathfrak{N}} \cap \mathrm{F}(G))$ and $A^{\mathfrak{N}}/(A^{\mathfrak{N}} \cap \mathrm{F}(G))$, it follows that $A/(\mathrm{F}(G) \cap A) \in \mathfrak{F}$. But this means that $G/\mathrm{F}(G) \in \mathfrak{F}$, that is, $G \in \mathfrak{N} \diamond \mathfrak{F}$.

Assume now that $G \in \mathfrak{N} \diamond \mathfrak{F}$. As in the previous paragraph, $A^{\mathfrak{N}} \cap \mathrm{F}(A) = A^{\mathfrak{N}} \cap \mathrm{F}(G)$ is a normal subgroup of G. Since $A/(A \cap \mathrm{F}(G)) \cong G/\mathrm{F}(G) \in \mathfrak{F}$ and $A/A^{\mathfrak{N}} \in \mathfrak{N}$, we deduce that $A/(A^{\mathfrak{N}} \cap \mathrm{F}(A)) \in \mathfrak{F}$ by the quasi-R_0 lemma [119, IX, 1.13]. This means, applying again the quasi-R_0 lemma, that $A \in \mathfrak{N} \diamond \mathfrak{F}$. This completes the proof of the theorem. □

The main purpose of the next result, due to Hauck, Martínez-Pastor, and Pérez-Ramos [147], is to show that the Fitting classes containing the class $\mathfrak{U}$ of all supersoluble groups which behave well with respect to two totally permutable subgroups behave also well with respect to products of finitely many pairwise totally permutable subgroups.

Theorem 5.3.7. *Let $\mathfrak{F}$ be a Fitting class containing the class $\mathfrak{U}$ of all supersoluble groups. Let the group $G = G_1 G_2 \cdots G_r$ be the product of the pairwise totally permutable subgroups $G_1, G_2, \dots, G_r$.*

1. *Assume that $\mathfrak{F}$ satisfies the following property:*

   If a group $G = AB$ is the product of the totally permutable subgroups A and B such that $A \in \mathfrak{F}$ and $B \in \mathfrak{F}$, then $G \in \mathfrak{F}$. (5.2)

   If $G_i \in \mathfrak{F}$ for all $i \in \{1, 2, \dots, r\}$, then $G \in \mathfrak{F}$.

2. *Assume that $\mathfrak{F}$ satisfies the following property:*

   If a group $G = AB$ is the product of the totally permutable subgroups A and B such that $G \in \mathfrak{F}$, then $A \in \mathfrak{F}$ and $B \in \mathfrak{F}$. (5.3)

   If $G \in \mathfrak{F}$, then $G_i \in \mathfrak{F}$ for all $i \in \{1, 2, \dots, r\}$.

Proof. Since the arguments needed to prove Statements 1 and 2 are very similar, we will present a common proof for them and we will indicate the differences when necessary.

Assume that the result is false. There exists a group $G = G_1G_2 \cdots G_r$ such that $G_1, G_2, \ldots, G_r$ are pairwise totally permutable subgroups of G. Then it satisfies one of following conditions:

Case 1: If Statement 1 is false, then $G_1, G_2, \ldots, G_r \in \mathfrak{F}$ and $G \notin \mathfrak{F}$.

Case 2: If Statement 2 is false, then $G \in \mathfrak{F}$, but not all $G_i \in \mathfrak{F}$ for $i \in \{1, 2, \ldots, r\}$.

Among all possible counterexamples, we consider one with least $|G| + |G_1| + |G_2| + \cdots + |G_r|$. We split the proof into the following independently stated steps from which a conclusion will hold.

1. *We may assume that $G_2, G_3, \ldots, G_r$ are nilpotent groups and G_1 is not nilpotent. We denote $A = G_1$ and $B = G_2 \cdots G_r$. Moreover, B is supersoluble and $[B, A^{\mathfrak{N}}] = 1$.*

If Case 2 holds, obviously not all G_i can be nilpotent. Assume that $G_i \in \mathfrak{N}$ for all $i \in \{1, 2, \ldots, r\}$ and that Case 1 holds. Then $G \in \mathfrak{U} \subseteq \mathfrak{F}$ by Corollary 4.1.38, which is a contradiction.

Assume now that there exists $i, j \in \{1, 2, \ldots, r\}, i \neq j$, such neither G_i nor G_j is nilpotent. By Theorem 4.2.7, it follows that $[G_k, G_t^{\mathfrak{N}}] = 1$ for all $k, t \in \{1, 2, \ldots, r\}$, $k \neq t$. Then $G_t \leq \mathrm{C}_G(G_i^{\mathfrak{N}})$ for every $t \neq i$ and $\mathrm{C}_G(G_i^{\mathfrak{N}})$ is a proper subgroup of G; $G_h \leq \mathrm{C}_G(G_j^{\mathfrak{N}})$ for every $h \neq j$ and $\mathrm{C}_G(G_i^{\mathfrak{N}})$ is also a proper subgroup of G. Hence

$$\mathrm{C}_G(G_i^{\mathfrak{N}}) = \Big(\prod_{t \in \{1,2,\ldots,r\} \setminus \{i\}} G_t\Big)(G_i \cap \mathrm{C}_G(G_i^{\mathfrak{N}}))$$

is a product of pairwise totally permutable subgroups. Suppose that Case 1 holds. Then $\mathrm{C}_G(G_i^{\mathfrak{N}})$ is a product of pairwise totally permutable subgroups in $\mathfrak{F}$ since $\mathrm{C}_{G_i}(G_i^{\mathfrak{N}})$ is a normal subgroup of the $\mathfrak{F}$-group G_i. We conclude that $\mathrm{C}_G(G_i^{\mathfrak{N}}) \in \mathfrak{F}$ by the choice of G. An analogous argument shows that $\mathrm{C}_G(G_j^{\mathfrak{N}}) \in \mathfrak{F}$. Then

$$G = \mathrm{C}_G(G_i^{\mathfrak{N}})\mathrm{C}_G(G_j^{\mathfrak{N}}) \in \mathrm{N}_0\,\mathfrak{F} = \mathfrak{F},$$

against the choice of G. Suppose now that Case 2 holds. By the choice of G, we obtain that $G_t \in \mathfrak{F}$ for all $t \neq i$. In a similar way, using G_j, we deduce that $G_l \in \mathfrak{F}$ for all $l \neq j$. Then $G_k \in \mathfrak{F}$ for all $k \in \{1, 2, \ldots, r\}$, which contradicts the choice of G.

Consequently, in both cases, there exists a unique $i \in \{1, 2, \ldots, r\}$ such that G_i is not nilpotent. Without loss of generality we may suppose $i = 1$. Now the conclusion is clear by Theorems 4.1.38 and 4.2.7.

2. *$\langle B^G \rangle \cap A \in \mathfrak{N}$ and $\langle B^G \rangle$ is supersoluble.*

Since $[B, A^{\mathfrak{N}}] = 1$, we can deduce that $\langle B^G \rangle \cap A^{\mathfrak{N}} \leq \mathrm{Z}(\langle B^G \rangle \cap A)$ and so $\langle B^G \rangle \cap A$ is nilpotent. Finally, since $\langle B^G \rangle = (\langle B^G \rangle \cap A)B$ is a product of pairwise totally permutable nilpotent subgroups, then $\langle B^G \rangle$ is supersoluble by Corollary 4.1.38.

3. *There exists a prime number* p *such that* $G = A^{\mathfrak{N}}A_p\langle B^G\rangle$, *with* A_p *a Sylow* p*-subgroup of* A.

Let A_q be a Sylow q-subgroup of A for a prime q. Note that $\langle B^G\rangle = AB \cap \langle B^G\rangle = (A \cap \langle B^G\rangle)B$ and so $A^{\mathfrak{N}}A_q\langle B^G\rangle = A^{\mathfrak{N}}A_q(A \cap \langle B^G\rangle)B$ is a product of pairwise totally permutable subgroups. Moreover, it is a normal subgroup of G. Suppose that $A^{\mathfrak{N}}A_q\langle B^G\rangle$ is a proper subgroup of G for all primes q.

Assume that Case 1 holds. Then $A^{\mathfrak{N}}A_q\langle B^G\rangle$ is an $\mathfrak{F}$-group for all primes q and so $G = \prod_{q\in\mathbb{P}}(A^{\mathfrak{N}}A_q\langle B^G\rangle) \in \mathrm{N}_0\mathfrak{F} = \mathfrak{F}$, contradicting the choice of G.

Assume that Case 2 holds. Then $A^{\mathfrak{N}}A_q\langle B^G\rangle$ is a normal subgroup of the $\mathfrak{F}$-group G and so it is an $\mathfrak{F}$-group. By the choice of G, we deduce that $A^{\mathfrak{N}}A_q(A\cap\langle B^G\rangle) \in \mathfrak{F}$. In particular, $A^{\mathfrak{N}}A_q \in \mathrm{S}_n\mathfrak{F} = \mathfrak{F}$ for all primes q, and so $A \in \mathrm{N}_0\mathfrak{F} = \mathfrak{F}$, against supposition.

4. *If Case* 1 *holds, then for all primes* $q \neq p$, $A^{\mathfrak{N}}A_q[A, B]$ *is a normal* $\mathfrak{F}$*-subgroup of* G, *where* A_q *denotes a Sylow* q*-subgroup of* A, *but* $A^{\mathfrak{N}}A_p[A_p, B] \notin \mathfrak{F}$.

We note first that $A^{\mathfrak{N}}A_q[A, B]$ is a normal subgroup of $G = AB$ contained in $A^{\mathfrak{N}}\langle B^G\rangle$ by Statement 3. Since $A^{\mathfrak{N}}$ is a normal subgroup of A, $A^{\mathfrak{N}} \in \mathfrak{F}$. Moreover, $\langle B^G\rangle \in \mathfrak{U} \subseteq \mathfrak{F}$ by Statement 2. This implies that $A^{\mathfrak{N}}\langle B^G\rangle \in \mathrm{N}_0\mathfrak{F} = \mathfrak{F}$.

Suppose now that $A^{\mathfrak{N}}A_p[A_p, B] \in \mathfrak{F}$. Since

$$\langle (A^{\mathfrak{N}}A_p)^G\rangle = \langle (A^{\mathfrak{N}}A_p)^B\rangle = A^{\mathfrak{N}}A_p[A^{\mathfrak{N}}A_p, B] = A^{\mathfrak{N}}A_p[A_p, B]$$

is a normal subgroup of G, then

$$\langle A^G\rangle = A[A, B] = (A^{\mathfrak{N}}A_p[A_p, B])\Big(\prod_{q\neq p} A^{\mathfrak{N}}A_q[A, B]\Big) \in \mathrm{N}_0\mathfrak{F} = \mathfrak{F}$$

by the previous paragraph. Consequently $G = \langle A^G\rangle\langle B^G\rangle \in \mathrm{N}_0\mathfrak{F} = \mathfrak{F}$, against the hypothesis.

5. *If Case* 1 *holds, then* $G = A^{\mathfrak{N}}A_pB$.

If $A^{\mathfrak{N}}A_pB$ is a proper subgroup of G, then $A^{\mathfrak{N}}A_pB \in \mathfrak{F}$ by the choice of G. But this contradicts Statement 4 because $\langle (A^{\mathfrak{N}}A_p)^G\rangle = A^{\mathfrak{N}}A_p[A_p, B]$ is a normal subgroup of $A^{\mathfrak{N}}A_pB$.

6. *If Case* 2 *holds, then for all primes* $q \neq p$, $A^{\mathfrak{N}}A_q \in \mathfrak{F}$, *where* A_p *is a Sylow* q*-subgroup of* A. *Moreover* $A^{\mathfrak{N}}A_p \notin \mathfrak{F}$.

We note that $A^{\mathfrak{N}}A_q$ is contained in $A \cap A^{\mathfrak{N}}\langle B^G\rangle = A^{\mathfrak{N}}(A \cap \langle B^G\rangle)$ by Statement 3. But $A^{\mathfrak{N}}(A \cap \langle B^G\rangle) \in \mathfrak{F}$ because $A^{\mathfrak{N}}$ is a normal subgroup of $G \in \mathfrak{F}$ and $A \cap \langle B^G\rangle$ is a supersoluble normal subgroup of A. Then $A^{\mathfrak{N}}A_q \in \mathrm{S}_n\mathfrak{F} = \mathfrak{F}$. Finally, if $A^{\mathfrak{N}}A_q \in \mathfrak{F}$, then $A = \prod_{q\in\mathbb{P}}(A^{\mathfrak{N}}A_q) \in \mathrm{N}_0\mathfrak{F} = \mathfrak{F}$, against the hypothesis.

7. *$A = A^{\mathfrak{N}} A_p$ and so $A/A^{\mathfrak{N}}$ is a p-group, and for all primes q dividing $|B|$, $p \leq q$. Moreover, $G = A^{\mathfrak{N}} A_p B_p B_{p'}$, where B_p is a Sylow p-subgroup of B and $B_{p'}$ is a Hall p'-subgroup of B. Furthermore, $B_{p'}$ is a normal subgroup of G.*

Clearly, $A^{\mathfrak{N}} A_p$ is a normal subgroup of A. Thus if Case 1 holds, $A^{\mathfrak{N}} A_p \in \mathfrak{F}$. Applying Statement 5 and the choice of $(G, G_1, G_2, \dots, G_r)$, it follows that $A = A^{\mathfrak{N}} A_p$. Assume that Case 2 holds. Then $\langle (A^{\mathfrak{N}} A_p)^G \rangle = A^{\mathfrak{N}} A_p [A_p, B] \in \mathfrak{F}$.

Suppose that $p \geq q$ for all primes q dividing $|B|$. Since $A_p B$ is a supersoluble group by Corollary 4.1.38, it follows that $A_p \mathrm{O}_p(B)$ is a normal subgroup of $A_p B$. Hence A_p is subnormal in $A_p B$. Therefore, by Theorem 4.2.7, $A^{\mathfrak{N}} A_p$ is a subnormal subgroup of $A^{\mathfrak{N}} A_p B$. If we are in Case 1, $A = A^{\mathfrak{N}} A_p$ is subnormal in G and so $G = A \langle B^G \rangle \in \mathrm{N}_0 \mathfrak{F} = \mathfrak{F}$, against the choice of $(G, G_1, G_2, \dots, G_r)$. If we are in Case 2, $A^{\mathfrak{N}} A_p$ is a subnormal subgroup of $A^{\mathfrak{N}} A_p [A_p, B] \in \mathfrak{F}$ and so $A^{\mathfrak{N}} A_p \in \mathfrak{F}$, contrary to Statement 6.

Therefore there exists a prime q dividing $|B|$ with $p < q$. Let

$$\pi(B) \cup \{p\} = \{p_1, p_2, \dots, p_t = p, p_{t+1}, \dots, p_n\}$$

with $p_1 < p_2 < \dots < p_t = p < p_{t+1} < \dots < p_n$. We let $\pi = \{p, p_{t+1}, \dots, p_n\}$ and $\pi' = (\pi(B) \cup \{p\}) \setminus \pi$. Then $B_{\pi'}$, a Hall π'-subgroup of B, normalises a Hall π-subgroup B_π of B since B is supersoluble. Since $A_p B$ is a supersoluble group by Corollary 4.1.38, $B_{\pi'}$ normalises $A_p B_\pi$. Hence $A^{\mathfrak{N}} A_p B_\pi$ is a normal subgroup of G. Suppose that $A^{\mathfrak{N}} A_p B_\pi$ is a proper subgroup of G. We note that $B_\pi = \mathrm{O}_\pi(G_2) \cdots \mathrm{O}_\pi(G_r)$ is the product of the pairwise totally permutable nilpotent subgroups $\mathrm{O}_\pi(G_i)$, each of which is totally permutable with A. Then $A^{\mathfrak{N}} A_p B_\pi \in \mathfrak{F}$ by the choice of G. Therefore $G = (A^{\mathfrak{N}} A_p B_\pi) \langle B^G \rangle \in \mathrm{N}_0 \mathfrak{F} = \mathfrak{F}$, a contradiction which implies that $G = A^{\mathfrak{N}} A_p B_\pi$.

By the choice of $(G, G_1, G_2, \dots, G_r)$, it follows that $A = A^{\mathfrak{N}} A_p$ and $B = B_\pi$. Thus $p \leq q$ for all primes $q \in \pi(B)$. Since $A_p B$ is supersoluble, it is clear that $B_{p'}$ is a normal subgroup of $G = A^{\mathfrak{N}} A_p B_p B_{p'}$, as desired.

8. *B is a normal p'-subgroup of G.*

We note that $B_{p'} = \mathrm{O}_{p'}(G_2) \cdots \mathrm{O}_{p'}(G_r)$. If $AB_{p'}$ is a proper subgroup of G, then $AB_{p'} \in \mathfrak{F}$ by the choice of $(G, G_1, \dots, G_r)$. Now, since $A^{\mathfrak{N}} B_{p'}$ is a normal subgroup of G and $G/A^{\mathfrak{N}} B_{p'}$ is a p-group by Statement 7, it follows that $AB_{p'}$ is a subnormal subgroup of G. This means that $G = (AB_{p'}) \langle B^G \rangle \in \mathrm{N}_0 \mathfrak{F} = \mathfrak{F}$, against the hypothesis. Hence $G = AB_{p'}$ and $B = B_{p'}$ by the choice of $(G, G_1, \dots, G_r)$. By Statement 7, B is normal in G.

9. *For all $j \in \{2, \dots, r\}$, $G_j = [G_j, A]$. Moreover, $\prod_{k=1}^{t} G_{j_k} = [\prod_{k=1}^{t} G_{j_k}, A]$ for each set of indices $\{j_1, \dots, j_t\} \subseteq \{2, \dots, r\}$. In particular, $\langle A^G \rangle = G$ and $B = [A, B]$ is a nilpotent group.*

First, we remark that for every $j \in \{2, \dots, r\}$, $[G_j, A] = [G_j, A_p] \leq G_j$ because $A = A^{\mathfrak{N}} A_p$, $[A^{\mathfrak{N}}, B] = 1$, and $A_p B$ is a supersoluble group, with p the smallest prime dividing its order. Now, since G_j is a p'-group, we know by [119, A, 12.5] that $G_j = [G_j, A_p]\mathrm{C}_{G_j}(A_p)$ for all $j \in \{2, \dots, r\}$. Then we have that

$$\langle A^G \rangle = \langle A^{G_2 \cdots G_r} \rangle \leq A\Big(\prod_{j=2}^{r} [G_j, A_p]\Big) \leq \langle A^G \rangle.$$

Then $\langle A^G \rangle = A(\prod_{j=2}^{r}[G_j, A_p])$ is a product of pairwise totally permutable subgroups. Suppose $\langle A^G \rangle$ is a proper subgroup of G. If we are in Case 1, since A and $[G_j, A_p]$ for $j \in \{2, \dots, r\}$ are $\mathfrak{F}$-subgroups, by the choice of G we conclude that $\langle A^G \rangle \in \mathfrak{F}$ and $G = B\langle A^G \rangle \in \mathrm{N}_0\,\mathfrak{F} = \mathfrak{F}$, against the hypothesis. If we are in Case 2, by the choice of G we conclude from $\langle A^G \rangle \in \mathfrak{F}$ that $A \in \mathfrak{F}$, once again contradicting the hypothesis. Consequently

$$G = \langle A^G \rangle = A\Big(\prod_{j=2}^{r} [G_j, A_p]\Big).$$

By the choice of $(G, G_1, G_2, \dots, G_r)$ we conclude that $G_j = [G_j, A_p] = [G_j, A]$ for all $j \in \{2, \dots, r\}$.

Then, for any $\{j_1, \dots, j_t\} \subseteq \{2, \dots, r\}$,

$$B = \prod_{k=1}^{t} G_{j_k} = \prod_{k=1}^{t} [G_{j_k}, A] \leq \Big[\prod_{k=1}^{t} G_{j_k}, A\Big] \leq \prod_{k=1}^{t} G_{j_k}.$$

In particular, $B = [A, B]$ is a nilpotent group by Lemma 4.2.12.

10. *For all $j \in \{2, \dots, r\}$, G_j is an abelian group and, moreover, A normalises each subgroup of G_j.*

Choose $j \in \{2, \dots, r\}$. We claim that A does not centralise any non-trivial Sylow subgroup of G_j. Assume that it does and let $(G_j)_q$ be a non-trivial Sylow q-subgroup of G_j for some prime q such that $[(G_j)_q, A] = 1$. Then $\langle A^G \rangle \leq A(\prod_{1 \neq i \neq j} G_i)(G_j)_{q'}$, where $(G_j)_{q'}$ denotes the Hall q'-subgroup of G_j, and, by Statement 9, we deduce that $G = A(\prod_{1 \neq i \neq j} G_i)(G_j)_{q'}$. Now, by the choice of $(G, G_1, G_2, \dots, G_r)$, we obtain that $G_j = (G_j)_{q'}$ and $(G_j)_q = 1$, against our assumption.

Since $A_p(G_j)_q$ is a product of totally permutable subgroups and it is a supersoluble group by Corollary 4.1.38, it follows that A_p normalises each subgroup of $(G_j)_q$ but does not centralise $(G_j)_q$ for all primes $q \in \pi(G_j)$. By Lemma 1.3.4, $(G_j)_q$ is an abelian group for all primes $q \in \pi(G_j)$, and hence G_j is an abelian group.

Again, since A_pG_j is a supersoluble group which is a product of two totally permutable subgroups and p is the smallest prime dividing its order, we deduce that A_p normalises each subgroup of G_j by Lemma 4.1.37. Now, since $[A^{\mathfrak{N}}, G_j] = 1$, the result follows by Statement 7.

11. *For every $j \in \{2, \dots, r\}$, G_j is a cyclic p_j-group for some prime p_j.*

Choose any $j \in \{2, \dots, r\}$. Since G_j is abelian by Statement 10, it is a direct product of cyclic subgroups of prime power orders. Let $G_j = \times_i T_{j_i}$, $T_{j_i} \cong C_{p_i^{\alpha_i}}$ for some primes $p_i > 2$ and some integers $\alpha_i \geq 0$ for each i. Then

$$G = A\Big(\prod_{1 \neq k \neq j} G_k\Big)\Big(\times_i T_{j_i}\Big)$$

is a product of pairwise totally permutable subgroups with

$$|A| + \sum_{1 \neq k \neq j} |G_k| + \sum_i |T_{j_i}| < |G_1| + \dots + |G_r|$$

because

$$\sum_i |T_{j_i}| < \prod_i |T_{j_i}|$$

unless $G_j = T_{j_i}$ for some index j_i, that is, G_j is a cyclic p_j-group for some prime p_j. If we are in Case 1, by the choice of $(G, G_1, G_2, \dots, G_r)$ we conclude that since A, G_k for $1 \neq k \neq j$, and T_{j_i} belong to $\mathfrak{F}$, then $G \in \mathfrak{F}$, against assumption. If we are in Case 2, since $G \in \mathfrak{F}$ we conclude that $A \in \mathfrak{F}$, against the hypothesis. In any case it follows that G_j must be a cyclic p_j-group for some prime p_j, as desired.

12. *$B = G_2 \cdots G_r$ is an abelian group.*

Since B is a nilpotent group by Statement 9, it suffices to show that every Sylow q-subgroup of B is abelian for all primes q. Consider a pair of indices $i, j \in \{2, \dots, r\}$ such that G_i and G_j are q-groups and denote $T_{i,j} = G_i G_j$. By Statement 9, $T_{i,j} = [T_{i,j}, A] = [T_{i,j}, A_p]$. Moreover, by Statement 11, $T_{i,j}$ is the product of two totally permutable cyclic A-invariant q-subgroups, where q is an odd prime (recall that $p < q$). Then $T_{i,j}$ is an abelian group by Lemma 3.1.10. This means that $[G_i, G_j] = 1$ for every pair of q-groups G_i and G_j. Consequently, we deduce that any Sylow q-subgroup of B is abelian, by Statement 11, and the result follows.

13. *The final contradiction.*

Since $B = G_2 \cdots G_r$ is an abelian group and A normalises each subgroup of G_r, it is clear that $AG_2 \cdots G_{r-1}$ is totally permutable with G_r. Assume that $AG_2 \cdots G_{r-1}$ is a proper subgroup of G. If we are in Case 1, then $AG_2 \cdots G_{r-1}$ is an $\mathfrak{F}$-group by minimality of $(G, G_1, G_2, \dots, G_r)$ and so $G = (AG_2 \cdots G_{r-1})G_r$ is the product of two totally permutable subgroups in $\mathfrak{F}$, which is an $\mathfrak{F}$-group by our assumption, contradicting the hypothesis. If we are in Case 2, then we can apply the hypothesis to $G = (AG_2 \cdots G_{r-1})G_r \in \mathfrak{F}$ to obtain that $AG_2 \cdots G_{r-1} \in \mathfrak{F}$, and by the choice of $(G, G_1, G_2, \dots, G_r)$, we conclude that $A \in \mathfrak{F}$, against our supposition. In any case, we reach a final contradiction which proves the theorem. □

Finally, we will study the behaviour of radicals and injectors of totally permutable products with respect to Fitting classes. We will follow Hauck, Martínez-Pastor, and Pérez-Ramos' original treatment of this material, [145], with only minor modifications.

Proposition 5.3.8. *Let $\mathfrak{F}$ be a Fitting class containing the class $\mathfrak{U}$ of all supersoluble groups and satisfying the following condition:*

> *If a group $G = AB$ is the product of the totally permutable subgroups A and B, then $G_{\mathfrak{F}} = A_{\mathfrak{F}} B_{\mathfrak{F}}$.* (5.4)

Then, for a group $G = AB$ which is the product of the totally permutable subgroups A and B, the following properties hold:

1. *If A and B belong to $\mathfrak{F}$, then G belongs to $\mathfrak{F}$.*
2. *If G belongs to $\mathfrak{F}$, then A and B belong to $\mathfrak{F}$.*

Proof. 1. It is clear.

2. Let $G = AB$ be an $\mathfrak{F}$-group which is the product of the totally permutable subgroups A and B. We argue by induction on $|G|$. Assume that $\langle B^G \rangle = G$. Since B centralises $A^{\mathfrak{N}}$ by Theorem 4.2.7, we obtain that $A^{\mathfrak{N}}$ is centralised by B and all its conjugates and so $A^{\mathfrak{N}} \leq Z(G)$. This implies that $A^{\mathfrak{N}} \leq Z(A)$. It follows that $A^{\mathfrak{N}} = 1$ and so $A \in \mathfrak{N} \subseteq \mathfrak{F}$. By the hypothesis, we have that $G = G_{\mathfrak{F}} = A_{\mathfrak{F}} B_{\mathfrak{F}} = AB_{\mathfrak{F}}$ and $B = B_{\mathfrak{F}}(A \cap B) \in \mathfrak{F}$. Therefore we can assume that $\langle B^G \rangle$ is a proper subgroup of G and analogously $\langle A^G \rangle$ is a proper subgroup of G. Since $B(\langle B^G \rangle \cap A) = \langle B^G \rangle \in \mathfrak{F}$, it follows that $B \in \mathfrak{F}$ by the inductive hypothesis. Analogously, $A \in \mathfrak{F}$. □

It is not known whether the converse of this result holds, but it does for Fischer classes as we will show in Theorem 5.3.10. We need a preliminary result.

Proposition 5.3.9. *Let $\mathfrak{F}$ be a Fitting class containing the class $\mathfrak{U}$ of all supersoluble groups and satisfying the following condition:*

> *For a group $G = AB$ which is the product of the totally permutable subgroups A and B, the following properties hold:*
>
> 1. *If A and B belong to $\mathfrak{F}$, then G belongs to $\mathfrak{F}$.*
> 2. *If G belongs to $\mathfrak{F}$, then A and B belong to $\mathfrak{F}$.*

Suppose that the group $G = G_1 G_2 \cdots G_r$ is the product of the pairwise totally permutable subgroups $G_1, G_2, \dots, G_r$. Then $(G_1)_{\mathfrak{F}} (G_2)_{\mathfrak{F}} \cdots (G_r)_{\mathfrak{F}}$ is a normal subgroup of G contained in $G_{\mathfrak{F}}$ and $(G_i)_{\mathfrak{F}} = G_i \cap G_{\mathfrak{F}}$ for all $i \in \{1, 2, \dots, r\}$.

Proof. For any $i \in \{1, 2, \ldots, r\}$, by Lemma 4.2.12, we have that

$$\Big[G_i, \prod_{j \in \{1,\ldots,r\} \setminus \{i\}} G_j\Big] \leq Z_{\mathfrak{U}}(G) = \prod_{i=1}^{r} Z_{\mathfrak{U}}(G_i) \leq \prod_{i=1}^{r} (G_i)_{\mathfrak{F}}.$$

In particular, $\prod_{i=1}^{r} (G_i)_{\mathfrak{F}}$ is a normal subgroup of G and hence $\prod_{i=1}^{r} (G_i)_{\mathfrak{F}} \leq G_{\mathfrak{F}}$ by Statement 1 and Theorem 5.3.7. Moreover, $\prod_{i=1}^{r} (G_i)_{\mathfrak{F}} \leq \prod_{i=1}^{r} (G_i \cap G_{\mathfrak{F}})$, which is a normal subgroup of $G_{\mathfrak{F}}$ by Corollary 4.1.22 (2). This implies that $\prod_{i=1}^{r} (G_i \cap G_{\mathfrak{F}}) \in \mathfrak{F}$ and so $G_i \cap G_{\mathfrak{F}} \in \mathfrak{F}$ for all $i \in \{1, 2, \ldots, r\}$ by Statement 2 and Theorem 5.3.7. Consequently, $G_i \cap G_{\mathfrak{F}} = (G_i)_{\mathfrak{F}}$ for all $i \in \{1, 2, \ldots, r\}$ and the result is proved. □

Theorem 5.3.10. *Let $\mathfrak{F}$ be a Fischer class containing the class $\mathfrak{U}$ of all supersoluble groups. Suppose that the group $G = G_1 G_2 \cdots G_r$ is the product of the pairwise totally permutable subgroups $G_1, G_2, \ldots, G_r$. Then $G_{\mathfrak{F}} = (G_1)_{\mathfrak{F}} (G_2)_{\mathfrak{F}} \cdots (G_r)_{\mathfrak{F}}$ and $(G_i)_{\mathfrak{F}} = G_i \cap G_{\mathfrak{F}}$ for all $i \in \{1, 2, \ldots, r\}$.*

Proof. We argue by induction on $|G| + |G_1| + \cdots + |G_r|$. We split the proof in the following independently stated steps:

1. *$(G_1)_{\mathfrak{F}} (G_2)_{\mathfrak{F}} \cdots (G_r)_{\mathfrak{F}}$ is a normal subgroup of G contained in $G_{\mathfrak{F}}$ and $(G_i)_{\mathfrak{F}} = G_i \cap G_{\mathfrak{F}}$ for $i \in \{1, 2, \ldots, r\}$.*

This follows from Theorem 5.3.3 and Proposition 5.3.9.

2. *$L = (G_1)_{\mathfrak{F}} (G_2)_{\mathfrak{F}} \cdots (G_r)_{\mathfrak{F}}$. Then $[G, G_{\mathfrak{F}}] \leq L$; in particular, $G_{\mathfrak{F}}/L$ is abelian.*

As in the proof of Proposition 5.3.9, we have $[G_i, \prod_{j \in \{1,\ldots,r\} \setminus \{i\}} G_j] \leq L$. Hence for all $i \in \{1, 2, \ldots, r\}$

$$[G_i L, G] = [G_i, G][L, G] \leq [G_i, G_i]\Big[G_i, \prod_{j \in \{1,\ldots,r\} \setminus \{i\}} G_j\Big][L, G] \leq G_i L$$

and so $T_i = G_i L$ is a normal subgroup of G and $G = \prod_{j=1}^{r} T_j$. If there exists some $i \in \{1, 2, \ldots, r\}$ such that $T_i = G$, then

$$G_{\mathfrak{F}} = \prod_{j \in \{1,\ldots,r\} \setminus \{i\}} (G_j)_{\mathfrak{F}} (G_i \cap G_{\mathfrak{F}}) = \prod_{i=1}^{r} (G_i)_{\mathfrak{F}}$$

and the result follows. Therefore we can assume that T_i is a proper normal subgroup of G for all $i \in \{1, 2, \ldots, r\}$ and $G = \prod_{j=1}^{r} T_j$. Since $(T_i)_{\mathfrak{F}} = G_{\mathfrak{F}} \cap G_i L = (G_{\mathfrak{F}} \cap G_i) L = L$ by Statement 1, we have

$$[G, G_{\mathfrak{F}}] = \Big[\prod_{i=1}^{r} T_i, G_{\mathfrak{F}}\Big] = \prod_{i=1}^{r} [T_i, G_{\mathfrak{F}}] \leq \prod_{i=1}^{r} (T_i)_{\mathfrak{F}} = L,$$

which proves Statement 2.

If L is a proper subgroup of $G_{\mathfrak{F}}$, since $G_{\mathfrak{F}}/L$ is abelian by Statement 2, we can consider a normal subgroup $T = L\langle x\rangle$ of $G_{\mathfrak{F}}$ such that $x^p \in L$, where x is a p-element and $x = x_1x_2\cdots x_r$ for some $x_i \in G_i$, x_i a p-element for all $i \in \{1, 2, \dots, r\}$ and a prime number p.

3. *We may assume that* $G_i = (G_i)_{\mathfrak{F}}\langle x_i\rangle$ *for all* $i \in \{1, 2, \dots, r\}$.

Let $i \in \{1, 2, \dots, r\}$ and $S = (G_i)_{\mathfrak{F}}\langle x_i\rangle(\prod_{j\in\{1,\dots,r\}\setminus\{i\}} G_j)$. Obviously $T \leq S_{\mathfrak{F}}$ because T is a subnormal subgroup of G, $T \leq S$, and $T \in \mathfrak{F}$. Let $R_i = (G_i)_{\mathfrak{F}}\langle x_i\rangle \leq G_i$. If R_i is a proper subgroup of G_i, by the inductive hypothesis, we have that $S_{\mathfrak{F}} = (\prod_{j\in\{1,\dots,r\}\setminus\{i\}} (G_j)_{\mathfrak{F}})(R_i)_{\mathfrak{F}}$. Then

$$\begin{aligned} x \in S_{\mathfrak{F}} \cap G_{\mathfrak{F}} &= \Big(\prod_{j\in\{1,\dots,r\}\setminus\{i\}} (G_j)_{\mathfrak{F}}\Big)((R_i)_{\mathfrak{F}} \cap G_{\mathfrak{F}}) \\ &\leq \Big(\prod_{j\in\{1,\dots,r\}\setminus\{i\}} (G_j)_{\mathfrak{F}}\Big)(G_i \cap G_{\mathfrak{F}}) \\ &= \prod_{i=1}^{r} (G_i)_{\mathfrak{F}} = L \end{aligned}$$

by Statement 1. Then $x \in L$, which is a contradiction to the choice of x. Hence we may assume that $R_i = G_i$ for all $i \in \{1, 2, \dots, r\}$, which proves Statement 3.

4. *Final contradiction.*

We claim that $G_1 \in \mathfrak{F}$. Let $a = x_1$ and $b = x_2\cdots x_r$. Note first that $\langle a, b\rangle \leq \langle x_1\rangle\langle x_2\rangle\cdots\langle x_r\rangle$, which is a p-group. Therefore, since $[G_1, \prod_{j=2}^{r} G_j] \leq \mathrm{F}(G)$ by Corollary 4.2.11 and $\prod_{j=2}^{r} G_j$ centralises $G_1^{\mathfrak{N}}$ by Theorem 4.2.7, it follows that

$$C = \langle [\langle a\rangle, \langle b\rangle]^{G_1^{\mathfrak{N}}}\rangle \leq \mathrm{O}_p(G)$$

and C is normalised by $\langle a, b\rangle$.

On the other hand, since $G_1^{\mathfrak{N}} \leq (G_1)_{\mathfrak{F}} \leq G_{\mathfrak{F}}$ by Statement 3 and $x \in G_{\mathfrak{F}}$, we have that

$$Y = G_1^{\mathfrak{N}} C\langle x\rangle \leq G_{\mathfrak{F}}.$$

Now $CY = G_1^{\mathfrak{N}} C\langle a\rangle\langle x\rangle = G_1^{\mathfrak{N}} C\langle a, b\rangle$ is a subgroup of $G_{\mathfrak{F}}$. Since $CY/G_1^{\mathfrak{N}} \cong C\langle a, b\rangle/(G_1^{\mathfrak{N}} \cap C\langle a, b\rangle)$ is a p-group, we conclude that $(CY)^{\mathfrak{N}}$ is contained in $G_1^{\mathfrak{N}}$. Moreover, $(CY)^{\mathfrak{N}}$ is normalised by $G_1^{\mathfrak{N}}$, which is a normal subgroup of G by Theorem 4.2.7. Consequently, $(CY)^{\mathfrak{N}}$ is a subnormal subgroup of $G_{\mathfrak{F}}$. Thus $CY \in \mathfrak{F}$ because $\mathfrak{F}$ is a Fischer class. Analogously, $G_1^{\mathfrak{N}}\langle a\rangle$ is an $\mathfrak{F}$-group. Moreover, $G_1^{\mathfrak{N}}\langle a\rangle$ is a subnormal subgroup of G_1 since it contains $G_1^{\mathfrak{N}}$. We conclude that

$$G_1 = (G_1)_{\mathfrak{F}}(G_1^{\mathfrak{N}}\langle a\rangle) \in \mathrm{N_0}\,\mathfrak{F} = \mathfrak{F}.$$

Finally, for every $i \in \{2, \dots, r\}$, we note that

$$x_i x_{i+1} \cdots x_r x_1 \cdots x_{i-1} = (x_1 x_2 \cdots x_r)^{x_1 \cdots x_{i-1}} \in G_{\mathfrak{F}}.$$

Then we can argue in a similar way to deduce that $G_i \in \mathfrak{F}$ for all $i \in \{2, \dots, r\}$. Consequently $G \in \mathfrak{F}$ and $G = G_{\mathfrak{F}} = \prod_{i=1}^r G_i = \prod_{i=1}^r (G_i)_{\mathfrak{F}} = L$, which contradicts our previous assumption and concludes the proof. □

If $\mathfrak{F}$ is a Fitting class containing $\mathfrak{U}$, the $\mathfrak{F}$-injectors of soluble totally permutable products are not prefactorised in general, even if the product is direct.

Example 5.3.11. We consider the Fitting class $\mathfrak{F} = (\mathfrak{N} \diamond \mathfrak{D}) \cap (\mathfrak{N}^2 \mathfrak{S}_2)$, where $\mathfrak{D} = \mathfrak{D}(5)$ is the class of all soluble groups such that $\prod_{i=1}^n \det(g \text{ on } M_i) = 1$ for all $g \in G$, the product being taken over the 5-chief factors $M_1, M_2, \dots, M_n$ of a given chief series of G (see [119, IX, 2.14 (a)]), and $\mathfrak{S}_2$ denotes the class of all 2-groups.

Let V_5 be a faithful and irreducible Sym(3)-module over GF(5), the finite field of 5 elements, here $\mathrm{Sym}(3) = \langle \beta, \alpha : \beta^3 = \alpha^2 = 1, \beta^\alpha = \beta^{-1} \rangle$ denotes the symmetric group of degree 3. Consider now the corresponding semidirect product $G = [V_5]\mathrm{Sym}(3)$. With the above notation, we obtain that in this group $\det(\alpha \text{ on } V_5) = -1$, but $\det(\beta \text{ on } V_5) = 1$. Now let V_3 be a faithful G-module over the field GF(3) of 3 elements and let $X = [V_3]G$ be the semidirect product with respect to the action of G over V_3.

Then the $\mathfrak{F}$-radical of the group $X \times X$ is $(X \times X)_{\mathfrak{F}} = X_{\mathfrak{F}} \times X_{\mathfrak{F}} = (V_3 V_5) \times (V_3 V_5)$, and so it is factorised. But $I = (X \times X)_{\mathfrak{F}} \langle (\alpha, \alpha) \rangle$ is an $\mathfrak{F}$-injector of $X \times X$. Consequently no $\mathfrak{F}$-injector of $X \times X$ is prefactorised.

We bring this section to a close by touching on the question of characterising the prefactorisation of the radical in terms of the prefactorisation of the injectors in the soluble universe. We record that, applying [47, 2.4.28], if $\mathfrak{F}$ is a Fitting class and G is a group with $G/G_{\mathfrak{F}}$ soluble (in particular, if G is soluble), then G has a unique conjugacy class of $\mathfrak{F}$-injectors. It is clear that, in this case, the core of an $\mathfrak{F}$-injector is precisely the $\mathfrak{F}$-radical. These facts will be needed at several points in the sequel.

Theorem 5.3.12. *Let $\mathfrak{F}$ be a Fitting class containing the class $\mathfrak{U}$ of all supersoluble groups. Let the soluble group $G = G_1 G_2 \cdots G_r$ be the product of the pairwise totally permutable subgroups $G_1, G_2, \dots, G_r$. Assume that there exists an $\mathfrak{F}$-injector I of G such that $I \cap G_i$ is an $\mathfrak{F}$-injector of G_i for all $i \in \{1, 2, \dots, r\}$ and $I = (I \cap G_1)(I \cap G_2) \cdots (I \cap G_r)$. Then the following hold:*

1. *$G_{\mathfrak{F}} = (G_1)_{\mathfrak{F}} (G_2)_{\mathfrak{F}} \cdots (G_r)_{\mathfrak{F}}$ and $(G_i)_{\mathfrak{F}} = G_i \cap G_{\mathfrak{F}}$ for all $i \in \{1, 2, \dots, r\}$.*
2. *If J is an $\mathfrak{F}$-injector of G, then $J \cap G_i$ is an $\mathfrak{F}$-injector of G_i for all $i \in \{1, 2, \dots, r\}$ and $J = (J \cap G_1)(J \cap G_2) \cdots (J \cap G_r)$.*
3. *If I_i is an $\mathfrak{F}$-injector of G_i for every $i \in \{1, 2, \dots, r\}$, then $J = I_1 I_2 \cdots I_r$ is an $\mathfrak{F}$-injector of G and $I_i = J \cap G_i$ for all $i \in \{1, 2, \dots, r\}$.*

Proof. 1. We begin by showing that if A_i and B_i are subgroups of G_i such that $Z_{\mathfrak{U}}(G_i) \leq A_i \cap B_i$ for each $i \in \{1, 2, \dots, r\}$, then $(A_1 A_1 \cdots A_r) \cap (B_1 B_2 \cdots B_r) = (A_1 \cap B_1)(A_2 \cap B_2) \cdots (A_r \cap B_r)$. Let $a_1 x = b_1 y \in (A_1 A_2 \cdots A_r) \cap (B_1 B_2 \cdots B_r)$, where $a_1 \in A_1$, $b_1 \in B_1$, $x \in A_2 \cdots A_r$, and $y \in B_2 \cdots B_r$. Then $b_1^{-1} a_1 = y x^{-1} \in G_1 \cap (G_2 \cdots G_r) \leq Z_{\mathfrak{U}}(G)$ by Lemma 4.2.12. In particular, $b_1^{-1} a_1 \in Z_{\mathfrak{U}}(G) \cap G_1 = Z_{\mathfrak{U}}(G_1) \leq A_1 \cap B_1$ by Corollary 4.2.14. Consequently, $a_1, b_1 \in A_1 \cap B_1$. The result follows by means of an argument with each component in the same way.

We claim now that $I^g = (I \cap G_1)^{g_1} (I \cap G_2)^{g_2} \cdots (I \cap G_r)^{g_r}$ if $g = g_1 g_2 \cdots g_r$ with $g_i \in G_i$ for every $i \in \{1, 2, \dots, r\}$. We recall that $\prod_{i=1}^r Z_{\mathfrak{U}}(G_i) = Z_{\mathfrak{U}}(G) \leq I$ by Corollary 4.2.14, and $[G_i, \prod_{j \in \{1,\dots,r\} \setminus \{i\}} G_j] \leq Z_{\mathfrak{U}}(G)$ by Lemma 4.2.12. Moreover, $Z_{\mathfrak{U}}(G_i) \leq I \cap G_i$ for all $i \in \{1, 2, \dots, r\}$. Then we have that

$$
\begin{aligned}
(I \cap G_1)^g (I \cap G_2)^g \cdots (I \cap G_r)^g &= (I \cap G_1)^{g_1} (I \cap G_2)^{g_2} \cdots (I \cap G_r)^{g_r} Z_{\mathfrak{U}}(G) \\
&= (I \cap G_1)^{g_1} (I \cap G_2)^{g_2} \cdots (I \cap G_r)^{g_r}.
\end{aligned}
$$

Therefore, it follows now that

$$
\begin{aligned}
G_{\mathfrak{F}} &= \bigcap_{g \in G} I^g = \bigcap \{ (I \cap G_1)^{g_1} (I \cap G_2)^{g_2} \cdots (I \cap G_r)^{g_r} : g_i \in G_i, 1 \leq i \leq r \} \\
&= \Big(\bigcap_{g_1 \in G_1} (I \cap G_1)^{g_1} \Big) \Big(\bigcap_{g_2 \in G_2} (I \cap G_2)^{g_2} \Big) \cdots \Big(\bigcap_{g_r \in G_r} (I \cap G_r)^{g_r} \Big) \\
&= (G_1)_{\mathfrak{F}} (G_2)_{\mathfrak{F}} \cdots (G_r)_{\mathfrak{F}}.
\end{aligned}
$$

In particular, $(G_i)_{\mathfrak{F}} \leq G_{\mathfrak{F}} \cap G_i$ for every $i \in \{1, 2, \dots, r\}$. But $G_{\mathfrak{F}} \cap G_i$ is a normal subgroup of the $\mathfrak{F}$-subgroup $I \cap G_i$. This implies that $G_{\mathfrak{F}} \cap G_i \leq (G_i)_{\mathfrak{F}}$ for every $i \in \{1, 2, \dots, r\}$, as desired.

2. Let J be an $\mathfrak{F}$-injector of G. There exists $g \in G$ such that $J = I^g$. If $g = g_1 g_2 \cdots g_r$, where $g_i \in G_i$ for each $i \in \{1, 2, \dots, r\}$, then we have as above that $J = I^g = (I \cap G_1)^{g_1} (I \cap G_2)^{g_2} \cdots (I \cap G_r)^{g_r}$. In particular, for every $i \in \{1, 2, \dots, r\}$ we have that $(I \cap G_i)^{g_i} \leq J \cap G_i$ and so $J \cap G_i = (I \cap G_i)^{g_i} ((J \cap G_i) \cap T)$, where $T = \prod_{j \in \{1,\dots,r\} \setminus \{i\}} (I \cap G_j)^{g_j}$. But $(J \cap G_i) \cap T \leq Z_{\mathfrak{U}}(G) \cap G_i = Z_{\mathfrak{U}}(G_i) \leq (I \cap G_i)^{g_i}$, which implies that $J \cap G_i = (I \cap G_i)^{g_i}$ is an $\mathfrak{F}$-injector of G_i.

3. Let I_i be an $\mathfrak{F}$-injector of G_i for every $i \in \{1, 2, \dots, r\}$. There exist elements $g_i \in G_i$ for $i \in \{1, 2, \dots, r\}$ such that $I_i = (I \cap G_i)^{g_i}$. Then $J = I_1 I_2 \cdots I_r = (I \cap G_1)^{g_1} (I \cap G_2)^{g_2} \cdots (I \cap G_r)^{g_r} = (I \cap G_1)^g (I \cap G_2)^g \cdots (I \cap G_r)^g = I^g$ is an $\mathfrak{F}$-injector of G for $g = g_1 g_2 \cdots g_r$. Finally, for every $i \in \{1, 2, \dots, r\}$, I_i is contained in $J \cap G_i$, which is an $\mathfrak{F}$-injector of G_i by Statement 2 and so $I_i = J \cap G_i$, which concludes the proof. □

Theorem 5.3.13. *Let $r \geq 2$ be an integer and let $\mathfrak{F}$ be a Fitting class containing the class $\mathfrak{U}$ of all supersoluble groups. Then the following statements are equivalent:*

1. *If* $G = G_1G_2\cdots G_r$ *is a soluble group which is the product of the pairwise totally permutable subgroups* $G_1, G_2, \dots, G_r$*, then* $G_{\mathfrak{F}} = (G_1)_{\mathfrak{F}}(G_2)_{\mathfrak{F}}\cdots(G_r)_{\mathfrak{F}}$.

2. *If* $G = G_1G_2\cdots G_r$ *is a soluble group which is the product of the pairwise totally permutable subgroups* $G_1, G_2, \dots, G_r$ *and* I *is an* $\mathfrak{F}$*-injector of* G*, then* $I \cap G_i$ *is an* $\mathfrak{F}$*-injector of* G_i *for all* $i \in \{1, 2, \dots, r\}$ *and* $I = (I \cap G_1)(I \cap G_2)\cdots(I \cap G_r)$*. Moreover, in this case, if* I_i *is an* $\mathfrak{F}$*-injector of* G_i *for every* $i \in \{1, 2, \dots, r\}$*, then* $I_1I_2\cdots I_r$ *is an* $\mathfrak{F}$*-injector of* G.

Proof. Statement 1 implies Statement 2.

Assume that Statement 1 holds and let $G = G_1G_2\cdots G_r$ be a soluble group which is the product of the pairwise totally permutable subgroups $G_1, G_2, \dots, G_r$. Let I be an $\mathfrak{F}$-injector of G. We argue by induction on $|G| + |G_1| + \cdots + |G_r|$ to prove that Statement 2 holds. We recall that $[G_i, \prod_{j\in\{1,\dots,r\}\setminus\{i\}} G_j] \leq \mathrm{F}(G) \leq G_{\mathfrak{F}}$ for every $i \in \{1, 2, \dots, r\}$ by Corollary 4.2.11. In particular,

$$\Big(G_i\Big(\prod_{j\in\{1,\dots,r\}\setminus\{i\}} G_j\Big)\Big)/G_{\mathfrak{F}} = (G_iG_{\mathfrak{F}}/G_{\mathfrak{F}})\Big(\Big(\prod_{j\in\{1,\dots,r\}\setminus\{i\}} G_j\Big)G_{\mathfrak{F}}/G_{\mathfrak{F}}\Big)$$

is a central product.

We split the proof into the following steps:

1. *$I \cap G_i$ is an $\mathfrak{F}$-injector of G_i for all $i \in \{1, 2, \dots, r\}$.*

We fix $i \in \{1, 2, \dots, r\}$. Since $G_iG_{\mathfrak{F}}$ is a normal subgroup of G, it follows that $I \cap (G_iG_{\mathfrak{F}})$ is an $\mathfrak{F}$-injector of $G_iG_{\mathfrak{F}}$. In particular, $I \cap (G_iG_{\mathfrak{F}}) = (I \cap G_i)G_{\mathfrak{F}} = (I \cap G_i)(\prod_{j\in\{1,2,\dots,r\}\setminus\{i\}} (G_j)_{\mathfrak{F}}) \in \mathfrak{F}$ and it is a product of pairwise totally permutable subgroups. Then $I \cap G_i \in \mathfrak{F}$ by Proposition 5.3.8 and Theorem 5.3.7. Moreover, we have that $I \cap G_i^{\mathfrak{N}} = (I \cap G_i) \cap G_i^{\mathfrak{N}}$ is an $\mathfrak{F}$-injector of $G_i^{\mathfrak{N}}$, because $G_i^{\mathfrak{N}}$ is a normal subgroup of G. We claim that $I \cap G_i$ is an $\mathfrak{F}$-maximal subgroup of G_i, which implies that $I \cap G_i$ is an $\mathfrak{F}$-injector of G_i by [119, IX, 1.6 (a)]. Assume that $I \cap G_i \leq W_i \leq G_i$ with $W_i \in \mathfrak{F}$. Then $I \cap (G_iG_{\mathfrak{F}}) \leq W_i(\prod_{j\in\{1,\dots,r\}\setminus\{i\}} (G_j)_{\mathfrak{F}}) \in \mathfrak{F}$ by Proposition 5.3.8 and Theorem 5.3.7, and $W_i(\prod_{j\in\{1,\dots,r\}\setminus\{i\}} (G_j)_{\mathfrak{F}}) \leq G_iG_{\mathfrak{F}}$. Since $I \cap (G_iG_{\mathfrak{F}})$ is an $\mathfrak{F}$-injector of $G_iG_{\mathfrak{F}}$, it is clear that $I \cap (G_iG_{\mathfrak{F}}) = W_i(\prod_{j\in\{1,\dots,r\}\setminus\{i\}} (G_j)_{\mathfrak{F}})$. Hence $W_i \leq (I \cap G_i)(\prod_{j\in\{1,\dots,r\}\setminus\{i\}} (G_j)_{\mathfrak{F}} \cap W_i)$ and so $W_i \leq (I \cap G_i)\,\mathrm{Z}_{\mathfrak{U}}(G_i) = I \cap G_i$ by Lemma 4.2.12 and Corollary 4.2.14, as desired.

2. *For any subgroup S of G containing $G_{\mathfrak{F}}$ we have that*

$$\pi_i(S) = \Big\{x_i \in G_i : \textit{there exists } x^{(i)} \in \prod_{j\in\{1,\dots,r\}\setminus\{i\}} G_j \textit{ such that } x_ix^{(i)} \in S\Big\}$$

is a subgroup of G_i and $S \cap G_i$ is normal in $\pi_i(S)$ for every $i \in \{1, 2, \dots, r\}$. Moreover, if $G_{\mathfrak{F}} \leq S \trianglelefteq T \leq G$, then $\pi_i(S)$ is a normal subgroup of $\pi_i(T)$ for all $i \in \{1, 2, \dots, r\}$.

Let $x_i, y_i \in \pi_i(S)$. There exist $x^{(i)}, y^{(i)} \in \prod_{j\in\{1,\dots,r\}\setminus\{i\}} G_j$ such that $x_i x^{(i)}$, $y_i y^{(i)} \in S$. Then $x_i x^{(i)} y_i y^{(i)} = x_i y_i x^{(i)} y^{(i)} c$ for some element $c \in G_{\mathfrak{F}} \le S$, which implies $x_i y_i \in \pi_i(S)$ and so $\pi_i(S)$ is a subgroup of G_i.

Let now $s \in S \cap G_i$ and $x_i \in \pi_i(S)$. Then $s^{x_i x^{(i)}} = s^{x_i} d \in S$ for some element $d \in G_{\mathfrak{F}} \le S$, and so $s^{x_i} \in S \cap G_i$. Therefore $S \cap G_i$ is normal in $\pi_i(S)$.

Finally, assume that $G_{\mathfrak{F}} \le S \trianglelefteq T \le G$ and let $x_i \in \pi_i(S)$ and $t_i \in \pi_i(T)$. Then $(x_i x^{(i)})^{t_i t^{(i)}} = (x_i)^{t_i} (x^{(i)})^{t^{(i)}} e$ for some $e \in G_{\mathfrak{F}} \le S$. Hence $x_i^{t_i} \in \pi_i(S)$. This proves that $\pi_i(S)$ is a normal subgroup of $\pi_i(T)$.

3. *We may assume that* $\pi_i(I) = G_i$ *for all* $i \in \{1, 2, \dots, r\}$.

Assume that $\pi_i(I)$ is a proper subgroup of G_i for some $i \in \{1, 2, \dots, r\}$. Since $I \le \prod_{i=1}^r \pi_i(I)$, we have that I is an $\mathfrak{F}$-injector of $\prod_{i=1}^r \pi_i(I)$ by [119, IX, 1.5 (c)]. Then we can deduce that $I = \prod_{i=1}^r (I \cap \pi_i(I)) \le \prod_{i=1}^r (I \cap G_i) \in \mathfrak{F}$ by the inductive hypothesis and applying Statement 1, Proposition 5.3.8, and Theorem 5.3.7. Then $I = \prod_{i=1}^r (I \cap G_i)$ and the result holds. Hence we may assume that $\pi_i(I) = G_i$ for all $i \in \{1, 2, \dots, r\}$.

4. $I \cap G_i = (G_i)_{\mathfrak{F}}$ *for all* $i \in \{1, 2, \dots, r\}$.

Consider $i \in \{1, 2, \dots, r\}$. From Statements 2 and 3, it follows that $I \cap G_i$ is a normal subgroup of $\pi_i(I) = G_i$. Then $I \cap G_i = (G_i)_{\mathfrak{F}}$ by step 1.

5. $I = G_{\mathfrak{F}}$. *Consequently,* $I = G_{\mathfrak{F}} = \prod_{i=1}^r (G_i)_{\mathfrak{F}} = \prod_{i=1}^r (I \cap G_i)$ *by Statement* 4.

Assume that $G_{\mathfrak{F}}$ is a proper subgroup of I. Then we can consider a subgroup T of G such that $G_{\mathfrak{F}}$ is a proper subgroup of a subnormal subgroup T of I and $|T : G_{\mathfrak{F}}| = p$ for a prime number p. Then $T = G_{\mathfrak{F}}\langle x \rangle$ for some p-element x such that $x^p \in G_{\mathfrak{F}}$. We can assume that $x = x_1 x_2 \cdots x_r$, with $x_i \in G_i$ and x_i a p-element for every $i \in \{1, 2, \dots, r\}$ (except for conjugacy).

We claim that $\pi_i(T) = G_i$ for all $i \in \{1, 2, \dots, r\}$. We note first that $\pi_i(T)$ is subnormal in $\pi_i(I) = G_i$ by Statement 2 and Statement 3. Consequently $G_{\mathfrak{F}} \le T \le \prod_{i=1}^r \pi_i(T)$, which is a subnormal subgroup of $G = \prod_{i=1}^r G_i$. Assume now that $\pi_i(T)$ is a proper subgroup of G_i for some $i \in \{1, 2, \dots, r\}$. Clearly $J = I \cap (\prod_{j=1}^r \pi_j(T))$ is an $\mathfrak{F}$-injector of $\prod_{j=1}^r \pi_j(T)$. Then, by the inductive hypothesis and Statement 4, we have that $T \le \prod_{j=1}^r (J \cap \pi_j(T)) \le \prod_{j=1}^r (I \cap G_j) = \prod_{j=1}^r (G_j)_{\mathfrak{F}} = G_{\mathfrak{F}}$, that is, $T \le G_{\mathfrak{F}}$, which contradicts the assumption.

Now we will prove that

$$\pi_i(T) \le \Big(\prod_{j\in\{1,\dots,r\}\setminus\{i\}} G_i\Big)(G_j)_{\mathfrak{F}}\langle x_i \rangle \quad \text{for all } i \in \{1, 2, \dots, r\}. \tag{5.5}$$

Take $i \in \{1, 2, \dots, r\}$ and let $t_i \in \pi_i(T)$. There exists $t^{(i)} \in \prod_{j\in\{1,\dots,r\}\setminus\{i\}} G_j$ such that $t_i t^{(i)} \in T$. Then $t_i t^{(i)} = x^l a = (x_1 x_2 \cdots x_r)^l a = x_i^l (x_1 \cdots x_{i-1} x_{i+1} \cdots x_r)^l c$

for suitable $l \in \mathbb{Z}, a, c \in G_{\mathfrak{F}}$ by the remark made at the beginning of the proof, and obviously $t_i \in \langle x_i\rangle(\prod_{j\in\{1,\dots,r\}\setminus\{i\}} G_j)G_{\mathfrak{F}}$. Hence $t_i \in (\prod_{j\in\{1,\dots,r\}\setminus\{i\}} G_j)(G_i)_{\mathfrak{F}}\langle x_i\rangle$ because $G_{\mathfrak{F}} = \prod_{j=1}^r (G_j)_{\mathfrak{F}}$.

Finally, we may suppose that $(G_i)_{\mathfrak{F}}\langle x_i\rangle = G_i$ for all $i \in \{1, 2, \dots, r\}$. Assume that there exists some $i \in \{1, 2, \dots, r\}$ such that $(G_i)_{\mathfrak{F}}\langle x_i\rangle$ is a proper subgroup of G_i. We note that

$$\begin{aligned} G = \prod_{j=1}^r G_j = G_i\Big(\prod_{j\in\{1,\dots,r\}\setminus\{i\}} G_j\Big) = \pi_i(T)\Big(\prod_{j\in\{1,\dots,r\}\setminus\{i\}} G_j\Big) \\ = (G_i)_{\mathfrak{F}}\langle x_i\rangle\Big(\prod_{j\in\{1,\dots,r\}\setminus\{i\}} G_j\Big). \end{aligned}$$

Then, by the inductive hypothesis and Statement 4, we obtain that

$$\begin{aligned} I = ((G_i)_{\mathfrak{F}}\langle x_i\rangle \cap I)\Big(\prod_{j\in\{1,\dots,r\}\setminus\{i\}} (G_j \cap I)\Big) \le (G_i \cap I) \prod_{j\in\{1,\dots,r\}\setminus\{i\}} (G_j \cap I) \\ = \prod_{j=1}^r (G_j)_{\mathfrak{F}} = G_{\mathfrak{F}} \end{aligned}$$

and the result would hold. Hence $(G_i)_{\mathfrak{F}}\langle x_i\rangle = G_i$ for all $i \in \{1, 2, \dots, r\}$.

Therefore we can deduce that $G/G_{\mathfrak{F}} = (\prod_{j=1}^r G_j)/\prod_{j=1}^r (G_j)_{\mathfrak{F}}$ is abelian, which implies that $I = G_{\mathfrak{F}}$ because the $\mathfrak{F}$-injectors would be normal in G. Now, by Theorem 5.3.12, the rest of the Statement 2 is true.

Statement 2 implies Statement 1 follows from Theorem 5.3.12. □

Remark 5.3.14. From Theorem 5.3.10, we see that the statements of Theorem 5.3.13 hold, in particular, for Fischer classes.

5.4 Mutually permutable products and other classes of groups

The topic of this section is the interplay between mutually permutable products and the classes of PST-, PT-, and T-groups studied in Chapter 2. It is known that soluble groups in these classes have a great deal of supersolubility built into their structure, so it is not surprising that they exhibit a good behaviour in mutually permutable products.

We know that the classes of PST-, PT-, and T-groups are not subgroup-closed. However, we have:

Theorem 5.4.1. *Let $\Theta \in \{\mathrm{PST}, \mathrm{PT}, \mathrm{T}\}$ and let the group $G = G_1G_2\cdots G_r$ be the product of the pairwise mutually permutable subgroups $G_1, G_2, \dots, G_r$. If G is a Θ-group, then G_1 is a Θ-group.*

Beidleman and Heineken [66], extending a previous result of Ballester-Bolinches and Cossey [37], proved the above theorem for mutually permutable products of two subgroups. Our proof of Theorem 5.4.1 is confined to an investigation of the behaviour of a local version of the class PST, namely the property n_p introduced in Section 2.1 of Chapter 2, in products of finitely many pairwise mutually permutable products (see [21]).

Theorem 5.4.2. *Let the group* $G = G_1G_2\cdots G_r$ *be the product of the pairwise mutually permutable subgroups* $G_1, G_2, \dots, G_r$. *If* G *is an* SC*-group and satisfies* n_p, *then* G_1 *satisfies* n_p.

Our proof of Theorem 5.4.2 depends heavily on the following lemma.

Lemma 5.4.3. *Let* N *be a normal subgroup of* G *such that* G/N *satisfies* n_p. *If* N *is either a non-abelian and simple or a* p'*-group, then* G *satisfies* n_p.

Proof. We proceed by induction on $|G|$. Clearly, we may assume that $N \neq 1$. Let g be a p'-element of G and let H be a subgroup of $\mathrm{O}_p(G)$. Then HN/N is a subgroup of $\mathrm{O}_p(G/N)$ and so gN normalises HN/N since G/N has n_p. Hence $HN = H^gN$. Since N is either non-abelian and simple or a p'-group, $H^g = \mathrm{O}_p(H^gN) = \mathrm{O}_p(HN) = H$. Therefore g normalises H. This means that p'-elements of G induce power automorphisms in $\mathrm{O}_p(G)$. Therefore, by Lemma 2.1.3, H is S-permutable in G.

Let T be a minimal normal subgroup of G. If $T = N$, then G/T satisfies n_p. Suppose that $T \neq N$ and T is contained in N. Then N/T is a non-trivial normal subgroup of G/T and N/T is a p'-group. By induction, G/T satisfies n_p. Assume that T is not contained in N. Then NT/T is a non-trivial normal subgroup of G/T which is G-isomorphic to N. Besides $(G/T)/(NT/T) \cong G/NT$ satisfies n_p. The induction hypothesis implies that G/T satisfies n_p.

Now if R is a non-trivial normal subgroup of G, there exists a minimal normal subgroup T of G such that $T \leq R$. Since G/T satisfies n_p, we conclude that G/R satisfies n_p. The induction argument is therefore complete. □

Proof of Theorem 5.4.2. Suppose, for a contradiction, that the result is false and let G be a counterexample of minimal order. Let N be a minimal normal subgroup of G. Then G/N is the product of the pairwise mutually permutable subgroups G_1N/N, $G_2N/N, \dots, G_rN/N$ by Lemma 4.1.10. In addition, G/N satisfies n_p. The minimal choice of G implies that $G_1N/N \cong G_1/(G_1 \cap N)$ satisfies n_p. In particular, $G_1 \cap N \neq 1$. Note that N is simple. By Lemma 4.3.8 and Lemma 5.4.3, N is a p-group. Hence $|N| = p$ and so $N \leq G_1$. In particular, $\mathrm{O}_{p'}(G) = 1$. Since G is an SC-group, we can apply Theorem 1.6.5 to conclude that the soluble residual D of G satisfies that $D/\mathrm{Z}(D)$ is a direct product of minimal normal subgroups of $G/\mathrm{Z}(D)$. Moreover, by Corollary 1.6.6, we have that $\mathrm{C}_G(D) = S$, where S denotes the soluble radical of G. This implies that $\mathrm{Z}(D)$ is contained in $\mathrm{Z}(S)$.

On the other hand, S is an n_p-group. By Lemma 2.1.15, S satisfies $\mathcal{U}_p^*$. Moreover, $O_{p'}(S) \leq O_{p'}(G) = 1$. Applying Theorem 2.2.27, we conclude that S is a p-group or the p-nilpotent residual $S(p)$ of S is an abelian Sylow p-subgroup of S on which S induces a group of power automorphisms.

Since G_1 is not an n_p-group, there exists a normal subgroup K of G_1 and a subgroup H/K of $O_p(G_1/K)$ such that H/K is not S-permutable in G_1/K. Let us choose H of the smallest possible order. It is rather clear that H is subnormal in G_1. Hence the soluble residual $H^{\mathfrak{S}}$ of H is subnormal in G_1 as well. Applying Corollary 4.3.7, the soluble residual $(G_1)^{\mathfrak{S}}$ of G_1 is a normal subgroup of G. Hence $H^{\mathfrak{S}}$ is subnormal in G. By Theorem 1.6.5, $D/Z(D)$ is a direct product of non-abelian minimal normal subgroups of $G/Z(D)$, which are simple. Since $H^{\mathfrak{S}}Z(D)/Z(D)$ is subnormal in $D/Z(D)$, it follows that $H^{\mathfrak{S}}Z(D)$ is normal in G. Since $(H^{\mathfrak{S}}Z(D))' = (H^{\mathfrak{S}})' = H^{\mathfrak{S}}$, we conclude that $H^{\mathfrak{S}}$ is normal in G. If $H^{\mathfrak{S}} \neq 1$, then $G_1/H^{\mathfrak{S}}$ satisfies n_p by minimality of G. It implies that G_1/K satisfies n_p as $H^{\mathfrak{S}}$ is a subgroup of K. This contradicts our choice of the pair (H, K). Consequently $H^{\mathfrak{S}} = 1$ and H is soluble. In this case, H is contained in the soluble radical of G_1, which is a subgroup of S by Theorem 4.3.14.

Assume that S is a p-group. Since G has n_p and $H \leq O_p(G)$, it follows that H is S-permutable in G and so is in G_1 by Lemma 1.2.7 (1). This contradiction implies that S is not a p-group. Then $O_p(G) = S(p)$ is the abelian Sylow p-subgroup of S and it does not contain central chief factors by [47, 4.2.19]. In particular, $Z(D) = Z(S) = 1$. Since all minimal normal subgroups of G have order p, it follows that $D = 1$ and $G = S$ is soluble. Applying Lemma 2.1.15, G_1 satisfies n_p. This final contradiction completes the proof of the theorem. □

We have been unable to prove the m_p- and t_p-versions (see Section 2.1) of Theorem 5.4.2.

We draw a conclusion from this theorem (note that PST-groups are those SC-groups satisfying n_p for all p by Lemma 2.1.3 and Corollary 2.1.20).

Corollary 5.4.4. *Let the group $G = G_1G_2 \cdots G_r$ be the product of the pairwise mutually permutable subgroups $G_1, G_2, \dots, G_r$. If G is a* PST-*group, then G_1 is a* PST-*group.*

An essential ingredient in our proof of Theorem 5.4.1 is the following nice property proved by Ballester-Bolinches and Cossey [37].

Lemma 5.4.5. *Let G be a* PST-*group. Every subgroup of the nilpotent residual of the soluble radical of G is normal in G.*

Proof. Let S be the soluble radical of G and let E be the nilpotent residual of S. It is clear that S is a soluble PST-group. Applying Theorem 2.1.8, we conclude E is an abelian Hall subgroup of S of odd order and every subgroup of E is normal in S. Moreover, every chief factor of S below E is non-central in S by [47, 4.2.19].

Therefore $E \cap Z(S) = 1$. Assume, by way a contradiction, that the result is false and let G be a counterexample with $|G|$ as small as possible. Then there exists a subgroup H of E such that H is not normal in G. Let us choose H of least order. Then every proper subgroup of H is a normal subgroup of G. This implies that H has no non-trivial proper subgroups. Since H is soluble, it follows that H is a cyclic group of primer order, p say.

Let D be the soluble residual of G. If $D = 1$, then $G = S$ is soluble and so every subgroup of E is normal in G. This contradiction yields $D \neq 1$ and then $C_G(D) = S$ by Corollary 1.6.6. Hence $Z(D)$ is contained in $Z(S)$ and so $E \cap Z(D) = 1$. Assume that $Z(D) \neq 1$ and let N be a minimal normal subgroup of G contained in $Z(D)$. Then N is an elementary abelian q-group for some prime q. Since E is a Hall subgroup of S, we have that $p \neq q$. The choice of G implies that HN is normal in G. Now, H is a normal Sylow subgroup of HN. Hence H is normal in G, a contradiction which yields $Z(D) = 1$. Therefore H is the soluble radical of HD by Theorem 2.1.19. Since, by minimality of G, HD is normal in G, it follows that so is H. This final contradiction completes the proof of the lemma. □

Proof of Theorem 5.4.1. By Corollary 5.4.4, we may assume that $\Theta \in \{\mathrm{PT}, \mathrm{T}\}$. We will say that a subgroup H of G is Θ-*well placed in* G when H is normal in G for $\Theta = \mathrm{T}$ and H is permutable in G for $\Theta = \mathrm{PT}$.

We suppose that the result is false and derive a contradiction. Let G be a Θ-group of minimal order such that $G = G_1 G_2 \cdots G_r$ is the product of the pairwise mutually permutable subgroups $G_1, G_2, \dots, G_r$. Suppose that $A = G_1$ does not belong to Θ. Let H be a subnormal subgroup of A which is not Θ-well placed in A and assume that H has minimal order. Let N be a minimal normal subgroup of G. Then G/N is the product of the pairwise mutually permutable subgroups AN/N, $G_2N/N, \dots,$ G_rN/N by Lemma 4.1.10. In addition, G/N is a Θ-group by Lemma 2.0.4. The minimal choice of G implies that $AN/N \cong A/(A \cap N)$ belongs to Θ. In particular, $A \cap N \neq 1$. Note that N is simple. By Lemma 4.3.8, N is contained in A. Moreover, $H_G = 1$. Arguing as in Theorem 5.4.2, the soluble residual of H is normal in G. Since H is core-free in G, it follows that H is soluble. In this case, H is contained in the soluble radical of A, which is a subgroup of soluble radical S of G by Theorem 4.3.14. By Lemma 5.4.5, $H \cap S^{\mathfrak{N}} = 1$ and so H is nilpotent. The minimality of H implies that H is a p-group for some prime p. Applying Corollary 5.4.4, A is a PST-group. Hence H is S-permutable in A. In particular, the p'-elements of A normalise each subgroup of H. Since H is not permutable in A, there exists a p-element $a \in A$ such that $H\langle a\rangle$ is not a subgroup of A. On the other hand, since H is not Θ-well placed in G, it follows that H is not subnormal in G. In particular, H is not subnormal in S and S is not nilpotent. Let M be a minimal normal subgroup of G contained in $S^{\mathfrak{N}}$. Then M is a cyclic q-group for some prime $q \neq p$. The minimal choice of G implies that $L = HM\langle a\rangle$ is a subgroup of A. Applying Theorem 1.1.19, $H\langle a\rangle$ is a Sylow p-subgroup of L. This contradiction proves the theorem for $\Theta = \mathrm{PT}$. Now if $\Theta = \mathrm{T}$, we have that HM is normal in A. Clearly H is a normal Sylow subgroup of HM, so that H is normal in A. It follows that our supposition in wrong and hence A is a Θ-group. □

One can also take the opposite point of view and ask about the consequences of a mutually permutable product, having factors belonging to PST, PT, or T. In this context, the following extension of a result proved by Ballester-Bolinches, Beidleman, Heineken, and Pedraza-Aguilera [35] is illustrative and definitive.

Theorem 5.4.6. *Let the group $G = G_1 G_2 \cdots G_r$ be the product of the pairwise mutually permutable subgroups $G_1, G_2, \dots, G_r$. If all G_i are* PST*-groups for $i \in \{1, 2, \dots, r-1\}$ and G_r is an* SC*-group, then G is an* SC*-group.*

The first tool needed to establish Theorem 5.4.6 is the analysis of the behaviour of the property $\mathcal{U}_p^*$, p a prime, studied in Section 2.1, in mutually permutable products.

Theorem 5.4.7. *Let $G = G_1 G_2 \cdots G_r$ be the product of the pairwise mutually permutable subgroups $G_1, G_2, \dots, G_r$. If G_i satisfies $\mathcal{U}_p^*$ for all $i \in \{1, 2, \dots, r-1\}$ and G_r is p-supersoluble, then G is p-supersoluble.*

Proof. Assume that the theorem is false, and among the counterexamples with least order we choose one $G = G_1 G_2 \cdots G_r$ such that the sum $|G_1| + |G_2| + \cdots + |G_r|$ is minimal. Applying Theorem 4.1.15, G is p-soluble. By Lemma 4.1.10, the hypotheses of the theorem are inherited by all proper quotients of G. Hence G contains a unique minimal normal subgroup, N say, and N is not cyclic, G/N is p-supersoluble, and $N = \mathrm{C}_G(N) = \mathrm{O}_p(G) = \mathrm{F}(G)$. Hence $\mathrm{O}_{p'}(G) = 1$. For every $i \in \{1, 2, \dots, r\}$, G_i is p-supersoluble and hence G_i' is p-nilpotent. By Corollary 4.1.26, we have that G_i' is a subnormal subgroup of G for each $i \in \{1, 2, \dots, r\}$. Thus G_i' is contained in N and so G_i is supersoluble for each $i \in \{1, 2, \dots, r\}$. By Corollary 4.1.30, G has a Sylow tower with respect to the reverse natural ordering of the prime numbers, p is the largest prime divisor of $|G|$, and N is the Sylow p-subgroup of G.

Let $i \in \{1, 2, \dots, r\}$ such that p divides $|G_i|$. Then $N \cap G_i$ is the non-trivial Sylow p-subgroup of G_i. Let $j \in \{1, 2, \dots, r\}$ such that $j \neq i$. Let $(G_j)_{p'}$ denote a Hall p'-subgroup of G_j. Then $G_i (G_j)_{p'}$ is a subgroup of G and $N \cap G_i$ is a Sylow p-subgroup of $G_i (G_j)_{p'}$. Since $G_i (G_j)_{p'}$ is a Sylow tower group with respect to the reverse natural ordering of the prime numbers, it follows that $N \cap G_i$ is normal in $G_i (G_j)_{p'}$. This implies that $N \cap G_i$ is a normal subgroup of G and so $N = N \cap G_i$ is contained in G_i.

Assume that there exists $j \in \{1, 2, \dots, r\}$ such that p does not divide $|G_j|$. Then $G_j' = 1$ and so G_j is an abelian p'-group. Let T denote $\prod_{i \neq j} G_i$. By the minimal choice of G, T is p-supersoluble. Let R be a minimal normal subgroup of T contained in N. Then $|R| = p$. Moreover $G_j R$ is a subgroup of G because N is contained in some of the factors G_l, $l \neq j$. Hence $G_j R \cap N = R$ is a normal subgroup of $G_j R$. Thus R is a normal subgroup of G and so $N = R$. This is a contradiction. Therefore

p divides the order of G_i for every $i \in \{1, 2, \dots, r\}$ and, consequently, N is contained in G_i for every $i \in \{1, 2, \dots, r\}$.

Consider now $W = G_2 G_3 \cdots G_r$. Then W is p-supersoluble by the minimal choice of G. Let X be a minimal normal subgroup of W contained in N. Then $|X| = p$. Recall that G_1 satisfies $\mathcal{U}_p^*$. Suppose that G_1 is p-nilpotent. Since $N = C_G(N)$, it follows that $G_1 = N$. Now suppose that G_1 satisfies Condition 2 of Theorem 2.2.27. Now, since N is self-centralising, $O_{p'}(G_1(p)) = 1$, where $G_1(p)$ denotes the p-nilpotent residual of G_1. Hence $G_1(p)$ is a Sylow p-subgroup of G_1. Consequently $N = G_1(p)$ and so every subgroup of N is normal in G_1. In both cases, we have that G_1 normalises X. This means that $N = X$. This final contradiction completes the proof of the theorem. □

The following result is an immediate consequence of Theorem 5.4.7 and it is a particular case of a more general result proved by Ballester-Bolinches, Beidleman, Heineken, and Pedraza-Aguilera [36].

Corollary 5.4.8. *Let $G = G_1 G_2 \cdots G_r$ be the product of the pairwise mutually permutable subgroups $G_1, G_2, \dots, G_r$. If G_i satisfies $\mathcal{U}_p^*$ for all $i \in \{1, 2, \dots, r\}$, then G is p-supersoluble.*

In the usual way we deduce from Theorems 5.4.7 and 2.1.8 the following consequence, which can be seen as the soluble version of Theorem 5.4.6.

Corollary 5.4.9. *Let the group $G = G_1 G_2 \cdots G_r$ be the product of the pairwise mutually permutable subgroups $G_1, G_2, \dots, G_r$. If all G_i are soluble* PST*-groups for $i \in \{1, 2, \dots, r-1\}$ and G_r is supersoluble, then G is supersoluble.*

Two additional facts are needed to prove Theorem 5.4.6. The first one was proved by Ballester-Bolinches, Beidleman, Heineken, and Pedraza-Aguilera in [35].

Lemma 5.4.10. *Let the group $G = AB$ be the product of the mutually permutable subgroups A and B. If A is an* SC*-group and B is a* PST*-group, then G is an* SC*-group.*

Proof. Suppose that the result is false and let G be a minimal counterexample to the claim that G is an SC-group. Note that B is an SC-group as it is a PST-group (Proposition 2.1.1). Let N be a minimal normal subgroup of G. Then $G/N = (AN/N)(BN/N)$ is a product of the mutually permutable subgroups AN/N and BN/N, by Lemma 4.1.10, which are an SC-group and a PST-group, respectively. Therefore, by minimality of G, G/N is an SC-group. Since the class of SC-groups is a formation by Theorem 1.6.3, N is the unique minimal normal subgroup of G and N is not simple. By Theorem 4.5.10, we may assume that $N \leq (A \cap B)_G$. Suppose that N is non-abelian. Then N is a direct product of finitely many copies of a simple group. Let L be a minimal normal subgroup of A contained in N. Then L is simple and $LZ(D)$ is a normal subgroup of B, where D is the soluble residual of B.

Since $L = (LZ(D))'$, it follows that L is normal in B. Consequently $L = N$ is simple, against the assumption. Therefore N is an elementary abelian p-group for some prime p. Denote by $C = C_G(N)$. Note that $C \neq G$ because N is not cyclic. Since A is an SC-group, we conclude that $A/C_A(N_i/N_{i+1})$ is abelian of exponent dividing $p-1$ for every chief factor N_i/N_{i+1} of A below N. Put $F = \bigcap C_A(N_i/N_{i+1})$. Hence A/F and then $(AC/C)/(FC/C)$ are abelian groups of exponent dividing $p-1$. Furthermore F stabilises a series of N. Thus $F/C_F(N)$ is a p-group by [119, A, 12.4 (a)] and so is FC/C. Then we have that AC/C and BC/C are normal extensions of a p-group by a group of exponent dividing $p-1$. By Corollary 4.1.30, G/C has a normal Sylow p-subgroup. But N is a minimal normal p-subgroup of G and so G/C contains no non-trivial normal p-subgroup by [119, A, 13.6 (b)]. In particular, G/C is a p'-group. Therefore AC/C and BC/C are abelian groups of exponent dividing $p-1$.

As A is an SC-group, every minimal normal subgroup T of A contained in N is cyclic of prime order. Moreover $B = B_p O^p(B)$, where B_p is a Sylow p-subgroup of B. By the above argument, B_p is contained in C, that is, it centralises N and, in particular, T. Moreover, since B is a PST-group, $O^p(B)$ induces power automorphisms in N by Proposition 2.1.1, hence it normalises T. Thus T is normal in G and from here $N = T$ is cyclic of prime order. This contradiction proves the lemma. □

Another key to Theorem 5.4.6 is:

Lemma 5.4.11 (Ballester-Bolinches et al. [35]). *Let the group $G = G_1 G_2 \cdots G_r$ be the product of the pairwise mutually permutable subgroups $G_1, G_2, \ldots, G_r$ and let H be a normal perfect subgroup of G_i. If $G_i G_j$ is an* SC*-group for every $j \neq i$, then H is a normal subgroup of G.*

Proof. By Lemma 4.1.26, G_i' is a subnormal subgroup of G. Moreover $H = H'$ is a normal subgroup of G_i'. Therefore H is a subnormal subgroup of G. Consider now the product $G_i G_j$ with $j \neq i$. By Corollary 4.3.7, $D_i D_j$ is the soluble residual of $G_i G_j$, where D_i and D_j denote the soluble residuals of G_i and G_j, respectively. Write $D_{i,j} = D_i D_j$. Since H is perfect, it follows that $H \leq D_{i,j}$ and $HZ(D_{i,j})/Z(D_{i,j})$ is a subnormal subgroup of $D_{i,j}/Z(D_{i,j})$, which is a direct product of simple non-abelian $G_i G_j$-invariant subgroups by Theorem 1.6.5. Consequently $HZ(D_{i,j})$ is a normal subgroup of $G_i G_j$. Since $H = (HZ(D_{i,j}))'$, it follows that H is normal in $G_i G_j$ and so H is normal in G. □

Recall that if G is a group, a *component* of G is a subnormal perfect subgroup C of G such that $C/Z(C)$ is simple (see [47, 2.2.18] or [159, X, 13.17]).

Proof of Theorem 5.4.6. Assume that the result is false. Let $G = G_1 G_2 \cdots G_r$ be a counterexample with least $|G| + |G_1| + |G_2| + \cdots + |G_r|$. By Lemmas 2.0.4 and 4.1.10, G/N inherits the hypotheses of the theorem for any non-trivial normal subgroup N of G. It follows that G/N is an SC-group. Therefore G has a unique minimal normal subgroup, N say, because the class of all SC-groups is a formation by Theorem 1.6.3. In addition, N is not simple. Assume that N is non-abelian. By Lemma 4.3.8, N is contained in at least one of the factors G_j, and $N \cap G_i \in \{1, N_i\}$ for $i \in \{1, 2, \dots, r\}$. We can suppose without loss of generality that $N \leq \bigcap_{j=1}^{s} G_j$ and $N \cap G_k = 1$ if $k \in \{s+1, \dots, r\}$. Let $N = N_1 \times N_2 \times \cdots \times N_l$ be a decomposition of N as a product of pairwise isomorphic simple groups. Since all G_j are SC-groups for $j \in \{1, 2, \dots, r\}$, all N_i are G_j-invariant for $j \in \{1, 2, \dots, s\}$. Now consider $k \in \{s+1, \dots, r\}$ and $T = G_j G_k$. By Lemma 4.1.37, we obtain that G_k induces a group of power automorphisms on N. Therefore all N_i are G_j-invariant for $i \in \{1, 2, \dots, l\}$ and $j \in \{1, 2, \dots, r\}$. Hence $l = 1$ and N is simple. This contradiction shows that N must be an elementary abelian p-group for some prime p. Let D be the soluble residual of G. Assume that $D \neq 1$. By Theorem 5.2.7, $D = D_1 D_2 \cdots D_r$, where D_i is the soluble residual of G_i for $i \in \{1, 2, \dots, r\}$. Let E be a component of G_i. Then E is a normal subgroup of G by Lemmas 5.4.10 and 5.4.11. Moreover E is contained in D_i and so in D. Applying [159, X, 13.18], we see that $E\mathrm{Z}(D)/\mathrm{Z}(D)$ is G-isomorphic to $E/\mathrm{Z}(E)$ and so $E\mathrm{Z}(D)/\mathrm{Z}(D)$ is a non-abelian chief factor of G and $E\mathrm{Z}(D)/\mathrm{Z}(D)$ is simple.

On the other hand, from Theorem 1.6.5 it follows that D_i is the product of the components of G_i, which are in fact components of G. Therefore D is a product of components of G each one contained in one of the factors G_i.

Assume now that N is not contained in G_k for some $k \in \{1, 2, \dots, r\}$. If $D_k \neq 1$, then $N \leq D_k$ since D_k is a normal subgroup of G by Corollary 4.3.7, contrary to assumption. Therefore $A = G_k$ is soluble. Let Y be a component of G contained in some $G_j = B$, $j \neq k$. Then N is contained in D_j. It follows that $A\mathrm{Z}(D)/\mathrm{Z}(D)$ and $B\mathrm{Z}(D)/\mathrm{Z}(D)$ are mutually permutable subgroups of $G/\mathrm{Z}(D)$ and $Y\mathrm{Z}(D)/\mathrm{Z}(D)$ is a minimal normal subgroup of $(A\mathrm{Z}(D)/\mathrm{Z}(D))(B\mathrm{Z}(D)/\mathrm{Z}(D))$. Since $A\mathrm{Z}(D)/\mathrm{Z}(D)$ is soluble, we obtain that $(A\mathrm{Z}(D)/\mathrm{Z}(D)) \cap (Y\mathrm{Z}(D)/\mathrm{Z}(D)) = 1$. Consequently $A\mathrm{Z}(D)/\mathrm{Z}(D)$ centralises $Y\mathrm{Z}(D)/\mathrm{Z}(D)$ by Lemma 4.3.3 (5). In particular, $[Y, A] \leq \mathrm{Z}(Y) \leq \Phi(Y)$. Since by [159, X, 13.19], $Y = [Y, A]\mathrm{C}_Y(A)$, it follows that $Y = \mathrm{C}_Y(A)$ and A centralises Y. In particular, A centralises N.

Assume that N is contained in $L = G_i$ for some $i \in \{1, 2, \dots, r\}$. Since G_i is an SC-group, every chief factor of G_i below N is cyclic. Let $F = \bigcap \mathrm{C}_L(N_t/N_{t+1})$, where the intersection is taken for all chief factors N_t/N_{t+1} of L below N. Then L/F is abelian of exponent dividing $p-1$. Let $C = \mathrm{C}_G(N)$. Since by [119, A, 12.4 (a)], $F/\mathrm{C}_F(N)$ is a p-group and $\mathrm{C}_L(F) \leq F$, it follows that $\mathrm{C}_L(N) = \mathrm{C}_F(N)$. Hence LC/C is a normal extension of a p-group by an abelian group of exponent dividing $p-1$. By Corollary 4.1.29, the group G/C is a normal extension of a p-group by a group of order divisible only by primes dividing $p-1$. In particular G/C has a normal Sylow p-subgroup. Since $\mathrm{O}_p(G/C) = 1$ by [119, A, 13.6 (b)], it follows that G/C is a p'-group. In particular, every Sylow p-subgroup of G centralises N.

Let R be a subgroup of N of order p. Let $i \in \{1, 2, \dots, r-1\}$. Then $G_i \leq \mathrm{C}_G(N) \leq \mathrm{C}_G(R)$ if N is not contained in G_i. Suppose that N is contained in G_i.

Since G_i is a PST-group, it follows that $O^p(G_i) \leq N_G(R)$ by Proposition 2.1.1. This implies that $G_i \leq N_G(R)$ for each $i \in \{1, 2, \ldots, r-1\}$. If G_r centralises N, then R is a normal subgroup of G and $R = N$ and if N is contained in G_r, we can consider R as a minimal normal subgroup of G_r contained in N. In both cases, we have a contradiction. Therefore, $D = 1$ and G is soluble. But in this case Corollary 5.4.9 shows that G is supersoluble and hence an SC-group. This completes the proof. □

Recall that the direct product of two copies of a symmetric group of degree 3 is not a PST-group, even though the factors are T-groups. This example shows clearly the limitations of the theory.

5.5 Concluding remarks

Our main object in this chapter has been to give an account of the relation of the structure of a factorised group as a product of pairwise totally and mutually permutable subgroups to the structure of these subgroups. The general background of the results was the following: let $G = G_1 G_2 \cdots G_r$ be the product of the pairwise permutable subgroups $G_1, G_2, \ldots, G_r$ such that the products $G_i G_j$ are either mutually or totally permutable for all i, j. If the G_i belong to a certain class $\mathfrak{F}$ of groups, can we describe in some satisfactory manner the class $\mathfrak{G}$ to which G belongs? In other words: what can we say about G if special properties of the factors $G_1, G_2, \ldots, G_r$ are known?

Another way to study the structure of factorised groups whose factors are connected by some particular properties has been looked at. As a point of departure consider the following observation: assume that G_1 and G_2 are totally permutable subgroups of a group G. Then for each $x \in G_1$ and $y \in G_2$, $\langle x, y\rangle = \langle x\rangle\langle y\rangle = \langle y\rangle\langle x\rangle$ is a supersoluble group by Lemma 3.1.1. This leads to the following definition due to Carocca [85]:

Definition 5.5.1. Let $\mathfrak{L}$ be a class of groups. Two subgroups G_1 and G_2 of a group G are said to be *connected with respect to the class* $\mathfrak{L}$, or simply $\mathfrak{L}$*-connected*, whenever for each $x \in G_1$ and $y \in G_2$ we have $\langle x, y\rangle \in \mathfrak{L}$.

In this language, two totally permutable subgroups are $\mathfrak{U}$-connected. However, there is no relation in general between totally permutable products and $\mathfrak{N}$-connected subgroups. The symmetric group of degree 3 is a totally permutable product which is not $\mathfrak{N}$-connected, while the Sylow 2-subgroups of the symmetric group of degree 4 are $\mathfrak{N}$-connected products of the Klein 4-group and a cyclic group of order 2, but this product is not totally permutable.

The special case $G = G_1 = G_2$ is of interest and it is closely related to the following general question:

> *Suppose that* $\mathfrak{L}$ *is a class of groups satisfying the following property: a group G belongs to* $\mathfrak{L}$ *if every 2-generator subgroup of G is in* $\mathfrak{L}$*. What can be said about* $\mathfrak{L}$*?*

A class $\mathfrak{L}$ of groups has been called *bigenetic* by Lennox [188] and 2-*recognisable* (2-*erkennbar*) by Brandl [77] if a group G is in $\mathfrak{L}$ if and only if every 2-generator subgroup of G is in $\mathfrak{L}$. Perhaps one of the most well-known examples of bigenetic classes that comes to mind is the class $\mathfrak{S}$ of all soluble groups. This is a consequence of Thompson's classification of minimal simple groups [241, 242, 243, 244, 245, 246] (see also Flavell [126] for an alternative proof). Many other well-known classes, as the class of all abelian, nilpotent or supersoluble groups are bigenetic as well. Applying [119, VII, 6.15], a saturated formation is bigenetic in the category of all soluble groups if it has a local definition composed of bigenetic formations. As a consequence, $\mathfrak{N}\mathfrak{A}$ or $\mathfrak{N}^2$ are bigenetic.

So far only the $\mathfrak{L}$-connectedness for the classes $\mathfrak{N}$, $\mathfrak{U}$, $\mathfrak{N}^2$, $\mathfrak{S}$, and for saturated formations contained in $\mathfrak{N}\mathfrak{A}$ have been studied. Carocca [85] showed that the saturated formations of full characteristic are closed under the formation of $\mathfrak{N}$-connected products. He also showed in [87] that $\mathfrak{S}$-connected products of soluble groups are soluble as well.

Structure and properties of $\mathfrak{N}$-connected products are well-understood (see Ballester-Bolinches and Pedraza-Aguilera [52], Ballester-Bolinches, Cossey, and Esteban-Romero [38], Beidleman and Heineken [64], Hauck, Martínez-Pastor, and Pérez-Ramos [146]). On the basis of the results of these papers, we may consider that, in some sense, totally permutable products are to supersolubility as $\mathfrak{N}$-connected subgroups are to nilpotency.

That products of $\mathfrak{N}^2$-connected subgroups are closely related to $\mathfrak{N}$-connected ones is made clear by the following beautiful theorem proved by Gállego, Hauck, and Pérez-Ramos in [135]: a soluble group $G = AB$ is the $\mathfrak{N}^2$-connected product of the subgroups A and B if and only if $G/\mathrm{F}(G) = (A\mathrm{F}(G)/\mathrm{F}(G))(B\mathrm{F}(G)/\mathrm{F}(G))$ is an $\mathfrak{N}$-connected product of $A\mathrm{F}(G)/\mathrm{F}(G)$ and $B\mathrm{F}(G)/\mathrm{F}(G)$. A similar result holds for products of $\mathfrak{N}\mathfrak{A}$-connected subgroups (see [134]).

Bibliography

[1] R. K. Agrawal, *The influence on a finite group of its permutable subgroups*, Canadian Math. Bull. 17 (1974), pp. 159–165.

[2] R. K. Agrawal, *Finite groups whose subnormal subgroups permute with all Sylow subgroups*, Proc. Amer. Math. Soc. 47 (1975), pp. 77–83.

[3] M. J. Alejandre, A. Ballester-Bolinches, and J. Cossey, *Permutable products of supersoluble groups*, J. Algebra 276 (2004), pp. 453–461.

[4] M. J. Alejandre, A. Ballester-Bolinches, J. Cossey, and M. C. Pedraza-Aguilera, *On some permutable products of supersoluble groups*, Rev. Mat. Iberoam. 20 (2004), pp. 413–425.

[5] M. J. Alejandre, A. Ballester-Bolinches, and M. C. Pedraza-Aguilera, *Finite soluble groups with permutable subnormal subgroups*, J. Algebra 240 (2001), pp. 705–722.

[6] B. Amberg, *Factorizations of Infinite Groups*, Habilitationsschrift, Universität Mainz, 1973.

[7] B. Amberg, A. Carocca, and L. Kazarin, *Criteria for the solubility and non-simplicity of finite groups*, J. Algebra 285 (2005), pp. 58–72.

[8] B. Amberg, S. Franciosi, and F. de Giovanni, *On normal subgroups of products of nilpotent groups*, J. Austral. Math. Soc. Ser. A 44 (1988), pp. 275–286.

[9] B. Amberg, S. Franciosi, and F. de Giovanni, *Products of groups*, Oxford Mathematical Monographs, The Clarendon Press Oxford University Press, New York, 1992, Oxford Science Publications.

[10] B. Amberg and A. Fransman, *Products of groups and group classes*, Israel J. Math. 87 (1994), pp. 9–18.

[11] B. Amberg and B. Höfling, *On finite products of nilpotent groups*, Arch. Math. (Basel) 63 (1994), pp. 1–8.

[12] B. Amberg and L. S. Kazarin, *On finite products of soluble groups*, Israel J. Math. 106 (1998), pp. 93–108.

[13] B. Amberg and L. S. Kazarin, *On the rank of a product of two finite p-groups and nilpotent p-algebras*, Comm. Algebra 27 (1999), pp. 3895–3907.

[14] B. Amberg and L. S. Kazarin, *On the product of a nilpotent group and a group with non-trivial center*, J. Algebra 311 (2007), pp. 69–95.

[15] B. Amberg and L. S. Kazarin, *Factorizations of groups and related topics*, Sci. China Ser. A 52 (2009), pp. 217–230.

[16] B. Amberg, L. S. Kazarin, and B. Höfling, *Finite groups with multiple factorizations*, Fundam. Prikl. Mat. 4 (1998), pp. 1251–1263.

[17] B. Amberg and D. J. S. Robinson, *Soluble groups which are products of nilpotent minimax groups*, Arch. Math. (Basel) 42 (1984), pp. 385–390.

[18] B. Amberg and W. R. Scott, *Products of abelian subgroups*, Proc. Amer. Math. Soc. 26 (1970), pp. 541–547.

[19] B. Amberg and Y. Sysak, *Radical rings and products of groups*, Groups St Andrews 1997 in Bath, I (Math. Proc. Cambridge Philos. Soc.) (C. M. Campbell, E. F. Robertson, N. Ruskuc, and G. C. Smith, eds.), London Math. Soc. Lecture Notes Ser., vol. 260, Cambridge Univ. Press, 1999, pp. 1–19.

[20] M. Arroyo-Jordá, P. Arroyo-Jordá, A. Martínez-Pastor, and M. D. Pérez-Ramos, *On finite products of groups and supersolubility*, J. Algebra 323 (2010), pp. 2922–2934.

[21] M. Asaad, A. Ballester-Bolinches, J. C. Beidleman, and R. Esteban-Romero, *Some classes of finite groups and mutually permutable products*, J. Algebra 319 (2008), pp. 3343–3351.

[22] M. Asaad and A. A. Heliel, *On permutable subgroups of finite groups*, Arch. Math. (Basel) 80 (2003), pp. 113–118.

[23] M. Asaad, A. A. Heliel, M. Ezzat Mohamed, and P. Csörgő, *Finite groups with some subgroups permutable with all Sylow subgroups*, JP J. Algebra Number Theory Appl. 4 (2004), pp. 437–446.

[24] M. Asaad and A. Shaalan, *On the supersolvability of finite groups*, Arch. Math. (Basel) 53 (1989), pp. 318–326.

[25] R. Baer, *Situation der Untergruppen und Struktur der Gruppe*, S. B. Heidelberg. Akad. Wiss. 2 (1933), pp. 12–17.

[26] R. Baer, *Engelsche Elemente Noetherscher Gruppen*, Math. Ann. 133 (1957), pp. 256–270.

[27] A. Ballester-Bolinches, *Totally and mutually permutable products of finite groups*, Mathematics and the 21st Century (Singapore) (A. A. Ashour and A.-S. F. Obada, eds.), World Scientific Publishing Corporation, 2001, pp. 159–167.

[28] A. Ballester-Bolinches, J. C. Beidleman, J. Cossey, R. Esteban-Romero, M. F. Ragland, and J. Schmidt, *Permutable subnormal subgroups of finite groups*, Arch. Math. (Basel) 92 (2009), pp. 549–557.

[29] A. Ballester-Bolinches, J. C. Beidleman, J. Cossey, and H. Heineken, *The structure of mutually permutable products of finite nilpotent groups*, Int. J. Algebra Comput. 17 (2007), pp. 895–904.

[30] A. Ballester-Bolinches, J. C. Beidleman, J. Cossey, H. Heineken, and M. C. Pedraza-Aguilera, *Mutually permutable products of two nilpotent groups*, Rend. Sem. Mat. Univ. Padova 115 (2006), pp. 273–279.

[31] A. Ballester-Bolinches, J. C. Beidleman, and R. Esteban-Romero, *On some classes of supersoluble groups*, J. Algebra 312 (2007), pp. 445–454.

[32] A. Ballester-Bolinches, J. C. Beidleman, and H. Heineken, *Groups in which Sylow subgroups and subnormal subgroups permute*, Illinois J. Math 47 (2003), pp. 63–69.

[33] A. Ballester-Bolinches, J. C. Beidleman, and H. Heineken, *A local approach to certain classes of finite groups*, Comm. Algebra 31 (2003), pp. 5931–5942.

[34] A. Ballester-Bolinches, J. C. Beidleman, H. Heineken, and M. C. Pedraza-Aguilera, *Prefactorized subgroups in pairwise mutually permutable products*, preprint.

[35] A. Ballester-Bolinches, J. C. Beidleman, H. Heineken, and M. C. Pedraza-Aguilera, *On pairwise mutually permutable products*, Forum Math. 21 (2009), pp. 1081–1090.

[36] A. Ballester-Bolinches, J. C. Beidleman, H. Heineken, and M. C. Pedraza-Aguilera, *Local classes and pairwise mutually permutable products of finite groups*, Doc. Math. 15 (2010), pp. 255–265.

[37] A. Ballester-Bolinches and J. Cossey, *Totally permutable products of finite groups satisfying SC or PST*, Monatsh. Math. 145 (2005), pp. 89–94.

[38] A. Ballester-Bolinches, J. Cossey, and R. Esteban-Romero, *On finite groups generated by strongly cosubnormal subgroups*, J. Algebra 259 (2003), pp. 226–234.

[39] A. Ballester-Bolinches, J. Cossey, and R. Esteban-Romero, *On totally permutable products of finite groups*, J. Algebra 293 (2005), pp. 269–278.

[40] A. Ballester-Bolinches, J. Cossey, and M. C. Pedraza-Aguilera, *On products of finite supersoluble groups*, Comm. Algebra 29 (2001), pp. 3145–3152.

[41] A. Ballester-Bolinches, J. Cossey, and M. C. Pedraza-Aguilera, *On mutually permutable products of finite groups*, J. Algebra 294 (2005), pp. 127–135.

[42] A. Ballester-Bolinches, J. Cossey, and X. Soler-Escrivà, *On a permutability property of subgroups of finite soluble groups*, Commun. Contemp. Math. 12 (2010), pp. 207–221.

[43] A. Ballester-Bolinches and R. Esteban-Romero, *Sylow permutable subnormal subgroups of finite groups II*, Bull. Austral. Math. Soc. 64 (2001), pp. 479–486.

[44] A. Ballester-Bolinches and R. Esteban-Romero, *Sylow permutable subnormal subgroups of finite groups*, J. Algebra 251 (2002), pp. 727–738.

[45] A. Ballester-Bolinches and R. Esteban-Romero, *On finite* $\mathcal{T}$*-groups*, J. Austral. Math. Soc. Ser. A 75 (2003), pp. 1–11.

[46] A. Ballester-Bolinches and R. Esteban-Romero, *On finite soluble groups in which Sylow permutability is a transitive relation*, Acta Math. Hungar. 101 (2003), pp. 193–202.

[47] A. Ballester-Bolinches and L. M. Ezquerro, *Classes of Finite Groups*, Mathematics and its Applications, vol. 584, Springer, New York, 2006.

[48] A. Ballester-Bolinches, X. Guo, and M. C. Pedraza-Aguilera, *A note on m-permutable products of finite groups*, J. Group Theory 3 (2000), pp. 381–384.

[49] A. Ballester-Bolinches, L. A. Kurdachenko, J. Otal, and T. Pedraza, *Infinite groups with many permutable subgroups*, Rev. Mat. Iberoam. 24 (2008), pp. 745–764.

[50] A. Ballester-Bolinches, L. A. Kurdachenko, J. Otal, and T. Pedraza, *Infinite groups with Sylow permutable subgroups*, Annali Mat. Pura Appl. 189 (2010), pp. 553–565.

[51] A. Ballester-Bolinches and M. C. Pedraza-Aguilera, *Mutually permutable products of finite groups, II*, J. Algebra 218 (1999), pp. 563–572.

[52] A. Ballester-Bolinches and M. C. Pedraza-Aguilera, *On finite soluble products of $\mathcal{N}$-connected groups*, J. Group Theory 2 (1999), pp. 291–299.

[53] A. Ballester-Bolinches, M. C. Pedraza-Aguilera, and M. D. Pérez-Ramos, *On finite products of totally permutable groups*, Bull. Austral. Math. Soc. 53 (1996), pp. 441–445.

[54] A. Ballester-Bolinches, M. C. Pedraza-Aguilera, and M. D. Pérez-Ramos, *Finite groups which are products of pairwise totally permutable products*, Proc. Edinburgh Math. Soc. (2) 41 (1998), pp. 567–572.

[55] A. Ballester-Bolinches and M. D. Pérez-Ramos, *A question of R. Maier concerning formations*, J. Algebra 182 (1996), pp. 738–747.

[56] J. C. Beidleman, B. Brewster, and D. J. S. Robinson, *Criteria for Permutability to Be Transitive in Finite Groups*, J. Algebra 222 (1999), pp. 400–412.

[57] J. C. Beidleman, A. Galoppo, H. Heineken, and M. Manfredino, *On certain products of soluble groups*, Forum Math. 13 (2001), pp. 569–580.

[58] J. C. Beidleman, P. Hauck, and H. Heineken, *Totally permutable products of certain classes of finite groups*, J. Algebra 276 (2004), pp. 826–835.

[59] J. C. Beidleman and H. Heineken, *Totally permutable torsion subgroups*, J. Group Theory 2 (1999), pp. 377–392.

[60] J. C. Beidleman and H. Heineken, *A note on I-automorphisms*, J. Algebra 234 (2000), pp. 694–706.

[61] J. C. Beidleman and H. Heineken, *On the hyperquasicenter of a group*, J. Group Theory 4 (2001), pp. 199–206.

[62] J. C. Beidleman and H. Heineken, *Finite soluble groups whose subnormal subgroups permute with certain classes of subgroups*, J. Group Theory 6 (2003), pp. 139–158.

[63] J. C. Beidleman and H. Heineken, *On the Fitting core of a formation*, Bull. Austral. Math. Soc. 68 (2003), pp. 107–112.

[64] J. C. Beidleman and H. Heineken, *Pairwise $\mathcal{N}$-connected products of certain classes of finite groups*, Comm. Algebra 32 (2004), pp. 4741–4752.

[65] J. C. Beidleman and H. Heineken, *Mutually permutable subgroups and group classes*, Arch. Math. (Basel) 85 (2005), pp. 18–30.

[66] J. C. Beidleman and H. Heineken, *Group classes and mutually permutable products*, J. Algebra 297 (2006), pp. 409–416.

[67] J. C. Beidleman, H. Heineken, and M. F. Ragland, *Strong Sylow bases and mutually permutable products*, J. Algebra 321 (2009), pp. 2022–2027.

[68] J. C. Beidleman and M. F. Ragland, *The intersection map of subgroups and certain classes of finite groups*, Ric. Mat. 10 (2007), pp. 205–210.

[69] T. R. Berger and F. Gross, *2-length and the derived length of a Sylow 2-subgroup*, Proc. London Math. Soc. (3) 34 (1977), pp. 520–534.

[70] J. Berkovich, *Groups of prime power order. Volume 1*, De Gruyter Expositions in Mathematics, vol. 46, Walter de Gruyter, Berlin, New York, 2008.

[71] E. Best and O. Taussky, *A class of groups*, Proc. Roy. Irish Acad. Sect. A 47 (1942), pp. 55–62.

[72] M. Bianchi, A. Gillio Berta Mauri, M. Herzog, and L. Verardi, *On finite solvable groups in which normality is a transitive relation*, J. Group Theory 3 (2000), pp. 147–156.

[73] J. Bochtler, *Radicals in mutually permutable products of finite groups*, J. Algebra 321 (2009), pp. 353–360.

[74] J. Bochtler, *Wechselseitig vertauschbare Produkte endlicher Gruppen*, Ph.D. thesis, Fakultät für Mathematik und Physik, Eberhard-Karls-Universität Tübingen, Tübingen, 2010.

[75] J. Bochtler and P. Hauck, *Mutually permutable products and Fitting classes*, Arch. Math. (Basel) 88 (2007), pp. 385–388.

[76] R. H. Bradway, F. Gross, and W. R. Scott, *The nilpotence class of core-free quasinormal subgroups*, Rocky Mountain J. Math. 1 (1971), pp. 375–382.

[77] R. Brandl, *Zur Theorie der untergruppenabgeschlossenen Formationen: Endlichen Varietäten*, J. Algebra 73 (1981), pp. 1–22.

[78] R. A. Bryce and J. Cossey, *The Wielandt subgroup of a finite soluble group*, J. London Math. Soc. 40 (1989), pp. 244–256.

[79] E. G. Bryukhanova, *The 2-length and 2-period of a finite solvable group*, Algebra Logic 18 (1979), pp. 5–20.

[80] E. G. Bryukhanova, *Connection between the 2-length and the derived length of a Sylow 2-subgroup of a finite solvable group*, Mat. Notes 29 (1981), pp. 85–90.

[81] G. Busetto and E. Jabara, *Some observations on factorized groups*, JP J. Algebra Number Theory Appl. 6 (2006), pp. 189–201.

[82] A. Carocca, *Products of mutually f-permutable subgroups in a finite group*, preprint.

[83] A. Carocca, *p-supersolvability of factorized finite groups*, Hokkaido J. Math. 21 (1992), pp. 395–403.

[84] A. Carocca, *Produtos de subgrupos mutuamente f-permutáveis em grupos finitos*, Tese de doutorado, Univ. de Brasília, Brasília, 1992.

[85] A. Carocca, *A note on the product of $\mathfrak{F}$-subgroups in a finite group*, Proc. Edinburgh Math. Soc. 39 (1996), pp. 37–42.

[86] A. Carocca, *On factorized finite groups in which certain subgroups of the factors permute*, Arch. Math. (Basel) 71 (1998), pp. 257–261.

[87] A. Carocca, *Solvability of factorized finite groups*, Glasgow Math. J. 42 (2000), pp. 271–274.

[88] A. Carocca and R. Maier, *Hypercentral embedding and pronormality*, Arch. Math. (Basel) 71 (1998), pp. 433–436.

[89] A. Carocca and R. Maier, *Theorems of Kegel-Wieland type*, Groups St Andrews 1997 in Bath, I (Cambridge) (C. M. Campbell, E. F. Robertson, and G. Smith, eds.), London Math. Soc. Lecture Notes Ser., vol. 260, Cambridge University Press, 1999, pp. 195–201.

[90] R. W. Carter, *Simple Groups of Lie Type*, second. ed., John Wiley & Sons, London, New York, Sydney, 1989.

[91] G. Casadio, *Costruzione di gruppi come prodotto di sottogruppi permutabili*, Rend. Mat., Univ. Roma, V V (1941), pp. 348–360.

[92] C. Casolo, *Subnormality in factorizable finite soluble groups*, Arch. Math. (Basel) 57 (1991), pp. 12–13.

[93] C. Casolo and U. Dardano, *Subnormality in the join of two subgroups*, J. Group Theory 7 (2004), pp. 507–520.

[94] M. R. Celentani, A. Leone, and D. J. S. Robinson, *The Maier-Schmid problem for infinite groups*, J. Algebra 301 (2006), pp. 294–307.

[95] N. S. Chernikov, *Groups that decompose into a product of permutable subgroups*, Naukova Dumka, Kiev, 1987, Russian.

[96] N. S. Chernikov, *On finite solvable groups decomposable into the product of two nilpotent groups*, Ukr. Math. J. 52 (2000), pp. 926–938.

[97] N. S. Chernikov, *On groups factorized by two subgroups with Chernikov commutants*, Ukr. Math. J. 52 (2000), pp. 457–463.

[98] N. S. Chernikov, *Factorization of periodic locally solvable groups by locally nilpotent and nilpotent subgroups*, Ukr. Math. J. 53 (2001), pp. 2072–2097.

[99] N. S. Chernikov, *Properties of a finite group representable as the product of two nilpotent groups*, Ukr. Math. J. 53 (2001), pp. 606–619.

[100] N. S. Chernikov, *Groups which are factorized by subgroups of finite exponents*, Acta Appl. Math. 85 (2005), pp. 81–92.

[101] N. S. Chernikov and A. P. Petravchuk, *Characterization of a periodic locally soluble group by its soluble Sylow π-subgroups of finite exponent*, Ukrain. Math. J. 39 (1987), pp. 619–624.

[102] M. Conder and I. M. Isaacs, *Derived subgroups of products of an abelian and a cyclic subgroup*, J. London Math. Soc. (2) 69 (2004), pp. 333–348.

[103] J. H. Conway, R. T. Curtis, S. P. Norton, R. A. Parker, and R. A. Wilson, *Atlas of Finite Groups*, Oxford Univ. Press, London, 1985.

[104] C. D. H. Cooper, *Power automorphisms of a group*, Math. Z. 107 (1968), pp. 335–356.

[105] J. Cossey, *Finite insoluble T-groups*, preprint.

[106] J. Cossey and L. M. Ezquerro, *Totally permutable products of finite groups*, preprint.

[107] J. Cossey and S. E. Stonehewer, *On the derived length of finite dinilpotent groups*, Bull. Lond. Math. Soc. 30 (1998), pp. 247–250.

[108] J. Cossey and S. E. Stonehewer, *Products of finite nilpotent groups*, Comm. Algebra 27 (1999), pp. 289–300.

[109] J. Cossey and S. E. Stonehewer, *Cyclic permutable subgroups of finite groups*, J. Austral. Math. Soc. 71 (2001), pp. 169–176.

[110] J. Cossey and S. E. Stonehewer, *The embedding of a cyclic permutable subgroup in a finite group*, Illinois J. Math. 47 (2003), pp. 89–111.

[111] J. Cossey and S. E. Stonehewer, *The embedding of a cyclic permutable subgroup in a finite group. II*, Proc. Edinburgh Math. Soc. 47 (2004), pp. 101–109.

[112] J. Cossey and Y. Wang, *Finite dinilpotent groups of small derived length*, J. Austral. Math. Soc. Ser. A 67 (1999), pp. 318–328.

[113] M. de Falco, F. de Giovanni, and C. Musella, *Locally finite products of totally permutable nilpotent groups*, Algebra Colloq. 16 (2009), pp. 535–540.

[114] F. de Giovanni, *Prodotti di gruppi nilpotenti*, Milan J. Math. 61 (1991), pp. 257–265.

[115] R. Dedekind, *Über Gruppen, deren sámtliche Teiler Normalteiler sind*, Math. Ann. 48 (1897), pp. 548–561.

[116] W. E. Deskins, *On quasinormal subgroups of finite groups*, Math. Z. 82 (1963), pp. 125–132.

[117] E. Detomi, *On some products of nilpotent groups*, Comm. Algebra 26 (1998), pp. 3239–3247.

[118] M. R. Dixon and S. E. Stonehewer, *The join of disjoint permutable subgroups*, Comm. Algebra 30 (2002), pp. 2967–2973.

[119] K. Doerk and T. Hawkes, *Finite Soluble Groups*, De Gruyter Expositions in Mathematics, vol. 4, Walter de Gruyter, Berlin, New York, 1992.

[120] K. Doerk and T. O. Hawkes, *On the residual of a direct product*, Arch. Math. 30 (1978), pp. 458–468.

[121] J. Douglas, *On the solvability of bicyclic groups*, Proc. Natl. Acad. Sci. USA 47 (1961), pp. 1493–1495.

[122] L. M. Ezquerro and X. Soler-Escrivà, *On mutually M-permutable products of finite groups*, Comm. Algebra 31 (2003), pp. 1949–1960.

[123] L. M. Ezquerro and X. Soler-Escrivà, *Some permutability properties related to $\mathfrak{F}$-hypercentrally embedded subgroups of finite groups*, J. Algebra 264 (2003), pp. 279–295.

[124] L. M. Ezquerro and X. Soler-Escrivà, *Some new permutability properties of hypercentrally embedded groups*, J. Austral. Math. Soc. Ser. A 79 (2005), pp. 243–255.

[125] E. Fisman, *Products of solvable products of finite groups*, Arch. Math. (Basel) 61 (1993), pp. 201–205.

[126] P. Flavell, *Finite groups in which every two elements generate a soluble subgroup*, Invent. Math. 121 (1995), pp. 279–285.

[127] T. Foguel, *Conjugate-permutable subgroups*, J. Algebra 191 (1997), pp. 235–239.

[128] T. Foguel, *Groups with all cyclic subgroups conjugate-permutable groups*, J. Group Theory 2 (1999), pp. 47–51.

[129] S. Franciosi, F. de Giovanni, H. Heineken, and M. L. Newell, *On the Fitting length of a soluble product of nilpotent groups*, Arch. Math. (Basel) 57 (1991), pp. 313–318.

[130] S. Franciosi, F. de Giovanni, and Y. P. Sysak, *An extension of the Kegel-Wielandt theorem to locally finite groups*, Glasgow Math. J. 38 (1996), pp. 171–176.

[131] S. Franciosi, F. de Giovanni, and Y. P. Sysak, *On locally finite groups factorized by locally nilpotent subgroups*, J. Pure Appl. Algebra 106 (1996), pp. 45–56.

[132] S. Franciosi, F. de Giovanni, and Y. P. Sysak, *On ascending and subnormal subgroups of infinite factorizable groups*, Ukr. Math. J. 49 (1997), pp. 943–949.

[133] S. Franciosi, F. de Giovanni, and Y. P. Sysak, *On subnormal subgroups of factorized groups*, J. Algebra 198 (1997), pp. 469–480.

[134] P. Gállego, P. Hauck, and M. D. Pérez-Ramos, *On 2-generated subgroups and products of groups*, J. Group Theory 11 (2008), pp. 851–867.

[135] P. Gállego, P. Hauck, and M. D. Pérez-Ramos, *Soluble products of connected subgroups*, Rev. Mat. Iberoam. 24 (2008), pp. 433–461.

[136] The GAP Group, *GAP – Groups, Algorithms, and Programming, Version 4.4.12*, 2008.

[137] W. Gaschütz, *Gruppen, in denen das Normalteilersein transitiv ist*, J. Reine Angew. Math. 198 (1957), pp. 87–92.

[138] J. D. Gillam, *A finite p-group $P = AB$ with Core(A) = Core(B) = $\{1\}$*, Rocky Mountain J. Math. 3 (1973), pp. 15–17.

[139] F. Gross, *Finite groups which are the product of two nilpotent subgroups*, Bull. Austral. Math. Soc. 9 (1973), pp. 267–274.

[140] F. Gross and T. R. Berger, *A universal example of a core-free permutable subgroup*, Rocky Mountain J. Math. 12 (1982), pp. 345–365.

[141] O. Grün, *Beiträge zur Gruppentheorie. I*, J. Reine Angew. Math. 174 (1936), pp. 1–14.

[142] P. Hall, *Theorems like Sylow's*, Proc. London Math. Soc, III. Ser. 6 (1956), pp. 286–304.

[143] P. Hall and G. Higman, *On the p-length of p-soluble groups and reduction theorems for Burnside's problem*, Proc. London Math. Soc. III Ser. 6 (1956), pp. 1–42.

[144] P. Hauck, A. Martínez-Pastor, and M. D. Pérez-Ramos, *Fitting classes and products of totally permutable groups*, J. Algebra 252 (2002), pp. 114–126.

[145] P. Hauck, A. Martínez-Pastor, and M. D. Pérez-Ramos, *Injectors and radicals in products of totally permutable groups*, Comm. Algebra 31 (2003), pp. 6135–6147.

[146] P. Hauck, A. Martínez-Pastor, and M. D. Pérez-Ramos, *Products of $\mathcal{N}$-connected groups*, Illinois J. Math. 47 (2003), pp. 1033–1045.

[147] P. Hauck, A. Martínez-Pastor, and M. D. Pérez-Ramos, *Products of pairwise totally permutable groups*, Proc. Edinburgh Math. Soc. 46 (2003), pp. 147–157.

[148] H. Heineken, *Produkte abelscher Gruppen und ihre Fittinggruppe*, Arch. Math. (Basel) 48 (1987), pp. 185–192.

[149] H. Heineken, *Products of finite nilpotent groups*, Math. Ann. 287 (1990), pp. 643–652.

[150] H. Heineken, *The product of two finite nilpotent group and its Fitting series*, Arch. Math. (Basel) 59 (1992), pp. 209–214.

[151] H. Heineken and J. C. Lennox, *A note on products of abelian groups*, Arch. Math. (Basel) 41 (1983), pp. 498–501.

[152] B. Höfling, *Produkte endlicher nilpotente Gruppen*, Diplomarbeit, Johannes Gutenberg-Universität Mainz, Mainz, 1993.

[153] T. W. Hungerford, *Algebra*, Graduate Texts in Mathematics, vol. 73, Springer-Verlag, New York, 2003, Reprint of the 1974 original.

[154] B. Huppert, *Über das Produkt von paarweise vertauschbaren zyklischen Gruppen*, Math. Z. 58 (1953), pp. 243–264.

[155] B. Huppert, *Über die Auflösbarkeit faktorisierbarer Gruppen. III*, Math. Z. 64 (1956), pp. 138–148.

[156] B. Huppert, *Zur Sylowstruktur auflösbarer Gruppen*, Arch. Math. 12 (1961), pp. 161–169.

[157] B. Huppert, *Endliche Gruppen I*, Grund. Math. Wiss., vol. 134, Springer Verlag, Berlin, Heidelberg, New York, 1967.

[158] B. Huppert and N. Blackburn, *Finite Groups II*, Grundlehren der Mathematischen Wissenschaften, vol. 242, Springer-Verlag, Berlin-Heidelberg-New York, 1982.

[159] B. Huppert and N. Blackburn, *Finite groups III*, Grundlehren der Mathematischen Wissenschaften [Fundamental Principles of Mathematical Sciences], vol. 243, Springer-Verlag, Berlin, 1982.

[160] B. Huppert and N. Itô, *Über die Auflösbarkeit faktorisierbarer Gruppen. II*, Math. Z. 61 (1954), pp. 94–99.

[161] N. Itô, *Remarks on factorisable groups*, Acta Sci. Math. (Szeged) 14 (1951), pp. 83–84.

[162] N. Itô, *Über das Produkt von zwei abelschen Gruppen*, Math. Z. 62 (1955), pp. 400–401.

[163] N. Itô, *Über das Produkt von zwei zyklischen 2-Gruppen*, Publ. Math. Debrecen 4 (1956), pp. 517–520.

[164] N. Itô and J. Szép, *Über die Quasinormalteiler von endlichen Gruppen*, Acta Sci. Math. (Szeged) 23 (1962), pp. 168–170.

[165] K. Iwasawa, *Über die endlichen Gruppen und die Verbände ihrer Untergruppen*, J. Fac. Sci. Imp. Univ. Tokyo Sect. I 4 (1941), pp. 171–199.

[166] E. Jabara, *A note on a class of factorized p-groups*, Czechoslovak Math. J. 55 (2005), pp. 993–996.

[167] P. M. Johnson, *A property of factorizable groups*, Arch. Math. (Basel) 60 (1993), pp. 414–419.

[168] L. S. Kazarin, *On the product of two groups that are close to being nilpotent*, Mat. Sb. 110 (1979), pp. 51–65.

[169] L. S. Kazarin, *On products of two nilpotent groups*, Vopr. teorii grupp i gomol. algebry (1981), pp. 62–66.

[170] L. S. Kazarin, *On products of two nilpotent groups*, Vopr. teorii grupp i gomol. algebry (1982), pp. 47–49.

[171] L. S. Kazarin, *On the product of finite groups*, Sov. Math. Dokl. 27 (1983), pp. 354–357, Translation from *Dokl. Akad. Nauk. SSSR*, 269, 528–531 (1983).

[172] L. S. Kazarin, *On the product of a 2-group and a group of odd order*, Vopr. Teor. Grupp Gomologicheskoj Algebry 1985 (1985), pp. 85–92, Russian.

[173] L. S. Kazarin, *On groups which are the product of two solvable subgroups*, Comm. Algebra 14 (1986), pp. 1001–1066.

[174] L. S. Kazarin, *On Burnside's* p^{α}*-lemma*, Mat. Zametki 44 (1990), pp. 749–751.

[175] L. S. Kazarin, *The product of a group with a cyclic Sylow* p*-subgroup and a group with a nontrivial center*, Ukr. Math. J. 44 (1992), pp. 697–702.

[176] L. S. Kazarin, *Soluble products of groups*, Infinite groups 1994. Proceedings of the International Conference held in Ravello, May 23–27, 1994 (Berlin) (F. de Giovanni and M. L. Newel, eds.), Walter de Gruyter, 1996, pp. 111–123.

[177] L. S. Kazarin, A. Martínez-Pastor, and M. D. Pérez-Ramos, *The product of two* π*-decomposable groups*, preprint.

[178] L. S. Kazarin, A. Martínez-Pastor, and M. D. Pérez-Ramos, *On the product of a* π*-decomposable group and a* π*-group*, J. Algebra 315 (2007), pp. 640–653.

[179] L. S. Kazarin, A. Martínez-Pastor, and M. D. Pérez-Ramos, *On the product of two* π*-decomposable soluble groups*, Publ. Mat. 53 (2009), pp. 439–456.

[180] O. H. Kegel, *Produkte nilpotenter Gruppen*, Arch. Math. (Basel) 12 (1961), pp. 90–93.

[181] O. H. Kegel, *Sylow-Gruppen und Subnormalteiler endlicher Gruppen*, Math. Z. 78 (1962), pp. 205–221.

[182] O. H. Kegel, *On the solvability of some factorised linear groups*, Illinois J. Math. 9 (1965), pp. 535–547.

[183] O. H. Kegel, *Zur Struktur mehrfach faktorisierter endlicher Gruppen*, Math. Z. 87 (1965), pp. 42–48.

[184] W. Knapp, *Cosubnormality and the Hypercenter*, J. Algebra 234 (2000), pp. 609–619.

[185] L. E. Knop, *Normal subgroups of groups which are products of two abelian subgroups*, Proc. Amer. Math. Soc. 40 (1973), pp. 37–41.

[186] V. N. Knyagina and V. S. Monakhov, *On the* p*-length of the product of two Schmidt groups*, Sib. Math. Zh. 45 (2004), pp. 329–333, Russian. Translation in *Sib. Math. J.*, 45 (2), 269–2772 (2004).

[187] L. A. Kurdachenko and I. Y. Subbotin, *Transitivity of normality and pronormal subgroups*, Combinatorial group theory, discrete groups, and number theory. A conference in honor of Gerhard Rosenberger's sixtieth birthday, Fairfield, CT, USA, December 8–9, 2004 and the AMS special session on infinite groups, Bard College, Annandale-on-Hudson, NY, USA, October 8–9, 2005 (Providence, RI) (B. Fine et al., eds.), Contemporary Mathematics, vol. 421, American Mathematical Society (AMS), 2006, pp. 201–212.

[188] J. C. Lennox, *Bigenetic properties of finitely generated hyper-(abelian-by-finite) groups*, J. Austral. Math. Soc. Ser. A 16 (1973), pp. 309–315.

[189] J. C. Lennox and S. E. Stonehewer, *Subnormal Subgroups of Groups*, Clarendon Press, Oxford, 1987.

[190] P. Longobardi, *Gruppi finite a fattoriali modulari*, Note Mat. II (1982), pp. 73–100.

[191] R. Maier, *Endliche metanilpotente Gruppen*, Arch. Math. (Basel) 23 (1972), pp. 139–144.

[192] R. Maier, *Um problema da teoria dos subgrupos subnormais*, Bol. Soc. Brasil. Math. 8 (1977), pp. 127–130.

[193] R. Maier, *Zur Vertauschbarkeit und Subnormalität von Untergruppen*, Arch. Math. (Basel) 53 (1989), pp. 110–120.

[194] R. Maier, *A completeness property of certain formations*, Bull. London Math. Soc. 24 (1992), pp. 540–544.

[195] R. Maier and P. Schmid, *The Embedding of Quasinormal Subgroups in Finite Groups*, Math. Z. 131 (1973), pp. 269–272.

[196] R. Maier and S. Sidki, *A note on subnormality in factorizable finite groups*, Arch. Math. (Basel) 42 (1984), pp. 97–101.

[197] A. Mann, *The derived length of p-groups*, J. Algebra 224 (2000), pp. 263–267.

[198] B. McCann, *On p-groups which are the product of a finite extra-special group and a nilpotent group of class two*, J. Group Theory 2 (1999), pp. 301–305.

[199] V. S. Monakhov, *The product of two groups, one of which contains a cyclic subgroup of index* ≤ 2, Math. Notes 16 (1974), pp. 757–762 (English. Russian original).

[200] V. S. Monakhov, *On the product of two groups with cyclic subgroups of index* 2, Izv. Akad. Nauk Belarusi, Ser. Fiz.-Mat. Nauk (1996), pp. 21–24 (Russian. English summary).

[201] M. Morigi, *A note on factorized (finite) p-groups*, Rend. Sem. Mat. Univ. Padova 98 (1997), pp. 101–105.

[202] K. H. Müller, *Schwachnormale Untergruppen: Eine gemeinsame Verallgemeinerung der normalen und normalisatorgleichen Untergruppen*, Rend. Sem. Mat. Univ. Padova 36 (1966), pp. 129–157.

[203] V. I. Mysovskikh, *Investigation of subgroup embeddings by the computer algebra package GAP*, Computer algebra in scientific computing—CASC'99 (Munich) (Berlin), Springer, 1999, pp. 309–315.

[204] V. I. Mysovskikh, *Subnormalizers and embedding properties of subgroups of finite groups*, Zapiski Nauchnikh Seminarov POMI 265 (1999), pp. 258–280.

[205] F. Napolitani, *Modularità e distributività nell'insieme dei sottogruppi subnormali*, Rend. Sem. mat. Univ. Padova 43 (1970), pp. 215–220.

[206] H. Neumann, *Varieties of Groups*, Springer Verlag, New York, 1967.

[207] O. Ore, *Contributions to the theory of groups of finite order*, Duke Math. J. 5 (1939), pp. 431–460.

[208] T. A. Peng, *Finite groups with pronormal subgroups*, Proc. Amer. Math. Soc. 20 (1969), pp. 232–234.

[209] T. A. Peng, *Pronormality in finite groups*, J. London Math. Soc. (2) 3 (1971), pp. 301–306.

[210] E. A. Pennington, *On products of nilpotent groups*, Math. Z. 134 (1973), pp. 81–83.

[211] L. Rédei, *Zur Theorie der faktorisierbaren Gruppen I*, Acta Math. Acad. Sci. Hungar. 1 (1950), pp. 74–98.

[212] D. J. S. Robinson, *Groups in which normality is a transitive relation*, Proc. Camb. Phil. Soc. 60 (1964), pp. 21–38.

[213] D. J. S. Robinson, *A note on finite groups in which normality is transitive*, Proc. Amer. Math. Soc. 19 (1968), pp. 933–937.

[214] D. J. S. Robinson, *A Course in the Theory of Groups*, second. ed., Springer-Verlag, New York, 1995.

[215] D. J. S. Robinson, *Permutability properties of subgroups*, Groups St Andrews 1997 in Bath, II (Cambridge) (C. M. Campbell, E. F. Robertson, and G. Smith, eds.), London Math. Soc. Lecture Notes Ser., vol. 261, Cambridge University Press, 1999, pp. 633–638.

[216] D. J. S. Robinson, *A survey of groups in which normality or permutability is a transitive relation*, Algebra. Some recent advances (Basel) (I. B. S. Passi, ed.), Trends in Mathematics, Birkäuser, 1999, pp. 171–189.

[217] D. J. S. Robinson, *The structure of finite groups in which permutability is a transitive relation*, J. Aust. Math. Soc. 70 (2001), pp. 143–159.

[218] D. J. S. Robinson, *Minimality and Sylow-permutability in locally finite groups*, Ukr. Math. J. 54 (2002), pp. 1038–1049.

[219] D. J. S. Robinson, *Finite groups whose cyclic subnormal subgroups are permutable*, Algebra Colloq. 12 (2005), pp. 171–180.

[220] J. S. Rose, *A Course on Group Theory*, Cambridge University Press, London, New York, Melbourne, 1978.

[221] P. Schmid, *Subgroups Permutable with All Sylow Subgroups*, J. Algebra 207 (1998), pp. 285–293.

[222] R. Schmidt, *Subgroup lattices of groups*, De Gruyter Expositions in Mathematics, vol. 14, Walter de Gruyter, Berlin, 1994.

[223] W. R. Scott, *Group theory*, Dover Publ., Mineola, NY, USA, 1987.

[224] V. G. Sementovskiĭ, *Pronormal subgroups of finite groups*, Vescī Akad. Navuk BSSR Ser. Fīz.-Mat. Navuk 135 (1966), pp. 12–16.

[225] V. I. Senashov, A. I. Sozutov, and V. P. Shunkov, *Investigation of groups with finiteness conditions in Krasnoyarsk*, Russ. Math. Surv. 60 (2005), pp. 805–848.

[226] E. I. Shatylo, *On groups of type CS*, Ukr. Math. J. 23 (1971), pp. 527–535.

[227] Z. Shen, W. Liu, and X. Kong, *Finite groups with self-conjugate-permutable subgroups*, Comm. Algebra 38 (2010), pp. 1715–1724.

[228] V. P. Shunkov, *On groups decomposable into a uniform product of their p-subgroups*, Sov. Math. Dokl. 5 (1964), pp. 147–149.

[229] S. E. Stonehewer, *Permutable subgroups of infinite groups*, Math. Z. 125 (1972), pp. 1–16.

[230] S. E. Stonehewer, *Permutable subgroups of some finite p-groups*, J. Austral. Math. Soc. Ser. A 16 (1973), pp. 90–97.

[231] S. E. Stonehewer, *On finite dinilpotent groups*, J. Pure Appl. Algebra 88 (1993), pp. 239–244.

[232] I. Y. Subbotin, *The hypercentral coradical of a KI-group*, Ukr. Math. J. 34 (1982), pp. 650–654.

[233] M. Suzuki, *Group theory I*, Grundlehren der Mathematischen Wischenschaften, vol. 247, Springer-Verlag, Berlin-Heidelberg-New York, 1982.

[234] J. Szép, *Über die als Produkt zweier Untergruppen darstellbaren endlichen Gruppen*, Comment. Math. Helv. 22 (1949), pp. 31–33.

[235] J. Szép, *On factorisable groups*, Acta Sci. Math. (Szeged) 13 (1950), pp. 235–238.

[236] J. Szép, *On factorisable, not simple groups*, Acta Sci. Math. (Szeged) 13 (1950), pp. 239–241.

[237] J. Szép, *On the structure of groups which can be represented as the product of two subgroups*, Acta Sci. Math. (Szeged) 12A (1950), pp. 57–61.

[238] J. Szép, *On factorisable simple groups*, Acta Sci. Math. (Szeged) 14 (1951), p. 22.

[239] J. Szép, *Zur Theorie der endlichen einfachen Gruppen*, Acta Sci. Math. (Szeged) 14 (1951), pp. 111–112.

[240] J. G. Thompson, *An example of core-free quasinormal subgroups of p-groups*, Math. Z. 96 (1967), pp. 226–226.

[241] J. G. Thompson, *Nonsolvable finite groups all of whose local subgroups are solvable*, Bull. Amer. Math. Soc. 74 (1968), pp. 383–437.

[242] J. G. Thompson, *Nonsolvable finite groups all of whose local subgroups are solvable. II*, Pac. J. Math. 33 (1970), pp. 451–536.

[243] J. G. Thompson, *Nonsolvable finite groups all of whose local subgroups are solvable. III*, Pac. J. Math. 39 (1971), pp. 483–534.

[244] J. G. Thompson, *Nonsolvable finite groups all of whose local subgroups are solvable. IV*, Pac. J. Math. 48 (1973), pp. 511–592.

[245] J. G. Thompson, *Nonsolvable finite groups all of whose local subgroups are solvable. V*, Pac. J. Math. 50 (1974), pp. 215–297.

[246] J. G. Thompson, *Nonsolvable finite groups all of whose local subgroups are solvable. VI*, Pac. J. Math. 51 (1974), pp. 573–630.

[247] V. V. Tsybulenko, *On the construction of finite groups with systems of commutative primary subgroups*, Ukr. Math. J. 25 (1973), pp. 701–705.

[248] V. N. Tyutyanov, *Products of π-nilpotent subgroups*, Mat. Sb. 187 (1996), pp. 97–102.

[249] R. W. van der Waall and A. Fransman, *On products of groups for which normality is a transitive relation on their Frattini factor groups*, Quaestiones Math. 19 (1996), pp. 59–82.

[250] V. G. Vasil'ev, *Strong uniform products of cyclic p-groups*, Algebra Logic 16 (1977), pp. 335–340.

[251] V. G. Vasil'ev, *Solvability length of uniform products of cyclic groups*, Algebra Logic 17 (1978), pp. 180–184.

[252] P. Wang, *A problem on products of two supersolvable subgroups of a finite group*, J. Math. Res. Exposition 13 (1993), pp. 279–282, Chinese.

[253] M. Weinstein (ed.), *Between Nilpotent and Solvable*, Polygonal Publishing House, Passaic, NJ, USA, 1982.

[254] H. Wielandt, *Über das Produkt von paarweise vertauschbaren nilpotenten Gruppen*, Math. Z. 55 (1951), pp. 1–7.

[255] H. Wielandt, *Über die Existenz von Normalteilern in endlichen Gruppen*, Math. Nachr. 18 (1958), pp. 274–280.

[256] H. Wielandt, *Über Produkte von nilpotenten Gruppen*, Illinois J. Math. 2 (1958), pp. 611–618.

[257] H. Wielandt, *Vertauschbarkeit von Untergruppen und Subnormalität*, Math. Z. 133 (1973), pp. 275–276.

[258] H. Wielandt, *Kriterien für Subnormalität in endlichen Gruppen*, Math. Z. 138 (1974), pp. 199–203.

[259] H. Wielandt, *Über das Erzeugnis paarweise kosubnormaler Untergruppen*, Arch. Math. (Basel) 35 (1980), pp. 1–7.

[260] J. S. Wilson, *A note on products of abelian by finite groups*, Arch. Math. (Basel) 54 (1990), pp. 117–118.

[261] R. Wilson, P. Walsh, J. Tripp, I. Suleiman, R. Parker, S. Norton, S. Nickerson, S. Linton, J. Bray, and R. Abbott, *Atlas of Finite Group Representations - Version 3: Suzuki group* Sz(8), `http://brauer.maths.qmul.ac.uk/Atlas/v3/exc/Sz8/`, visited 13th November, 2009.

[262] T. Yoshida, *Character-theoretic transfer*, J. Algebra 52 (1978), pp. 1–38.

[263] G. Zacher, *Caratterizzazione dei t-gruppi risolubili*, Ricerche Mat. 1 (1952), pp. 287–294.

[264] G. Zacher, *I gruppi risolubili finiti in cui i sottogruppi di composizione coincidono con i sottogrupi quasi-normali*, Atti Accad. Naz. Lincei Rend. cl. Sci. Fis. Mat. Natur. (8) 37 (1964), pp. 150–154.

[265] G. Zappa, *Sulla costruzione dei gruppi prodotto di due dati sottogrupi permutabilit tra loro*, Atti Secondo Congreso Un. Mat. Ital., Bologna, 1940 (Perella, Rome), Edizioni Cremonense, 1942, pp. 119–125.

List of symbols

Symbols
$X < Y$, X is a proper subgroup of Y
$A \cong_G B$, A is G-isomorphic to B
A^{φ}, $A\varphi$, $\varphi(A)$, image of the subset A by φ
G/N, quotient group of G by N
$G \cong H$, G and H are isomorphic
$G \times H$, direct product of G and H
$L^{\natural}$, base group of the wreath product $L \wr H$
$Q_1 \curlyvee Q_2 \curlyvee \cdots \curlyvee Q_r$, central product, 46
$V_1 \otimes V_2$, tensor product, 75
$X \cap Y$, intersection of the sets X and Y
$X \cup Y$, union of the sets X and Y
$X^y = y^{-1}Xy$, conjugate of X by y
$[K]H$, semidirect product of a normal subgroup K by a subgroup H
$\mathfrak{F} \circ \mathfrak{G}$, formation product or Gaschütz product of $\mathfrak{F}$ and $\mathfrak{G}$
$\mathfrak{F} \diamond \mathfrak{G}$, Fitting product of $\mathfrak{F}$ and $\mathfrak{G}$
$\Phi(G)$, Frattini subgroup of G
$\bigcap_{i \in I} X_i$, intersection of X_i, $i \in I$
$\times_{i \in I} X_i$, direct product of X_i, $i \in I$
$\langle X, Y \rangle$, join of X and Y
$\langle X^G \rangle$, normal closure of X in G, 3
$\langle X^Y \rangle$, 3
$\langle X \rangle$, subgroup generated by X
$\lceil x \rceil$, 36
$X \leq Y$, X is a subgroup of Y
$X \not\leq Y$, X is not a subgroup of Y
$|X : Y|$, index of Y in X
$|X|$, order of X
$\prod_{i \in I} X_i$, product of X_i, $i \in I$
$X \subseteq Y$ or $Y \supseteq X$, X is a subset (subclass) of Y
$N \lhd G$, N is a proper normal subgroup of G
$N \trianglelefteq G$, N is a normal subgroup of G
$G \wr H$, wreath product of G and H
a^{φ}, $a\varphi$, $\varphi(a)$, image of the element a by φ
$f_1 \subseteq f_2$, 11
$x^y = y^{-1}xy$, conjugate of x by y
$[X, Y]$, commutator subgroup of X and Y
$[g_1, g_2, \dots, g_r]$, iterated commutator
$[x, y] = x^{-1}y^{-1}xy$, commutator of x and y
$X \setminus Y$, set difference between X and Y
$V_1 \oplus V_2$, direct sum of the modules V_1 and V_2

A
$\mathfrak{A}$, class of all abelian groups
${}^2A_l(q)$, simple group, 86
A^T, transpose of the matrix A, 85
$A_B = \bigcap_{b \in B} A^b$, largest B-invariant subgroup of A
A_G, core of A in G
$\mathfrak{A}_n$, 130
AB, product of A and B, 1
$A_l(q)$, simple group, 86
$\mathrm{Alt}(n)$, alternating group of degree n
$\mathrm{Aut}(G)$, automorphism group of G

Index

R

S

Index of names

Bei Fragen zur Produktsicherheit wenden Sie sich bitte an:
If you have any questions regarding product safety,
please contact:

Walter de Gruyter GmbH
Genthiner Straße 13
10785 Berlin
productsafety@degruyterbrill.com